ELECTRON AND PHOTON INTERACTIONS WITH ATOMS

PHYSICS OF ATOMS AND MOLECULES

Series Editors: P.G. Burke *Queen's University of Belfast, Northern Ireland*
and
H. Kleinpoppen *University of Stirling, Scotland*

Editorial Advisory Board:
R.B. Bernstein *(Austin, U.S.A.)*
J.C. Cohen-Tannoudji *(Paris, France)*
W. Hanle *(Giessen, Germany)*
W.E. Lamb, Jr. *(Tucson, U.S.A.)*
M.R.C. McDowell *(London, U.K.)*

1976: ELECTRON AND PHOTON INTERACTIONS WITH ATOMS
Edited by H. Kleinpoppen and M.R.C. McDowell

In preparation:
THEORY OF ELECTRON–ATOM COLLISIONS
By P.G. Burke and C.J. Joachain

PROGRESS OF ATOMIC SPECTROSCOPY
Methods and Applications
Edited by W. Hanle and H. Kleinpoppen

A Continuation Order Plan is available for this series. A continuation order will bring delivery of each new volume immediately upon publication. Volumes are billed only upon actual shipment. For further information please contact the publisher.

ELECTRON AND PHOTON INTERACTIONS WITH ATOMS

Festschrift for Professor Ugo Fano

Edited by

H. Kleinpoppen

University of Stirling
Stirling, Scotland

and

M. R. C. McDowell

Royal Holloway College
Surrey, England

PLENUM PRESS · NEW YORK AND LONDON

Library of Congress Cataloging in Publication Data

Main entry under title:

Electron and photon interactions with atoms.

Includes bibliographies and index.
1. Electromagnetic interactions—Addresses, essays, lectures. 2. Fano, Ugo, I. Fano, Ugo. II. Kleinpoppen, Hans. III. McDowell, M. R. C.
QC794.8.E4E38   539.7'54   75-37555
ISBN 0-306-30846-0

© 1976 Plenum Press, New York
A Division of Plenum Publishing Corporation
227 West 17th Street, New York, N.Y. 10011

United Kingdom edition published by Plenum Press, London
A Division of Plenum Publishing Company, Ltd.
Davis House (4th Floor), 8 Scrubs Lane, Harlesden, London, NW10 6SE, England

All rights reserved

No part of this book may be reproduced, stored in a retrieval system, or transmitted, in any form or by any means, electronic, mechanical, photocopying, microfilming, recording, or otherwise, without written permission from the Publisher

Printed in the United States of America

The International Symposium

on

ELECTRON AND PHOTON INTERACTIONS WITH ATOMS

was granted

Sponsorship by the European Physical Society

and financial support was given by

The Royal Society

I.B.M. United Kingdom Ltd.

The University of Stirling

Scientific Organizing Committee of the Symposium

B. Bederson, P.G. Burke, J. Kessler, G.V. Marr, M.R.C. McDowell,
I.C. Percival, H. van Regemorter, and H. Kleinpoppen (Chairman)

Local Organizing Committee of the Symposium

K. Blum, D. Hils, H. Kleinpoppen, I. McGregor,
T.W. Ottley, M.C. Standage

International Symposium on Electron and Photon Interactions with Atoms in honor of Ugo Fano, University of Stirling, July, 16-19, 1974

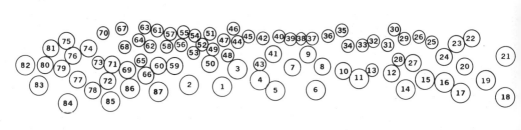

1. U. Fano
2. M. Lambropoulos
3. P.S. Farago
4. A. Temkin
5. D.E. Golden
6. S. Ormonde
7. C.J. Joachain
8. H.S. Taylor
9. G.V. Marr
10. K.H. Winters
11. H.G.M. Heideman
12. A.C.H. Smith
13. P. Jaegle
14. J.A.R. Samson
15. F. Combet Farnoux
16. A.E. Kingston
17. D.J. Burns
18. F.J. de Heer
19. M.R.C. McDowell
20. T.M. Miller
21. J.C. Zorn
22. D. Hils
23. T. Suzuki
24. L. Parcell
25. G. Nienhuis
26. K. Takayanagi
27. A.D. Stauffer
28. F. Wuilleumier
29. V.L. Jacobs
30. S. Geltman
31. E.H.A. Granneman
32. M. Mittleman
33. Mrs. Mittleman
34. K. Ross
35. T. Heindorff
36. C. Bottcher
37. E. Reichert
38. W.C. Lineberger
39. M.R. Flannery
40. W. Eissner
41. K.R. Lea
42. M. Willmers
43. F.W. Byron
44. K.C. Mathur
45. D. O'Connell
46. M. Eminyan
47. T.W. Ottley
48. H. Suzuki
49. R.G. Keesing
50. B. Bederson
51. M. le Dorneuf
52. K. Taylor
53. R. Bruch
54. J. Peresse
55. R.J. Tweed
56. R. Browning
57. F. Bely-Dubau
58. K. Persy
59. K. Rubin
60. P. Lambropoulos
61. F. Brouillard
62. M. Rudge
63. J. Dubau
64. V. Myerscough
65. D.L. Moores
66. J.F. Williams
67. P. Defrance
68. A.J. Duncan
69. J.W. McConkey
70. S.J. Smith
71. D.W. Walker
72. J. Slevin
73. I.C. Malcolm
74. M.C. Standage
75. D. Andrick
76. J. Macek
77. H.D. Zeman
78. A.F. Starace
79. G.F. Hanne
80. P.G. Burke
81. K.T. Lu
82. K.V. Codling
83. I. Hertel
84. R.S. Berry
85. K.H. Nygaard
86. G.W.F. Drake
87. H. Kleinpoppen

Foreword

The quantum description of electromagnetic interactions among electrons, atoms, and photons, after initial qualitative success 50 years ago, is now finding quantitative success in a period of exciting theoretical and experimental progress. Theory and experiment tend to advance hand in hand, and early experimental success in observing the full details of collision dynamics in nuclear physics attracted much of the leading theoretical effort in this direction. As a result of recent atomic physics progress on both theoretical and experimental fronts, in areas such as angular distributions and correlations, photon and electron spin polarization, cross section threshold studies, and phase shift analysis, the full details of atomic physics interactions are being laid open to view. The influence of dynamic and geometric effects are being sorted out and powerful new tools for the analysis of a great variety of electron–photon processes are emerging.

A central contributor to the development of these theoretical concepts, as well as to the stimulation of fruitful directions for experimental research, has been Professor Ugo Fano.

Fano's success in introducing important new points of view in atomic physics no doubt results in part from the fact that he has not restricted his interest to this one area of research. His major concern during an earlier stage of his career was stimulated by the practical problem of the interaction of high-energy particles with condensed matter. Indeed, his contributions in applied physics are as impressive as those in scattering theory, spectroscopy, and atomic structure, and follow a similar pattern. He actively collaborates with experimenters and theorists interested in a broad range of problems, from basic to applied and spanning such fields as nuclear physics, biology, medicine, and radiation chemistry.

The breadth of Fano's work is illustrated by the fact that the term "Fano effect" has at least two different meanings. As discussed in this volume, the "Fano effect" refers to the polarization of photoelectrons ejected from heavy alkali atoms when photoionized near the minimum in the cross section. This effect is illustrative of his contributions to the subject of this volume. Another "Fano effect," however, refers to the deviation from a Poisson distribution of the average number of ion pairs produced by a monochromatic particle beam due to the discrete nature of the ionization process. Still another "Fano effect" may be found in the literature describing autoionization processes in ultraviolet and soft x-ray spectroscopy.

Atomic physicists are well familiar with his work on the theory of angular momentum coupling, much of it done in collaboration with his cousin, G. Racah. Vacuum ultraviolet spectroscopists are also well aware of his role in sorting out the interpretation of multiple electron transitions in atomic absorption, so beautifully revealed in synchrotron light experiments. His continuing interest in the many-body problem as applied to atomic collision theory is also demonstrated by significant work on the interpretation of energetic, multiply charged ion collisions with atoms.

Atomic physicists are doubtless less familiar with his work on the theory of diffraction gratings and on Stokes parameters, which also typifies his wide diversity of interests within the context of atomic physics.

The idea of a "Festschrift" for a scientist as vigorous and productive as Ugo Fano will not seem strange to the many colleagues who have been stimulated and educated through collaboration and discussion with him. Ugo Fano's Debye sphere has included not only a great many scientists, but also a great many vital scientific ideas. A sufficient fraction of these ideas has been bold and controversial enough to ensure an exciting and rewarding experience for anyone engaged in collaborative research with him.

During his long career at the National Bureau of Standards, a great many postdoctoral research associates and more experienced scientists from the NBS, universities, and industrial laboratories have been influenced by collaboration with Ugo Fano on specific projects in one of the many fields of his interest. This spirit of "learning together" is still very much in evidence in his group at the University of Chicago, where it is reflected in the excellence and the diversity of research workers and published papers. Observing the quality of this output, one is tempted to suggest that the significance of this Festschrift rests less on Fano's past achievements than on the promise of things yet to come.

An examination of the work reported in this volume will indeed show that recent progress, both experimental and theoretical, sets the stage for further acceleration in the rate of new understanding of basic atomic processes. We can confidently predict that much of that progress will come from Ugo Fano and his collaborators.

John W. Cooper
National Bureau of Standards
Washington, D.C.

Lewis M. Branscomb
IBM Corporation
Armonk, New York

Contents

Introduction xv
    H. Kleinpoppen and M. R. C. McDowell

1. Effects of Configuration Interaction on Electron and Photon Interactions with Atoms 1
    P. G. Burke

2. Low-Energy Electron Scattering and Attachment by C, N, and O Atoms 27
    R. K. Nesbet and L. D. Thomas

3. The Ejection of Electrons from Autoionizing States in Atoms .. 35
    C. Bottcher

4. The Study of Photoionization Processes Using Synchrotron Radiation 39
    G. V. Marr

5. Photoabsorption of the 4d Electrons in Xenon and Barium: A Comparison 69
    D. L. Ederer, T. B. Lucatorto, E. B. Saloman, R. P. Madden, M. Manalis, and Jack Sugar

6. Effects of Anisotropic Electron–Ion Interaction on the Photoelectron Angular Distribution of Open-Shell Atoms .. 83
    Dan Dill, Anthony F. Starace, and Steven T. Manson

7. Study of Atomic Subshell Properties by Electron Spectrometry . 89
    M. O. Krause and F. Wuilleumier

8. Quantum Defect Theory: Photoionization 99
    J. Dubau

9. Electron–Alkali Scattering and Photodetachment of Alkali Negative Ions 109
    D. L. Moores

10. Photodetachment Threshold Processes 125
    W. C. Lineberger, H. Hotop, and T. A. Patterson

11. Multiple Photoionization of the Rare Gases 133
    James A. R. Samson and G. N. Haddad

12. Nonstatistical Branching Ratios in Atomic Processes 141
    A. R. P. Rau

13. Study of Atomic Structure by Means of (e, 2e) Impulsive
    Reactions 149
    A. Giardini Guidoni, G. Missoni, R. Camilloni, and
    G. Stefani

14. Phase Shift Analysis and Dispersion Relations 161
    B. H. Bransden

15. Corrections for Forward Scattering to Positron–Helium Total
    Cross Section Measurements 181
    P. G. Coleman, T. C. Griffith, G. R. Heyland, and
    T. L. Killeen

16. Testing of Classical and Quantum-Mechanical Criteria for Elastic
    Scattering of Electrons by Noble Gases 185
    B. van Wingerden, F. J. de Heer, R. H. J. Jansen, and
    J. Los

17. Spin Polarization in Electron–Atom Scattering 191
    B. Bederson and T. M. Miller

18. Conservation of Total Spin in Electron–Atom Collisions 203
    D. W. Walker

19. Spin Polarization of Electrons by Resonance Scattering 215
    E. Reichert

20. Influence of Spin Polarization on Resonance Scattering by Neon 229
    T. Suzuki, H. Tanaka, M. Saito, and H. Igawa

21. Asymmetry in the Single Scattering of Electrons from One-
    Electron Atoms 235
    P. S. Farago

22. Intense Source for Highly Polarized Electrons Using the Fano
    Effect 241
    W. von Drachenfels, U. T. Koch, T. M. Müller, and
    W. Paul

Contents

23. Electron Impact Excitation of Light Atoms at Intermediate Energies.................................... 245
    M. R. C. McDowell

24. The Multichannel Eikonal Treatment of Electron–Atom Collisions.. 275
    M. R. Flannery and K. J. McCann

25. Recent Progress in the Application of Eikonal–Born Series Methods in Atomic Physics...................... 285
    F. W. Byron, Jr.

26. *Ab Initio* Optical Theory of Elastic Electron–Atom Scattering.. 299
    F. W. Byron, Jr. and Charles J. Joachain

27. The Scattering of Electrons from Hydrogen Atoms........... 309
    J. F. Williams

28. On the Anisotropy of the Quenching Radiation from Metastable Hydrogen and Deuterium Atoms 339
    G. W. F. Drake, A. van Wijngaarden, and P. S. Farago

29. New Measurements of Differential and Integral Cross Sections for Electron Impact Excitation of the $n = 2$ States of Helium.. 349
    G. Joyez, A. Huetz, F. Pichou, and J. Mazeau

30. Coherent Electron Impact Excitation of Different L States in the $n = 3$ Shell of Atomic Hydrogen................... 365
    Stephen J. Smith

31. Details of Collision Dynamics from the Electron Scattering by Laser-Excited Sodium Atoms...................... 375
    I. V. Hertel

32. Excitation and Ionization in the Coulomb-Projected Born Approximation................................ 387
    Sydney Geltman

33. Calculations of Triple Differential Cross Sections............. 397
    D. H. Phillips and M. R. C. McDowell

34. Differential Cross Sections for Electron Impact Ionization of Helium 411
    V. L. Jacobs

35. The Polarized Frozen Target and Polarized Frozen Core Methods in Low-Energy Electron Scattering and in Atomic Structure Calculations.......................... 415
   M. LeDourneuf, H. van Regemorter, and Vo Ky Lan

36. Progress Report on the Use of the Many-Body Theory in Inelastic Scattering from Atoms.......................... 435
   H. S. Taylor, A. Chutjian, and L. D. Thomas

37. Direct Observation of Exchange Scattering by Spin Flip of Polarized Electrons in Excitation of Mercury........ 445
   G. F. Hanne and J. Kessler

38. Electron–Photon Coincidence Technique for Electron Impact on Atoms.. 455
   M. Eminyan, H. Kleinpoppen, J. Slevin, and M. C. Standage

39. Theory of Measurement of Impact Radiation on Atoms....... 485
   Joseph Macek

40. On the Theory of Electron–Photon Coincidence Experiments .. 501
   K. Blum and H. Kleinpoppen

41. Spatial Asymmetries in Atomic Collisions................... 515
   H. G. Berry, L. J. Curtis, D. G. Ellis, and R. M. Schectman

42. Atoms in Intense Electromagnetic Fields.................... 525
   P. Lambropoulos and Melissa Lambropoulos

43. The Effect of Multimode Laser Operation on Multiphoton Absorption by Atoms........................... 553
   Joel Gersten and Marvin Mittleman

44. Two-Photon Processes 559
   R. Stephen Berry

45. Electron Spin Polarization from Multiple Photoionization Processes 581
   Herbert D. Zeman

46. Total Cross Section for Elastic Scattering of Fast Charged Particles by a Neutral Atom...................... 595
   Michio Matsuzawa and Mitio Inokuti

47. Recent Developments in Variational Principles and Variational Bounds: A Road Map........................... 601
   Robert Blau, A. R. P. Rau, L. Rosenberg, and Larry Spruch

Contents

48. Backscattering of Slow Electrons by Positive Ions 609
    G. Drukarev

49. Quantum Defect Theory of Excited $^1\Sigma_u^+$ Levels of H_2 613
    O. Atabek and C. Jungen

50. Auger Spectroscopy of Foil-Excited Beryllium Ions........... 621
    R. Bruch, G. Paul, J. Andrä, and B. Fricke

51. The Perturbed Series, $2p^5(^2P_{3/2,1/2})ns$, of Ions along the Ne I Isoelectronic Sequence 627
    K. T. Lu and M. D. W. Mansfield

52. Effect of LS Term Dependence on Some Rare Gas Transition Probabilities................................. 633
    P. L. Altick and Jack R. Woodyard, Sr.

53. Resonances and Cusps in Electron Impact on Atoms 639
    D. E. Golden

54. High-Resolution Studies of Electron–Atom Collisions 651
    F. H. Read

55. Electron Excitation of Xenon near Threshold............... 661
    N. Swanson, R. J. Celotta, and C. E. Kuyatt

56. Resonances in the Excitation of Ne by Electrons at Energies between 40 and 50 eV 669
    H. G. M. Heideman and T. van Ittersum

57. Resonance Series in Helium............................. 671
    D. W. O. Heddle

Index.. 679

During a coffee break at the Symposium. *Front row*: Ugo Fano (center), P. S. Farago (left), and J. Kessler (right). *Back row*: M. R. C. McDowell (left) and H. van Regemorter (right).

Introduction

As a result of recent activities in experimental and theoretical physics on angular distribution and correlations, spin polarization, threshold behavior, and phase shift analysis in electron–atom scattering and atomic photoionization, a new understanding of the details of these processes has emerged. The members of the Scientific Organizing Committee of the International Symposium on Electron and Photon Interactions with Atoms, therefore, felt it appropriate to bring together theoreticians and experimentalists for a workshop-type International Symposium in order to provide an opportunity to survey the present state of investigations on the analysis of the above topics and to provide an opportunity for discussions on future experimental and thereotical studies. To illustrate some of the basic concepts taken up at the Symposium, it might be appropriate to sketch the most recent developments and their future prospects.

1. *Geometric and Dynamic Effects in Impact Excitation*

A theoretical formulation has been presented by Fano and Macek for the angular distribution and polarization of light excited by atomic and electronic collisions and modulated in time by the action of internal and external fields. This formulation disentagles the geometric and dynamic effects from which alignment tensors and orientation vectors of radiating atoms can be derived. These vector and tensor parameters have simple relations to angular momentum quantities involved in the collision process (e.g., an orientation vector which is proportional to the average angular momentum, $\langle \mathbf{J} \rangle$, and an alignment tensor whose components are proportional to the mean value of quadratic expressions in J_x, J_y, J_z, such as $J_x^2 - J_y^2$).

The theory shows how the emission of line radiation averaged over all directions depends on dynamic factors such as line strength, while the anisotropy and polarization depend only on the alignment and orientation of the excited atoms. The new theory of Fano and Macek is also applicable to processes of time dependence of alignment and orientation between excitation and emission. This concept of studying time dependence in the collision process extends beyond previous theories (e.g., Percival–Seaton theory on the polarization of line radiation). Applications of the new theory can be made for beam foil spectroscopy (beats and spatial asymmetries), for angular electron–photon correlations in delayed photon–electron coincidence experiments of excitation impact processes, and also to the polarization of impact radiation. Recent measurements of electron–photon correlation from electron impact excitation permitted measurement of complex excitation amplitudes and their phase differences. These measurements determine the Fano–Macek alignment and orientation parameters, and they also determine the amount of orbital angular momentum transferred to the atom during the excitation process.

2. *Analysis of Spin and Polarization Effects*

Much progress has been made in the analysis of spin effects in both the fields of electron–atom collision and atomic photoionization processes.

(i) *Electron–Atom Processes.* New theoretical and experimental developments have recently taken place: direct and exchange scattering cross sections could be measured by means of spin analysis of atoms and electrons. Triplet and singlet scattering amplitudes and phase differences between amplitudes are soon to be measured. Electron scattering experiments using lasers to produce oriented atoms in excited short-lived states with well-defined quantum numbers were recently successful. New measurements on spin effects in resonance scattering processes have also been reported (University of Mainz). Detailed theoretical analysis of spin effects were given. New theoretical developments of spin effects in spin–orbit interactions for one-electron atoms have been published. A generalized theory of spin effects in polarized electron–atom scattering with arbitrary atomic spin has been developed based upon the density matrix formalism.

(ii) *Photoionization Processes.* Of course, the important role the Fano effect plays dominates a large part of the interest and applications in the above field. However, theories of photoionization measurements for general atomic systems have been developed in terms of the density matrix formalism which determines the ejected electron and residual ion polarization as a function of the target atom and incident photon polarization states. The new types of parameters set out by these theories for photoionization with polarized photons and atoms represent a challenge for the application of new experimental methods. The physics of multiple photon ionization processes represents a further new field with an impressive range and variety of polarization effects to be studied by experiments and theory.

3. *Phase Shift Analysis*

Important progress has been made in the field of phase shift and dispersion relation analysis of electron–atom scattering processes. The present results were critically reviewed and discussed in the light of future applications. New suggestions for alternative methods of phase shift analysis were considered.

4. *Resonance Structure, Electron–Electron Correlation, Threshold Behavior, Cusps, etc.*

A variety of interesting effects may be summarized under this heading. New results on resonance structure in electron–atom and ion collisions were reported by a considerable number of theoreticians and experimentalists. This situation reflects, on the one hand, the high standard of the technology of high-resolution measurements in electron–atom scattering, and, on the other hand, the advanced status of approximations available. Electron–electron correlation measurements of electron ionization processes for low- and high-momentum transfer revealed important new information on threshold behavior of ionization processes and momentum distributions of electrons in atomic states. Important experimental investigations have been carried out in connection with cusps and threshold laws in photoionization, photodetachment measurements, and electron scattering processes. These experiments were important tests of different types of cusps and threshold laws predicted by theory.

5. *Subshell Photoionization Processes*

Advances in experimental techniques over the last few years have begun to stimulate direct measurements of the distribution of photoelectrons in energy and angle which offers the possibility of studying the photoionization cross section not only of outer but also of individual electronic subshells (photon–electron spectrometry by including the use of synchroton ultraviolet radiation). Measurements of subshell photon ionization cross sections revealed interesting structure at the subshell energies. Ratios for subshell cross sections and the asymmetry parameter β of angular distributions from different subshells were determined both by theory and experiment. Theoretical models for the description of the subshell photoionization processes were developed whereby in particular intrachannel couplings and multielectron correlation effects were included.

6. Theoretical Approximations

Significant progress has been made in the theory of electron impact excitation of atoms. Improved variational methods have produced well-converged calculations above the lowest excitation threshold in a number of atoms. Related studies on the application of configuration interaction wave functions in photoionization and in scattering processes have elucidated the resonance structure in complex atoms.

A variety of distorted wave calculations is leading to improved predictions of total and differential excitation cross sections over a wide range of energies for light atoms and producing models of the electron–photon coincidence processes discussed above. At higher energies the study of the effects produced by distortion and polarization has clarified our understanding of the range of validity of the first Born approximation and its nonprivileged status as only one of an infinite range of zero-order approximations.

Important improvements have been made in applying perturbation theory, by analytic investigation of the energy dependence of successive terms of the Born and Glauber series, appropriate combination of these approaches allowing the derivation of results correct to a given order in an inverse power expansion of the energy.

H. Kleinpoppen
M. R. C. McDowell

1

Effects of Configuration Interaction on Electron and Photon Interactions with Atoms

P. G. Burke

A review is presented of the effects of configuration interaction on photoionization of atoms and electron–atom scattering. The emphasis is on experimental and theoretical studies carried out since the basic paper by Fano (1961) on this subject.

1. Introduction

It is a very great pleasure for me to have the opportunity to give a paper at this symposium in honor of my friend and mentor Ugo Fano. I had the very good fortune to first meet Ugo at Berkeley when he had just completed the work for his famous 1961 paper on the effects of configuration interaction (CI) on intensities and phase shifts (Fano, 1961). This paper, which developed earlier work by Fano in the 1930s (Fano, 1935), laid the framework for developments in this subject over the past 13 years. Later, in the summer of 1965, I had the privilege of working with Ugo and with John Cooper at the National Bureau of Standards in Washington. This was a period of intense experimental activity, with Madden and Codling (1963, 1965) producing exciting new absorption spectra in a whole variety of gases using radiation from the 180 MeV synchrotron and with Simpson, Kuyatt, Mielczarek, Chamberlain, and Heideman producing complementary electron scattering data with their newly perfected high-resolution electron

P. G. BURKE • Department of Applied Mathematics and Theoretical Physics, The Queen's University of Belfast, Belfast, BT7 1NN, Northern Ireland.

spectrometer (e.g., Simpson, 1964; Simpson et al., 1964; Kuyatt et al., 1965; Chamberlain and Heideman, 1965). Again the influence of Fano in laying down the basic theoretical framework for our understanding of these experiments in papers with Cooper, Prats, and Simpson (Cooper et al., 1963; Simpson and Fano, 1963; and Fano and Cooper, 1965) was of fundamental importance.

In this talk my plan is to take this topic of the effects of CI on electron and photon interactions with atoms and to trace its development from this period in the early 1960s through to the present day. Although this represents only a part of Fano's contribution to the subject matter of this symposium, I hope to show you that it is a subject which deserves and is receiving an increasing degree of attention, and where future important developments may be expected.

2. Basic Processes and Theory

The basic processes that will be considered in this paper are elastic and inelastic scattering of electrons by atoms or ions

$$e + A_i \rightarrow e + A_j \tag{1}$$

and photoionization or photodetachment

$$h\nu + A_i \rightarrow e + A_j^+ \tag{2}$$

We are also interested in this symposium in the situation where there is more than one electron in the final state or where there is more than one photon incident in the initial state. However, partly for reasons of time and partly because, broadly speaking, CI effects have not yet been studied in such detail in these cases, although they certainly play an important role, I will not be considering them further in this paper.

From an experimental point of view the above processes may be studied by measuring the total cross section as a function of the incident electron or photon energy and spin state, by observing the energy, angular distribution, and spin state of the ejected electron, and, finally, by observing the quantum state of the final atom or ion, most usually by looking at the polarization and angular distribution of the decay photon, in the most recent work, in coincidence with the scattered electron. These measurements give us progressively more information about the basic scattering process, which in turn can be compared with theory.

It is important to note at this point that the two processes considered above are not independent and measurements of one can give information about the other. To see this we observe that the final state in the photo-

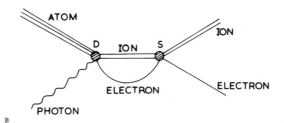

Figure 1. Relationship between photoionization and electron–ion scattering.

ionization process (2) corresponds to scattering of an electron by an ion as illustrated diagrammatically in Figure 1. In this case the photon interacts with the atom through the dipole operator **D**, producing an electron and an ion. These can then scatter, either resonantly or nonresonantly, giving rise to the observed final state. Clearly the scattering process denoted by S in the figure corresponds to Eq. (1). Because of the conservation laws of parity and angular momentum, and because we are usually only concerned with dipole photons, only some of the possible scattering states occur in photoionization.

We can show that the cross section for photoionization in the dipole length approximation is

$$\frac{d\sigma}{d\Omega_f} = 4\pi^2 \alpha a_0^2 \omega |\langle \Psi_{jE}^-|\hat{\varepsilon} \cdot \mathbf{D}|\Psi_i\rangle|^2 \quad (3)$$

where $\hat{\varepsilon}$ is the photon polarization vector, **D** is the dipole moment operator, and ω is the photon energy in atomic units. In addition, Ψ_i is the initial atomic state and Ψ_{jE}^- is the final scattering state with ingoing waves in all channels and an outgoing wave in the jth channel of interest. The final state in addition is normalized to unit energy interval in the continuum

$$\langle \Psi_{jE}^-|\Psi_{j'E'}^-\rangle = \delta(E - E')\delta_{jj'} \quad (4)$$

which defines the normalization factor in Eq. (3). Equivalent expressions can also be obtained using the velocity and the acceleration forms of the matrix element. Since the scattering state Ψ_{jE}^- contains, through its asymptotic form, complete information about the collision, we see from Eq. (3) the connection mentioned earlier between processes (1) and (2).

Our basic problem is to calculate Ψ_i and Ψ_{jE}^-. The method proposed by Fano (1961) and by Fano and Prats (1964) and developed further by Altick (1968) and Bates and Altick (1973) is to expand these functions in terms of the eigenstates of some suitably chosen zero-order Hamiltonian H_0. We define

$$H_0\phi_j = \varepsilon_j\phi_j$$
$$H_0\psi_{i\varepsilon} = \varepsilon\psi_{i\varepsilon} \qquad (5)$$

where ϕ_j are discrete eigenfunctions and $\psi_{i\varepsilon}$ are continuum eigenfunctions in the ith channel. Some calculations stop at this point. For example, in the work of Cooper (1962) a model potential was used to represent the atomic shells and the resultant zero-order eigenfunctions used directly in (3) to calculate the photoionization cross section. This approach gives a broad picture of the distribution of oscillator strength in the continuum, but details, as expected, tend to be inaccurate. A more refined form to take for H_0 is the Hartree–Fock Hamiltonian. This gives better results than the model potential method, as expected, but fails to allow for channel coupling effects.

To proceed further we can use the eigenfunctions defined by Eq. (5) as a basis for expanding the complete wave function,

$$\Psi_{jE} = \sum_i \int d\varepsilon \, a_{jE,i\varepsilon}\psi_{i\varepsilon} + \sum_i b_{jE,i}\phi_i \qquad (6)$$

and a similar expansion can be used for Ψ_j. We have omitted the ingoing wave boundary condition since we usually need to determine a complete set of linearly independent solutions. The index i in the first expansion refers to the channels and in any practical calculation is truncated to a finite number of terms. The second expansion is included for completeness. The coefficients $a_{jE,i\varepsilon}$ and $b_{jE,i}$ are now determined by diagonalizing the complete Hamiltonian

$$\langle\Psi_{jE}|H|\Psi_{j'E'}\rangle = E\,\delta(E - E')\,\delta_{jj'} \qquad (7)$$

in this basis. In the work of Altick the continuum integral is represented by quadrature and (7) reduces to the diagonalization of a finite-dimensional matrix. The **R**-matrix method considered recently by Burke *et al.* (1971), Allison *et al.* (1972), Fano and Lee (1973), and Lee (1974) can be incorporated into this framework by introducing a boundary condition operator into H_0 (Block, 1957; Lane and Robson, 1966; Robson and Lane, 1967). This ensures that the eigenfunctions (5) form a discrete and complete set over some inner region $r \leqslant r_a$.

It is important to note at this point that the traditional close coupling approach to electron–atom scattering discussed for example by Seaton (1953), Burke and Smith (1962), Burke and McVicar (1965), Smith and Morgan (1968), and Burke and Seaton (1971) is completely equivalent to the above. In this case expansion (6) is replaced in configuration space by

$$\Psi_{jE} = \mathscr{A}\sum_i \Phi_i F^E_{ij}(r) + \sum_i b_{jE,i}\phi_i \qquad (8)$$

where Φ_i are channel functions representing the target states and the angular and spin parts of the motion of the scattered electron. Again the second expansion is included for completeness. The diagonalization of H in this case yields coupled integrodifferential equations for the $F_{ij}^E(r)$. Provided that the same representation of the target states are used in (6) and (8) and the same number of channels are retained in the expansion, the same results will be obtained.

The matrix variational methods discussed by Harris and Michels (1971) and by Nesbet (1973) can also be shown to be equivalent provided that the second expansion in (6) and (8) is augmented to include 1-electron, 2-electron, etc., excitations of the Bethe–Goldstone type. In this case the expansion basis is formed from $(N + 1)$-electron determinants constructed from non-orthogonal Slater orbitals, together with sine and cosine functions which give the asymptotic form. This basis then defines a subspace of the $(N + 1)$-electron Hilbert space in which H is diagonalized.

A problem which is inherent in all of the methods discussed so far is how to represent the N-electron target states. The usual method is to pre-diagonalize the N-electron Hamiltonian in the discrete part of the basis defined by Eq. (5). This implies a redefinition of what we mean by the zero-order Hamiltonian H_0 since the new zero-order continuum eigenfunctions are linear combinations of the old ones (Mahaux and Weidenmüller, 1969). We will then need to study the convergence of the cross sections as more terms are included in the representation of the target states and as more channels are included in expansions (6) and (8).

Another approach which has proved very successful in treating CI effects in photoionization of closed shell atoms is the random-phase or time-dependent Hartree–Fock approximation considered by Altick and Glassgold (1964), Dalgarno and Victor (1966), Amusia et al. (1971, 1973), Amusia (1973), and Wendin (1971, 1972, 1973). This approach is based on the assumption that the electrons in an atom can be treated as a dense electron gas (Gell-Mann and Brueckner, 1957). In this case the main correction to the independent particle or Hartree–Fock approximation arises from the ring diagrams where particle–hole pairs are excited from the Fermi sea. Amusia et al. (1971, 1973) show that exchange effects are also important for atoms. An important aspect of this approach is that both the initial and final state are improved at the same time and thus the length and velocity forms of the photoionization matrix element are equivalent for closed-shell atoms.

In the case of light atoms and open shell systems the random phase approximation omits diagrams which can be important. In this case it is necessary to use the full many-body perturbation treatment of Brueckner (1959) and Goldstone (1957). A very successful program using this approach has been carried out by Kelly (e.g., Kelly, 1973), and photoionization results for a number of closed- and open-shell systems were obtained. More recently the method has been extended to calculate the optical potential describing elastic electron–atom scattering (Pindzola and Kelly, 1974), and good agreement has been obtained with experiment. It however remains to be seen how effective this method will be in describing inelastic electron–atom collisions.

Finally, in concluding this very brief survey of theoretical methods, we note that the Green's function method of Martin and Schwinger (1959) has been used with remarkable success by Yarlagadda et al. (1973) and Thomas et al. (1973) for elastic and inelastic scattering of electrons by helium atoms and will be discussed further by Taylor (1974) in this symposium.

3. Effects of Configuration Interaction in Photoionization

The role of CI in the formation and decay of resonances was brought sharply into focus by the work of Fano (1961). The theory was developed further by Shore (1967, 1968) and by Mies (1968). This is well illustrated by the photoionization of helium atoms

$$h\nu + \mathrm{He} \rightarrow \mathrm{e} + \mathrm{He}^+(1s) \qquad (9)$$
$$\searrow \quad \nearrow$$
$$\mathrm{He}(2s2p\,^1P)$$

Photoionization can proceed directly, as indicated by the upper arrow in this equation, or, if the incident energy is appropriate, via one of an infinite number of intermediate resonant states. These resonances are members of Rydberg series converging to an excited state of the He^+ ion and are known as Feshbach (1958, 1962), core-excited, or closed-channel resonances (Schulz, 1973). The coupling between channels, in this case the $n = 1$ and $n = 2$ channels, is the mechanism for decay or autoionization of the resonance. This is illustrated in Figure 2, taken from Fano and Cooper (1968), which shows the channels occurring in the spectrum of He. Photon absorption in the ground state can lead to an excited He atom or to a He^+ ion in its ground state or to an infinite number of excited states or to a He^{++} ion. The coupling between the final state channels can give rise to transitions as indicated by arrows in the figure.

An important measure of the effect of CI is the line shape and width. Fano (1961) and Fano and Cooper (1965) showed that for an isolated resonance the total absorption cross section is given by

$$\sigma(E) = \sigma_a \frac{(\varepsilon + q)^2}{1 + \varepsilon^2} + \sigma_b \qquad (10)$$

Effects of Configuration Interaction

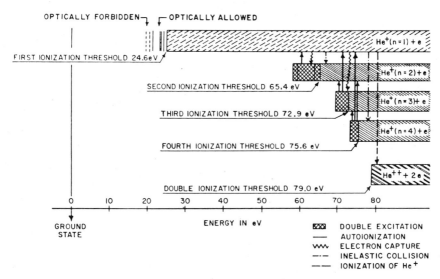

Figure 2. Diagram showing several channels in the spectrum of helium and their interconnection by autoionization processes [from Fano and Cooper (1968), Figure 26].

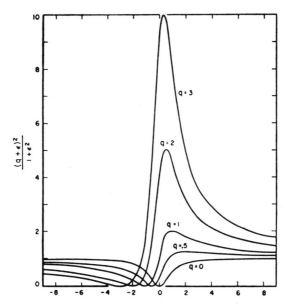

Figure 3. Natural line shapes for different values of q with $\rho^2 = 1$. (Reverse the scale of the abscissas for negative q) [from Fano (1961), Figure 1].

where σ_a and σ_b are the parts of the cross section which interact and are orthogonal to the resonance and are slowly varying functions of energy, $\varepsilon = (E - E_r)/\frac{1}{2}\Gamma$, where E_r is the resonance energy and Γ the resonance width, and q, which has been called the line profile index by Fano, defines the line shape. In addition a correlation coefficient $\rho^2 = \sigma_a/(\sigma_a + \sigma_b)$ has been defined by Fano and Cooper (1965). Figure 3 shows the dependence of the line shape for different values of q when $\rho^2 = 1$, and Figure 4 shows examples of the profiles of autoionization lines in the rare gases. Case (a) corresponds

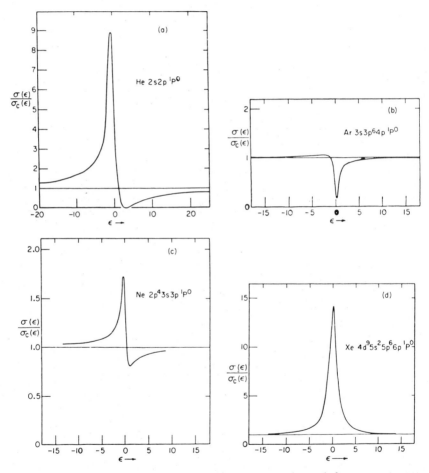

Figure 4. Profiles of autoionization lines in the rare gases: (a) $2s2p\,^1P^0$ in He ($q = -2.8$, $\rho^2 = 1$); (b) $3s3p^64p\,^1P^0$ in Ar ($q = -0.22$, $\rho^2 = 0.86$); (c) $2p^4(^3P)3s3p\,^1P^0$ in Ne ($q = -2.0$, $\rho^2 = 0.17$); (d) $4d^95s^25p^66p\,^1P^0$ in Xe ($q = 200$, $\rho^2 = 0.0003$) [from Fano and Cooper (1968), Figure 27].

to $2s2p\,^1P$ in He, case (b) to $3s3p^64p\,^1P$ in Ar, case (c) to $2p^4(^3P)3s3p\,^1P$ in Ne, and case (d) to $4d^95s^25p^66p\,^1P$ in Xe. Accurate calculations are still awaited for all cases except He, but such studies can be expected to be very fruitful and are being carried out.

A considerable body of knowledge has now been accumulated about the role of CI in determining resonance widths. One of the first and still most interesting examples is in He, where Cooper *et al.* (1963) pointed out that the appropriate zero-order states are

$$(2snp \pm 2pns)^1P \tag{11}$$

The "+" and "−" series are respectively broad and narrow and correspond to electrons moving radially in phase and out of phase. This interpretation satisfactorily explained the absorption spectrum obtained by Madden and Codling (1965) and shown in Figure 5. More detailed calculations by Burke and McVicar (1965), Altick and Moore (1966), and Lipsky and Russek (1966) have confirmed this picture and shown in addition a third $2pnd\,^1P$ series converging to the $n = 2$ threshold of He^+ (Burke, 1968; Macek, 1968). The correlated motion of the two electrons, so well illustrated in this case, may have important implications elsewhere. Recently Fano and Lin (1973, 1974) and Lin (1974) have raised the possibility that the same effect may be important in understanding the Wannier (1953) threshold law of ionization. The (e, 2e) coincidence experiments by Ehrhardt *et al.* (1972a,b) and by Cvejanović and Read (1973) are of fundamental importance in helping to solve this problem.

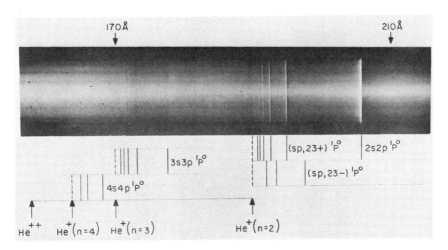

Figure 5. The absorption spectrum of helium between 160 and 215 Å showing the many resonances due to two-electron excitations [from Madden and Codling (1965), Figure 3].

Progress has also been made in understanding CI effects on resonances in the photoionization of complex atoms (e.g., Conneely *et al.*, 1970; Kelly and Simons, 1973). Some of the most interesting recent work has been carried out on Be and Mg by Dubau and Wells (1973) and Bates and Altick (1973), extending earlier work of Moores (1967) and Burke and Moores (1968). Figure 6 shows the results for Mg obtained by Dubau and Wells compared with absolute measurements of Ditchburn and Marr (1953) and relative measurements of Mehlman-Balloffet and Esteva (1969). The calculations included the coupling between the ground and first excited state of Mg^+ and used a multiconfiguration ground state for Mg. The main features are the $3pns$ resonances with widths comparable to separation and the narrow $3pnd$ resonances. Calculations have also been carried out for Ca by Wells (1973). In this case good agreement is only obtained with the experiments of Garton and Codling (1965) and of Newson (1966) if the dipole and quadrupole polarizabilities of the core and a dielectric correlation term, introduced by Chisholm and Öpik (1964), which allows for the effect on one electron due to the polarization of the core by the other, are included. These terms have also been shown to be important in calculations of alkali negative ion bound states by Norcross (1974).

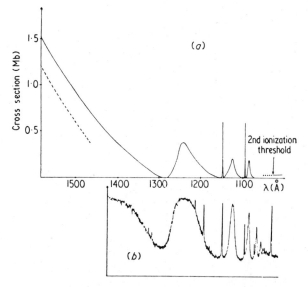

Figure 6. Photoionization of Mg: (a) theoretical curve (dipole length) compared with absolute measurement made by Ditchburn and Marr; (b) experimental curve of Mehlman-Balloffet and Esteva [from Dubau and Wells (1973), Figure 2].

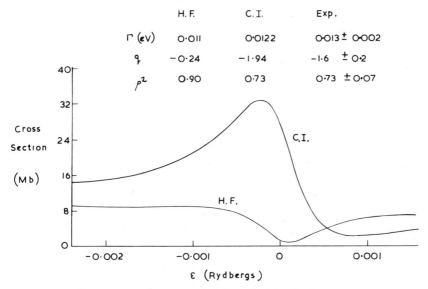

Figure 7. Profile of the $2p^53s^1P^0$ autoionizing line in neon.

Finally, some new studies of the effect of CI in the shape of resonances in the rare gases have been made by Luke (1973) and by Burke and Taylor (1975). We show in Figure 7 the results of two calculations for the $1s^22s2p^63p^1P$ resonance in Ne. Both calculations have included the coupling between the ^2P ground states and ^2S excited state of Ne$^+$, but while the first represents these states by single configurations the second includes 9 configurations for the ground state, 6 for the excited state, and 18 for the ground state of Ne. In both cases the width of the resonance agrees well with the measurements of Codling et al. (1967), but only the CI calculation gives a result in agreement with experiment for the line profile index. Since this quantity depends on the relative signs and magnitudes of the three amplitudes corresponding to direct photoionization into the nonresonant continuum, photoionization of the resonant state, and decay of this resonance by autoionization it can be a very sensitive test of CI effects.

The variation of photoelectron angular distributions through autoionizing resonances is also a sensitive probe of the dynamics of the photoionization process (Fano and Dill, 1972; Dill, 1973). The differential cross section for linearly polarized light is

$$\frac{d\sigma}{d\Omega} = \frac{\sigma}{4\pi}(1 + \beta P_2(\cos\delta)) \qquad (12)$$

where σ is the integrated cross section and β is the asymmetry parameter

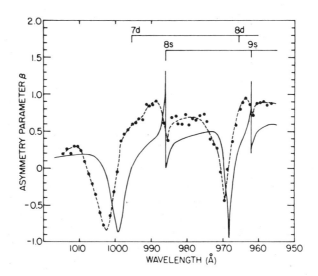

Figure 8. Asymmetry parameter β as a function of wavelength for the autoionization structure lying between the $^2P_{3/2}$ and $^2P_{1/2}$ ionization thresholds of xenon; the solid line represents the theoretical values; the dots and the dashed line represent the experimental data [from Samson and Gardner (1973), Figure 3].

which lies in the range $-1 \leqslant \beta \leqslant 2$ (Cooper and Zare, 1968). Samson and Gardner (1973) have measured this parameter for the autoionization structure lying between the $^2P_{3/2}$ and $^2P_{1/2}$ ionization thresholds of xenon. Figure 8 compares the measurements with a theory by Dill, who expressed β in terms of the reduced dipole transition amplitudes which are determined semiempirically using the multichannel quantum defect theory described by Seaton (1966) and Fano (1970). The development of an *ab initio* theory to describe these amplitudes in a similar way to the recent work by Fano and Lee (1973) and Lin (1974) on argon would now be of value. Further discussion of the role of photoelectron angular distributions is given by Dill *et al.* (1974) in a contribution to this symposium.

We conclude our discussion of photoionization by considering CI effects in ejection of an electron leaving the ion in an excited state. For the simplest systems H$^-$ and He the degeneracy of the final states means that coupling effects are particularly important. We show in Figure 9 results of calculations carried out by Hyman *et al.* (1972) using a 1s–2s–2p final state wave function. The probability of leaving the ion in a 2p state is much larger than that of leaving it in a 2s state close to threshold, although as expected this situation is reversed at very high energies. On the other hand, the frozen core Hartree–

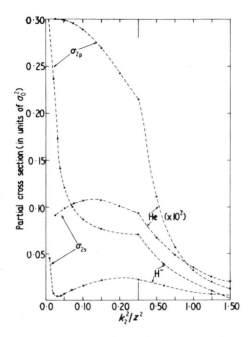

Figure 9. $n = 2$ photoionization cross sections for H^- and He; the change in energy scale at $k_2^2/z^2 = 0.25$ is indicated by the vertical bars [from Hyman et al. (1972), Figure 1].

Fock calculation of Bell et al. (1973), which neglects the 2s–2p interaction, gives a σ_{2p}/σ_{2s} ratio close to unity at low energies. This ratio can be deduced from the experiments of Krause and Wuilleumier (1972, 1974), who measured the ejected electron intensity at 54° 44' and at 90°. Since β is different for the 2s and 2p states (from conservation of angular momentum it must be 2 for the 2s state), the ratio can be deduced if β is known. In addition, the relative intensity of forming He^+ in the $n = 2$ state was also measured by Krause and Wuilleumier and their result and that of Samson (1969) are compared with theory in Figure 10. The latter also shows the asymptotic limit calculated by Salpeter and Zaidi (1962).

In the heavier rare gases Krause and Wuilleumier (1974) have measured the components of the photoionization cross section of Ne, and Lynch et al. (1973) have measured the cross section for ejection of the 3s electron in Ar. The latter measurements are compared with the Hartree–Fock calculations of Kennedy and Mansen (1972) and the random-phase approximation with exchange (RPAE) calculations of Amusia et al. (1972). In Figure 11 the experiments strongly favor the RPAE results, which takes account of CI

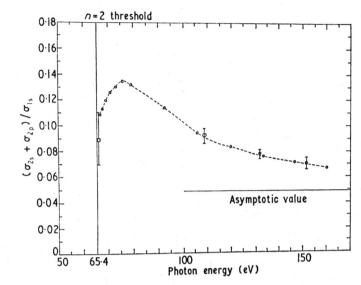

Figure 10. The $(\sigma_{2s} + \sigma_{2p})/\sigma_{1s}$ ratio measured by Krause and Wuilleumier compared with theory from Jacobs and Burke (1972).

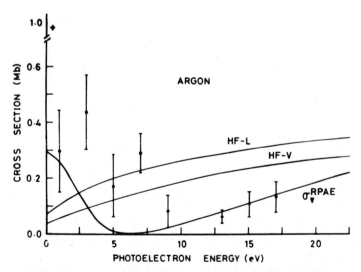

Figure 11. The Ar $3s$ subshell photoionization cross section; the solid curves show theoretical Hartree–Fock length and velocity results and random phase approximation with exchange results [from Lynch *et al.* (1973), Figure 1].

effects between the $3s3p^6\varepsilon p$ and $3s^23p^5\varepsilon d$ channels. This allows for the virtual oscillation of the outer $3p$ shell which is out of phase with the incident electromagnetic wave and which almost completely screens the inner $3s$ shell, giving rise to the minimum in the cross section. A similar situation is found in Kr and Xe, although not in Ne, where the intershell interaction is weaker.

We have not had time to discuss the situation in any detail for heavier atoms where the spin–orbit interaction is important, giving rise to the well-known Fano effect (Fano, 1969a,b). In this case further parameters in addition to σ and β can be measured (Jacobs, 1972), and a study of the role of CI in such situations, while remaining for the most part a challenge for the future, is one where new insights can be expected.

4. Effects of Configuration Interaction in Electron Scattering

We have now seen that an important problem in electron–atom collision theory is to disentangle, as far as possible, CI effects involving the electrons in the target or N-electron correlations from CI effects caused by the interaction of the incident electron with the target or $(N + 1)$-electron correlations. One way of doing this is to make use of the very different energy dependences which these effects have. Thus while the $(N + 1)$-electron correlation effect can dominate at low incident electron energies, it becomes less important at high energies, where the Born or eikonal approximations are appropriate. On the other hand, the N-electron correlation effect can play a significant role at all energies, as can easily be seen in the case of an allowed transition, which at high energy is just proportional to the oscillator strength which in turn depends on the atomic wave function used. A considerable amount of experimental effort is now being given to this high-energy region (e.g., Luyken *et al.*, 1972; Weigold *et al.*, 1973; and Tau and McConkey, 1974) and a significant understanding has now been obtained which is dealt with by McDowell and by Byron in this volume. I will therefore concentrate on the low-energy region, where both effects can be simultaneously important for complex atomic systems.

Before looking at the difficulties in electron scattering by a complex atom, it is helpful first to consider briefly electron scattering by atomic hydrogen and hydrogen-like ions. In this case exact target states are available and only $(N + 1)$-electron correlation effects can occur. In the case of hydrogen the situation is now fairly well understood for excitation of the $2s$ and $2p$ states below the $n = 3$ excitation threshold. The 3-state $1s$–$2s$–$2p$ close coupling approximation gives results which are within 10–20% of experiment, and including in addition 20 Hylleraas type terms in the wave

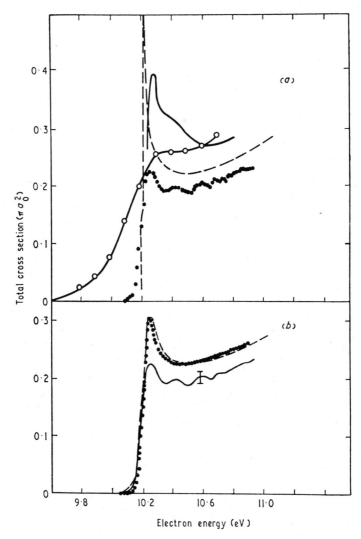

Figure 12. The excitation of the 2p states of atomic hydrogen by electron impact close to the 2p threshold. (a) (– – –) Taylor and Burke (1967); (———) Marriot and Rotenburg (1968); (○—○) Chamberlain et al. (1964); (● ● ●) Williams and McGowan (1968); (b) (———) Williams and McGowan (1968); (● ● ●) Williams and Willis (1974); (– – –) Taylor and Burke (1967), folded with the electron energy distribution function of Williams and Willis [from Williams and Willis (1974), Figure 1].

function, as was done by Taylor and Burke (1967), brings the results into accord with the most recent experiments of Williams and Willis (1974) as shown in Figure 12. However, above the $n = 3$ threshold, loss of flux into these and higher channels plays an important role in the 2s and 2p state excitation, and no really satisfactory theory has been devised, although limited success has been achieved by representing the infinity of omitted channels by a few pseudostate channels in the expansion of the wave function (Burke and Webb, 1970; Callaway and Wooten, 1974).

In the case of 2s and 2p state excitation in He^+ the situation is apparently more complicated. We show in Figure 13 experimental results obtained by Dance et al. (1966) and by Dolder and Peart (1973) compared with the 3-state close coupling approximation, a Coulomb–Born calculation (Burgess et al., 1970), and a distorted wave polarized orbital calculation (McDowell et al., 1973). The results of a 3-state plus 20 correlation term calculation close to threshold by Burke and Taylor (1969), which are not shown in Figure 13, lie a little higher than the Coulomb–Born calculations. Theoretically one might expect this last approximation to be the most accurate one below the $n = 3$ threshold, and in fact it gives excellent results for the positions and widths

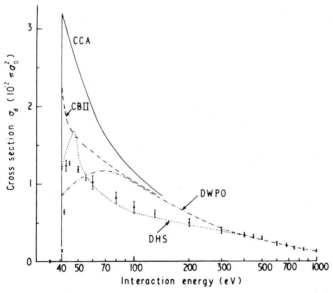

Figure 13. The excitation of the 2s states of He^+ by electron impact. Solid points, Dolder and Peart (1973); dotted curve (DHS), Dance et al.; CCA, 3-state close coupling; CB II, Coulomb–Born approximation; DWPO, distorted-wave polarized orbital [from Dolder and Peart (1973), Figure 4].

of the resonances below the $n = 2$ threshold which involve essentially the same matrix elements. The disagreement with experiment is thus difficult to explain, although it may be due to inaccurate allowance for cascade and the failure of the Coulomb–Born approximation to give reliable results except at very high energies. Further light is thrown on this by calculations carried out by Damburg and Propin (1972), which indicate that at least in the case of $1s$–$2p$ excitation in atomic hydrogen strong coupling between the $2s$ and $2p$ channels can have a significant effect even at 1000 eV.

The effects of CI have been studied recently for electron impact excitation of complex positive ions by Seaton (1973), by Saraph and Seaton (1974), and by Robb (1975). This work extended the earlier work of Ormonde et al. (1973) by including N-electron correlation effects as well as $(N + 1)$-electron correlation effects. We discuss here the low-energy cross sections for the transitions

$$1s^2 2s^2 2p^2\ {}^3P \to {}^1D, \qquad 1s^2 2s^2 2p^2\ {}^3P \to {}^1S \qquad (13)$$

in e^-–N^+ scattering collisions. The following approximations were used:

(i) The $1s^2 2s^2 2p^2\ {}^3P^e$, ${}^1D^e$, and ${}^1S^e$ states included in the close coupling expansion using a single configuration to represent each target state. This approximation is the same as that used by Henry et al. (1969).
(ii) The ${}^3P$, ${}^1D$, and ${}^1S$ target states are again included, but they are now represented by the two configuration wave functions $c_1\ 1s^2 2s^2 2p^2 + c_2\ 1s^2 2p^4$.
(iii) The same as (i), but with the additional inclusion of the $1s^2 2s 2p^3\ {}^3D^0$ ${}^3P^0$ states.
(iv) The same as (ii), but with the additional inclusion of the $1s^2 2s 2p^3\ {}^3D^0$ and ${}^3P^0$ states.

Results at an incident electron energy of 0.4 Ry are given in Table 1. We see that columns (a) and (d), which represent essentially the same physical approximations carried out using completely different numerical methods, agree very well with each other. On the other hand, calculation (iii) of Ormonde et al. is not in line with these. If we believe the results of Saraph and Seaton and of Robb we can infer that the effect of higher closed channels is small and can be neglected, a result which is consistent with the results for H but not those for He^+ (if we believe experiment). We also note that the effect of including CI in the target is small but not completely negligible.

We show in Figure 14 the behavior of the collision strengths obtained by Saraph and Seaton. We note Rydberg series of resonances converging to each of the higher thresholds of the N^+ ion, which strongly affect the collision strength in their neighborhood. It is important to be able to predict the

Table 1. Contributions to the e⁻–N⁺ Collision Strengths from $SL\pi = {}^2P^0$ and ${}^2D^0$ for $[k({}^3P)]^2 = 0.4\,\text{Ry}$

(taken from Robb, 1975)

Approximation (see text)	$SL\pi = {}^2P^0$				$SL\pi = {}^2D^0$			
	(a)	(b)	(c)	(d)	(a)	(b)	(c)	(d)
(i)	2.67	2.6	2.597	2.651	0.311	0.37	—	0.368
(ii)	2.78	—	2.685	2.703	0.359	—	0.389	0.384
(iii)	2.50	1.4	—	2.307	0.299	0.12	—	0.315
(iv)	2.40	—	2.306	2.330	0.303	—	0.338	0.328

(a) From Saraph and Seaton (1974).
(b) From results given in graphical form by Ormonde et al. (1973).
(c) From Robb (1975) using all calculated energy differences.
(d) From Robb (1975) using experimental energy differences.

position and width of these resonances accurately, particularly if any fall close to threshold. Clearly, higher thresholds are important if they support resonances in the energy range of interest.

Perhaps, an even more complex situation is the low-energy scattering of electrons by atomic oxygen. The ground state in this case is $1s^2 2s^2 2p^4 \, {}^3P^e$ and the dominant large-range potential comes from the atomic quadrupole moment and dies off asymptotically as r^{-3}. In addition there is an important contribution arising from the polarization of the atom by the incident electron which behaves asymptotically as r^{-4}. To include part of the effect of polarization Rountree et al. (1974) included a number of real and pseudostates in a close coupling expansion where the $2p$ electron was excited into $3s$ and pseudo-$\overline{3d}$ orbitals. Their results for the total elastic scattering cross section, obtained including terms of the ground state configuration $1s^2 2s^2 2p^4({}^3P, {}^1D, {}^1S)$, the first excited triplet states $1s^2 2s^2 2p^3 3s({}^3S, {}^3D, {}^3P)$, and the pseudostates $1s^2 2s^2 2p^3 \overline{3d}({}^3S, {}^3P, {}^3D)$ are shown in Figure 15. They added the non-s-wave partial cross sections obtained by Henry et al. (1969) to the pseudostate s-wave calculations (for LS = ${}^4P$ and ${}^2P$) to obtain curve A. Circles, triangles, and squares represent, respectively, the experimental results of Sunshine et al. (1967), Neynaber et al. (1961), and Lin and Kivel (1959). In addition curve B corresponds to the 3-state $1s^2 2s^2 2p^4({}^3P, {}^1D, {}^1S)$ results of Henry et al. (1969), curve C to the polarized orbital results of Henry (1967), and curve D to results of Saraph (1973), who included N-electron correlation effects in the target and retained the three ground state terms as in Henry et al.

More recently Nesbet and Thomas (1974), in a paper contributed to this conference, have carried out matrix variational calculations for low-energy

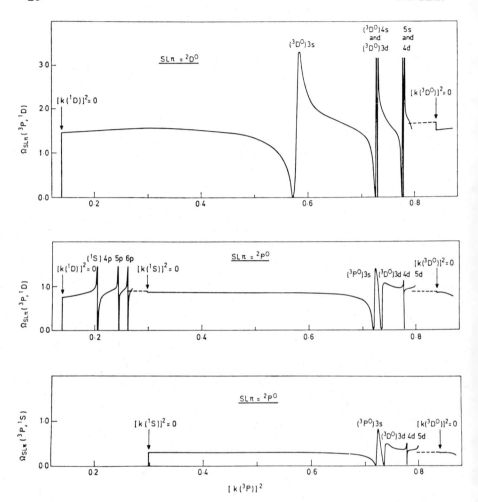

Figure 14. The $SL\pi = {}^2D^0$ contribution to the collision strength $\Omega({}^3P-{}^1D)$ and the $SL\pi = {}^2P^0$ contribution to the collision strength $\Omega({}^3P-{}^1D)$ and $({}^3P-{}^1S)$ for e^--N^+ scattering [from Saraph and Seaton (1974), Figure 1].

electron scattering by atomic oxygen. In their most detailed calculations near-degenerate correlations, represented by the target atom configurations $1s^22s2p^5$ and $1s^22p^6$, were included as well as $2p$ orbital virtual excitation and CI target atom effects. Their results lie somewhat lower than those of Rountree *et al.* except at the very lowest energy, agreeing somewhat better with the experimental data. However, the scatter in the data makes any firm conclusion impossible at this stage, and it is clearly necessary to carry out further experiments.

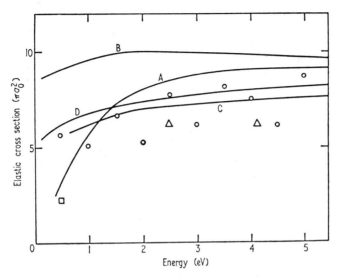

Figure 15. The total elastic scattering cross section for e⁻–O scattering. A, Rountree *et al.* (1974); B, Henry *et al.* (1969); C, Henry (1967); D, Saraph (1973); ○, Sunshine *et al.* (1967); △, Neynaber *et al.* (1961); □, Lin and Kivel (1959) [from Rountree *et al.* (1974); Figure 2].

Finally, we discuss the very interesting situation in the low-energy scattering of electrons by atomic nitrogen. We show in Figure 16 the energy levels corresponding to the $1s^2 2s^2 2p^3(^4S, {}^2D, {}^2P)$ states of N and the $1s^2 2s^2 2p^4(^3P, {}^1D, {}^1S)$ states of N⁻. The results shown for the latter were calculated by Henry *et al.* (1969) retaining just the ground state terms of N in the expansion of the total wave function. The ^3P state gives rise to a

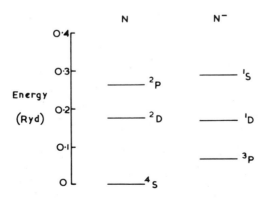

Figure 16. The energy levels of N and N⁻.

p-wave-shape resonance in the low-energy elastic cross section, and the ^1D and ^1S states, while not contributing to scattering on the ground state because of spin conservation, contribute to elastic and inelastic scattering on the ^2D and ^2P excited states of N. Recent calculations by Berrington et al. (1975) and by Thomas et al. (1974) show that the relative position of the ^3P and ^4S states is a very sensitive function of amount of CI included in the calculation, with of course very significant implications for the low-energy cross section. The best indications are that the ^3P state lies just in the continuum, perhaps by as little as 0.1 eV according to recent calculations by Le Dourneuf, Vo Ky Lan, and Burke reported by Van Regemorter (1974) in this symposium. The width of the associated resonance is very narrow and may be impossible to observe with present experimental techniques. The importance of using accurate atomic wave functions in variational calculations of scattering lengths has already been stressed by Peterkop and Rabik (1971), and the situation in nitrogen is an example where the low-energy cross section is also a sensitive function of the atomic wave functions used.

We conclude by noting that increasing emphasis is being given to scattering of electrons by heavy atoms, where the spin–orbit interaction plays an important role. In this case scattering of spin-polarized electrons by atoms, which has been reviewed recently by Drukarev (1973), can be expected to play an even more important role than hitherto in unraveling the basic mechanisms involved in the collision process. This is discussed further by Bedersen (1974) and by Reichert (1974) in this symposium.

Acknowledgments

This work has been supported by the Science Research Council and by the U.S. Office of Naval Research under Contract No. N00014-69-C-0035.

References

Allison, D. C. S., Burke, P. G., and Robb, W. D. (1972). *J. Phys. B.*, **5**, 55.
Altick, P. L. (1968). *Phys. Rev.*, **169**, 21.
Altick, P. L., and Glassgold, A. E. (1964). *Phys. Rev.*, **133**, A632.
Altick, P. L., and Moore, E. N. (1966). *Phys. Rev.*, **147**, 59.
Amusia, M. Ya. (1973). *The Physics of Electronic and Atomic Collisions.* (Invited Lectures and Progress Reports.) Eds. B. C. Čobić and M. V. Kurepa, p. 172. Institute of Physics, Yugoslavia.
Amusia, M. Ya., Cherepkov, N. A., and Chernysheva, L. V. (1971). *Sov. Phys.—JETP.*, **60**, 160.
Amusia, M. Ya., Ivanov, V. K., Cherepkov, N. A., and Chernysheva, L. A. (1972). *Phys. Letters*, **40A**, 361.

Amusia, M. Ya., Cherepkov, N. A., Janev, R. K., Shelftel, S. I., and Živanović, Dj. (1973). *J. Phys. B.*, **6**, 1028.
Bates, G. N., and Altick, P. L. (1973). *J. Phys. B.*, **6**, 653.
Bederson, B. (1974). Contribution to this volume.
Bell, K. L., Kingston, A. E., and Taylor, I. R. (1973). *J. Phys. B.*, **6**, 1228.
Berrington, K. A., Burke, P. G., and Robb, W. D. (1975). *J. Phys. B.*, to be published.
Block, C. (1957). *Nucl. Phys.*, **4**, 503.
Brueckner, K. A. (1959). *The Many Body Problem*, p. 47. New York: Wiley.
Burgess, A., Hummer, D. G., and Tully, J. A. (1970). *Phil. Trans. Roy. Soc.*, **A266**, 225.
Burke, P. G. (1968). *Adv. Atom. Molec. Phys.*, **4**, 173.
Burke, P. G., and McVicar, D. D. (1965). *Proc. Phys. Soc.*, **86**, 989.
Burke, P. G., and Moores, D. L. (1968). *J. Phys. B.*, **1**, 575.
Burke, P. G., and Seaton, M. J. (1971). *Meth. Comp. Phys.*, **10**, 1.
Burke, P. G., and Smith, K. (1962). *Rev. Mod. Phys.*, **34**, 458.
Burke, P. G., and Taylor, A. J. (1969). *J. Phys. B.*, **2**, 44.
Burke, P. G., and Taylor, K. T. (1974). *J. Phys. B.*, to be published.
Burke, P. G., and Webb, T. G. (1970). *J. Phys. B.*, **3**, L131-4.
Burke, P. G., Hibbert, A., and Robb, W. D. (1971). *J. Phys. B.*, **4**, 153.
Byron, F. W. Jr. (1974). Contribution to this volume.
Callaway, J., and Wooten, J. W. (1974). *Phys. Rev.*, **A9**, 1924.
Chamberlain, G. E., and Heideman, H. G. M. (1965). *Phys. Rev. Letters*, **15**, 337.
Chamberlain, G. E., Smith, S. J., and Heddle, D. W. O. (1964). *Phys. Rev. Letters*, **12**, 647.
Chisholm, C. D. H., and Öpik, U. (1964). *Proc. Phys. Soc.*, **83**, 541.
Codling, K., Madden, R. P., and Ederer, D. L. (1967). *Phys. Rev.*, **155**, 26.
Conneely, M. J., Smith, K., and Lipsky, L. (1970). *J. Phys. B.*, **3**, 493.
Cooper, J. W. (1962). *Phys. Rev.*, **128**, 681.
Cooper, J., and Zare, R. N. (1968). *Lectures in Theoretical Physics.* Vol. X1-C, p. 317. Eds. S. Geltman, K. T. Mahanthappa, and W. E. Brittin. New York: Gordon and Breach.
Cooper, J. W., Fano, U., and Prats, F. (1963). *Phys. Rev. Letters*, **10**, 518.
Cvejanović, S., and Read, F. H. (1973). *Electronic and Atomic Collisions.* (Abstracts of Papers at the VIII ICPEAC.) Eds. B. C. Čobić and M. V. Kurepa, p. 442. Institute of Physics, Yugoslavia.
Dalgarno, A., and Victor, G. A. (1966). *Proc. Roy. Soc.*, **A291**, 291.
Damburg, R., and Propin, R. (1972). *J. Phys. B.*, **5**, 533.
Dance, D. F., Harrison, M. F. A., and Smith, A. C. H. (1966). *Proc. Roy. Soc.*, **A270**, 74.
Dill, D. (1973). *Phys. Rev.*, A **7**, 1976.
Dill, D., Starace, A. F., and Mansen, S. T. (1974). Contribution to this volume. (See also *Phys. Rev.* **A11**, 1596.)
Ditchburn, R. W., and Marr, G. V. (1953), *Proc. Phys. Soc.*, **A66**, 655.
Dolder, K. T., and Peart, B. (1973). *J. Phys. B.*, **6**, 2415.
Drukarev, G. (1973). *The Physics of Electronic and Atomic Collisions.* (Invited Lectures and Progress Reports.) Eds. B. C. Čobić and M. V. Kurepa, p. 597. Institute of Physics, Yugoslavia.
Dubau, J., and Wells, J. (1973). *J. Phys. B.*, **6**, L31.
Ehrhardt, H., Hesselbacher, K. H., Jung, K., and Willmann, K. (1972a). *J. Phys. B.*, **5**, 1559.
Ehrhardt, H., Hesselbacher, K. H., Jung, K., Schulz, M., and Willmann, K. (1972b). *J. Phys. B.*, **5**, 2107.
Fano, U. (1935). *Nuovo Cimento*, **12**, 156.
Fano, U. (1961). *Phys. Rev.*, **124**, 1866.
Fano, U. (1969a). *Phys. Rev.*, **178**, 131.

Fano, U. (1969b). *Phys. Rev.*, **184**, 250.
Fano, U. (1970). *Phys. Rev.*, A **2**, 353.
Fano, U., and Cooper, J. W. (1965). *Phys. Rev.*, **137**, A1364.
Fano, U., and Cooper. J. W. (1968). *Rev. Mod. Phys.*, **40**, 441.
Fano, U., and Dill, D. (1972). *Phys. Rev.*, A **6**, 185.
Fano, U., and Lee, C. M. (1973). *Phys. Rev. Letters*, **31**, 1573.
Fano, U., and Lin, C. D. (1973). *The Physics of Electronic and Atomic Collisions.* (Invited Lectures and Progress Reports.) Eds. B. C. Čobić and M. V. Kurepa, p. 229. Institute of Physics, Yugoslavia.
Fano, U., and Lin, C. D. (1974). Invited paper at the IV ICAP Heidelberg.
Fano, U., and Prats, F. (1964). *J. Natl Acad. Sci. Ind.*, **33**, Pt IV, 553.
Feshbach, H. (1958). *Ann Phys.* (*N.Y.*), **5**, 357.
Feshbach, H. (1962). *Ann Phys.* (*N.Y.*), **19**, 287.
Garton, W. R. S., and Codling, K. (1965). *Proc. Phys. Soc.*, **86**, 1067.
Gell-Mann, M., and Brueckner, K. A. (1957). *Phys. Rev.*, **106**, 364.
Goldstone, J. (1957). *Proc. Roy. Soc.* (*Lond.*), **A239**, 267.
Harris, F. E., and Michels, H. H. (1971). *Meth. Comp. Phys.*, **10**, 143.
Henry, R. J. W. (1967). *Phys. Rev.*, **162**, 56.
Henry, R. J. W., Burke, P. G., and Sinfailam, A. L. (1969). *Phys. Rev.*, **178**, 218.
Hyman, H. A., Jacobs, V. L., and Burke, P. G. (1972). *J. Phys. B.*, **5**, 2282.
Jacobs, V. L. (1972). *J. Phys. B.*, **5**, 2257.
Kelly, H. P. (1973). *The Physics of Electronic and Atomic Collisions.* (Invited Lectures and Progress Reports.) Eds. B. C. Čobić and M. V. Kurepa, p. 240. Institute of Physics, Yugoslavia.
Kelly, H. P., and Simons, R. L. (1973). *Phys. Rev. Letters*, **30**, 529.
Kennedy, D. J., and Mansen, S. T. (1972). *Phys. Rev.*, A **5**, 227.
Krause, M. O., and Wuilleumier, F. (1972). *J. Phys. B.*, **5**, L143.
Krause, M. O., and Wuilleumier, F. (1974). Contribution to this volume.
Kuyatt, C. E., Simpson, J. A., and Mielczarek, S. R. (1965). *Phys. Rev.*, **138**, A385.
Lane, A. M., and Robson, D. (1966). *Phys. Rev.*, **151**, 774.
Lee, C. M. (1974). *Phys. Rev.*, A**10**, 584.
Lin, C. D. (1974). *Phys. Rev.*, A**10**, 1986.
Lin, S. C., and Kivel, B. (1959). *Phys. Rev.*, **114**, 1026.
Lipsky, L., and Russek, A. (1966). *Phys. Rev.*, **142**, 59.
Luke, T. M. (1973). *J. Phys. B.*, **6**, 30.
Luyken, B. J. F., de Heer, F. J., and Baas, R. C. (1972). *Physica*, **61**, 200.
Lynch, M. J., Gardner, A. B., Codling, K., and Marr, G. V. (1973). *Phys. Letters*, **43A**, 237.
Macek, J. H. (1968). *J. Phys. B.*, **1**, 831.
Madden, R. P., and Codling, K. (1963). *Phys. Rev. Letters*, **10**, 516.
Madden, R. P., and Codling, K. (1965). *Astrophys. J.*, **141**, 364.
Mahaux, C., and Weidenmüller, H. A. (1969). *Shell Model Approach to Nuclear Reactions*, Ch. 3. Amsterdam: North Holland.
Marriott, R., and Rotenberg, M. (1968). *Phys. Rev. Letters*, **21**, 722.
Martin, P. C., and Schwinger, J. (1959). *Phys. Rev.*, **115**, 1342.
McDowell, M. R. C. (1974). Contribution to this volume.
McDowell, M. R. C., Morgan, L. A., and Myerscough, V. P. (1973). *J. Phys. B.*, **6**, 1441.
Mehlman-Balloffet, G., and Esteva, J. M. (1969). *Astrophys. J.*, **157**, 945.
Mies, F. H. (1968). *Phys. Rev.*, **175**, 164.
Moores, D. L. (1967). *Proc. Phys. Soc.*, **91**, 830.
Nesbet, R. K. (1973). *Comput. Phys. Commun.*, **6**, 275.
Nesbet, R. K., and Thomas, L. D. (1974). Contribution to this volume.

Newson, G. H. (1966). *Proc. Phys. Soc.*, **87**, 975.
Neynaber, R. H., Marino, L. L., Rothe, E. W., and Trujillo, S. M. (1961). *Phys. Rev.*, **123**, 148.
Norcross, D. W. (1974). *Phys. Rev. Letters*, **32**, 192.
Ormonde, S., Smith, K., Torres, B. W., and Davies, A. R. (1973). *Phys. Rev.*, A **8**, 262.
Peterkop, R., and Rabik, L. (1971). *J. Phys. B.*, **4**, 1440.
Pindzola, M. S., and Kelly, H. P. (1974). *Phys. Rev.*, A **9**, 323.
Reichert, E. (1974). Contribution to this volume.
Robb, W. D. (1975). *J. Phys. B.*, **8**, L46.
Robson, D., and Lane, A. M. (1967). *Phys. Rev.*, **161**, 992.
Rountree, S. P., Smith, E. R., and Henry, F. J. W. (1974). *J. Phys. B.*, **7**, L167.
Salpeter, E. E., and Zaidi, M. H. (1962). *Phys. Rev.*, **125**, 248.
Samson, J. A. R. (1969). *Phys. Rev. Letters*, **22**, 693.
Samson, J. A. R., and Gardner, J. L. (1973). *Phys. Rev. Letters*, **31**, 1327.
Saraph, H. (1973). *J. Phys. B.*, **6**, L243.
Saraph, H. E., and Seaton, M. J. (1974). *J. Phys. B.*, **7**, L36.
Schulz, G. J. (1973). *Rev. Mod. Phys.*, **45**, 378.
Seaton, M. J. (1953). *Phil. Trans. Roy. Soc.*, **A245**, 469.
Seaton, M. J. (1966). *Proc. Phys. Soc.*, **A88**, 801.
Seaton, M. J. (1973). *The Physics of Electronic and Atomic Collisions*. (Invited Lectures and Progress Reports.) Eds. B. C. Čobić and M. V. Kurepa, p. 253. Institute of Physics, Yugoslavia.
Shore, B. W. (1967). *Rev. Mod. Phys.*, **39**, 439.
Shore, B. W. (1968). *Phys. Rev.*, **171**, 43.
Simpson, J. A. (1964). *Rev. Sci. Instr.*, **35**, 1698.
Simpson, J. A., and Fano, U. (1963). *Phys. Rev. Letters*, **11**, 158.
Simpson, J. A., Mielczarek, S. R., and Cooper, J. W. (1964). *J. Opt. Soc. Am.*, **54**, 269.
Smith, K., and Morgan, L. A. (1968). *Phys. Rev.*, **165**, 110.
Sunshine, G., Aubrey, B. B., and Bedersen, B. (1967). *Phys. Rev.*, **154**, 1.
Tau, K.-H., and McConkey, J. W. (1974). *J. Phys. B.*, **7**, L185.
Taylor, A. J., and Burke, P. G. (1967). *Proc. Phys. Soc.*, **92**, 336.
Taylor, H. S. (1974). Contribution to this volume.
Thomas, L. D., Yarlagadda, B. S., Csanak, G., and Taylor, H. S. (1973). *Comput. Phys. Commun.*, **6**, 316.
Thomas, L. D., Oberoi, R. S., and Nesbet, F. K. (1974). *Phys. Rev.*, A **10**, 1605.
Van Regemorter, H. (1974). Contribution to this volume.
Wannier, G. H. (1953). *Phys. Rev.*, **90**, 817.
Weigold, E., Hood, S. T., and Teubner, P. J. O. (1973). *Phys. Rev. Letters*, **30**, 475.
Wells, J. (1973). Ph.D. thesis, University of London.
Wendin, G. (1971). *J. Phys. B.*, **4**, 1080.
Wendin, G. (1972). *J. Phys. B.*, **5**, 110.
Wendin, G. (1973). *J. Phys. B.*, **6**, 42.
Williams, J. F., and McGowan, J. W. (1968). *Phys. Rev. Letters*, **21**, 719.
Williams, J. F., and Willis, B. A. (1974). *J. Phys. B.*, **7**, L61.
Yarlagadda, B. S., Csanak, G., Taylor, H. S., Schneider, B., and Yaris, R. (1973). *Phys. Rev.*, A **7**, 146.

2

Low-Energy Electron Scattering and Attachment by C, N, and O Atoms

R. K. NESBET AND L. D. THOMAS

> *Recent theoretical calculations provide a consistent description of low-energy electron scattering by C, N, and O atoms and of the corresponding negative ion states. This work is summarized and reviewed. New results are reported of e–O scattering calculations by the matrix variational method. These calculations include near-degenerate correlation and virtual excitation effects, which must be taken into account in order to locate resonances and negative ions states correctly in energy. They also include the principal residual effect of target atom polarizability, which produces a large reduction in the low-energy electron scattering cross section, consistent with earlier polarized orbital calculations.*

1. Introduction

Theoretical methods applicable to the quantitative description of low-energy electron scattering by complex atoms have recently been reviewed (Nesbet, 1975). In the case of atomic C, N, and O, experimental measurements are difficult because of the need for molecular dissociation. The few experiments that have been carried out are reviewed by Bederson and Kieffer (1971). No experiment of high precision has been performed, but e–N (Neynaber *et al.*, 1963) and e–O (Neynaber *et al.*, 1961; Sunshine *et al.*, 1967) scattering experiments give the general trend of the total ground state cross sections for low impact energies. There are no comparable experiments on e–C scattering. Nevertheless, because of the abundance of these

---

R. K. NESBET and L. D. THOMAS • IBM Research Laboratory, San Jose, California 95193, U.S.A.

light elements, the electron–atom scattering cross sections are important for dynamical models in astrophysical applications and in the study of plasmas containing dissociated atoms. The negative ion states are also of considerable interest. Several states of the expected ground configurations of C^- and N^- have not yet been identified experimentally, either as bound states or as scattering resonances.

The present paper will discuss some very recent theoretical results that should greatly clarify the low-energy properties of the e–C,N,O systems. The principal results are contained in a study of e–C,N,O scattering and resonances by Thomas *et al.* (1974), in the accompanying paper by Van Regemorter (1974), referring to calculations by Le Dourneuf and Vo Ky Lan (1974), and in improved e–O total cross section calculations to be presented here. This recent work quantitatively confirms earlier calculations of negative ion energies by bound state methods that included effects of three-electron correlation (Moser and Nesbet, 1971).

The scattering calculations considered here differ from previous work by inclusion of the short-range correlation effect due to interactions among the $2s$ and $2p$ valence orbitals and by inclusion of the full effect of the electric dipole polarizability of the target atom. The results to be discussed here indicate the importance of both of these effects in e–C,N,O scattering and in the corresponding negative ion states.

2. Resonances and Negative Ion States

The ground state electronic configurations of C, N, and O are of the form $1s^2 2s^2 2p^q$, with a large contribution from the near-degenerate configuration $1s^2 2p^{q+2}$ in states of compatible LS quantum numbers. Negative ion states of the configuration $1s^2 2s^2 2p^{q+1}$ may be bound with respect to the neutral ground state or may occur as electron–neutral scattering resonances if not bound. Not all of these expected negative ion states have been observed experimentally.

The $^4S^0$ ground state of C^- is bound by 1.24 eV with respect to $C(^3P)$, and $O^-(^2P^0)$ is bound by 1.47 eV with respect to $O(^3P)$ (Edlén, 1960). The excited state $C^-(^2D^0)$ has recently been identified as bound by 0.035 eV (Ilin, 1973). The excited state $C^-(^2P^0)$ of configuration $2s^2 2p^3$ has not been observed. None of the corresponding states of N^- has been observed: ^3P, ^1D, ^1S of configuration $2s^2 2p^4$.

Electron affinities of these atoms and negative ion excitation energies were computed several years ago using a variational Bethe–Goldstone formalism that included three-electron correlation effects (Moser and Nesbet, 1971). The electron affinities computed for C and O (1.29 and 1.43 eV, respectively) were in good agreement with experimental values and with the extrapolated values of Edlén (1960). The computed energy of $N^-(^3P)$, 0.12 eV above threshold, disagreed with Edlén's extrapolation (-0.05 eV). Excitation energies of the first excited states of C^- and N^- were computed (Moser and Nesbet, 1971), giving $C^-(^2D^0)$ bound (-0.02 eV) and $N^-(^1D)$ at 1.47 eV above the neutral threshold. The latter state is metastable since it cannot interact except through spin–orbit coupling with the e–$N(^4S^0)$ continuum, and it lies below the $N(^2D^0)$ threshold at 2.38 eV.

Recent scattering theory calculations, to be discussed here, appear to confirm the results of these bound state variational calculations. Total cross sections for e–C,N,O scattering were computed by Thomas et al. (1974) using the matrix variational method (Nesbet, 1973) in two approximations: single configuration (SC) and near-degenerate configuration interaction (CI). Both approximations neglect the principal effects of target atom electric dipole polarizability. The SC calculations are the variational equivalent of earlier close-coupling calculations including all states of the ground configuration $2s^2 2p^q$. The CI calculations augment this with configuration $2s 2p^{q+1}$ (which gives the $2s \rightarrow 2p$ contribution to dipole polarizability) and with $2p^{q+2}$. The short-range correlations introduced by the CI calculations have dramatic effects on resonance structures found at low energies in SC calculations. In general, resonances are moved by the CI calculations to locations compatible with known experimental data and with the bound state variational calculations mentioned above. The qualitative conclusions of this work (Thomas et al., 1974) are that O^- should have no low-energy resonances (the $^2P^0$ state is well below threshold), that $N^-(^3P)$ is either very weakly bound or a very narrow resonance just above threshold, and that $C^-(^2D^0)$ should be bound while $C^-(^2P^0)$ should be a resonance near 0.6 eV.

In the case of $N^-(^3P)$, the conclusion that this state lies very near the neutral ground state has been confirmed and made more precise by close-coupling calculations that include both the near-degenerate CI correlation effects and the full effect of target atom polarizability. Details of this work are given by Van Regemorter (1974). The $N^-(^3P)$ state is found to be a narrow resonance at 0.13 eV, in remarkable agreement with the earlier bound state calculation of Moser and Nesbet (1971), which located this state at 0.12 eV. As a check on the reliability of this result, the $O^-(^2P^0)$ ground state was computed with the same procedure, referred to as the "polarized frozen core" approximation, and found to be bound by 1.48 eV (Le Dourneuf and Vo Ky Lan, 1974), very close to the experimental value (1.47 eV) and to the bound state variational result (1.43 eV).

3. Scattering Cross Sections

The importance of negative ion resonances in low-energy electron scattering by complex atoms is illustrated in Figure 1, which compares SC and CI matrix variational calculations of the e–C total ground state cross section (Thomas *et al.*, 1974) with a polarized orbital calculation (Henry, 1968) that includes the full effect of target atom electric dipole polarizability but omits the wave function structure required to describe resonances. In comparison with the SC calculation, the $^2D^0$ state of C$^-$ in the CI calculation has moved below threshold, and the $^2P^0$ state appears as a prominent resonance. The polarized orbital calculation of Henry (1968) indicates that the general effect of target atom polarizability is to decrease the total cross section. From these results, it is expected that the true e–C cross section would be well approximated by superimposing the $^2P^0$ resonance structure near 0.6 eV on the polarized orbital curve.

Similar results for e–N scattering are shown in Figure 2, which compares SC and CI calculations (Thomas *et al.*, 1974) with a polarized orbital calculation (Henry, 1968) and with available experimental data (Neynaber *et al.*, 1963). In this case, the near-degeneracy correlation and polarization effects included in the CI wave function have moved the N$^-$(^3P) state just below threshold. There is no rigorous variational bound on the relative location of this state, so that further improvement of the wave function may displace it

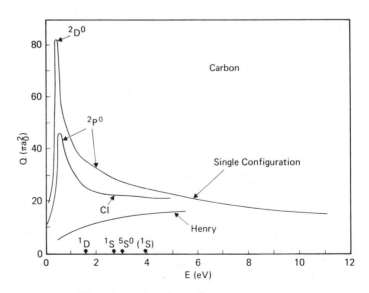

Figure 1. e–C total ground state cross section.

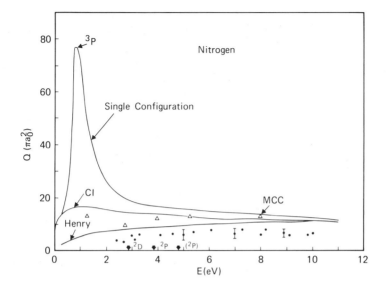

Figure 2. e–N total ground state cross section.

either up or down. It can be concluded, however, that this state is either very loosely bound or a narrow resonance just above threshold. The polarized orbital calculation of Henry (1968) indicates that a substantial reduction of the cross section should result from the full inclusion of target atom polarization.

These conclusions are supported by the e–N scattering calculations reported by Van Regemorter (1974). In the polarized frozen core approximation, the CI wave function is combined with a pseudostate expansion that describes the full effect of target electric dipole polarization. The computed cross section (Van Regemorter, 1974; Le Dourneuf and Vo Ky Lan, 1974) can be described qualitatively as that which would be obtained by superimposing a narrow ^3P resonance structure at 0.13 eV on the polarized orbital cross section curve. It can be concluded that the e–N cross section obtained in the polarized frozen core approximation must be close to the true cross section in the low-energy range considered here.

We report here preliminary results of matrix variational calculations of the e–O total cross section that include polarization effects in a multi-channel formalism capable of accurate description of threshold effects and resonances.

Calculations and experimental data for the total e–O ground state are compared in Figure 3. The SC and CI calculations follow the pattern described above for C and N scattering. The curve labeled BG is obtained

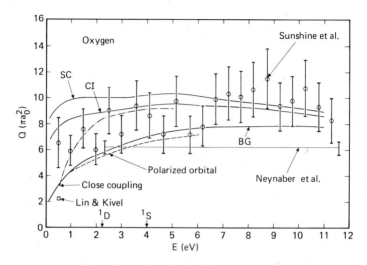

Figure 3. e–O total ground state cross section.

by variational Bethe–Goldstone (BG) equations for [2p] virtual excitations, in addition to the $2s \to 2p$ effects included in the CI wave function. These results include all open-channel states arising from s, p, d, and f continuum orbitals.

Comparison with the polarized orbital calculation of Henry (1967), shown in Figure 3, indicates that the BG calculation includes a very good representation of target atom polarizability. Because the BG calculation can represent all relevant aspects of the electronic wave function, it should be close to the true cross section. Since there is no low-energy resonance structure, in contrast to C and N, this cross section is well approximated by the polarized orbital result.

A recent close-coupling calculation, including polarization pseudostates (Rountree *et al.*, 1974), also shown in Figure 3, represents [2p] virtual excitations less completely than does the BG calculation. These results approach the CI curve at energies above the ^1D threshold.

The BG cross section follows the general trend of the lower limit of $\pm 20\%$ error bars shown on the rather scattered data of Sunshine *et al.* (1967). The BG cross section is in reasonable agreement with the shock tube result of Lin and Kivel (1959) and with the least-squares value used by Neynaber *et al.* (1961) to represent their data. Measurements of the ratio of forward to backward e–O scattering have recently been reported (Dehmel *et al.*, 1974). As shown in Figure 4, the ratio obtained by integrating the BG differential cross section over the experimental angular range lies within the experimental error bars.

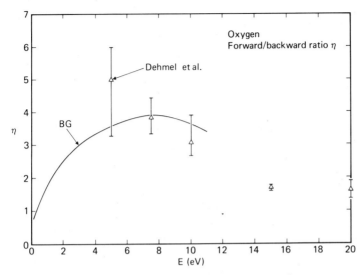

Figure 4. e–O forward/backward scattering ratio.

4. Discussion

The results presented here indicate that a quantitative theory of low-energy electron scattering by complex atoms requires two distinct elements: negative ion states or resonances must be correctly located, and the full effect of target atom polarizability must be taken into account. Recent calculations of e–N scattering, reported by Van Regemorter (1974) in this volume, and of e–O scattering, reported here, apparently meet these criteria. The correct location of resonances is verified by comparison with bound state calculations (Moser and Nesbet, 1971; Le Dourneuf and Vo Ky Lan, 1974), and the inclusion of polarization effects is verified by comparison with polarized orbital calculations (Henry, 1967, 1968).

It has been shown that resonances are correctly located in matrix variational calculations that include only the near-degenerate correlation and polarization effects of $2s \to 2p$ or $2s^2 \to 2p^2$ virtual excitations (Thomas *et al.*, 1974). The residual effect of target atom polarization results primarily in a general reduction of the electron scattering cross section. It would be of interest to verify this conclusion by carrying out close-coupling calculations with wave functions of the same structure as those denoted by CI here.

As applied to negative ion bound states (Le Dourneuf and Vo Ky Lan, 1974), the polarized frozen core method is computationally much simpler than the variational Bethe–Goldstone method (Moser and Nesbet, 1971), which had to include three-electron correlation effects in order to obtain comparable agreement with known electron affinities. The structural difference between these methods is that relaxation and correlation of the neutral core is completely omitted from the negative ion calculation in the polarized frozen core method, whereas such effects are computed for both neutral atom and ion in the Bethe–Goldstone method, and subtracted to obtain a net differential effect. The results considered here indicate a definite computational advantage of the polarized frozen core approach, which corresponds to direct computation of the one-electron Green's function for an electron added to the neutral atom.

Acknowledgment

This work has been supported in part by the Office of Naval Research, Contract No. N00014-72-C-0051.

References

Bederson, B., and Kieffer, L. J. (1971). *Rev. Mod. Phys.*, **43**, 601.
Dehmel, R. C., Fineman, M. A., and Miller, D. R. (1974). *Phys. Rev.*, A **9**, 1564.
Edlén, B. (1960). *J. Chem. Phys.*, **33**, 98.
Henry, R. J. W. (1967). *Phys. Rev.*, **162**, 56.
Henry, R. J. W. (1968). *Phys. Rev.*, **172**, 99.
Ilin, R. N. (1973). In *Atomic Physics 3*. Eds. S. J. Smith and G. K. Walters, p. 309. New York: Plenum Press.
Le Dourneuf, M., and Vo Ky Lan (1974). *Abstract, IV Intl. Conf. on Atomic Physics, Heidelberg*, 154.
Lin, S. C., and Kivel, B. (1959). *Phys. Rev.*, **114**, 1026.
Moser, C. M., and Nesbet, R. K. (1971). *Phys. Rev.*, A **4**, 1336.
Nesbet, R. K. (1973). *Comput. Phys. Commun.*, **6**, 275.
Nesbet, R. K. (1975). *Adv. Quantum Chem.*, **9**, 215.
Neynaber, R. H., Marino, L. L., Rothe, E. W., and Trujillo, S. M. (1961). *Phys. Rev.*, **123**, 148.
Neynaber, R. H., Marino, L. L., Rothe, E. W., and Trujillo, S. M. (1963). *Phys. Rev.*, **129**, 2069.
Rountree, S. P., Smith, E. R., and Henry, R. J. W. (1974). *J. Phys. B.*, **7**, L167.
Sunshine, G., Aubrey, B. B., and Bederson, B. (1967). *Phys. Rev.*, **154**, 1.
Thomas, L. D., Oberoi, R. S., and Nesbet, R. K. (1974). *Phys. Rev.*, A **10**, 1605.
Van Regemorter, H. (1974). Contribution to this volume.

3

The Ejection of Electrons from Autoionizing States in Atoms

C. BOTTCHER

Experimentalists are increasingly interested in the information that can be obtained from the angular and energy distributions of electrons ejected from autoionizing states. We consider two problems in this area: (a) the relation of the angular distributions to the symmetry of the autoionizing state; (b) the effect of interactions between the ejected and scattered electrons when the autoionizing state is excited by low-energy electrons.

We are studying the mechanisms for exciting autoionizing states in atoms and molecules and the information which can be obtained from the energy and angular distributions of the ejected electrons. It is a pleasure to acknowledge the contributions made to this subject by Professor Ugo Fano. This paper outlines progress in two areas of interest to experimentalists.

1. Angular Distribution of Ejected Electrons

When electrons are ejected from an autoionizing state excited by photoionization or electron impact one would expect the angular distribution of the ejected electrons to yield valuable information on the symmetry of the states involved. This problem is more complicated than that associated with resonances in elastic scattering, since in photoionization, say, the photon can couple a resonance to more than one outgoing partial wave. For the same reason the variation in angular distribution between states of different

---

C. BOTTCHER • Department of Theoretical Physics, The University, Manchester M13 9PL, England.

symmetry is less striking than in elastic scattering. It can be shown that if one asks what is the correlation between the directions of the electric vector and the ejected electron, one has a standard problem in angular correlation theory, to which the answer may be found in the nuclear literature (Goldfarb, 1959). Special cases have recently been considered by Dill et al. (1974).

As an illustration consider the process

$$hv + A(S_0 L_0) \to e(\mathbf{k}_i) + A^+(S_i L_i) \tag{1}$$

Let the cross section be proportional to the square of a matrix element

$$M_\mu^{(\sigma)} = \langle \Psi_0 | H_\mu^{(\sigma)} | \Psi_i \rangle \tag{2}$$

where $H^{(\sigma)}$ is a tensor of rank σ, a function of spatial coordinates only. If we introduce the function

$$M(\mathbf{k}_0, \mathbf{k}_i) = \sum_\mu M_\mu^{(\sigma)} Y_{\sigma\mu}(\mathbf{k}_0) \tag{3}$$

it is readily shown that

$$M(\mathbf{k}_0, \mathbf{k}_i) \sim \sum_{Ll} \langle LM_L | L_i M_{L_i} l m_l \rangle \langle LM_L | L_0 M_{L_0} \sigma \mu \rangle$$
$$\times Y_{lm_l}(\mathbf{k}_i)^* Y_{\sigma\mu}(\mathbf{k}_0) \langle SM_{S_0} | S_i M_{S_i} s m_s \rangle \langle S_i L_i s l S_0 L \| H^{(\sigma)} \| S_0 L_0 \rangle \tag{4}$$

which has the form of a scattering amplitude with the reactance matrix

$$i^{l-\sigma} T_L(L_i l \| L_0 \sigma) = \langle S_0 M_{S_0} | S_i M_{S_i} s m_s \rangle \langle S_i L_i s l S_0 L \| H^{(\sigma)} \| S_0 L_0 \rangle$$

These standard theorems give the result that (averaging over spin quantum numbers),

$$\overline{|M(\mathbf{k}_0, \mathbf{k}_i)|^2} \sim \sum_k \mathscr{L}_k P_k(\cos \vartheta) \tag{5}$$

where ϑ is the angle between $\mathbf{k}_0$ and $\mathbf{k}_i$ and

$$\mathscr{L}_k = \sum_{LiL'l'} A_k(lLl'L'; L_i) A_k(\sigma L \sigma L'; L_0) B_{0i}(lL) B_{0i}(l'L') \tag{6}$$

We have written $B_{0i}(lL) = \langle S_i L_i s l S_0 L \| H^{(\sigma)} \| S_0 L_0 \rangle$ for brevity. The result (5), (6) is easily extended to more complicated coupling schemes in atoms, and to molecules. Numerical calculations suggest that fairly accurate experiments will be necessary to assign quantum numbers to the resonances and disentangle the background effects.

2. Shifted Thresholds in Ejection by Low-Energy Electrons

Hicks et al. (1974) have studied the process

$$e_s(E) + \text{He} \to e_s(E - E_r) + \text{He}^{**}(2s^2)$$
$$\to e_S + e_i(\varepsilon) + \text{He}^+ \quad (7)$$

as a function of E. They found that ejected electrons did not appear until $E = 58.7$ eV while E_r is known to be 57.8 eV. The energy of the ejected electrons is also increased by 0.9 eV, suggesting that E_r is effectively shifted by this amount. As E is increased E_r tends to its usual value. The effect has now been studied for several resonances: the larger the width Γ the greater the shift. Hicks et al. have advanced a classical model in which the outgoing electron e_s is at a distance $r_1 = vt(t \sim \hbar/\Gamma)$ when the atom decays, so that the ejected electron is repelled by a barrier l^2/r_1. This accounts fairly well for some of the observed results, but the total picture is very complicated and classical theory is probably only valid if $r_1 \gg a_0$. Thus we are examining a quantal description adapting the formalism used for resonance processes in molecules (Bottcher, 1974a,b).

We introduce a projection operator P onto those states of the $e_s + $ He system in which the He$^+$ core is unexcited and its complement $Q = 1 - P$. This enables us to isolate the interaction of e_s with the autoionizing atom He**. Formally, the transition matrix element is

$$T_{if} = \langle X_{ir}|H_{QP}|\psi_f\rangle \quad (8)$$

where

$$(E - \mathcal{H}) = H_{QP}\psi_i \quad (9)$$

$\mathcal{H}$ is the complex optical Hamiltonian in Q-space and X_{ir} has an outgoing wave in the $e_s + $ He** channel. Since He** is unstable the wave number k_c is a complex number given by

$$\tfrac{1}{2}k_c^2 = E - E_r + \frac{i}{2}\Gamma \quad (10)$$

and X_{ir} decays like $\exp(-r_1/vt)$. In the final state the interaction of e_i with e_s is averaged over X_{ir}^2, leading to a barrier in agreement with the classical model. The physical significance of $|X_{ir}(r_1)|^2$ is that it is the probability that He** decays when e_s is at a distance r_1. Clearly this probability should tend to zero as $r_1 \to \infty$. K. R. Schneider has now developed a technique for solving (9) with some approximations and has obtained energy shifts of the correct order, though rather larger than those observed. Hopefully agreement can be improved. We expect that the energy shift should be a strong function of the angle between e_i and e_s and this should be investigated theoretically in anticipation of coincidence experiments.

References

Bottcher, C. (1974a). *J. Phys. B.*, **7**, L221.
Bottcher, C. (1974b). *Proc. Roy. Soc.*, **A340**, 301.
Dill, D., Manson, S. T., and Starace, A. F. (1974). *Phys. Rev. Letters*, **32**, 971.
Goldfarb, L. J. B. (1959). Angular correlations and polarisation. In *Nuclear Reactions*, Vol. 1, p. 159. Eds. P. M. Endt and M. Demeur, Amsterdam: North-Holland.
Hicks, P. J., Cvejanovic, S., Comer, J., Read, F. H., and Sharp, J. M. (1974). *Vacuum*, **24**, 573.

4

The Study of Photoionization Processes Using Synchrotron Radiation

G. V. MARR

> The characteristics of synchroton radiation as a UV and x-ray source of photons are reviewed, monochromator design problems are stated, and experimental rigs in use at the Daresbury Synchrotron Radiation Facility are discussed. The requirements for a systematic study of atomic photoionization processes are outlined through energy level analysis, cross section measurements, and angular distribution studies of photoelectrons. Progress made is reviewed and new data presented for total cross section measurements on argon and krypton and the partial cross section for 3s electrons ejected from argon. Angular distribution studies of the asymmetry parameter are discussed and inert gas data compared with theory which provides evidence of multielectron correlation effects. An analysis of spin–orbit interaction effects on the angular distribution of s electrons from alkali metal vapors is reported.

1. Introduction

The processes involved in the interaction of ultraviolet radiation with atoms has been actively studied for many years. Review articles have been published on specific areas of activity (e.g., Biberman and Norman, 1967; Marr, 1967; Hudson and Kieffer, 1971), particular techniques have assumed sufficient significance to warrant conferences (e.g., Shirley, 1972, on electron spectroscopy; Marr and Munro, 1973, on synchrotron radiation), and it is timely, therefore, to examine the progress and prospects for the study of

G. V. MARR • The J. J. Thomson Physical Laboratory, The University, Whiteknights, Reading RG6 2AF, Berkshire, England.

photoionization processes at this symposium in honor of Professor Fano, who has contributed so much to our understanding of the subject.

The interaction of ultraviolet photons with atoms is important in determining the energy balance in situations to be found in the fields of aeronomy, astrophysics, plasma physics, and gas breakdown. An understanding of the behavior of these systems must be based on a detailed knowledge of their atomic interaction kinetics specified in absorption by photo- and autoionization. A systematic study requires three types of data:

(a) Discrete energy level analysis of atoms and ions.
(b) Transition probabilities for the photoionization processes as a function of photon energy.
(c) Information concerning the directional nature of the interaction.

Under (a) we are concerned with the observation and recording of absorption spectra under conditions of high resolution, and this is often most effectively carried out with photographic detectors. In this spectral region, the vacuum ultraviolet, term analysis provides information not only about the valence electrons, but also about inner core electrons. Data under (b) are obtained by absolute cross section measurements associated with photons of a given energy through oscillator strength distributions and lifetimes of various excited states of the ions. As a result of the continuous dependence on photon energy, cross sections for ionization into individual ionic states overlap, and the total absorption cross section has to be unfolded into a series of "partial cross sections" showing the probability for ionization into each ionic state available at a given photon energy. Data under (c) can acquire considerable significance in a low-density system. In a dense gas, the direction of the ejected photoelectron would soon lose significance because of the short time needed to produce an isotropic distribution through electron–atom collisions. However, in, for example, the upper reaches of the atmosphere, photoelectrons could travel an appreciable distance before losing energy to the medium by collision. A photoelectron ejected from atomic oxygen might spiral along magnetic field lines to deposit energy near the poles. The efficiency of this process would depend on the distribution of photoelectron velocities with respect to the magnetic field lines.

This is not to say that the reasons for study of the photoionization processes are only of interest in that they provide data for the elucidation of problems in other areas of activity. The data are important in understanding the basic processes. In the past, absorption cross section data showed conclusively the domination of autoionization over straight photoionization. More recently, the investigation of multiple ionization is providing an insight into "shakeup" and "shake off" processes, and the angular distribution measurements on photoelectrons could provide a significant contribution to the analysis of transition integrals for inner electrons.

2. Ultraviolet Photon Sources

Photoionization of ground state atoms begins with the threshold for cesium at 3180 Å (3.89 eV). At progressively shorter wavelengths, all the atomic species successively photoionize down to and beyond the thresholds for the innermost shells of the transuranium elements at about 0.04 Å (3×10^5 eV). Double ionization, that is, the ejection of two electrons by a single incident photon, begins at 815 Å (15.21 eV) for barium down to the threshold at 153 Å (81.01 eV) for lithium. Photon sources are required, therefore, to cover the range of 3.89 3V to about 10^6 eV (3180 Å to ~0.01 Å), in order that all aspects of the subject can be covered for all atomic species.

Photon sources available for localized regions within this very extensive spectral range are limited. The more usual sources are discussed in detail by Samson (1967), and it is sufficient here to indicate the categories into which they fall. The inert gases and hydrogen provide limited molecular emission continua down to ~600 Å. Optimum intensities, irrespective of whether continuous wave or pulsed spectra are involved, are limited to 10^8–10^9 photons $s^{-1} Å^{-1}$ bandwidth. Source designs vary considerably, and any particular gas filling is only useful over a few hundred angstroms. (H_2 is an exception and provides a continuum from the near-ultraviolet down to around 1650 Å.) To shorter wavelengths, flash tubes involving high current densities and operating over microsecond time scales can provide line and continuum emissions down into the x-ray region. However, these sources have a limited life before requiring maintenance or adjustment (~5×10^3 pulses with a uranium electrode are typical). While these are extremely useful for photographic work, reproducibility per flash is limited, and they tend to contaminate the system with electrode debris.

It is, of course, possible to provide dc discharges producing atomic resonance lines, He I (584 Å) being the most intense with 10^{10}–10^{11} photons s^{-1} intensity. Other lines have lower oscillator strengths and are not as favored. He II (304 Å) can be two orders of magnitude less in intensity. While it is not possible to cover large regions of the spectral range with these lamps, they do provide very stable sources which, because of their isolated sharp emission line characteristics, can be used with low-resolution monochromators or filters to provide high-resolution studies of selected wavelengths. If such lamps are used with undispersed radiation, care has to be taken to ensure that weak lines also in the spectrum do not provide spurious effects (e.g., Mitchell and Wilson, 1969).

3. Synchrotron Radiation

It is against the background of limited availability of ultraviolet photon sources that the electron synchrotron has to be considered. Given the right operation conditions, ultraviolet and x-radiation can be made available for photon interaction experiments in a clean high-vacuum system. The radiation emitted as the major energy loss mechanism from the orbiting electrons forms an intense continuum extending from the infrared to the x-ray region; it is highly directional and essentially plane-polarized. In the vacuum far-ultraviolet and x-ray region the divergence of the radiation is comparable to that of a conventional laser. The properties of synchrotron radiation are detailed in a number of texts and publications, and a review of its use has recently been compiled by Codling (1973). For the purpose of this paper, a simplified discussion will suffice.

Classical physics tells us that the instantaneous power radiation by an accelerated charge can be expressed by

$$P = \frac{e^2}{6\pi\varepsilon_0 m^2 c^3} \frac{dp^2}{dt} \tag{1}$$

where p is the momentum and the other symbols have their usual meaning. For radial acceleration, dp/dt changes with direction and so can be significant even if the change in energy is small or zero. Because of the mass difference between protons and electrons, only electron accelerators produce this power loss effectively at attainable energies. Radial accelerator machines are vastly more effective than linear ones, and the electron synchrotron or storage ring is an efficient source of this power loss called synchrotron radiation. In these machines the electrons are moving at speeds approaching that of light, and the relativistic form of Eq. (1) can be written

$$P = \frac{e^2 \gamma^2 \omega^2 p^2}{6\pi\varepsilon_0 m_0^2 c^3} \tag{2}$$

where

$$\gamma = (1 - v^2/c^2)^{-1/2} = \left(\frac{E}{m_0 c^2}\right) \tag{3}$$

where ω is the orbital frequency, approximately equal to c/R for an electron with velocity $v \to c$ and energy E moving over a circular orbit of radius R. For the Daresbury electron synchrotron NINA, E is ~ 5 GeV, R is ~ 20 m, and the power loss to synchrotron radiation is measured in megawatts.

Figure 1a shows the radiation pattern to be expected from a radiating electron in the rest frame of the electron. In the laboratory frame of reference, however, relativity theory shows that the radiation emitted in a direction θ'

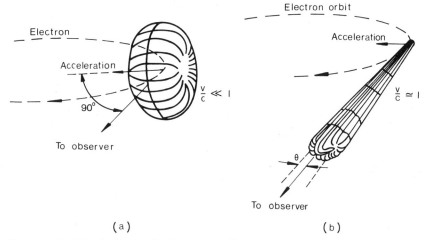

Figure 1. Spatial radiation distribution from a radially accelerated electron: (a) nonrelativistic case $v \ll c$; (b) relativistic case $v \approx c$.

to the direction of motion of the electron has to be transferred from the electron frame to that of the laboratory via

$$\tan \theta = \left(\frac{m_0 c^2}{E}\right) \frac{\sin \theta'}{v/c + \cos \theta'} \qquad (4)$$

The importance of the transformation is apparent when we let $\theta' = 90°$. The radiation node in the electron frame is moved to θ in the laboratory frame of reference, where $\tan \theta \approx (m_0 c^2)/E$. For $E = 5 \text{ GeV}$, $\tan \theta \approx \theta = 10^{-1}$ mrad. The situation is illustrated in Figure 1b, where the distortion is such that we can even "see" radiation from the backside of the radiation pattern, since angles greater than 90° will be projected in the forward direction. The radiation pattern for relativistic electrons is pushed forward into a narrow cone of half angle $\sim 1/\gamma$. On axis the radiation will be polarized with the E vector in the plane of the electron orbit. Above and below this plane there will be a perpendicular component of plane-polarized radiation produced as part of the relativistic distortion of the radiation pattern. Observations limited to the plane of the electron orbit will experience 100% plane-polarized radiation.

The narrow cone of radiation confined to the tangent to the electron orbit necessarily affects the frequency spread of the short pulse of radiation received by a detector. This is illustrated in Figure 2, in which we assume that the detector can receive radiation only while the electron is between the points A and B of its orbit. The time of observation is therefore limited

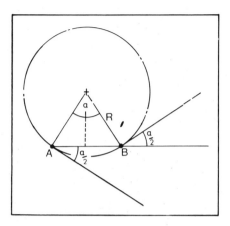

Figure 2. Part AB of the total electron orbit shown schematically. Light with an angular spread α is emitted so that it reaches a detector placed tangentially to the orbit.

to the time taken for the electron to move from A to B minus the time taken for the light to get from A to B, or

$$\Delta t = \frac{R}{v}\alpha - \frac{2R \sin(\alpha/2)}{c} \qquad (5)$$

By expanding $\sin(\alpha/2)$ and putting $v/c = 1$, we get

$$\Delta t \approx \frac{1}{3}\frac{R}{c}\alpha^3 \sim \frac{1}{\omega}\left(\frac{m_0 c^2}{E}\right)^3 \qquad (6)$$

A Fourier analysis of the pulse shows that we can expect to observe a spectrum of harmonics of ω up to $\sim 1/\Delta t$, which for 5 GeV electrons gives wavelengths down to <1 Å. Indeed, because of the variations in the electron orbits, the harmonics in practice are smoothed out and we get a pure continuum stretching from the infrared to the x-ray region. More precise calculations enable the exact details of the wavelength and angular distributions of the photons to be established. The peak of the radiation shifts to shorter wavelengths as E is increased and essentially adds to the low-wavelength (high-energy) end of the spectrum. Figure 3 shows some typical flux distributions for the NINA synchrotron.

To the properties of the light source already mentioned we can add one other property, that the light is 100% modulated. The synchrotron is

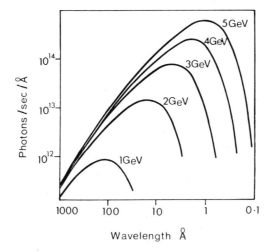

Figure 3. The absolute intensity spectrum falling onto a diffraction grating 30 mm × 50 mm ruled surface, placed 35 m from the NINA ring. The curves are for different values of electron energy.

filled by injecting bunches of electrons into the ring at finite intervals; radio frequency power then boosts the electrons into their required energy, and this provides further bunching of the electrons. The radiation seen at the detector is therefore modulated at various machine parameter frequencies. For NINA we have an injection frequency of 53 Hz, an orbital frequency of 1.36 MHz, and an acceleration frequency of 407.88 MHz.

A survey of the various synchrotron installations available as light sources is given in Table 1 in chronological order of the date of commencement of synchrotron radiation experiments. There is now an increasing requirement for electron storage rings rather than accelerators, the reason being the increased stability and higher circulating currents available in the electron beam orbit. With this improvement goes, however, a requirement of high vacuum conditions (10^{-10} torr, or 1.33×10^{-8} Pa) to minimize beam losses by collision and so provide a long half-life of the electron beam for synchrotron radiation. Figure 4 shows the layout of the synchrotron radiation facility at NINA and Figure 5 is a view of the northern beam line with its associated apparatus for atomic photoionization measurements. Three spectrographic instruments are in operation at any one time, suitably arranged so that each diffraction grating is exposed to radiation from the synchrotron ring.

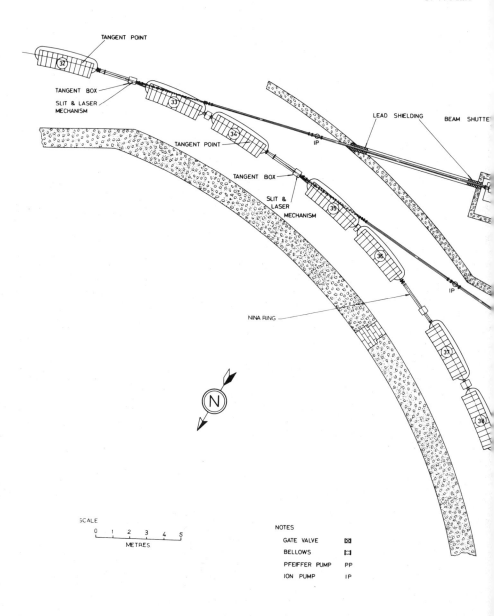

Figure 4. A general layout of the Synchrotron Radiation Facility at the Daresbury Laboratory. To the left is shown part of the NINA electron orbit ring. Radiation is accepted from points tangential to two focusing magnets. The distance from these points to the experimental hall is about 35 m.

Atomic Photoionization Processes

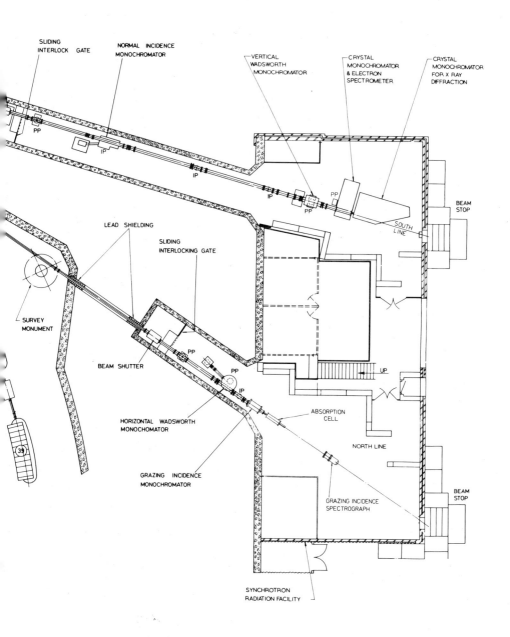

Table 1. Synchrotrons and Storage Rings (Present and Projected), as Sources of Synchrotron Radiation

λ_c is the "critical wavelength" which is related to the peak of the photon flux expected at the electron energies indicated.

Project	E(GeV)	R(m)	I(mA)	λ_c(Å)
Synchrotrons				
NBS (Washington)†	0.18	0.84	1	795
INS-SOR (Tokyo)	1.3	4.0	30	10
Frascati	1.1	3.6	15	15
DESY (Hamburg)	7.5	31.7	20	0.42
Bonn I	2.3	7.65	30	3.5
II	0.5	1.7	30	76
Moscow	0.68	2	–	35
Erevan (USSR)	6.0	24.7	20	0.64
DL (Daresbury)	5.0	20.8	20	0.93
PTB (Braunschweig)	0.14	0.46	–	937
Storage Rings				
Wisconsin	0.24	0.54	10	220
ACO (Orsay)	0.55	1.1	100	20
DCI (Orsay)	1.8	3.8	250	3.8
SLAC (Stanford)	2.5	12.7	250	4.5
INS-SOR (Tokyo)	0.3	1	100	200
DESY (Hamburg)	1.75	12.1	1000	12.7
	3.5	12.1	200	1.6
DL (Daresbury)	2.0	5.5	1000	3.9

† To be converted to a storage ring.

4. Monochromators

There are two main problems associated with the design of suitable monochromators for use with synchrotron radiation in the vacuum ultraviolet and x-ray regions. One is that the light source is immovable and most experiments require the monochromatized radiation to remain fixed in direction as the photon energy is varied. Figure 6 illustrates a design of the University of Reading group used at the Glasgow synchrotron in which a horizontal beam from the synchrotron is transmitted through a vertical Rowland circle concave diffraction grating instrument to provide a horizontal, monochromatized beam at the exit. Mirror–knife edge combinations replace the conventional entrance and exit slits, and the wavelength scan is accomplished by driving the grating round the Rowland circle. The mirror–knife edges are arranged to rotate through one-quarter the angle moved

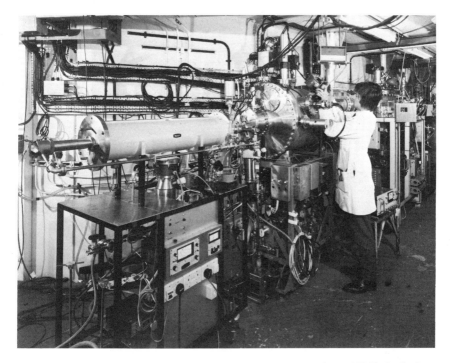

Figure 5. The northern beam line looking toward the synchrotron ring at NINA. In the foreground is the grazing incidence monochromator (West *et al.*, 1974a) and a heat pipe absorption cell for total cross section measurement. In the background is the horizontal-dispersion Wadsworth monochromator used for photoelectron spectroscopy.

through by the grating and so maintain the incident and diffracted beams in the horizontal plane. This instrument has been described in detail by Codling and Mitchell (1970).

The second problem to be overcome is that when a diffraction grating is set to transmit a particular wavelength in the first order, second order radiation at half the wavelength is almost invariably transmitted. The problem is enhanced with synchrotron radiation because the form of the spectral distribution is such that the intensities of the second- and higher-order radiations are often greater than that of the first order. Figure 7 shows a design (West *et al.*, 1974a), again from the Reading group, which attempts to solve both problems. Radiation from an entrance slit situated at the electron tangent point falls onto a plane diffraction grating G and then onto one of two mirrors M1 or M2 which can each be positioned in one of two locations to focus the monochromatized radiation at the exit slits S1 or S2. The wavelength is changed by a simple rotation of the diffraction grating, and provided

Figure 6. Photograph of the Reading University constant-deviation concave grating monochromator (Codling and Mitchell, 1970). The instrument is situated inside a vacuum can with remote drive control of the grating G and mirror-knife edge entrance and exit slit mechanisms, M1 and M2, respectively.

the first negative order of diffraction is used, the angles of diffraction are such as to form order sorting over limited wavelengths. This instrument is effective over the range 400 Å to 40 Å (30 eV to 300 eV). Various designs of monochromator have been used or suggested to minimize problems associated with the synchrotron facility; for details, see Marr and Munro (1973).

5. *Photo- and Autoionization Processes*

In the following sections we look briefly at the various experiments involving photoionization phenomena which have benefited by the use of synchrotron radiation. The survey is necessarily limited, but a comprehensive bibliography has been compiled which includes data up to February 1974 (Marr *et al.*, 1973, 1974). Of course, work has been carried out using the conventional ultraviolet and x-ray sources and it is not intended that we should minimize the importance of data gleaned from these sources. Details of this work which are to be found in the reviews cited at the beginning of

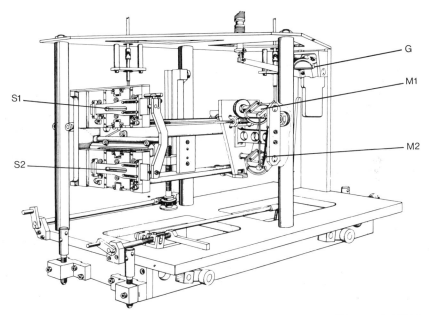

Figure 7. The Reading University grazing-incidence monochromator (West *et al.*, 1974a), providing a constant-deviation monochromatic beam through two exit slits S1 and S2, one above the other. Plane grating G and concave mirrors M1 and M2 focusing radiation onto slits S1 and S2. The instrument is mounted inside a vacuum tank with remote control of grating mirrors and exit slit widths.

the present paper and contributions to this symposium involve more recent data. The complementary aspect of electron scattering in many instances parallels the photon interaction work, and, again, this lies outside the scope of the present paper.

6. Energy Level Analysis

The early work in this field, carried out in the 1960s by workers at the National Bureau of Standards (NBS), Washington, using the 180 MeV electron synchrotron, has been effectively reviewed by Madden and Codling (1966), who also report their work on the rare gases. One- and two-inner-electron transitions to excited discrete states adjacent to photoionization continua are analyzed into Rydberg series, which provide beautiful examples of the asymmetric line shapes now described by the well-known Fano line shapes, attributed by Fano (1961) to configuration interactions between these excited quasi-discrete states and the underlying photoionization continua.

As the excitation energy increases, Auger type interactions become significant in which the atomic core may rearrange itself and eject an electron without requiring the involvement of the initially excited electron. The effect makes itself known by the study of absorption line halfwidths through a Rydberg series. Thus the lifetime for the internal Auger effect is essentially constant over the small energy interval of the Rydberg series. The members of the low principal quantum numbers of a series are dominated by the autoionization process involving the outer electron. However, as this lifetime increases with increasing principal quantum number (as $\sim n^3$), there comes a point where the internal Auger effect starts to dominate the lifetime of the excited state. At this point in the series, the widths of the resonances cease to decrease as n increases. Since the oscillator strength of a hydrogenic state decreases as $\sim n^{-3}$, the series loses contrast in relation to the photoionization background and becomes lost to experimental observation. This washout occurs at lower n values as the average excitation energy is increased due to the increased probability of Auger transitions. The understanding of the observed Rydberg series owed much to the work of Professor Fano and his colleagues (Cooper et al., 1963) at the NBS.

More recently, this aspect of identification and analysis has been extended to shorter wavelengths by the use of more energetic electron synchrotrons. Fine structure of x-ray absorption edges of Ar and other inert gases has been reported by Nakamura et al. (1968) using the Tokyo synchrotron and by Watson and Morgan (1969) using the Glasgow synchrotron. Metal vapor absorption spectra have been studied by Ederer et al. (1970) on lithium with the NBS machine, by Wolff et al. (1972) on sodium with the DESY synchrotron at Hamburg, and by Connerade et al. (1972) and Connerade and Mansfield (1973), of the Imperial College group, on cadmium and mercury with the Bonn synchrotron. With the extension to other atoms and higher energies, centrifugal barrier effects have been observed and analyzed (Cooper, 1964) for high-angular-momentum transitions. A considerable amount of data on autoionizing transition states is being steadily accumulated. Problems of wavelength calibration are considerable for precise work, and while some reliance has been placed on comparison of spectra with known gaseous absorption features, more recently low-inductance vacuum sparks have been installed between the furnace absorption cell and the spectrograph (Connerade and Mansfield, 1973).

7. Absolute Cross Section Measurements

With the requirement of accurate total cross section data, the electron synchrotron is becoming more used in this area of photoionization studies.

Essentially, the absorption cross section σ_v is determined by the attenuation of a photon flux from Φ_v^0 to Φ_v on passage through a low-density column of gas or vapor of length l and pressure p under conditions where

$$\Phi_v = \Phi_v^0 \exp\left| -\sigma_v \frac{N_0 T_0}{P_0} \int \frac{P\, dl}{T} \right| \qquad (7)$$

N_0 is Loschmidt's number, P_0 and T_0 are STP pressure and temperature, and $\int (P/T)\, dl$ represents the integration of the absorbing atoms in the line of sight of the cell. Experimental errors in σ_v are largely determined by the evaluation of the integral, and with the development of present experimental techniques of long low-pressure cells and photon counting, errors in σ_v can be reduced to $<5\%$ for permanent gases. For reactive vapors, vapor pressure data usually increase the error uncertainty.

The inert gases have again been the initial subjects for study. Watson (1972) and Lang (1973) have reported data using the Glasgow synchrotron, while Carlson *et al.* (1973) have examined argon with the Wisconsin storage ring. The latest of these experiments have been those carried out at the NINA synchrotron (West and Marr, 1975). Figure 8 shows the layout of the absorption cell and control system. A photon counting system is employed, with photomultiplier and sodium salicylate combinations for both the monitor detector and signal detector. This system is gated; it opens not only when the electron beam is circulating, but also for a short period after beam extraction to give an estimate of the background count level.

A differential pumping system is required to maintain a monochromator pressure of 2×10^{-6} torr (2.66×10^{-4} Pa) when the absorption cell pressure is in the range 0.2 to 1 torr (2.66 to 133 Pa). A zapon film is used as the first stage, as shown, with low-conductance slits placed next to the monochromator slit acting as the second stage. The cell is partially baked to a temperature of 150°C for 12 h before use. Research grade gases are introduced into the cell by means of a servo-operated needle valve. This is controlled by a capacitance manometer (MKS Baratron), previously calibrated against an inclined-plane oil manometer at the high pressure end of its range. This system maintained the pressure constant within 0.001 torr (0.133 Pa). Because of slight leakage through the zapon film, the gas under study is slowly replenished. Pressure differences along the cell are assumed to be negligible because of its large diameter (50 mm). The ratio of monitor count to signal count is measured at several pressures in the range 0.02 to 1 torr (2.66 to 133 Pa), and from this the cross section is deduced for a particular wavelength. From such data the statistical error in the cross section is 2%, and in addition to this a systematic error of 2% is possible, derived mainly from uncertainty in the measurement of gas pressure. Figure 9

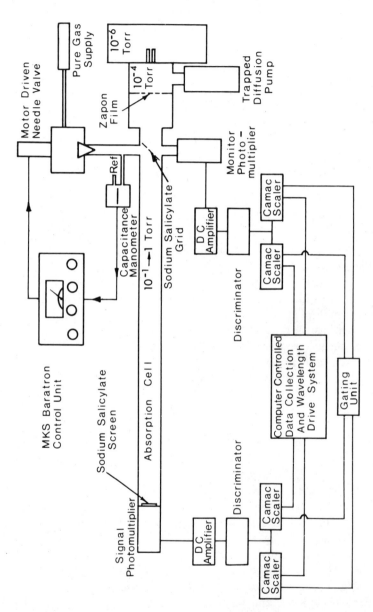

Figure 8. Schematic layout of absorption cell and associated control system.

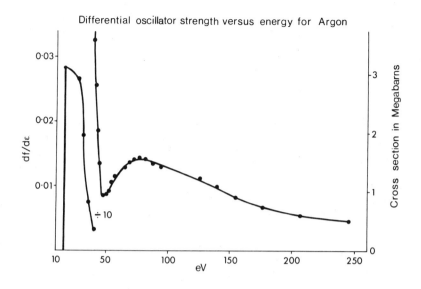

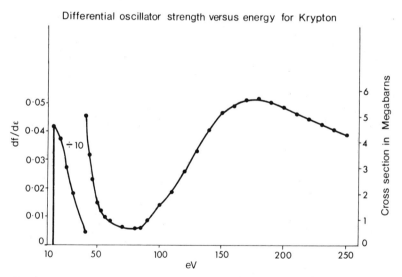

Figure 9. Cross section data for argon and krypton in the 50–350 Å region taken at the NINA synchrotron. (West and Marr, 1975.)

shows the photoionization cross sections of argon and krypton between 50 Å and 350 Å.

With these accurate cross section data, it has been possible to calculate the contribution from each shell of the atom to the total oscillator strength and compare this with theory. A one-electron model is used and the contribution of Pauli-forbidden transitions in each shell is calculated. For neon, the agreement with this theory is found to be good. For argon, the agreement is not so good; in particular, the experimental value for the contribution from the $n = 1$ and $n = 2$ shells is less than that predicted theoretically. The recently completed measurements on krypton, to which a similar analysis is applied, extend this discrepancy. It seems possible that the breakdown of the one-electron model is indicated, and the multielectron correlations have to be taken into account, as described, for example, by Amusia *et al.* (1971).

For the metal vapor cross section measurements, the absorption cell (Figure 8) is replaced with a heat pipe. To date, the total cross section of sodium has been measured in the same wavelength region, and measurements are being made on cadmium. The error in the present cross sections is larger than that for the rare gases because of uncertainty in the length of the vapor column, but it should not exceed 10%. The data will then be available for comparison with the relative cross section data from DESY (Wolff *et al.*, 1972).

8. Autoionization Line Shapes

The energy level analysis and the photoionization cross section data meet in the study of autoionization line shapes. Configuration interaction between the various discrete ($\phi_i \rightarrow \phi, \Phi$) and continuum ($\psi$) states result in absorption profiles which can be represented by (Fano and Cooper, 1965)

$$\sigma = \sigma_a \frac{(q + \varepsilon)^2}{1 + \varepsilon^2} + \sigma_b \tag{8}$$

The line shape parameter q is defined by (Fano, 1961)

$$q = \frac{\langle \Phi|r|\phi_i \rangle}{\pi \langle \psi|H|\phi \rangle \langle \phi|r|\psi \rangle} \tag{9}$$

and the energy parameter by

$$\varepsilon = \frac{E - E_r}{\frac{1}{2}\Gamma} \tag{10}$$

where

$$\Gamma = 2\pi |\langle \phi|H|\psi \rangle|^2 \tag{11}$$

is the halfwidth of the autoionizing feature, and E_r is the resonance energy. Mies (1968) has taken the autoionization theory a step further to include additional terms involving interaction between individual resonances by way of a coupling interaction matrix to the same continuum. This overlap matrix is defined by

$$\theta_{km} = \left\{\frac{(2\pi)^2}{\Gamma_k \Gamma_m}\right\}^{1/2} \sum_\beta V^*_{k\beta} V_{m\beta} \qquad (12)$$

where $V_{k\beta}$ represents the matrix elements linking the discrete states k to the continuum β replacing $\langle\psi|H|\phi\rangle$ in the Fano theory. Each resonance has a shape parameter defined by

$$q_k = \frac{t^b_k}{\pi \sum_\beta V_{k\beta} t^c_\beta} \qquad (13)$$

where t^b_k and t^c_β are now column vectors, in place of $\langle\Phi|r|\phi_i\rangle$ and $\langle\psi|r|\phi_i\rangle$ respectively.

Figure 10 shows data from Madden et al. (1969) which provides an illustration of the situation as $q \to 0$. These window resonances, which approach the threshold for the removal of a 3s subshell electron in argon, suggest that there should be a decrease in the absorption edge rather than a net increase as is usually the case. For argon the window resonances indicate that the small changes through the threshold can be accounted for by configuration interaction, provided the direct photoabsorption by the s subshell amounts to no more than $\sim 1\%$ of the p shell absorption at the appropriate photon energy (Fano and Cooper, 1968).

9. Photoelectron Spectroscopy

In order to gain further insight into the details of this configuration interaction phenomenon, we need to study the ejected electrons themselves through photoelectron spectroscopy. Consider the interaction of 350 Å (34.5 eV) photons on an assembly of Ar atoms. Two different ionic states are possible:

(a) $Ar(3s^2 3p^6\ {}^1S_0) + h\nu \to Ar^+(3s^2 3p^5\ {}^2P) + e\,(19.6\,\text{eV}\ p\ \text{electron})$
(b) $Ar(3s^2 3p^6\ {}^1S_0) + h\nu \to Ar^+(3s3p^6\ {}^2S) + e\,(6.2\,\text{eV}\ s\ \text{electron})$

Because of the large mass difference between the ion and the electron, the excess energy [19.6 eV in (a) and 6.2 eV in (b)] appears as kinetic energy of the photoelectron. A measurement of the relative numbers of electrons with the two different kinetic energies determines the partial cross section ratio, or branching ratio, for $Ar^+(^2P)$ and $Ar^+(^2S)$ at a photon energy of 35.4 eV.

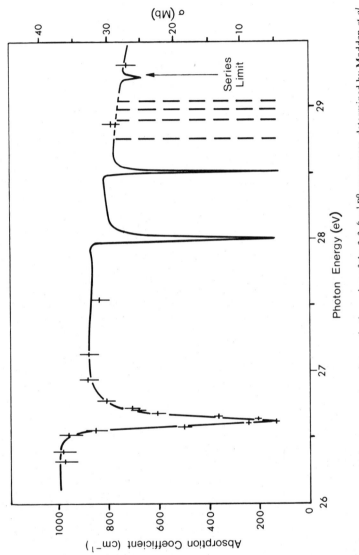

Figure 10. Absorption cross section of argon in the region of the $3s3p^6np\,{}^1P_1^0$ resonances, determined by Madden *et al.* (1969) at the NBS synchrotron.

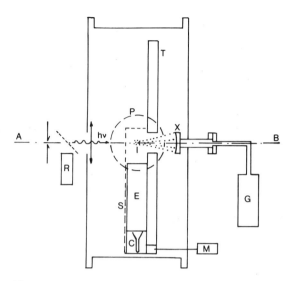

Figure 11. Schematic diagram of the 127° analyzer in plan view. C, channeltron; E, analyzer; G, gas supply; M, motor for rotation of the analyzer; P, pumping port below the analyzer; R, reference photomultiplier; S, mu metal shield; T, rotating table; X, focusing capillary array.

While a number of experiments have been carried out at specific photon energies, very little work has been done which makes effective use of the variation of branching ratio with photon energy. This is where synchrotron radiation has a very important part to play, and the first attempt to study the s/p ratio in argon was carried out at the Glasgow synchrotron (Lynch et al., 1973b). It has since been substantiated at the more powerful synchrotron radiation facility at NINA (West et al., 1974b). The absolute cross section data were used to obtain an estimate of the 3s cross section, which is only 2% of the total absorption at the 3s threshold. Figure 11 is a representation of the 127° electron analyzer system employed at NINA in conjunction with a horizontal-dispersion Wadsworth monochromator. Figure 12 shows the 3s cross section curve, which confirms the Glasgow low-flux results that the cross section decreases to a minimum at $\sim 10\,\text{eV}$ above threshold. Moreover, in agreement with Amusia et al. (1972b) and later papers (see Figure 12), this minimum is zero within the limits of experimental error. The 3s partial cross section is modified substantially from the independent particle by coupling with the $3p \rightarrow \varepsilon d$ virtual transitions.

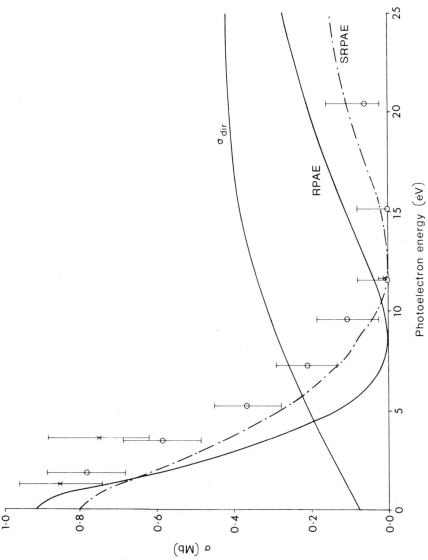

Figure 12. Partial cross section curve for the argon 3s photoelectron. O, experimental data from NINA; X, experimental data of Samson and Gardner (1974); σ_{dir}, calculated Hartree–Fock photoionization curve from Cooper and Manson (1969); RPAE, Random phase approximation with exchange calculations by Amusia (1973); SRPAE, simplified

10. The Angular Distribution of Photoelectrons

A photoionizing event contains more information than is displayed in either the wavelength dependence of total cross sections or electron-energy-resolved partial cross sections between particular molecular and ionic states. Additional insight into the coupling of orbital electrons with one another and with the interacting radiation field is potentially available from measurements of the trajectory direction of the photofragments.

The angular distribution of photoelectrons ionized by plane-polarized radiation is given by

$$N(\gamma) = \frac{\Phi n l \sigma_j}{4\pi}[1 + \beta P_2(\cos \gamma)] \tag{14}$$

where $P_2(\cos \gamma)$ is $(3\cos^2 \gamma - 1)/2$, Φ is the intensity of the radiation, l the interaction length, n the density of atoms and σ_j the photoionization cross section into the ionic state j. γ is defined by the direction of the photoelectrons with respect to the E vector of the radiation. Equation (14) depends upon the absorption taking place via an electric dipole process and not upon the details of the wave functions of the various states. The asymmetry parameter β, however, does depend on the details of the calculation, and for LS coupling of antisymmetrized products of single particle functions the β parameter becomes (Cooper and Zare, 1968)

$$\beta(E) = \frac{l(l-1)R_{l-1}^2 + (l+1)(l+2)R_{l+1}^2 - 6l(l+1)R_{l-1}R_{l+1}\cos[\delta_{l+1} - \delta_{l-1}]}{(2l+1)[lR_{l-1}^2 + (l+1)R_{l+1}^2]} \tag{15}$$

where

$$R_{l\pm 1} = \int_0^\infty P_{nl}(r) r P_{El\pm 1}(r) \, dr \tag{16}$$

are the radial dipole matrix elements and δ are the phase shifts associated with the asymptotic wave function of the continuum state.

β is generally a function of photon energy and has values between $+2$ and -1. Only for s electrons ($l = 0$) does $\beta = 2$ and not vary with energy. For p electrons ($l = 1$), photoionization can take place both through $p \to s$ and $p \to d$, and the continuum s and d waves interfere as in Eq. (15); β will therefore be expected to vary with energy. If only one of the $l \pm 1$ transitions can occur in a given photoionization channel, there will be no interference and β will be constant, e.g., for $p \to d$ alone, $\beta = 1$.

Recently, Walker and Waber (1973) have developed the theory to include relativistic wave functions and show that the β parameter can deviate from values predicted by Eq. (15). The heavy alkali atoms exhibit a number of anomalous effects which are attributed to the spin–orbit interaction process. Fermi (1930) showed that it was responsible for their anomalous doublet line strength ratios, Seaton (1951) used similar arguments to explain the nonzero minimum in the photoionization cross section curves, and Fano (1969) used the interaction to predict that spin-polarized electrons should be obtainable from the photoionization of cesium atoms if circular polarized radiation of the right wavelength is used. This prediction (the Fano effect) has since been verified by Lubell and Raith (1969) and independently by Heinzmann et al. (1970).

The spin–orbit interaction draws in the nodes of the $\psi(E^2 P_{1/2})$ wave function and expels outward those for $\psi(E^2 P_{3/2})$. The radial matrix elements in Eq. (16) are therefore j' dependent and give rise to two separate matrix elements instead of the single value obtained when $l = 0$. Writing these as R_1 and R_3, corresponding to $R(j' = \frac{1}{2})$ and $R(j' = \frac{3}{2})$ respectively, we get (Marr, 1974)

$$\beta(E) = \frac{2R_3^2 + 4R_3 R_1}{2R_3^2 + R_1^2} = \frac{2(\chi^2 - 1)}{\chi^2 + 2} \quad (17)$$

where χ is the spin–orbit perturbation function (Fano, 1969)

$$\chi(E) = \frac{2R_3 + R_1}{R_3 - R_1} \quad (18)$$

for the photoionization of the s electrons.

Figure 13 shows the evaluated β variation for cesium and compares it with theoretical data from different sources. Figure 14 uses the theoretical χ data to examine the variation of β through the alkali metal atomic species. Only where β departs significantly from 2 does the angular distribution depart from the normal $\cos^2 \gamma$ distribution expected from an s electron. From these data we see that a study of β can provide significant information concerning the matrix element which is complementary to the cross section data. When $\beta = +1$, $R_1 = 0$, and when $\beta = 0$, $R_3 = 0$. In the region $\chi = 0$, where the spin–orbit interaction is greatest, $\beta = -1$. A study of the β variation with photon energy can be expected to show up small changes in the matrix elements which would otherwise go undetected. In particular, the turning points in the β curve can be a sensitive test of the theoretical model under consideration. So far, direct measurements of β for the alkali metal atoms have not been made. However, it is only because of their low ionization thresholds that this information is available from other sources. For other atoms the most convenient method is to measure $\beta(E)$.

This type of measurement has been carried out at the Glasgow synchrotron (Mitchell and Codling, 1972; Lynch et al., 1972, 1973a) on

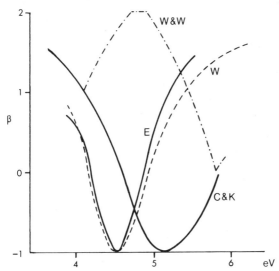

Figure 13. The angular distribution asymmetry parameter β for cesium as a function of photon energy (eV). E, based on experimental data, Baum *et al.* (1969); W, theoretical data, Weisheit (1972); C & K, theoretical data, Chang and Kelly (1972); W & W, theoretical data, Walker and Waber (1973).

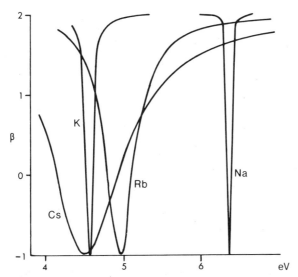

Figure 14. β values versus photon energy (eV), based on theoretical data from Weisheit (1972) for the alkali atoms starting from the ionization thresholds.

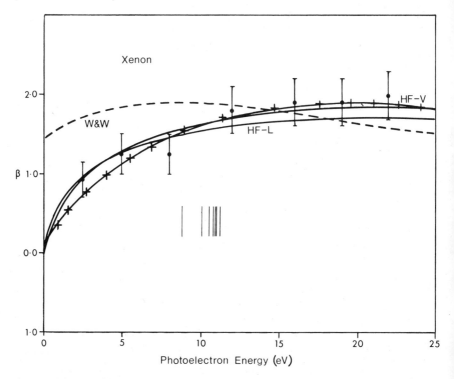

Figure 15. The asymmetry parameter β for xenon. Experimental data of Lynch et al. (1973a) compared with Hartree–Fock (———) calculations of Kennedy and Manson (1972). The Dirac–Slater (— — —) calculations of Walker and Waber (1974) and the many-electron correlation data (+ + +) of Amusia et al. (1972a).

the inert gases. No attempt has been made in these experiments to separate out spin components, and measurements by Carlson and Jonas (1971) at 584 Å show no significant variation for argon. Nevertheless, for other atoms it can be important in the neighborhood of autoionization features, and for heavy atoms where LS coupling breaks down, intensity deviations can be expected (Harrison, 1970).

Figure 15 shows data on xenon taken by Lynch et al. (1973a) and compared with various theoretical models. While the data are insensitive in comparison between different theoretical approaches for the general shape, nevertheless, the position of the turning point can be useful. Thus the data can distinguish between the data of Walker and Waber (1974) and the others. The data on argon are even more discriminating. Figure 16 compares the recent β measurements on argon (West et al., 1974b) taken at the NINA synchrotron, with the calculations by Amusia et al. (1972a). This extends

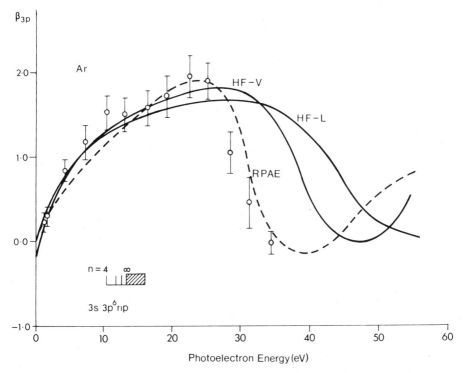

Figure 16. The asymmetry parameter β for argon experimental data from the NINA synchrotron compared with the many electron correlation data (———) of Amusia *et al.* (1972a) and the Hartree-Fock (———) calculations of Kennedy and Manson (1972).

the data of the Glasgow synchrotron and verifies the departure from the Hartree–Fock calculation suggested by Mitchell and Codling (1972).

11. Acknowledgments

Much of the recent experimental work reported in this paper has been made possible by financial support provided by the U.K. Science Research Council at the Glasgow and Daresbury synchrotrons. This support is gratefully acknowledged. My thanks are also due to colleagues in the atomic research group at Reading and Daresbury (Drs. K. Codling and J. B. West; R. Houlgate and J. Hamley, research students supported by the Gassiot Committee of the Royal Society and S.R.C. respectively) and to the Daresbury Laboratory staff.

References

Amusia, M. Ya. (1973). VIII ICPEAC: Invited lectures and progress reports, Belgrade, p. 172.
Amusia, M. Ya., Cherepkov, N. A., and Chernysheva, L. V. (1971). *Sov. Phys.—JETP*, **33**, 90.
Amusia, M. Ya., Cherepkov, N. A., and Chernysheva, L. V. (1972a). *Phys. Letts.*, **40A**, 15.
Amusia, M. Ya., Ivanov, K. V., Cherepkov, N. A., and Chernysheva, L. A. (1972b). *Phys. Letts.*, **40A**, 361.
Baum, G., Lubell, M. S., and Raith, W. (1972). *Phys. Rev.*, **A 5**, 1073.
Biberman, L. M., and Norman, G. E. (1967). *Sov. Phys. Usp*, **10**, 52.
Carlson, T. A., and Jonas, A. E. (1971). *J. Chem. Phys.*, **55**, 4913.
Carlson, R. W., Judge, D. L., Ogawa, M., and Lee, L. C. (1973). *Appl. Optics*, **12**, 409.
Chang, J. J., and Kelly, H. P. (1972). *Phys. Rev.*, **A 5**, 1713.
Codling, K. (1973). *Rep. Prog. Phys.*, **36**, 541.
Codling, K., and Mitchell, P. (1970). *J. Phys. E.*, **3**, 685.
Connerade, J. P., and Mansfield, M. W. D. (1973). *Proc. Roy. Soc.*, **A335**, 87.
Connerade, J. P., Mansfield, M. W. D., and Thimm, K. (1972). *Phys. Rev.*, **A 6**, 1955.
Cooper, J., and Zare, R. N. (1968). *J. Chem. Phys.*, **48**, 942.
Cooper, J., and Zare, R. N. (1968). *J. Chem. Phys.*, **49**, 4292.
Cooper, J. W. (1964). *Phys. Rev. Letts.*, **13**, 762.
Cooper, J. W., Fano, U., and Pratts, F. (1963). *Phys. Rev. Letts.*, **10**, 518.
Cooper, J. W., and Manson, S. T. (1969). *Phys. Rev.*, **177**, 157.
Ederer, D. L., Lucatorto, T., and Madden, R. P. (1970). *Phys. Rev. Letts.*, **25**, 1537.
Fano, U. (1961). *Phys. Rev.*, **124**, 1866.
Fano, U. (1969). *Phys. Rev.*, **178A**, 131.
Fano, U., and Cooper, J. W. (1965). *Phys. Rev.*, **137A**, 1364.
Fano, U., and Cooper, J. W. (1968). *Rev. Mod. Phys.*, **40**, 441.
Fermi, E. (1930). *Z. Phys.*, **59**, 680.
Harrison, H. (1970). *J. Chem. Phys.*, **52**, 901.
Heinzmann, U., Kessler, J., and Lorenz, J. (1970). *Z. Phys.*, **240**, 42.
Hudson, R. D., and Kieffer, L. J. (1971). *Atomic Data*, **2**, 205.
Kennedy, D. J., and Manson, S. T. (1972). *Phys. Rev.*, **A 5**, 227.
Lang, J. (1973). *Proc. Int. Symp. Synchrotron Radiation Users, Daresbury*. Eds. G. V. Marr and I. H. Munro, Daresbury Laboratory report DNPL/R 26, 240.
Lin, C. D. (1974). *Phys. Rev.*, **A 9**, 171.
Lubell, M. S., and Raith, W. (1969). *Phys. Rev. Letts.*, **23**, 211.
Lynch, M. J., Gardner, A. B., and Codling, K. (1972). *Phys. Letts.*, **40A**, 349.
Lynch, M. J., Codling, K., and Gardner, A. B. (1973a). *Phys. Letts.*, **43A**, 213.
Lynch, M. J., Gardner, A. B., Codling, K., and Marr, G. V. (1973b). *Phys. Letts.*, **43A**, 237.
Madden, R. P., and Codling, K. (1966). In *Autoionization*, p. 129. Ed. A. Temkin. Baltimore: Mono Book Co.
Madden, R. P., Ederer, D. L., and Codling, K. (1969). *Phys. Rev.*, **177**, 136.
Marr, G. V. (1967). *Photoionization Processes in Gases*. New York: Academic Press.
Marr, G. V. (1974). *J. Phys. B.*, **7**, L47.
Marr, G. V., and Munro, I. H., Eds. (1973). *Proc. Int. Symp. Synchrotron Radiation Users, Daresbury*. Daresbury Laboratory report, DNPL/R 26.
Marr, G. V., Munro, I. H., and Sharp, J. C. C. (1973). *Synchrotron Radiation: A Bibliography*. Daresbury Laboratory report, DNPL/R 24.
Marr, G. V., Munro, I. H., and Sharp, J. C. C. (1974). *Synchrotron Radiation: A Bibliography, Supplement, August 1972 to February 1974*. Daresbury Laboratory Internal report, DL/TM 127.

Mies, F. H. (1968). *Phys. Rev.*, **175**, 164.
Mitchell, P., and Codling, K. (1972). *Phys. Letts.*, **38A**, 31.
Mitchell, P., and Wilson, M. (1969). *Chem. Phys. Letts.*, **3**, 389.
Nakamura, M., Sasanuma, M., Sato, S., Watanabe, M., Yamashita, H., Iguchi, Y., Ejiri, A., Nakai, S., Yamaguchi, S., Sagawa, T., Nakai, Y., and Oshio, T. (1968). *Phys. Rev. Letts.*, **21**, 1303.
Samson, J. A. R. (1967). *Technique of Vacuum Ultraviolet Spectroscopy.* New York: Wiley.
Samson, J. A. R., and Gardner, J. L. (1974). *Phys. Rev. Letts.*, **33**, 671.
Seaton, M. J. (1951). *Proc. Roy. Soc.*, **A208**, 418.
Shirley, D. A., Ed. (1972). *Electron Spectroscopy.* Amsterdam: North-Holland.
Walker, T. E. H., and Waber, J. T. (1973). *J. Phys. B.*, **6**, 1165.
Walker, T. E. H., and Waber, J. T. (1974). *J. Phys. B.*, **7**, 674.
Watson, W. S. (1972). *J. Phys. B.*, **5**, 2292.
Watson, W. S., and Morgan, F. J. (1969). *J. Phys. B.*, **2**, 277.
Weisheit, J. C. (1972). *Phys. Rev.*, **A5**, 1621.
West, J. B., Codling, K., and Marr, G. V. (1974a). *J. Phys. E.*, **7**, 137.
West, J. B., Houlgate, R. G., Codling, K., and Marr, G. V. (1974b). *J. Phys. B.*, **7**, L470.
West, J. B., and Marr, G. V. (1975). In press.
Wolff, H. W., Radler, K., Sonntag, B., and Haensel, R. (1972). *Z. Phys.*, **257**, 353.

5
Photoabsorption of the 4d Electrons in Xenon and Barium: A Comparison

D. L. EDERER, T. B. LUCATORTO, E. B. SALOMAN,
R. P. MADDEN, M. MANALIS, AND JACK SUGAR

The 4d absorption spectra of xenon and barium show very different behavior. In xenon the cross section has been measured for two series of resonances converging to the $4d^9(^2D_{5/2,3/2})$ limit. By a parameterization technique, the amplitudes and widths of these resonances have been obtained. The widths of the resonances are essentially constant, but the widths of the resonances converging to the $4d^9(^2D_{5/2})$ limit (N_V edge) are somewhat narrower than those converging to the $4d^9(^2D_{3/2})$ limit (N_{IV} edge). The oscillator strength in these $4d^9(^2D_{5/2,3/2})np$ series is small (0.06) compared to the total continuum oscillator strength integrated over open p and f channels. The barium spectrum is very different from the xenon spectrum because the 4f orbital contracts in excited barium and overlaps the 4d orbit. This contraction produces two terms, ^3P and ^3D, of the $4d^94f$ configuration below the 4d ionization limits, while the strong electrostatic exchange interaction drives the ^1P term of this configuration some 10 eV above the limit. Furthermore, extensive mixing of the $4d^96s^26p$ configuration with $4d^95d^26p$ and $4d^95d6s6p$ produces many weak resonances near the $4d^{10} \rightarrow 4d^9(^2D_{5/2})6s^26p(^2P_{3/2})$ resonance. A suggested classification of these features is given with the aid of known features of the La I spectrum. Finally, from the known x-ray splitting of the $N_{IV,V}$ threshold and the energy interval between the $6s^26p(^2P)$ and $6s^2(^1S)$ levels in La I and La II, the ionization thresholds of the 4d electron were determined to be 814,800(1000) cm^{-1} and 792,500(1000) cm^{-1}.

D. L. EDERER, T. B. LUCATORTO, E. B. SALOMAN, R. P. MADDEN, M. MANALIS, and JACK SUGAR
• Optical Physics Division, Institute for Basic Standards, National Bureau of Standards, Washington, D.C. 20234, U.S.A. M. Manalis's present address is Physics Department, University of California at Santa Barbara, California.

Introduction

During the past decade there has been extensive growth in the use of vacuum ultraviolet (VUV) radiation as a research tool. One of the most interesting and intriguing phenomena in the field of VUV photoabsorption spectroscopy was the photoabsorption of $4d$ electrons in medium- to high-Z elements. Professor Fano's early enthusiastic involvement in this work encouraged experiments and calculations that eventually led to a detailed understanding of $4d$ photoabsorption. In this paper we shall present measurements in xenon and barium gas that deal with several aspects of the behavior of the $4d$ electron upon photoionization.

Ten years ago the problem of $4d$ absorption spectra began to unfold. Photon-absorption measurements of the fine structure preceding the $N_{IV,V}$ thresholds in xenon gas were made by Madden's group at the National Bureau of Standards (NBS) (Codling and Madden, 1964) and by the Lukirskii group (Lukirskii et al., 1964b) in Leningrad. These measurements were followed later that year by measurements of the $4d$ continuum cross section (Lukirskii et al., 1964a; Ederer, 1964). The surprising features of these spectra were: (1) only one series of resonances (associated with an excitation to an outer p orbital) was observed to converge to the N_{IV} and N_V threshold; (2) the oscillator strength associated with a transition to an f orbital was extremely small near the $N_{IV,V}$ threshold; and (3), the continuum cross section, although small at threshold (2×10^{-18} cm^2, or 2 Mb), increased rapidly with increasing photon energy to a maximum of 30 Mb at a photon energy 20 eV above the $N_{IV,V}$ threshold. Another 90 eV above the maximum, the cross section decreased by an order of magnitude. In an accompanying paper, J. W. Cooper (1964) showed that the results could be understood qualitatively in terms of the behavior of the dipole matrix element as a function of photon energy. He suggested that the centrifugal barrier prevented significant overlap between the outgoing f wave electron and the initial d state until the outgoing electron had sufficient energy to overcome this potential barrier. Thus the cross section should be small at threshold and increase rapidly. The cross section subsequently decreased rapidly because the $4d$ wave function has a node and the dipole matrix element changes sign 80 eV above threshold causing a "Cooper minimum" (Cooper, 1962) to occur in the absorption cross section. Although qualitatively the idea of the potential barrier was of fundamental importance, rough calculations suggested that the electron–electron correlations were not properly accounted for. Thus the $4d$ photoabsorption cross section provided a stimulus for the use of many-body theory to account for the electron correlations.

Since $4d$ photoabsorption in solids is atom-like in character (Ederer, 1964; Codling et al., 1966), the publication in 1967 of a comprehensive set of measurements of 4d absorption structure of the rare earth metals in solid form (Zimkina et al., 1967; Formichev et al., 1967) was relevant to atomic physics. The measurements were important to future atomic theoretical developments for two reasons: firstly, it was a systematic study covering all the rare earths as well as tin, tellurium, and xenon, and secondly, the measurements covered a broad energy range with sufficient emphasis on the fine structure preceding the $N_{IV,V}$ threshold. Several significant facts emerged from these measurements. In the first place, the continuum cross section peaks up more sharply in the rare earths than in xenon. Secondly, the peak decreases in amplitude as a function of Z until it disappears completely at an atomic number coinciding with the filling of the 4f shell. Finally, the fine structure below the 4d ionization threshold is sharp, of low oscillator strength, and quite different from element to element. Formichev et al. (1967) offered a suggestion, verified later, that the rare earth fine structure was due to transitions of the type $4d^{10}4f^N \rightarrow 4d^9 4f^{N+1}$ for positive ions embedded in the crystal lattice.

In a review article, Professor Fano and J. W. Cooper (1968) laid the ground work for later calculations that would deal with the rare earths by developing the concept of a resonance near threshold. Concurrently, Professor Fano and his students (Starace, 1970, 1971; Dehmer and Fano, 1970) were developing the technique to properly account for exchange and electron correlations in the photoionization process. Several other groups (Amusia et al., 1971; Wendin, 1973a; Kelley and Simons, 1973) were using many-body formulations with great success to account for electron correlation. The measurement of the 4d absorption spectrum of some rare earth metals and their oxides (Haensel et al., 1970) served as a catalyst and prompted a brilliant series of papers by Professor Fano, his students, and NBS colleagues (Dehmer et al., 1971; Starace, 1972; Sugar, 1972; Dehmer and Starace, 1972). The answers to nagging details were obtained, and more important, the general photoabsorption process was finally understood.

It is with this rich background that we began the absorption measurements on barium vapor to learn more about the behavior of the potential barrier at $Z \leqslant 57$, i.e., before the contraction of the 4f electron occurs at cerium. Since barium is a closed-shell atom, one might expect a xenon-like behavior. However, the results shown in Figure 1 indicate that such a simple interpretation is not possible.

Experimental Procedure

Xenon

Xenon was contained between the entrance and exit slits of a specially built (Madden et al., 1967) 3-m-radius grazing-incidence monochromator of

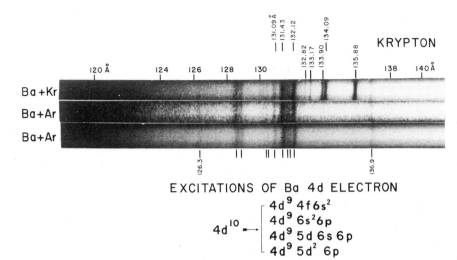

Figure 1. Photoabsorption in barium vapor (88–105 eV). The top spectrum was taken using krypton as a buffer gas in the heat pipe to provide a calibration spectrum in this region. The two lower spectral plates were taken using argon as a buffer gas; argon is structureless in this region. The pressure–path length product for all three exposures was 7 torr-cm (reduced to standard temperature).

0.065 Å spectral slit width. Continuum radiation from the NBS 180 MeV synchrotron was used as a source. Repeated scans at pressure–path length products between 1 torr-cm and 10 torr-cm in the 170–200 Å wavelength range produced a series of transmission scans which were reduced to a cross section by using a method described elsewhere (Ederer, 1969). Each xenon structure was modeled as a Lorentzian of variable width and amplitude, superimposed on a constant background cross section. The observed transmission curve was compared to the calculated transmission curve by convoluting the calculated transmission of the model cross section with a gaussian spectrometer window function. The Lorentzian line-shape parameters were adjusted until a best least-squares fit was obtained between the observations and the parameterized transmission. The final parameters listed in Table 1 are the weighted means of the parameters obtained in individual runs. The error assigned to each parameter is the standard deviation of the mean and does not include directly an estimate of the magnitude of systematic errors, which we estimate are of the same order or magnitude.

Barium

It was necessary to use an entirely different technique to obtain the barium absorption spectrum. The barium was contained in a heat pipe

Table 1. Cross Section Parameters for 4d Inner Shell Excitations to Final States of the Type $4d^95s^25p^6(^2D_{5/2,3/2})np$ in Xenon

n	λ (Å)	E (eV)	n*	σ_{np} (cm^{-1})	Γ_n (eV)	$f_n^d \times 10^4$	$f_n n^{*3}$
[A] Parameters for final states $4d^95s^25p^6(^2D_{5/2})np$							
6	190.41[a]	65.11	2.36	350(30)[b]	0.114(8)	209(25)	0.280(50)
7	186.81	66.37	3.40	90(15)	0.09(2)	42(12)	0.165(40)
8	185.47	66.84	4.41	53(20)	0.09(2)	25(12)	0.21(10)
9	184.84	67.08	5.41	25[c]	0.11[c]	14	0.22
∞	183.55	67.55		$\sigma_{5/2,\infty} = 28(5)$			
[B] Parameters for final states $4d^95s^25p^6(^2D_{3/2})np$							
6	184.94	67.04	2.34	240(40)	0.13(3)	166(50)	0.212(60)
7	181.42	68.34	3.39	60(10)	0.11(3)	38(12)	0.148(50)
8	180.14	68.82	4.42	29(5)	0.11(3)	17(5)	0.147(40)
9	179.53	69.06	5.42	20(5)	0.11(3)	13(4)	0.207(50)
∞	178.33	69.52		$\sigma_{3/2,\infty} = 20(5)$			

[a] Data from Codling and Madden (1964).
[b] The number in parentheses is the standard deviation of the parameters from the mean obtained from several sets of data.
[c] These parameters were fixed according to the prescription outlined in the test.
[d] The oscillator strength f_n for a Lorentzian line is calculated from the formula: $f_n = 5.31 \times 10^{-4} \sigma_{np} \Gamma$ when Γ_n is expressed in eV and σ_{np} in cm^{-1}(STP).

oven of the type used to contain lithium vapor (Ederer et al., 1970). Thin-film plastic windows were used to contain the argon or krypton buffer gas, and the oven was inserted in the light beam between the synchrotron and the spectrograph. The heat pipe was operated in the pressure range 1.0–2.0 torr, and the hot vapor produced an optical path of about 10 cm. The absorption spectrum was observed on high-speed 101 plates with a 3-m grazing-incidence spectrograph of 0.05 Å spectral slit width (Madden et al., 1967). Typically, an exposure between 30 min and 60 min was long enough to obtain a plate of adequate density.

Results

Xenon

The earlier investigations of the structure preceding the $N_{IV,V}$ threshold (Codling and Madden, 1964; Lukirskii et al., 1964b; Haensel et al., 1969) dealt primarily with the energy position of the resonances and the relative cross section values, rather than absolute cross section measurements. In the fitting procedure, a Lorentzian line shape was assumed, and the width and peak cross section of the resonances were adjustable parameters, but the energy positions, taken from Codling and Madden (1964), were fixed.* The background cross section due to the open channels of the outer $5s$ and $5p$ shells was an adjustable parameter, as were the continuum cross sections due to $(^2D_{5/2})\varepsilon p, \varepsilon f$ and $(^2D_{3/2})\varepsilon p, \varepsilon f$ channels. The analysis proceeded in three distinct steps. The $4d^9(^2D_{5/2})6p$ resonance was analyzed by itself. The group of resonances between 66 eV and 68 eV comprised the next interval. This region included the $4d^9(^2D_{3/2})6p$ resonance and the other members of $4d^9(^2D_{5/2})np$ converging to the limit at 67.55 eV. Finally the remaining $(^2D_{3/2})np$ resonances converging to the limit at 69.52 eV were analyzed as a group. The first region presented no problem because it was straightforward to determine, rather accurately, the parameters of the single resonance involved. In the second group, however, the $4d^9(^2D_{3/2})6p$ resonance and higher members of the $4d^9(^2D_{5/2})np$ resonances blended together to form a continuum. The width of the $4d^9(^2D_{5/2})9p$ resonance was fixed and set equal to the width of the $4d^9(^2D_{5/2})6p$ resonance. The amplitude of this resonance (also fixed) was determined by assuming that the average oscillator strength density of the series (averaged over a period of the Rydberg series) is constant (Fano and Cooper, 1965). The superposition of the $n = 10$ and higher members of the $4d^9(^2D_{5/2})np$ series was accounted for by fitting an approximate cross section (Watanabe, 1965) of the form

$$\sigma_\Sigma = \sum_{n=10}^{\infty} \sigma_{np} = \tfrac{1}{2}\sigma_{5/2,\infty}\left[1 + \frac{2}{\pi}\arctan\left(\frac{E - E_0}{\tfrac{1}{2}\Gamma}\right)\right]$$

to the data. The magnitude of the continuum threshold absorption cross section for the $4d^9(^2D_{5/2})$ core, $\sigma_{5/2,\infty}$, was an adjustable parameter. The quantity Γ, however, was fixed at 0.11 eV, the width of the $4d^9(^2D_{5/2})6p$ resonance, and E_0 was taken as the energy position of the $4d^9(^2D_{5/2})10p$ resonance. The analysis of the final group of resonances between 68 eV and 70 eV proceeded in a straightforward manner since there were no superimposed resonances from other series. The superposition of resonances in the $4d^9(^2D_{3/2})np$ series for $n \geqslant 10$ was accounted for in the same manner as the $4d^9(^2D_{5/2})$ series. The parameter values obtained from this fitting process are tabulated in Table 1, and the absorption coefficient synthesized from these parameters is shown in Figure 2.

The widths of the autoionizing resonances involving the $5s$ and $5p$ electrons of xenon decrease with n, indicating that these lifetimes grow larger with increasing n and hence depend on the interaction time of the excited electron in the field of outer core electrons (Fano and Cooper, 1965).

*Because of the rather large spectral slit function (0.94 Å), the data of Lukirskii et al. (1964b) were subject to greater uncertainty than those of Codling and Madden (1964).

Photoabsorption of 4d Electrons

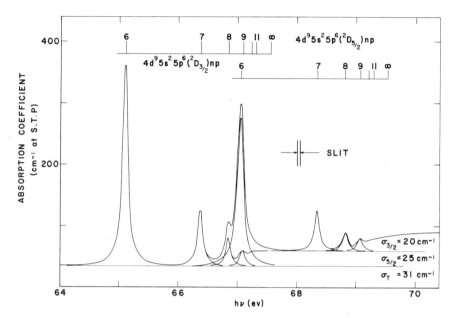

Figure 2. Total absorption coefficient of xenon between 64 and 69 eV as synthesized from the parameters in Table 1. Series members are indicated and the limits identified. Contributions of the different open channels are individually shown.

However, the width of the resonances involving the 4d electron is essentially constant, although there is a tendency for the first resonance to be somewhat broader than other members of the series. Thus the lifetime of the d hole is dominated primarily by the Auger process, where an outer-shell electron fills the d-shell hole, ejecting another outer-shell electron in the process. Autoionization of the outer electron, manifested by decreasing widths, is a comparatively weak process, contributing at most 20% to the width of the first member of the $N_{IV,V}$ series. From the results, it is consistent to state that the widths of both series of resonances are the same, although there is a tendency for the width of the $4d^9(^2D_{3/2})np$ resonances to be somewhat larger than the width of the $4d^9(^2D_{5/2})np$ resonances.

Lukirskii et al. (1964b), from deconvolution of their data, obtained widths of 0.18(4) eV and 0.11(4) eV for the N_{IV} and N_V levels respectively. Since the spectral slit width of the monochromator used was 0.94 Å (compared to 0.065 Å in the present experiment) it is surprising that the agreement with our experiment is as good as it is: 0.11(3) eV for the width of resonances converging to the N_{IV} limit, and 0.09(3) eV for the resonances converging to the N_V limit.

The oscillator strength density in the continuum region is a smoothly varying function which can be related to the average oscillator strength density in the discrete spectrum. The oscillator strength f_n of each series number divided by ΔE_n, the energy spacing between adjacent series members, is the appropriate average. Since $\Delta E_n \propto 1/n^{*3}$, the average oscillator strength density is proportional to $f_n n^{*3}$, which should be nearly constant over the series. In the last column of Table 1 the product $f_n n^{*3}$ has been tabulated for each series member that was analyzed. From these data, the continuum oscillator strength density near threshold can be interpreted as a slowly varying function since the averaged discrete oscillator strength density is nearly constant.

The cross section at the $4d^9(^2D_{5/2})$ limit $(\sigma_{5/2,\infty})$ was compared with the cross section at the $4d^9(^2D_{3/2})$ limit $(\sigma_{3/2,\infty})$ and found to have a ratio of 1.4(4), which is consistent with 1.5, the value of the ratio if the statistical weights of the $4d^9(^2D_{5/2})$ and $4d^9(^2D_{3/2})$ levels account for the number of open channels available to the $4d$ electron. Lukirskii et al. (1964a) obtained a value of 1.42 for this ratio. In other words, the cross section at threshold is dominated by final states of p symmetry rather than f symmetry. While transitions to final states of f symmetry are negligible compared to transitions to final states of p symmetry at threshold, the oscillator strength summed over the discrete p channel transitions is small (approximately 0.06 in the series) compared to the total oscillator strength (approximately 11) arising from the $d \to f$ and $d \to p$ transitions (Ederer, 1964).

The background absorption coefficient due to transitions of the type $5s^2 5p^6 \to 5s^2 5p^5 \varepsilon s, \varepsilon d$ and $5s 5p^6 \to 5s 5p^6 \varepsilon p$ was $31(5)\,\mathrm{cm}^{-1}$. This result is somewhat lower than early measurements of the continuum cross section (Ederer, 1964) but consistent with the more recent measurement of Cairns et al. (1969).

Barium

While the interpretation of the xenon spectrum, shown in Figure 2, was straightforward, such simplicity cannot be found in the barium spectrum shown in Figure 1. The wavelength position of the discrete structures involving the excitation of the $4d$ electron are listed in Table 2. The rather large estimated probable error in the measurements is due to the weakness and diffuseness of the barium features and not the calibration. Although the barium cross section was not measured, densitometer traces giving a qualitative measure of the trend of this quantity are presented in Figure 3. Some of the discrete structure in Figure 3A is due to excitation in the krypton buffer gas to final states of the form $3d^9(^2D_{5/2,3/2})np$; these lines were used in calibration. In Figure 3B, argon was used as a buffer, and the excitations in barium vapor are visible without interfering structure. The pressure-path length product reduced to room temperature was about 7 torr-cm for both cases.

Table 2. Wavelength and Classification of 4d Transitions in Barium Vapor

Wavelength (Å)	Energy (eV)	Classification	
136.9(1)[a]	90.56(8)[a]	$4d^9 4f(^3P_1)$	0.04% $^1P_1$
132.2(1)	93.78(8)	$4d^9(^2D_{5/2})6s^26p(^2P_{3/2})$	46% $(5/2, 3/2)_1$
131.9(1)	94.00(8)		
131.7(1)	94.14(8)		
131.4(1)	94.35(8)	$4d^9 4f\,^3D_1$	0.41% $^1P_1$
130.9(1)	94.71(8)	$4d^9(^2D_{5/2})6s^26p(^2P_{3/1})$	19% $(5/2, 3/2)_1$
130.5(2)	95.00(16)		
130.2(2)	95.22(16)		
128.9(1)	96.18(8)	$4d^9(^2D_{3/2})6s^26p(^2P_{1/2})$	28% $(3/2, 1/2)_1$
128.6(1)	96.41(8)	$4d^9(^2D_{3/2})6s^26p(^2P_{1/2})$	35% $(3/2, 1/2)_1$
126.3(1)	98.16(8)	$4d^9(^2D_{3/2})6s6p(^2P_{1/2})$	27% $(3/2, 1/2)_1$

$d^9(^2D_{5/2})6s^2(^1S_0)$ limit = 98.4(2) eV or 792,500(1000) cm^{-1}
$d^9(^2D_{3/2})6s^2(^1S_0)$ limit = 101.0(2) eV or 814,800(1000) cm^{-1}

[a] The uncertainty listed in parentheses is the estimated probable error.

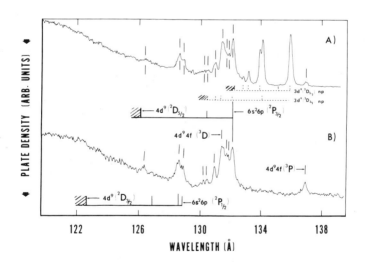

Figure 3. Densitometer traces of the absorption spectrum of barium. (A) Krypton used as buffer gas providing known lines for calibration. (B) Argon used as a buffer gas. All lines are due to barium. The relative fraction of $6s^26p(^2P)$ obtained from the calculation of Stein (1967) is shown by the vertical lines attached to the term identification underneath the spectra.

The Ba structure can be divided into two groups separated by about 3 Å (or 2 eV) and a single absorption feature below and above these groups. From the densitometer traces shown in Figure 3, the plate density is observed to decrease as the wavelength decreases, indicating an increasing absorption cross section. Other reports (Rabe et al., 1974; Connerade et al., 1974) have shown that this cross section rises to a maximum corresponding to the maximum in the $4d^{10} \rightarrow 4d^9\varepsilon f$ transition probability occurring at 110 Å. The lack of similarity of the Ba structure to xenon below the $4d$ ionization threshold and the very weak feature in Ba at 136.9 Å suggest that this spectrum is more similar to the discrete $4d^{10} \rightarrow 4d^9 4f$ spectrum (Sugar, 1972) of triply ionized lanthanum than to xenon. To test this interpretation, a calculation of energy levels of $4d^9 4f$ was carried out utilizing scaled Hartree–Fock (HF) values for the radial energy integrals. The scaling was the same as was indicated by Sugar (1972) for the lanthanides; 75% for the Slater parameters F^K, 67% for the G^K, and 100% for the spin–orbit parameters. The calculated intervals between the accessible upper levels, $^3P_1$, $^3D_1$, and $^1P_1$, were 3.6 eV and 15.9 eV. If one interprets the lowest-energy line observed (at 90.6 eV) as the transition to the $^3P_1$ level, the strong feature at 94.4 eV as the transition to the $^3D_1$ level, and the large resonance (Rabe et al., 1974) above threshold at 112 eV as the transition to the $^1P_1$ level, then the observed intervals of 3.80(15) eV and 17.6(20) eV agree well with the predictions. Furthermore, the predicted relative intensities of 1, 10, and 2500 for the $^3P_1$, $^3D_1$, and $^1P_1$ lines are also in good agreement with the observations. The close correspondence of the calculations to the experimental results indicates that the $4f$ electron orbit in excited barium contracts and overlaps the $4d$ orbit. In such a case it is more consistent to speak of the autoionization of a bound state than to speak of the outgoing $4d$ electron surmounting the centrifugal barrier.

The remaining structure is interpreted as arising from $4d^{10}6s^2 \rightarrow 4d^9 6s^2 6p$ transitions. An HF calculation of the radial energy integrals for the excited configuration and a subsequent diagonalization of the energy matrices show nearly pure J, j coupling in $4d^6 6s^2 6p$. The eigenvectors for the $J = 1$ levels indicate that two absorption lines of relative intensities 3 to 2 should be observed, the first associated with the $4d^9(^2D_{5/2})$ threshold and the second associated with the $4d^9(^2D_{3/2})$ threshold, with a separation of about 2.4 eV. Experimentally, the $4d^9$ spin–orbit splitting has been determined from x-ray data to be 2.66(15) eV (Bearden, 1967). We interpret the prominent lines at 132.2 Å and 128.6 Å, having an energy separation of 2.63(15) eV, as these transitions. We suggest that the remaining structure may be accounted for by configuration interaction among the outer electron groups $6s^2 6p$, $5d6s6p$, and $5d^2 6p$. The complexity of an exact calculation led us to choose a more qualitative approach to the interpretation of these features using neutral lanthanum as a model. Stein (1967) has analyzed the configuration interaction for the above electron groups in lanthanum and has given the eigenvectors for their mixture. The difference between lanthanum and barium in the present experiment is the presence of the $4d^9$ hole in barium. If we assume weak coupling between these outer electrons and the $4d^9$ hole, the mixing among the three outer electron groups should be similar to their mixture in lanthanum.

The allowed dipole transitions of this type in barium having significant oscillator strength are $4d^{10}6s^2 \rightarrow 4d^9(^2D_{5/2})6s^26p(^2P_{3/2})$ and $4d^{10}6s^2 \rightarrow 4d^9(^2D_{3/2})6s^26p(^2P_{1/2})$. These upper states have $^1P_1$ components of 60% and 40%, respectively, in our calculation. Stein's calculation for lanthanum shows that $6s^26p(^2P_{3/2})$ is distributed mainly among two levels and $6s^26p(^2P_{1/2})$ among three. The structure associated with these terms is shown in Figure 3 between 130 and 126 Å. The relative fraction of $6s^26p(^2P_{3/2,1/2})$ in these terms is shown by the vertical lines attached to the term identification under the spectra. This comparison shows the probable origin of the main barium resonances. While nearly all (90%) the $^2P_{1/2}$ of lanthanum is identified by Stein, he has given only 65% of the $^2P_{3/2}$. The rest, presumably, is distributed among a number of levels and may account for the remaining small peaks in the vicinity of the $5d^9(^2D_{5/2})6s^26p(^2P_{3/2})$ resonances. The discrepancies between the barium spectrum and the lanthanum model are likely to be due to deviations from our assumption of weak coupling between the core hole and the outer electrons.

Table 2 contains the wavelengths and energies of the observed resonances of barium and their classifications. These classifications show the major component of the eigenvectors for two of the three $4d^94f$ levels. Since the oscillator strength is proportional to the percentage of $^1P_1$ in the eigenvector, this quantity is also given. The remaining classifications are proposed transitions to $4d^96s^26p$. The notation is for J_1J_2 coupling and includes the component giving rise to the transition and its percentage in the eigenvector.

This interpretation is intended to illustrate the role final state configurations of the type $5d^26p$ and $5d6s6p$ play in the $4d$ excitation spectrum of barium. A more exact accounting would require a detailed calculation that included configuration mixing between these configurations.

By subtracting the energy of $6s^26p(^2P_{3/2})$ and $6s^26p(^2P_{1/2})$ in La I from the energy of $6s^2(^1S)$ in La II (Garten and Wilson, 1966), one can determine the binding energies of the $6p$ electron on a ^1S core to be 36,096 cm^{-1} (4.47 eV) and 37,155 cm^{-1} (4.60 eV). From the known position of $4d^9(^2D_{5/2})6s^26p(^2P_{3/2})$ and $4d^9(^2D_{3/2})6s^26p(^2P_{1/2})$, we can estimate the energy of the $4d^9(^2D_J)6s^2(^1S)$ core (the $N_{IV,V}$ edges) by using the $6p$ binding energies derived from La. By this procedure, we arrive at values for the two limits $4d^9(^2D_{5/2})$ and

$4d^9(^2D_{3/2})$ of 792,500(1000) cm^{-1} and 814,800(1000) cm^{-1} respectively. These values correspond to 98.4(2) eV and 101.0(2) eV, respectively. The error is an estimate of error arising from the use of lanthanum to compute the 6p binding energy, rather than barium. A recent report (Rabe et al., 1974) gives the binding energy of the $N_{IV,V}$ edges in barium as 100 eV and 102 eV, which compares reasonably well with the present results, and a calculation of the $N_{IV,V}$ threshold in atomic barium yields 98.6 eV (Wendin, 1973b), in good agreement with the present results. The $N_{IV,V}$ edges determined from x-ray measurements are 89.9 eV and 92.5 eV (Bearden and Burr, 1967). This disagreement is somewhat artificial because solid x-ray targets were used, and thus the ionization threshold is expected to be lower.

Acknowledgments

The authors gratefully acknowledge many helpful discussions with A. W. Weiss. The heat pipe oven and much of the monochromator used in the experiments were built by Mr. David C. Morgan; we wish to express our sincere thanks to him for his dedicated efforts, and also to Mrs. Patricia Fritz for her expert handling of the manuscript.

References

Amusia, M. Ya., Cherepkov, N. A., and Chernysheva, L. V. (1971). *Zh. Eksp. Teor. Fiz.*, **60**, 160.
Bearden, J. A. (1967). *Rev. Mod. Phys.*, **39**, 78.
Bearden, J. A., and Burr, A. F. (1967). *Rev. Mod. Phys.*, **39**, 125.
Brandt, W., and Lundqvist, S. (1967). *J. Quant. Spectrosc. Radiat. Transfer*, **7**, 441.
Cairns, R. B., Harrison, H., and Schoen, R. I. (1969). *Phys. Rev.*, **183**, 52.
Codling, K., and Madden, R. P. (1964). *Phys. Rev. Letters*, **12**, 106.
Codling, K., Madden, R. P., Hunter, W. R., and Angel, D. W. (1966). *J. Opt. Soc. Am.*, **56**, 189.
Connerade, J. P., Mansfield, M. W. D., Thimm, K., and Tracy, D. (1974). *Proc. IV Int. Conf. on VUV Rad. Phys.*, Contrib. No. 167, Hamburg, W. Germany.
Cooper, J. W. (1962). *Phys. Rev.*, **128**, 681.
Cooper, J. W. (1964). *Phys. Rev. Letters*, **13**, 762.
Dehmer, J. L., and Fano, U. (1970). *Phys. Rev.*, A **2**, 304.
Dehmer, J. L., and Starace, A. F. (1972). *Phys. Rev.*, B **5**, 1792.
Dehmer, J. L., Starace, A. F., Fano, U., Cooper, J. W., and Sugar, J. (1971). *Phys. Rev. Letters*, **26**, 1521.
Ederer, D. L. (1964). *Phys. Rev. Letters*, **13**, 760.
Ederer, D. L. (1969). *Appl. Opt.*, **8**, 2315.
Ederer, D. L., Lucatorto, T., and Madden, R. P. (1970). *Phys. Rev. Letters*, **25**, 1537.
Fano, U., and Cooper, J. W. (1965). *Phys. Rev.*, **137**, A1364.
Fano, U., and Cooper, J. W. (1968). *Rev. Mod. Phys.*, **40**, 441.

Formichev, V. A., Zimkina, T. M., Gribovskii, S. A., and Zhukova, I. I. (1967). *Soviet Physics-Solid State*, **9**, 1163.
Garton, W. R. S., and Wilson, M. (1966). *Astrophys. J.*, **145**, 333.
Haensel, R., Keitel, G., Schreiber, P., and Kunz, C. (1969). *Phys. Rev. Letters*, **22**, 398.
Haensel, R., Rabe, P., Sonntag, B., and Kunz, C. (1970). *Solid State Commun.*, **8**, 1845.
Kelley, H. P., and Simons, R. L. (1973). *Phys. Rev. Letters*, **30**, 529.
Lukirskii, A. P., Britov, I. A., and Zimkina, T. M. (1964a). Opt. i. *Spectroskopiya*, **17**, 438.
Lukirskii, A. P., Zimkina, T. M., and Britov, I. A. (1964b). *Izv. Akad. Nauk SSSR, Ser. Fiz.*, **28**, 772.
Madden, R. P., Ederer, D. L., and Codling, K. (1967). *Appl. Opt.*, **6**, 31.
Rabe, P., Radler, K., and Wolff, H. W. (1974). *Proc. IV Int. Conf. on VUV Rad. Phys.*, Contrib. No. 169, Hamburg.
Starace, A. F. (1970). *Phys. Rev.*, A **2**, 118.
Starace, A. F. (1971). *Phys. Rev.*, A **3**, 1242.
Starace, A. F. (1972). *Phys. Rev.*, B **5**, 1773.
Stein, J. (1967). Thesis, Hebrew University, Jerusalem.
Sugar, J. (1972). *Phys. Rev.*, B **5**, 1785.
Watanabe, T. (1965). *Phys. Rev.*, **139**, A1747.
Wendin, G. (1973a). *J. Phys., B.*, **6**, 42.
Wendin, G. (1973b). *Phys. Letters*, **46A**, 101.
Zimkina, T. M., Formichev, V. A., Gribovskii, S. A., and Zhukova, I. I. (1967). *Soviet Physics—Solid State*, **9**, 1128.

6

Effects of Anisotropic Electron–Ion Interaction on the Photoelectron Angular Distribution of Open-Shell Atoms

DAN DILL, ANTHONY F. STARACE, AND
STEVEN T. MANSON

The photoelectron asymmetry parameter β in LS coupling is obtained, starting from the representation of β as an expansion into contributions from alternative angular momentum transfers j_t, each of which has a characteristic angular distribution. For open-shell atoms the photoelectron–ion interaction is generally anisotropic: photoelectron phase shifts and electric dipole matrix elements depend on both the multiplet term of the residual ion and the total orbital momentum of the outgoing photoelectron channel. Consequently β depends on the term levels of the residual ion and contains contributions from all allowed values of j_t. These findings contradict the independent particle model theory for β, to which our expressions reduce only in the limiting cases of (1) spherically symmetric atoms (e.g., closed-shell atoms) and (2) open-shell atoms for which the electron–ion interaction is isotropic (e.g., very light elements). Numerical results for atomic oxygen and atomic sulfur are shown.

For open-shell atoms the photoelectron–ion interaction is generally anisotropic, i.e., photoelectron phase shifts and electric dipole matrix elements depend in general on both the multiplet term of the residual ion and the total orbital momentum of the photoelectron–ion final state. Consequently the photoelectron asymmetry parameter β is expected to be dependent on the term level of the residual ion. To explore this effect of

DAN DILL • Department of Chemistry, Boston University, Boston, Massachusetts 02215, U.S.A. ANTHONY F. STARACE • Behlen Laboratory of Physics, The University of Nebraska, Lincoln, Nebraska 68508, U.S.A. STEVEN T. MANSON • Department of Physics, Georgia State University, Atlanta, Georgia 30303, U.S.A.

anisotropic electron–ion interactions, we have derived the LS coupling form for β starting from the representation of β as an expansion into contributions from alternative angular momentum transfers j_t (Fano and Dill, 1972; Dill and Fano, 1972; Dill, 1973). Our expression for β differs considerably from that of the independent particle model theory for β developed by Cooper and Zare (1969), who ignored final-state electron–ion interaction. Whereas we find β to depend on the ionic term levels and to have contributions from all angular momentum transfers j_t, the model of Cooper and Zare gives a β independent of the ionic term level and dependent on only the single angular momentum transfer $j_t = l_0$, where l_0 is the photoelectron's initial orbital angular momentum. Our expressions for β do reduce to those of Cooper and Zare, however, in the following two limiting cases: (1) spherically symmetric atoms, e.g., closed-shell atoms, and (2) open-shell atoms for which the electron–ion interaction is isotropic, e.g., very light elements. This explains why the Cooper–Zare theory has so successfully predicted the photoelectron angular distributions of the rare gases. For most open-shell atoms, however, we expect anisotropic electron–ion interactions to have a large influence on the photoelectron angular distribution. A brief sketch of the theory is presented below, and illustrative calculations for atomic oxygen and atomic sulfur are discussed. A more detailed presentation will be given elsewhere (Dill *et al.*, 1975).

A general photoionization process in LS coupling may be represented as

$$A(J_0\pi_0) + \gamma(j_\gamma = 1, \pi_\gamma = -1) \rightarrow A^+(J_c\pi_c) + e^-(lsj, \pi_e = (-1)^l) \tag{1}$$

Provided the atom is unpolarized and no measurements are made of the orientation of the ion or of the photoelectron's spin, the differential cross section and the asymmetry parameter may be represented as incoherent sums over allowed angular momentum transfers j_t (Dill, 1973):

$$\frac{d\sigma}{d\Omega} = \sum_{j_t} \frac{\sigma(j_t)}{4\pi} [1 + \beta(j_t)P_2(\cos\theta)] \tag{2}$$

$$\beta = \sum_{j_t} \sigma(j_t)\beta(j_t) \Big/ \sum_{j_t} \sigma(j_t) \tag{3}$$

where $\mathbf{j}_t = \mathbf{j}_\gamma - \mathbf{l} = \mathbf{L}_c - \mathbf{L}_0$ and the allowed values of j_t are determined from conservation of angular momentum and parity. Explicit expressions for the partial cross sections $\sigma(j_t)$ and asymmetry parameters $\beta(j_t)$ are given by Dill (1973) in terms of scattering amplitudes $S_l(j_t)$, whose form in

LS coupling is (Dill et al., 1974)

$$S_l(j_t) = \frac{4\pi}{\lambda}\left(\frac{\pi\alpha h\nu}{3}\right)^{1/2} i^{-l} \exp(i\sigma_{\varepsilon l})\hat{J}_0\hat{l}_0 \begin{pmatrix} l & 1 & l_0 \\ 0 & 0 & 0 \end{pmatrix}(l_0^n L_0 S_0\{|l_0^{n-1}L_cS_c)$$

$$\times \sum_L \exp(i\delta_{\varepsilon l}^{L_cS_cL})R_{\varepsilon l}^{L_cS_cL}\hat{L}^2 \begin{Bmatrix} L_0 & L_c & j_t \\ l & 1 & L \end{Bmatrix}\begin{Bmatrix} L_0 & L_c & l_0 \\ l & 1 & L \end{Bmatrix} \quad (4)$$

Here $\sigma_{\varepsilon l}$ is the Coulomb phase shift, dependent on the photoelectron orbital momentum l and kinetic energy ε, $\hat{x} \equiv (2x + 1)^{1/2}$, $\nu\lambda = c$, $R_{\varepsilon l}^{L_cS_cL}$ is the radial dipole matrix element, and $\delta_{\varepsilon l}^{L_cS_cL}$ is the photoelectron phase shift relative to Coulomb waves.

The sum over L in Eq. (4) is dynamically weighted by phase shifts and dipole matrix elements that are dependent on L as well as on the ion core level L_cS_c. These phase shifts and matrix elements are manifestations of anisotropic (i.e., L-dependent) electron–ion interactions. The Cooper–Zare model of angular distributions results from Eqs. (2)–(4) only in the limiting cases of (1) isotropic, i.e., L-independent, electron–ion interactions and (2) isotropic target states. In the first case, the dynamic factors may be taken out of the summation, since

$$\exp(i\delta_{\varepsilon l}^{L_cS_cL})R_{\varepsilon l}^{L_cS_cL} \xrightarrow[\text{no interaction}]{} \exp(i\delta_{\varepsilon l})R_{\varepsilon l} \quad (5)$$

The sum over L may then be performed analytically to yield for the scattering amplitude an expression independent of L, L_c, and S_c and nonzero only for the single angular momentum transfer, $j_t = l_0$. In the second case, $L_0 = 0$, e.g., as in closed-shell atoms, and the $6j$ symbols collapse, leaving L and j_t single-valued: $L = 1$ and $j_t = l_0$.

We conclude that anisotropic electron–ion interactions may be studied experimentally and theoretically by investigating the photoelectron angular distribution of open-shell atoms. A measure of the strength of these interactions is the difference between different photoelectron phase shifts $\delta_{\varepsilon l}^{L_cS_cL}$. When these differences are large our results for β differ from those of Cooper and Zare (1969) in two respects: we find (1) that β is dependent on the term level of the residual ion, L_cS_c, and (2) that β has contributions from all allowed angular momentum transfers and not just the single value $j_t = l_0$.

Photoelectron angular distribution calculations using the theory presented above have been carried out for the following reactions in atomic oxygen (Starace et al., 1974) and atomic sulfur (Dill et al., 1974):

$$O[2p^4(^3P)] + \gamma \rightarrow O^+(2p^3 L_cS_c) + e^- \quad (l = 0, 2)$$

and

$$S[3p^4(^3P)] + \gamma \rightarrow S^+(3p^3 L_cS_c) + e^- \quad (l = 0, 2)$$

Table 1. Allowed Values for the Ion Core Level ($L_c S_c$), the Photoelectron Orbital Angular Momentum (l), Angular Momentum Transfer (j_t), and Total Orbital and Spin Angular Momenta (LS) for the Photoionization Reactions $np^4(^3P) + h\nu \to np^3(^4S, {}^2D, {}^2P) + e^-$, where $n = 2$ for Oxygen and $n = 3$ for Sulfur

$L_c S_c$	l	j_t	LS
^4S	0	1	^3S
^4S	2	1	^3D
^2D	0	1	^3D
^2D	2	1	$^3D, {}^3P, {}^3S$
^2D	2	2	$^3D, {}^3P, {}^3S$
^2D	2	3	$^3D, {}^3P, {}^3S$
^2P	0	1	^3P
^2P	2	1	$^3D, {}^3P$
^2P	2	2	$^3D, {}^3P$

In Table 1 we list the allowed values of the angular momentum transfer j_t for each ionic term level $L_c S_c$ and electron orbital momentum l. Also shown are allowed total orbital and spin momenta for the electron–ion final state system. Note that whereas the ^4S ionic term level has only a single value of j_t, the ^2D and ^2P ionic term levels have more than one allowed value of j_t.

In Figure 1 we have plotted d-wave photoelectron phase shifts $\delta_{\varepsilon d}^{L_c S_c L}$ for oxygen as a function of kinetic energy ε. Of the many such phase shifts that are possible, we have plotted those leading to the ^2D ionic term level

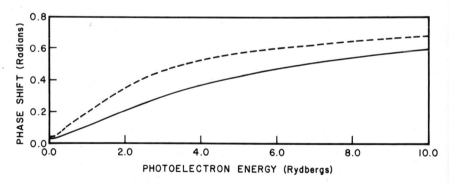

Figure 1. Hartree–Fock d-wave phase shifts $\delta_{\varepsilon d}^{L_c S_c L}$ for the ^2D oxygen ion term versus photoelectron kinetic energy ε for two alternative allowed values of L. Solid line corresponds to $L = 0$, i.e., the state $2p^3(^2D)\varepsilon d(^3S)$, and the dashed line to $L = 2$, i.e., the state $2p^3(^2D)\varepsilon d(^3D)$.

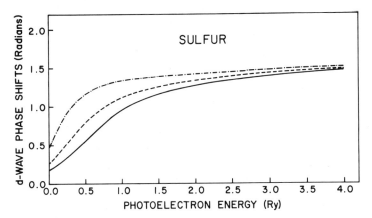

Figure 2. Hartree–Fock d-wave phase shifts $\delta_{\varepsilon d}^{L_c S_c L}$ for the ^2D sulfur ion term versus photoelectron kinetic energy ε for alternative allowed values of L. Solid line corresponds to $L = 0$, i.e., the state $3p^3(^2D)\varepsilon d(^3S)$; dashed line corresponds to $L = 1$, i.e., the state $3p^3(^2D)\varepsilon d(^3P)$; dot-dashed line corresponds to $L = 2$, i.e., the state $3p^3(^2D)\varepsilon d(^3D)$.

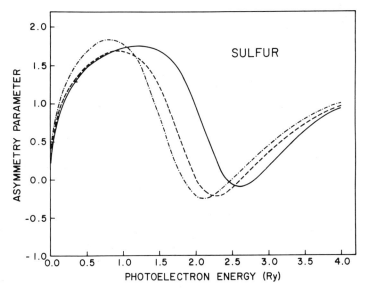

Figure 3. Asymmetry parameters $\beta(^3P \to L_c S_c)$ for the photoionization reactions $S[3p^4(^3P)] \to S^+(3p^3 L_c S_c) + e^-$ as a function of photoelectron kinetic energy. Solid line, ^4S ionic term; dashed line, ^2D; dot-dashed line, ^2P.

for two of the allowed values of L listed in Table 1. Three of the corresponding phase shifts in atomic sulfur are plotted in Figure 2. Recalling that the phase shift differences are measures of anisotropic electron–ion interaction, we note that whereas the oxygen phase shifts differ by ~ 0.2 rad at most, the sulfur phase shifts differ by as much as ~ 0.6 rad. The small phase shift differences in oxygen, which are typical only of the lightest elements, indicate weak anisotropic electron–ion interactions. Hence Hartree–Fock calculations for β show little dependence on the ionic term level. Furthermore, β may be obtained accurately using only the single angular momentum transfer $j_t = l_0$ (Starace et al., 1974).

The large phase-shift differences in sulfur, however, which are typical of most open-shell atoms, indicate a sizable anisotropic electron–ion interaction. Figure 3 shows that this leads to clear differences between the asymmetry parameters β appropriate to different ionic term levels (Dill et al., 1974). Note also that these differences are large enough to be experimentally measurable at energies where the cross section is large enough to give experimentally adequate photoelectron intensities. Additional theoretical calculations are being carried out for atomic chlorine and possibly for other open-shell elements for which these effects of anisotropic electron–ion interactions might be experimentally measured.

References

Cooper, J., and Zare, R. N. (1969). *Lectures in Theoretical Physics. Vol. XI-C*, Eds. S. Geltman, K. T. Mahanthappa, and W. E. Britten, p. 317. New York: Gordon and Breach.
Dill, D. (1973). *Phys. Rev.*, A **7**, 1976.
Dill, D., and Fano, U. (1972). *Phys. Rev. Letters*, **29**, 1203.
Dill, D., Manson, S. T., and Starace, A. F. (1974). *Phys. Rev. Letters*, **32**, 971.
Dill, D., Starace, A. F., and Manson, S. T. (1975). *Phys. Rev.*, A **11**, 1596.
Fano, U., and Dill, D. (1972). *Phys. Rev.*, A **6**, 185.
Starace, A. F., Manson, S. T., and Kennedy, D. J. (1974). *Phys. Rev.*, A **9**, 2453.

7

Study of Atomic Subshell Properties by Electron Spectrometry

M. O. KRAUSE AND F. WUILLEUMIER

> Atomic properties and excitation processes that can be studied by electron spectrometry are briefly described. A discussion is given of photon–atom interactions differentiated, process by process and subshell by subshell, by photoelectron analysis. Recent results on helium and neon are emphasized and their implications on future work indicated.

With the advent of high-resolution electron spectrometry, many atomic properties and excitation processes implicitly contained in such measurable quantities as photoattenuation coefficients, charge states, or similarly gross observables can now be differentiated and investigated directly and in detail. Other properties that are not the exclusive domain of electron spectrometry, such as binding energies and natural level widths, can be obtained with ease and accuracy. Detailed and reliable experimental data allow us to test theoretical models and approximations rigorously and sensitively at a level of increased differentiation. Information on subshell properties is essential for refinement of the theory of atomic structure and dynamics and also for improvement of analytical methods that rely on emission of electrons or photons following excitation of inner- and, in some instances, outer-shell electrons to bound or continuum states.

General treatment of the capability of electron spectrometry in the study of atomic properties and processes at a highly differentiated level is given in several works (Krause, 1969, 1971; Siegbahn et al., 1969; Rudd and Macek, 1972; Sevier, 1972; Wuilleumier and Krause, 1974; Krause, 1975).

M. O. KRAUSE • Transuranium Research Laboratory, Oak Ridge National Laboratory, Oak Ridge, Tennessee.   F. WUILLEUMIER • Laboratoire de Chimie Physique, l'Université Paris VI and L.U.R.E., Bâtiment 350, 91405 Orsay, France.

The following are representative cases of a variety of experiments using the photoelectron, Auger electron, characteristic electron energy loss, or electron–electron coincidences as a probe: (a) multiple-electron excitation by photons as exhibited in photoelectron spectra or by electrons as exhibited in Auger spectra; (b) multiple electron excitation in an Auger process; (c) subshell binding energies; (d) subshell photoionization cross sections; (e) channel identification for simultaneous excitation–ionization processes in photoeffect; (f) angular distributions of photoelectrons from different subshells; (g) natural widths of atomic levels; (h) energies and branching ratios for nonradiative transitions; (i) parameters of autoionization profiles; (j) energies of and transition rates to excited levels obtained from characteristic energy loss spectra; and (k) momentum distribution of wave functions and reaction mechanisms of electron–atom collisions from e,2e coincidence experiments. Electron spectrometric experiments that use synchrotron radiation and lasers as excitation sources and aim, *inter alia*, at determinations of subshell photoionization cross sections, energy levels of negative ions, and spin polarization are described elsewhere in this volume. Investigations have just begun of multiplet structure of atoms (ions) with partially filled shells, as experimentalists move from the noble gases to atomic vapors.

Two special applications extend the use of electron spectrometry. First, x-rays can be investigated by means of the photoelectron ejected from some converter atom. Based on the photoelectric effect, the *p*hotoelectron serves for the *a*nalysis of *x*-rays (PAX) and makes x-ray processes in atoms accessible for study by electrons (Krause, 1973). Thus, energies, line widths, and relative intensities, or branching ratios, of x-ray lines in singly and multiply ionized atoms can be measured (Krause *et al.*, 1972; Keski-Rahkonen and Krause, 1974). Second, photoionization can be simulated by a coincidence measurement of ions and high-energy electrons that have lost energy in small-angle scattering in the electron–atom interaction (Van der Wiel and Wiebes, 1971). Since the energy loss, or momentum transfer, of the incoming electron is related to the oscillator strength, or photoionization cross section, the same experiments can be performed that can be done by observing photoelectrons created by the continuously variable radiation from a synchrotron or storage ring.

It is beyond the scope of this paper to discuss all representative electron spectrometry experiments of atomic properties. We shall, instead, emphasize how photoelectrons enable us to distinguish single and multiple photoionization processes in the various atomic subshells, and to derive absolute partial cross sections.

In the photoelectron spectrum of Figure 1, the Cu L_α x-rays produce photolines that correspond to the ejection of a single orbital electron from

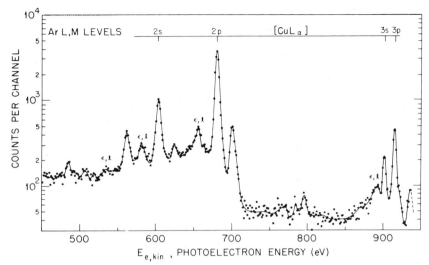

Figure 1. Photoelectron spectra of argon excited by Cu L_α x-rays. Shake up lines are designated by ε, l. Undesignated lines are produced by Cu L photons other than Cu L_α.

either the 2s, 2p, 3s, or 3p subshell. The energy $E_{e,\text{kin}}$ of a photoline is given by

$$E_{e,\text{kin}} = h\nu - E_{nlj} \qquad (1)$$

where $h\nu$ is the photon energy and E_{nlj} the binding energy of the nlj electron. For an electrostatic analyzer, the intensity of the line $N(e)$ is given by

$$N(e) = KE_{e,\text{kin}}\, d\sigma_{nlj}/d\Omega \qquad (2)$$

where $d\sigma_{nlj}/d\Omega$ is the partial differential photoionization cross section and K a factor which varies little in most energy ranges and which contains instrumental parameters, source density, and incoming photon intensity. The width of a photoline is obtained from the Voigt integral

$$S = \int G(x - \xi) L(\xi)\, d\xi \qquad (3)$$

where G is the Gaussian distribution of the spectrometer function and L is the sum of the Lorentzian distributions of the atomic level nlj and the x-ray line.

The photoelectron spectrum provides quantitative information on the energy, the differential cross section, and the width of a subshell characterized by the quantum numbers nlj. In addition, the anisotropy parameter β_{nlj} can be obtained from a measurement of $d\sigma_{nlj}/d\Omega$ at different angles θ

between photoelectron and photon propagation directions according to

$$\frac{d\sigma_{nlj}}{d\Omega} = \frac{\sigma_{nlj}}{4\pi}\left[1 - \frac{\beta_{nlj}}{4}(3\cos^2\theta - 1)\right] \quad (4)$$

Finally, the peaks designated by εl in Figure 1 evince processes in which one electron from an nlj subshell is ionized and another one from an $n'l'j'$ subshell is simultaneously excited to some bound, discrete level of the ion. These $\varepsilon l, nl'$ processes, often called "shake up," are better resolved in the spectrum shown in Figure 2, clearly demonstrating the occurrence of an $\varepsilon l, nl'$ transition in which a 1s electron of neon goes into an εp continuum channel and a 2p electron is concomitantly excited into an nl' level, $n \geqslant 3$, whereby spin exchange leads to a splitting of each of these n states. The spectrum also shows that this type of process may also involve a 1s, 2s electron pair, and, by the continuum below about 47 eV, that two electrons can be simultaneously ionized.

Wuilleumier and Krause (1974) made a systematic study of all aspects of photoionization of neon as a function of photon energy in the range from 100 to 2000 eV. As demonstrated in Figures 1 and 2, all processes that take place could be identified and their differential cross sections obtained with the aid of Eq. (2). However, $d\sigma_{nlj}/d\Omega$ was obtained only on a relative basis, since it is very involved to determine it absolutely on the basis of Eq. (2). It was then mandatory to determine the relative cross sections for all processes taking place, in order to derive the absolute partial cross sections by normalization to the known, absolute total photoionization cross section. The procedure employed can be sketched by the following scheme:

$$I_x(\text{photo})_{\text{rel}} \propto d\sigma_x/d\Omega_{\text{rel}} \xrightarrow{\beta_x} \sigma_{x,\text{rel}} \xrightarrow{\sigma_{\text{tot}}} \sigma_{x,\text{abs}} \quad (5)$$

where x designates the various subshells and subshell combinations. The resulting partitioning of σ_{tot} into its components is displayed in Figure 3. With these data, current theoretical models could be tested, subshell by subshell and process by process. As an example, the absolute 2s cross section is compared in Figure 4 with theoretical predictions based on (a) the independent-particle central-potential model which uses the Herman–Skillman wave functions in one case and Hartree–Fock wave functions for the continuum electron in the other case, and (b) the random-phase approximation with spin exchange (RPAE) which uses the Hartree–Fock wave functions and includes intra- and intershell electron–electron interactions (Amusia et al., 1972). As this comparison indicates, the sophisticated RPAE calculation agrees well with these experimental data; and in view of similarly good agreement with data on argon at very low photon energies (Lynch et al.,

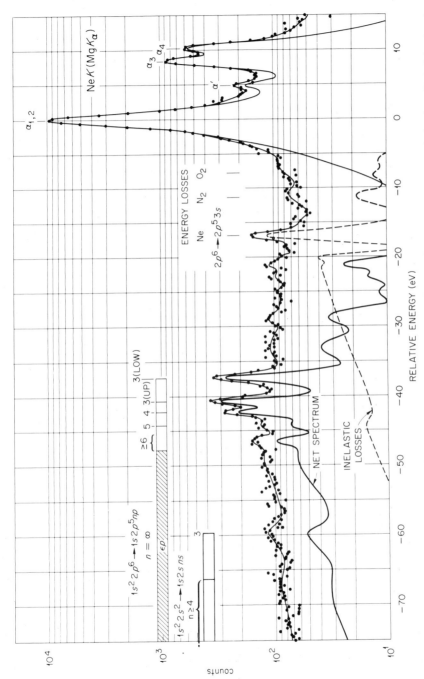

Figure 2. Photoelectron spectrum of neon produced by Mg K_α and exhibiting photolines due to single-electron emission from K shell ($\alpha_{1,2}$) and to two-electron excitation, $\varepsilon l, nl'$ processes, involving the emission of a K electron and excitation of a $2p$ or $2s$ electron (Carlson et al., 1971).

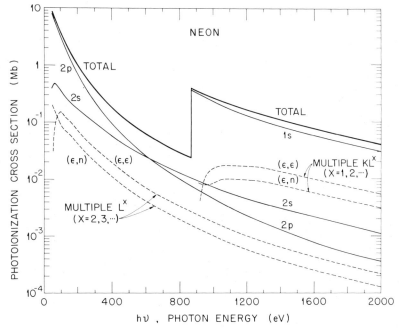

Figure 3. Complete partitioning of the total photoionization cross section of neon into its components (Wuilleumier and Krause, 1974).

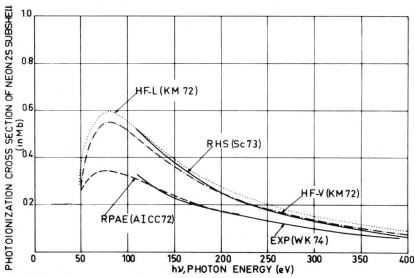

Figure 4. Comparison of the experimental 2s cross section of single-electron emission (Wuilleumier and Krause, 1974) with the theoretical results of Amusia *et al.* (1972), Kennedy and Manson (1972), and Scofield (1973).

1973), this model might be regarded to be generally dependable in the non-relativistic regime. On the other hand, the Herman–Skillman calculations generally overestimate the partial cross sections by about 10–40%. According to Fadley (1974), this overestimate is due to the inclusion of multiple-electron processes in the partial cross section of the single-electron process.

It would be desirable to extend this study on neon to other, more complex atoms, and especially to atoms with open shells, to test the theory of photoionization under more severe conditions. However, if for an element like xenon we were to proceed as in the case of neon, the experiment would become very tedious. An alternate, simpler procedure is to determine the partial cross sections of interest not by normalization to the total cross section of the atom under study but by normalization to the $2p$ or $1s$ subshell cross section of neon, which are now known to an accuracy of better than $\pm 5\%$ in the respective regions of interest (Wuilleumier, 1973; and Krause, 1974).

According to the more important of the two-electron selection rules, there may be an excitation of two electron states if one electron changes only its n and the other electron changes n by an arbitrary amount and its l by ± 1 (Goudsmit and Gropper, 1931). In the case of helium this would mean an $\varepsilon l, nl'$ process could leave the He^+ with an electron in an s or a p state. For the most intense transitions to the $n = 2$ level, $1s^2 \to 2s, \varepsilon p$ and $1s^2 \to 2p, \varepsilon s$ (or $2p, \varepsilon d$) are allowed. An experimental determination of the probabilities of these two alternate transitions at different photon energies is of basic importance to the understanding of photon–atom interaction. As has been shown by Krause and Wuilleumier (1972), this can be realized by a photoelectron spectrometric measurement in which the degeneracy of the two final states $2s, \varepsilon p$ and $2p, \varepsilon s$ is "lifted" by a measurement of the angular distribution of the emitted photoelectrons. It follows from symmetry considerations that $1s^2 \to 2s, \varepsilon p$ would yield a photoelectron having the same anisotropy parameter β_s as $1s^2 \to 1s, \varepsilon p$, which is a single-electron transition. The ratio of the photoelectron intensities of the two photolines is then independent of observation angle, according to Eq. (4). If, however, the photoline corresponding to an $\varepsilon l, nl'$ event is due to $1s^2 \to 2p, \varepsilon s$ or contains a fraction of $1s^2 \to 2p, \varepsilon s$ transitions, the intensity of the photoline referred to the $1s^2 \to 1s, \varepsilon p$ line would vary with angle, since its β_p would usually be different from β_s.

Figure 5 shows that the intensity ratio of $He^+(n = 2)/He^+(n = 1)$ varies with observation angle, indicating that both processes take place in the energy range studied. Data suggest that the $2p$ state is populated preferentially at low energies and the $2s$ state preferentially at high energies. These data support the results of the calculation of Jacobs and Burke (1972), who have used the best available bound state and continuum wave functions.

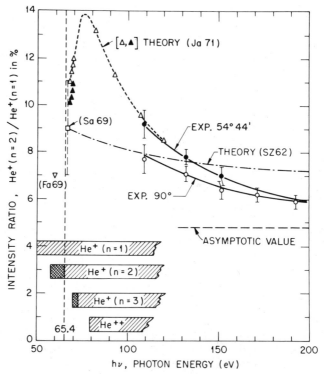

Figure 5. Selection of the processes that leave the He$^+$ ion in the $n = 2$ and $n = 1$ states and their intensity ratios at two different angles. Experiment at 54° 44′ can be directly compared with theoretical partial cross sections (Wuilleumier and Krause, 1975).

Partially because of the efforts needed to identify all pertinent processes and partially because of intensity limitations, absolute subshell cross sections near thresholds have often been derived from a measurement of the relative probabilities for single-electron emission from the various subshells and normalization to the total cross section. This procedure is faulty since the εl, nl' double-electron transitions are active even at threshold, as demonstrated for helium (see Figure 5) and for neon (Wuilleumier and Krause, 1974). In fact, the relative intensity of these processes can be substantial at threshold and may rise and fall rapidly somewhat above threshold. Only the double-ionization process, an εl, $\varepsilon' l'$ event, has zero intensity at threshold.

Acknowledgment

Research for this study was sponsored by the U.S. Atomic Energy Commission under contract with Union Carbide Corporation.

References

Amusia, M. Ya., Ivanov, V. K., Cherepkov, N. A., and Chernysheva, L. V. (1972). *Phys. Letts.*, A **40**, 361.
Carlson, T. A., Krause, M., and Moddeman, W. E. (1971). *J. Phys. (Paris)*, **32**, C4–76.
Fadley, C. S. (1974). *Chem. Phys. Letts.*, **25**, 225.
Goudsmit, S., and Gropper, L. (1931). *Phys. Rev.*, **38**, 225.
Jacobs, V. L., and Burke, P. G. (1972). *J. Phys. B*, **4**, L67.
Kennedy, D. J., and Manson, S. T. (1972). *Phys. Rev.*, A **5**, 227.
Keski-Rahkonen, O., and Krause, M. O. (1974). *Physica Fennica*, **9**, S1–261.
Krause, M. O. (1969). *Phys. Rev.*, **177**, 151.
Krause, M. O. (1971). *J. Phys. (Paris)*, **32**, C4–67.
Krause, M. O. (1973). *Adv. X-Ray Anal.*, **16**, 74.
Krause, M. O. (1975). In *Atomic Innershell Processes*, Ed. B. Crasemann, Chap. 6.2. New York: Academic Press.
Krause, M. O., and Wuilleumier, F. (1972). *J. Phys. B*, **5**, L143.
Krause, M. O., Wuilleumier, F., and Nestor, C. W., Jr (1972). *Phys. Rev.*, **6**, 871.
Lynch, M. J., Gardner, A. B., Codling, K., and Marr, G. V. (1973). *Phys. Letts.*, A **43**, 237.
Rudd, M. E., and Macek, J. H. (1972). *Case Studies in Atomic Physics*, **3**, 47. Amsterdam: North-Holland.
Scofield, J. (1973). *Lawrence Livermore Laboratory, Report No. UCRL 51326*. Available from National Information Center, National Bureau of Standards, Springfield, Va.
Sevier, K. D. (1972). *Low Energy Electron Spectrometry*. New York: Wiley, Interscience.
Siegbahn, K. *et al.* (1969). *ESCA Applied to Free Molecules*. Amsterdam and London: North-Holland.
Van der Wiel, M. J., and Wiebes, G. (1971). *Physica*, **53**, 225.
Wuilleumier, F. (1973). *Adv. X-Ray Anal.*, **16**, 63.
Wuilleumier, F., and Krause, M. O. (1974). *Phys. Rev.*, A **10**, 242.
Wuilleumier, F., and Krause, M. O. (1975). To be published. [See also Krause and Wuilleumier (1972).]

8
Quantum Defect Theory: Photoionization

J. Dubau

> Quantum defect theory is used to express the photoionization cross section as a function of quantities which vary slowly with energy. A parametric form is obtained for the resonances structure. Direct calculations are compared with extrapolated results in the resonance region. The numerical case of Be is considered as an illustration of the techniques employed.

1. Introduction

The quantum defect theory (Seaton, 1966, 1969) has proved to be a very useful approximation for studying the excitation cross sections of ions by electron collision in the resonance region.

This paper extends the quantum defect theory method to the study of photoionization cross sections. The data required for the extrapolation, from above the excitation threshold of Be^+, are obtained by solving a set of coupled integrodifferential equations (Dubau and Wells, 1973a).

Fano (1970) and Lu (1971) developed a theory of similar nature, but since they were mainly concerned with the analysis of experimental data, their approach was completely different from that of this paper.

J. Dubau • Department of Physics and Astronomy, University College, Gower Street, London, WC1E 6BT, England.

2. Wave Functions

The numerical calculations were done using the dipole-length approximation. The total photoionization cross section is then

$$Q(E) = A \frac{(I + E)}{\omega_i} \sum_j |(\Psi_B(x)\|r\|\Psi_j(E, x))|^2 \tag{1}$$

where $\Psi_B(x)$ and $\Psi_j(E, x)$ are the wave functions corresponding to the initial bound state $[2s^2(^1S)]$ and the final free state $[2skp(^1P)]$ respectively. The energy, E, is measured from the first ionization threshold; I is the energy necessary to ionize the initial bound state, ω_i the statistical weight of the initial bound state, and A a coefficient dependent on the normalization of the wave functions.

As in the case of electron–ion collision, we approximate the true free wave function by a finite sum of antisymmetrical products

$$\Psi_j(E, x) = \mathscr{A} \sum_i \phi_i \frac{F_{ij}(r)}{r} \tag{2}$$

where ϕ_i incorporates the normalized target ion functions as well as the spin and the angular part of the orbital of the added electron and j represents some particular solution when there is an energy degeneracy.

Using the Kohn variational method, the best approximated solution of the Schrödinger wave equation is obtained when the radial functions $F_{ij}(r)$ satisfy the following coupled integrodifferential equations:

$$\left(\frac{d^2}{dr^2} - \frac{l(l + 1)}{r^2} + \frac{2}{r} + k^2 \right) \mathbf{F} + \mathbf{UF} = 0 \tag{3}$$

The convention used is: bold type for nondiagonal matrices, $\mathbf{F} = (F_{ij}(r))$, and italic type, without subscripts, for diagonal matrices. The energy of the outgoing electron is $k^2 : k_i^2 = E - E_i$, where E_i is the energy in rydbergs of the target state corresponding to ϕ_i. $\mathbf{U}$ is the usual integrodifferential operator used in the collision-coupled equation method (Percival and Seaton, 1957).

The solutions will satisfy the physical boundary condition

$$\mathbf{F}(0) = 0 \tag{4}$$

Let us now consider in detail the case of Be. The energy band of interest lies between the first and second ionization thresholds. The numerical data required are extrapolated from above the second ionization threshold. We shall include in the summation (2) only three terms corresponding to the two first target states of Be^+, $2s$, $2p$. Therefore we have three equations, the second and third having degenerate energies:

$$k_2^2 = k_3^2 = k_1^2 - (E_{2p} - E_{2s}) \tag{5}$$

The asymptotic boundary conditions are different above ($k_2^2 > 0$) and below ($k_2^2 < 0$) the second threshold.

For $k_2^2 > 0$,

$$\mathbf{F}(\mathbf{S}, r) \underset{r \to \infty}{\sim} \frac{1}{\sqrt{k}} [\exp(-i\varphi) - \exp(i\varphi)\mathbf{S}] \tag{6}$$

where

$$\varphi = kr + \frac{1}{k} \ln(2kr) - \frac{l\pi}{2} + \arg \Gamma\left(l + 1 - \frac{i}{k}\right) \tag{7}$$

and $\mathbf{S}$ is a $(3, 3)$ matrix corresponding to the three independent solutions $\Psi_j(E, \mathbf{S}, x)$.

For $k_2^2 < 0$,

$$\left.\begin{aligned}
F_{11}(\mathscr{S}, r) &\underset{r \to \infty}{\sim} \frac{1}{\sqrt{k}} [\exp(-i\varphi_1) - \exp(i\varphi_1)\mathscr{S}_{11}] \\
F_{i1}(\mathscr{S}, r) &\underset{r \to \infty}{\sim} d_i \left(\frac{2r}{v_2}\right)^{v_2} e^{-r/v_2} \quad i = 2, 3
\end{aligned}\right\} \tag{8}$$

where $k_2^2 = -1/v_2^2$, $\mathscr{S}$ has only one element, $\mathscr{S}_{11}$, and there exists only one independent physical solution, $\Psi_1(E, \mathscr{S}, x)$.

With the former boundary conditions, (6) and (8), the solutions can be normalized:

$$\int dx \, \Psi_i(E, \mathbf{S}, x)\Psi_j(E', \mathbf{S}', x) = \delta_{ij} 4\pi \delta(E - E') \tag{9}$$

We shall define the following matrix elements:

$$\left.\begin{aligned}
\mathscr{H}_j(\mathbf{S}) &= (\Psi_B(x) \| r \| \Psi_j(E, \mathbf{S}, r)) \\
\mathscr{H}_1(\mathscr{S}) &= (\Psi_B(x) \| r \| \Psi_1(E, \mathscr{S}, r))
\end{aligned}\right\} \tag{10}$$

$\mathscr{H}(\mathbf{S})$ will be the row matrix containing the elements $\mathscr{H}_j(\mathbf{S})$.

We now consider another solution of (3), $\mathbf{F}(E, r)$, still satisfying (4) but also analytic in the neighborhood of $r = 0$. It has been proved (Ham, 1955) that $\mathbf{F}(E, r)$ is analytic in E for any finite r:

$$\mathbf{F}(E, r) = \sum_{p=0}^{\infty} E^p \mathbf{F}_{(p)}(r) \tag{11}$$

The corresponding solutions $\Psi_j(E, r)$ are also analytic in energy. It can be proved that the reduced tensor element

$$\mathcal{H}_j(E) = (\Psi_B(x)\|r\|\Psi_j(E, x)) \tag{12}$$

is analytic in E, provided that the integrals on the right-hand side of (12) are uniformly convergent.

The main hypothesis of the quantum defect theory is that there exists some finite radius r_0 such that for $r \geq r_0$ the potential $\mathbf{U}$ is zero. In practice we have nonzero long-range potentials. For this case the theory is still valid as an approximate extrapolation procedure.

If we consider two independent analytic Coulomb solutions for $r \geq r_0$, we have

$$\mathbf{F}(E, r) = f(E, r)\mathbf{A}(E) + g(E, r)\mathbf{B}(E) \tag{13}$$

$\mathbf{A}(E)$ and $\mathbf{B}(E)$ are also analytic.

In this paper, the functions $f(E, r)$ and $g(E, r)$ are taken to be the analytic continuation of the usual regular and irregular Coulomb functions. For an energy $k^2 > 0$ they behave asymptotically as

$$f(k^2, r) \underset{r \to \infty}{\sim} \frac{1}{\sqrt{k}} \sin(\varphi) \qquad g(k^2, r) \underset{r \to \infty}{\sim} \frac{1}{\sqrt{k}} \cos(\varphi) \tag{14}$$

and for $k^2 < 0$,

$$\left.\begin{aligned} f(k^2, r) &= \xi \sin \pi v - \theta \cos \pi v \\ g(k^2, r) &= \xi \cos \pi v + \theta \sin \pi v \end{aligned}\right\} \tag{15}$$

where

$$\left.\begin{aligned} \xi &\underset{r \to \infty}{\sim} \alpha(v, l)\left(\frac{2r}{v}\right)^{-v} e^{r/v} \\ \theta &\underset{r \to \infty}{\sim} \beta(v, l)\left(\frac{2r}{v}\right)^{v} e^{-r/v} \end{aligned}\right\} \tag{16}$$

($k^2 = -1/v^2$); the functions $\alpha(v, l)$ and $\beta(v, l)$ have been defined by Eissner et al. (1969).

Using (6), (10), (12), (13), and (14) we obtain

$$\left.\begin{aligned} \mathbf{F}(S, r) &= \mathbf{F}(E, r)\frac{2}{i}(\mathbf{A} - i\mathbf{B})^{-1} \\ \mathcal{H}(S, r) &= \mathcal{H}(E, r)\frac{2}{i}(\mathbf{A} - i\mathbf{B})^{-1} \\ \mathbf{S} &= (\mathbf{A} + i\mathbf{B})(\mathbf{A} - i\mathbf{B})^{-1} \end{aligned}\right\} \tag{17}$$

From (17) we see the well-known result that the scattering matrix **S** is slowly varying above the threshold. Also, $\mathcal{H}(\mathbf{S})$ and, consequently, the photoionization cross section are smoothly continuous above the excitation threshold.

Under the threshold let us define the analytic continuation of **S** and $\mathcal{H}(\mathbf{S})$

$$\left. \begin{aligned} \chi &= (\mathbf{A} + i\mathbf{B})(\mathbf{A} - i\mathbf{B})^{-1} \\ \mathcal{H}(\chi) &= \mathcal{H}(E)\frac{2}{i}(\mathbf{A} - i\mathbf{B})^{-1} \end{aligned} \right\} \quad (18)$$

Following Gailitis (1963) and Seaton (1969), we can decompose χ and $\mathcal{H}(\chi)$ in submatrices; for $Be^+ + e^-$, we obtain

$$\left. \begin{aligned} \chi_{00} &= (\chi_{11}) \qquad \chi_{0c} = (\chi_{12}, \chi_{13}) \\ \chi_{c0} &= \begin{pmatrix} \chi_{21} \\ \chi_{31} \end{pmatrix} \qquad \chi_{cc} = \begin{pmatrix} \chi_{22}, \chi_{23} \\ \chi_{32}, \chi_{33} \end{pmatrix} \\ \mathcal{H}_0(\chi) &= (\mathcal{H}_1(\chi)) \qquad \mathcal{H}_c(\chi) = (\mathcal{H}_2(\chi), \mathcal{H}_3(\chi)) \end{aligned} \right\} \quad (19)$$

Then, using (8), (10), (12), (13), and (15), we have

$$\mathcal{H}(\mathcal{S}) = \mathcal{H}_0(\chi) + \mathcal{H}_c(\chi)(e^{-2\pi i v} - \chi_{cc})^{-1}\chi_{c0} \quad (20)$$

This is the general formula for $\mathcal{H}(\mathcal{S})$. All the matrices on the right-hand side of (20) are practically constant over a resonance, except $e^{-2\pi i v}$ [which is a (2, 2) matrix].

The second and third equations of (3) having the same energy, $k_2^2 = k_3^2$, $e^{-2\pi i v}$ is a multiple of the unit matrix. We are interested in diagonalizing χ_{cc} (which is a symmetric matrix)

$$\left. \begin{aligned} \chi_{cc} &= \mathbf{X}\chi_{cc}'\mathbf{X}^T \\ \mathbf{X}\mathbf{X}^T &= \mathbf{X}^T\mathbf{X} = 1 \end{aligned} \right\} \quad (21)$$

Then

$$\left. \begin{aligned} \mathcal{H}(\mathcal{S}) &= \mathcal{H}_0(\chi) + \mathcal{H}_c'(\chi)(e^{-2\pi i v_2} - \chi_{cc}')^{-1}\chi_{c0}' \\ \mathcal{H}_c'(\chi) &= \mathcal{H}_c(\chi)\mathbf{X} \qquad \chi_{c0}' = \mathbf{X}^T\chi_{c0} \end{aligned} \right\} \quad (22)$$

$\mathcal{H}(\mathcal{S})$ can therefore be written as

$$\mathcal{H}_1(\mathcal{S}) = a\left(1 + \sum_{p=2}^{3} \frac{b_p}{e^{-2\pi i v_2} - \chi_{pp}'}\right) \quad (23)$$

where a, b_p, and χ_{pp}', are complex numbers practically constant over a resonance. They are the profile parameters of the resonance.

Table 1. Profile Parameters for Be

Energy	a real	a imag.	b_2 real	b_2 imag.	b_3 real	b_3 imag.	χ'_{22} real	χ'_{22} imag.	χ'_{33} real	χ'_{33} imag.
0.09	1.183	−0.0496	−0.00151	0.00661	0.536	−0.627	0.635	−0.767	−0.349	−0.160
0.13	1.143	−0.1127	−0.00267	0.00270	0.543	−0.580	0.633	−0.771	−0.363	−0.155
0.17	1.110	−0.172	−0.00299	0.00034	0.550	−0.532	0.632	−0.774	−0.376	−0.150
0.22	1.061	−0.241	−0.00250	−0.00094	0.558	−0.469	0.630	−0.776	−0.389	−0.145
0.23	1.053	−0.254	−0.00231	−0.00102	0.559	−0.456	0.630	−0.776	−0.391	−0.144
0.24	1.044	−0.267	−0.00210	−0.00105	0.561	−0.444	0.630	−0.776	−0.393	−0.142
0.25	1.036	−0.280	−0.00187	−0.00103	0.562	−0.431	0.630	−0.776	−0.396	−0.141
0.26	1.028	−0.292	−0.00162	−0.00097	0.563	−0.418	0.630	−0.777	−0.398	−0.140
0.27	1.021	−0.305	−0.00135	−0.00087	0.565	−0.405	0.630	−0.777	−0.400	−0.139
0.285	1.009	−0.325	−0.00092	−0.00065	0.566	−0.386	0.630	−0.777	−0.403	−0.138
0.30	0.998	−0.343	−0.00047	−0.00036	0.568	−0.366	0.630	−0.777	−0.406	−0.136

Using (22) it may be proved that the average cross section over a resonance is continuous, through the threshold, with the total photoionization cross section (Baz, 1959) and (Dubau and Wells, 1973):

$$\int_{v_0}^{v_0+1} |\mathscr{H}_1(\mathscr{S})|^2 \, dv = \sum_{i=1}^{3} |\mathscr{H}_i(\chi)|^2 \qquad (24)$$

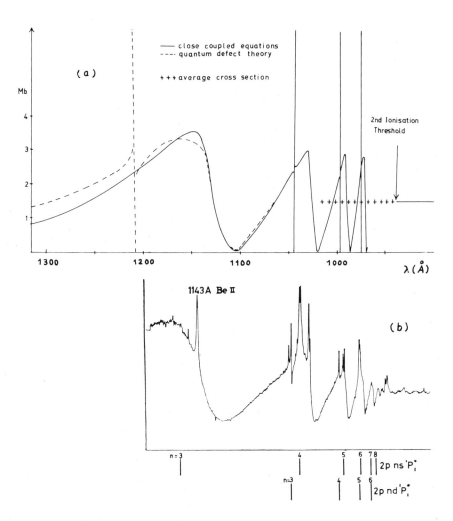

Figure 1. Photoionization cross section of beryllium. (a) Theoretical curves; (b) experimental curve of Mehlman-Balloffet and Esteva (1969).

3. Numerical Application

The target Be^+ functions were calculated using the computer program of Eissner and Nussbaumer (1969) and the radial functions for the added electron were calculated using the close-coupling method program of Seaton and Wilson (1972). The extrapolations were done from $E_1 = 0.285508$ Ry (excitation threshold) and $E_2 = 0.30$ Ry, where the $\mathbf{R}$ and $\mathcal{H}(\mathbf{R})$ matrices had been calculated:

$$\mathbf{R} = \mathbf{B}\mathbf{A}^{-1} \qquad \mathcal{H}(\mathbf{R}) = \mathcal{H}(E)\mathbf{A}^{-1} \qquad (25)$$

$\mathbf{R}$ and $\mathcal{H}(\mathbf{R})$ are smoothly continuous as $\mathbf{A}(E)$, $\mathbf{B}(E)$, $\mathcal{H}(E)$ and can be extrapolated. A linear extrapolation was done.

Using the following formulas χ and $\mathcal{H}(\chi)$ were obtained from $\mathbf{R}$ and $\mathcal{H}(\mathbf{R})$ [see Eq. (18)]:

$$\chi = (1 + i\mathbf{R})(1 - i\mathbf{R})^{-1}, \qquad \mathcal{H}(\chi) = \mathcal{H}(\mathbf{R})\frac{2}{i}(1 - i\mathbf{R})^{-1} \qquad (26)$$

The parameters a, b_2, b_3, χ'_{22}, χ'_{33} were then calculated (Table 1). From $\mathcal{H}(\mathbf{S})$ and $\mathcal{H}(\mathcal{S})$ the total photoionization cross section was derived using Eq. (1) (Figure 1).

The total photoionization cross section is plotted in megabarns (10^{-18} cm^2) as a function of wavelength in angstroms. The solid curve was obtained by solving the coupled equations directly and the dashed curve by extrapolation of $\mathbf{R}$ and $\mathcal{H}(\mathbf{R})$ from E_1 and E_2.

For the two first resonances the two theoretical curves are slightly different and show a spurious "2p2d" resonance where the uniform convergence criterion also breaks down [Eq. (12)]. The extrapolated curve is in good agreement for the other resonances.

For comparison we also plot the experimental curve of Mehlman-Balloffet and Esteva (1969).

Acknowledgments

This work was done in collaboration with Dr. J. Wells. We wish to thank Professor M. J. Seaton for his valuable advice.

References

Baz, A. I. (1959). *Soviet Physics–J.E.T.P.*, **9**, 1256.
Dubau, J., and Wells, J. (1973a). *J. Phys. B.*, **6**, L31.
Dubau, J., and Wells, J. (1973b). *J. Phys. B.*, **6**, 1452.
Eissner, W., and Nussbaumer, H. (1969). *J. Phys. B.*, **2**, 1028.

Eissner, W., Nussbaumer, H., Saraph, H. E., and Seaton, M. J. (1969). *J. Phys. B.*, **2**, 341.
Fano, U. (1970). *Phys. Rev.*, **A2**, 353.
Gailitis, M. (1963). *Soviet Physics–J.E.T.P.*, **17**, 1328.
Ham, F. S. (1955). *Solid State Physics*, **1**, 127. Eds. F. Seitz and D. Turnbull. New York: Academic Press.
Lu, K. T. (1971). *Phys. Rev.*, **A4**, 579.
Mehlman-Balloffet, G., and Esteva, J. M. (1969). *Astrophys. J.*, **157**, 945.
Percival, I. C., and Seaton, M. J. (1957). *Proc. Camb. Phil. Soc.*, **53**, 654.
Seaton, M. J. (1966). *Proc. Phys. Soc.*, **88**, 801.
Seaton, M. J. (1969). *J. Phys. B.*, **2**, 5.
Seaton, M. J., and Wilson, P. M. H. (1972). *J. Phys. B.*, **5**, L1.

9
Electron–Alkali Scattering and Photodetachment of Alkali Negative Ions

D. L. MOORES

> The relationship between the scattering of electrons by neutral alkali metal atoms and the photodetachment of the alkali negative ions is discussed, with particular reference to the behavior at the first excitation threshold. Elaborate calculations of the photodetachment cross sections and angular distribution of the electrons ejected in photodetachment of Li^-, Na^-, and K^- are described. Agreement between length and velocity forms and with experiment is good in the case of the first two ions but not as good for K^-; it is shown that improved results are obtained if a different model is chosen for the ground state of K^-. Structure in the cross sections for these ions is shown to be due to a combination of the effects of a resonance and the threshold behavior. Similar effects in the photodetachment of Rb^- and Cs^- which take the form of a "window-type" resonance just below threshold are described and interpreted. In general, the Wigner threshold law is found to hold only within microvolts of the threshold.

Introduction

The alkali negative ions are formed by attaching an electron to a neutral alkali metal atom and form bound states with a ^1S configuration. The energy required to detach the electron (the electron affinity of the alkali metals) is about 0.5 eV.

In the photodetachment process, the attached electron is ejected by impact of incident photons. After photodetachment, the negative ion breaks up into a neutral alkali plus a scattered electron; since the initial negative

---

D. L. MOORES • Department of Physics and Astronomy, University College, London, England.

ion state was ^1S, then, by the dipole selection rules, the final state of this system must be ^1P.

The wave function of this final state may thus be obtained by solving the electron–alkali scattering problem for the ^1P partial wave. According to theory (Wigner, 1949), for a P wave the partial elastic scattering cross section should exhibit a sharp discontinuity in slope at the excitation threshold. This should thus also appear in the photodetachment cross section at that photon energy for which the atom can just be left in the first excited state, following photodetachment. If discontinuity is observed at some wavelength λ, then the electron affinity will be given from energy conservation by

$$I = \frac{1}{\lambda} - E_1 \tag{1}$$

where E_1 is the (well-known) excitation energy of the atom. Following its theoretical prediction (Norcross and Moores, 1973) the existence of this effect was rapidly confirmed experimentally (Hotop et al., 1972) making use of a dye laser technique and the electron affinities of the alkali metals measured to much greater accuracy than before (Patterson et al., 1974). Subsequently, extensive calculations and measurements of the photodetachment cross sections, the subject of this paper, have been made, in a joint program of experimental and theoretical work by H. Hotop, A. Kasdan, W. C. Lineberger, D. L. Moores, D. W. Norcross, and T. A. Patterson.

Threshold Behavior

We consider the scattering of electrons by an alkali atom, the ground state of which is $ns(^2S)$ and the first excited state $np(^2P)$. The total energy of the system atom plus electron is

$$E = E(ns) + k_1^2 = E(np) + k_2^2 \tag{2}$$

At the threshold for excitation of np, $k_2^2 = 0$.

In order to understand the threshold behavior of the cross section, we describe the collision in terms of a partial-wave close-coupling formalism. According to the many-channel effective-range theory (Ross and Shaw, 1961), provided that the potentials of the problem are of short range, the reactance matrix may be written in the form

$$\mathbf{R} = \mathbf{k}^{l+1/2} \mathbf{M}^{-1} \mathbf{k}^{l+1/2} \tag{3}$$

where $\mathbf{k}^{l+1/2}$ is a matrix with elements $k_i^{l_i+1/2} \delta_{ij}$ and where $\mathbf{M}$ is so slowly varying as a function of energy that it may be regarded as constant over a small energy range. Considering the case when the total angular momentum of the system $L = 1$, we have a three-channel problem (nsk_1p, npk_2s, npk_2d). However, at energies of interest, effects of the channel npk_2d may be neglected, and we are left with a nondegenerate two-channel problem, in which the diagonal potentials fall off exponentially at large distances while the off-diagonal potential falls of as r^{-2}. Under these circumstances it may be shown (Bardsley and Nesbet, 1973) that the reactance matrix elements behave as (3). We then obtain, at sufficiently small $|k_2|$, for the P-wave partial cross sections

$$k_1^2 Q(ns \to ns) = \begin{cases} C_0\left(1 - \dfrac{2B}{C}|k_2|\right) & k_2^2 < 0 \\ C_0(1 - 2Bk_2) & k_2^2 > 0 \end{cases} \quad (4)$$

$$k_1^2 Q(ns \to np) = 3Bk_2 \quad k_2^2 > 0 \quad (5)$$

where B, C, and C_0 are constant, and C_0 and B are both positive. Hence the elastic cross section should be falling and the inelastic one rising immediately above threshold. As a function of k_1^2, the elastic cross section has an infinite derivative at threshold in the form of a cusp ($C > 0$) or a step ($C < 0$). For all other partial waves, the cross sections have continuous derivatives at threshold.

This behavior has been verified in numerical calculations such as those of Burke and Taylor (1969), Bardsley and Nesbet (1973), and Moores and Norcross (1973). In the elastic ^1P cross section for scattering of electrons by Li and by Na, a cusp is obtained; in the case of K, a downward step. In the alkalies, the effect in the ^1P partial wave is enhanced by the occurrence of a resonance just below threshold; this is apparent from the behavior of the elastic scattering phases calculated by Moores and Norcross for Li, Na, and K in a three-state close-coupling approximation, shown in Figure 1. For K, values of the constants B and C were obtained by fitting to the calculated cross section and used to give the straight lines in Figure 2. It is clear that the threshold law is only valid over a very narrow range. For both Li and Na, it was not found possible to determine constants B and C satisfying (4) and (5), even within $\pm 10^{-5}$ Ry of threshold. The explanation lies in the fact that the elements of **M** are only constant over a very narrow range of energy, owing to resonant behavior introduced by the off-diagonal inverse-square potential. The structure observed in the cross section is primarily due to the resonance, rather than the effects of the threshold law. This conclusion is supported by calculations for the ^3P partial wave, in which resonant behavior is absent, and for which, although a cusp is present, the Wigner threshold law has an almost negligible effect on the partial cross section.

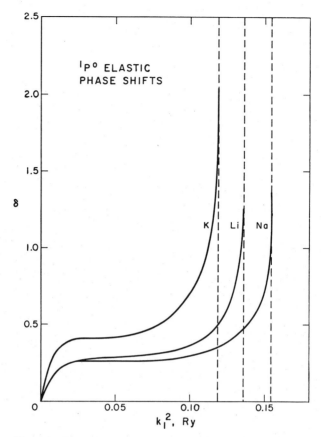

Figure 1. Phase shifts for electron–alkali atom scattering in the ^1P partial wave, calculated in a three-state ($ns, np, 3d$) close-coupling approximation and showing resonance structure.

Andrick *et al.* (1972) and Eyb and Hoffmann (1974) have measured the differential elastic scattering cross sections as a function of energy for Na and K in the vicinity of the excitation threshold, and observe what appears to be a cusp at 90° and above, but which vanishes at around 60°. However, according to theory, the cusp appears in the P waves, contributions from which must vanish at 90°. They remark that the structure observed shows predominantly a D wave character. The calculations for Na (Moores and Norcross, 1972) reveal that the ^1D phase shift exhibits resonant behavior, and the partial cross section peaks, just below threshold. The experimental results may be explained quite satisfactorily by looking at individual partial-wave contributions. The situation is rather complicated, since at least five singlet and five triplet partial waves contribute to the total. Around 90°, one

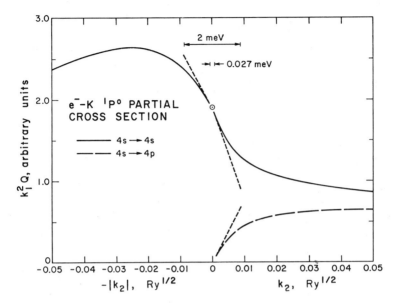

Figure 2. ^1P partial cross sections for elastic (———) and inelastic (— — —) electron–potassium scattering as a function of wave number in the vicinity of the excitation threshold. The dashed straight lines are the cross sections given by the Wigner threshold law.

finds that the behavior is dominated by the ^1D wave and that the structure observed is due to this resonance. Experimental energy resolution is so far insufficiently fine to distinguish the ^1D resonance from the ^1P threshold effect. Calculations are in progress to compute the differential cross section at a finer set of energy intervals in order to determine the form of the structure in more detail. At 60°, however, the ^3P partial wave dominates, the ^1S, ^1P, ^1D, and ^1F, tending to cancel. As we have seen, there is no resonance in this partial wave, and although a cusp is present, its effect is insignificant. The cross section thus appears to vary smoothly, in agreement with observation.

Clearly, a better way to investigate the cusp is in photodetachment, since here the relevant partial wave is picked out by the selection rules.

Photodetachment of Li^-, Na^-, and K^-

Elaborate calculations of the photodetachment cross sections of Li^-, Na^-, and K^- have been made by Moores and Norcross (1974), the details of which are given in their paper. It is shown that, near the excitation threshold, the photodetachment cross section $\kappa(nl)$ for which the atom is left in the state

nl has the form

$$\kappa(ns) = A_0(1 + C_1|k_2|) \qquad (6)$$

$$\left.\begin{array}{r}\kappa(ns) = A_0(1 - C_2 k_2) \\ \kappa(np) = C_3 k_2\end{array}\right\} \quad k_2^2 > 0 \qquad (7)$$

where A_0, C_1, C_2, C_3 are constant and the total cross section is equal to $\kappa(ns) + \kappa(np)$ above threshold. The photodetachment cross section thus also has an infinite slope at threshold.

In the calculations, configuration interaction wave functions and their corresponding eigenenergies, calculated by Weiss (1968), were used for the ground state of the negative ions, and solutions of three-state close-coupling equations for the final state of atom plus electron. Both dipole length and dipole velocity results were calculated. The results are shown in Figures 3, 4, and 5. A sharp upward cusp is obtained for both Li$^-$ and Na$^-$, for which the length and velocity calculations are in good agreement, while for K$^-$, for which this agreement is not as good, the cross section at the neutral excitation threshold appears as a step.

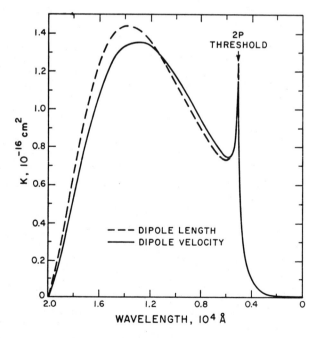

Figure 3. Photodetachment cross section for Li$^-$, using dipole length (— —) and dipole velocity (——) matrix elements.

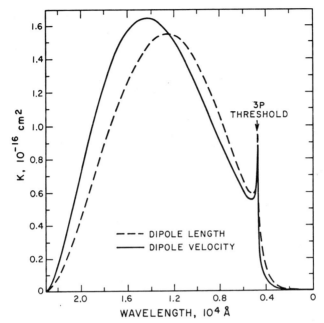

Figure 4. As Figure 3, for Na^-.

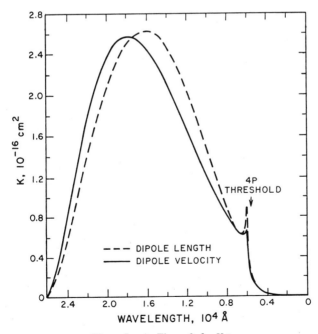

Figure 5. As Figure 3, for K^-.

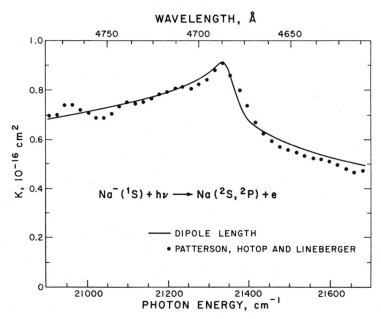

Figure 6. Photodetachment cross section of Na⁻ (dipole length) near 3p excitation threshold together with experimental data of Patterson, Hotop, and Lineberger (private communication). Normalizations described in text.

In Figure 6 are shown, on a larger scale, the experimental results (Patterson, Hotop, and Lineberger, private communication) for Na⁻, obtained using a tunable dye laser. The shape of the measured curve in the region of the cusp is in excellent agreement with the present result, the peak being observed at 4687 (± 7) Å, compared with the Weiss value of 4692 Å. Also shown in Figure 6 are the present dipole length results, folded with the experimental wavelength resolution and normalized in energy to make the peaks coincide. The experimental data, which are not absolute, have been normalized in turn to the theoretical results at about 4725 Å. This double normalization results in a good fit between the two curves (the fit using the dipole velocity results is even better) and leads to the conclusion that the experimentally observed feature is indeed the 3p threshold. The electron affinity of Na is then deduced, using (1), to be 0.543 (± 0.010) eV.

In Figure 7 are shown the results obtained for K⁻ in the vicinity of threshold, together with experimental results (Patterson, Hotop, and Lineberger, private communication). The shape of the experiment is in good agreement with the dipole velocity result, but not with the dipole length results. The same normalization procedure (of the dipole velocity calculation) was adopted as in Figure 6. Additional structure is observed in the experi-

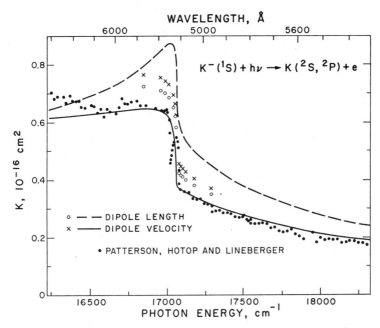

Figure 7. Photodetachment cross section of K⁻ near the 4p excitation threshold using dipole length (– – –) and dipole velocity (———) matrix elements. Experimental data of Patterson, Hotop, and Lineberger (●) normalized as described in text. Open circles and crosses are, respectively, dipole length and dipole velocity calculations, using a close-coupling wave function for the negative ion.

mental results at the center of the step, which is presumably the effect of the splitting of the 4p state into the $j = 1/2$ and $j = 3/2$ components. The calculations do not include the spin–orbit interaction, and hence no such structure is obtained. Assuming that the first minimum represents the $4p_{1/2}$ threshold, an affinity of 0.5012 (± 0.0015) eV is obtained for K compared with the value 0.472 eV of Weiss.

The partial K⁻ photodetachment cross sections are plotted against electron momentum relative to the 4p state in Figure 8, over the same range of momentum as the partial scattering cross sections (Figure 2). The partial cross section for the $4pk_2d$ channel, not shown, is more than two orders of magnitude less than that for the $4pk_2s$ channel. We see that, like the partial cross sections for electron scattering, the partial cross sections for photodetachment depart from the linear behavior demanded by the threshold law (6) and (7) for very small energies relative to the 4p threshold.

Another quantity of interest is the asymmetry parameter β for the angular distribution of those electrons (the "slow photoelectrons") which

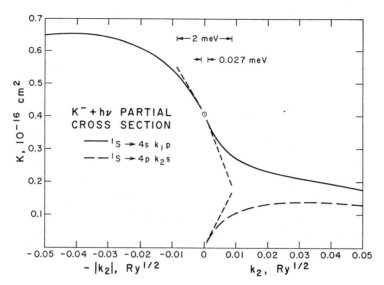

Figure 8. Partial photodetachment cross sections for K^- near the $4p$ threshold (dipole velocity result) for $4sk_1p$ (———) and $4pk_2s$ (— —) channels. Dashed straight lines represent the threshold law.

are ejected, leaving the neutral alkali atom in its excited state. This parameter, defined in Moores and Norcross's (1974) paper, is plotted against energy in Figure 9. The results are clearly sensitive to the choice of dipole operator, but for Li^- and Na^- the length and velocity results are still in reasonable agreement. For K^-, however, very large differences are obtained above 1.0 eV. The experimental value (Kasdan and Lineberger, 1974) at 0.425 eV is in better agreement with the velocity calculation.

The results of these calculations indicate that the wave functions used for K^- are less accurate than those used for Li^- and Na^-, the inaccuracy being such as to affect the length results more than the velocity results. Norcross (1974) has used improved versions of the atomic model potentials and included the so-called dielectric terms (which take into account the effect which the dipole moments induced in the core by the valence and scattered electrons have on each other) to compute electron affinities directly by solving the ^1S close-coupling equations with all channels closed. Results of calculations of the photodetachment of K^-, using the wave functions

Figure 9. Asymmetry parameter β for Li^-, Na^-, and K^- using dipole length (— —) and dipole velocity (———) matrix elements. Also shown (●) is an experimental point of Kasdan and Lineberger (1974). Results of improved calculations are given by open circles (length) and crosses (dipole velocity).

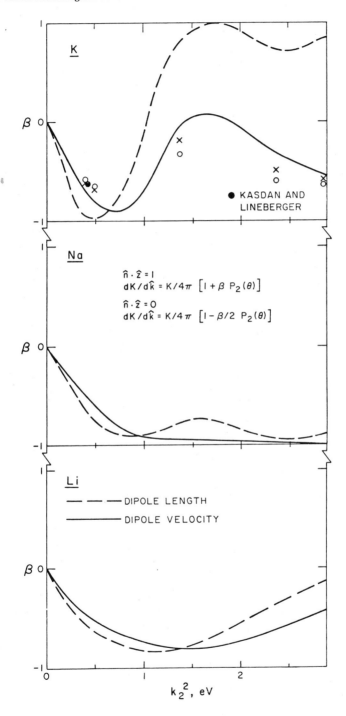

from this work for the ground state and including dielectric terms in the final state, are included in Figures 7 and 9. These calculations have the advantage that the same approximation is used for both initial and final states. The same number of terms were included in the close-coupling expansion as in the earlier calculations, but the improvement is quite marked. It is interesting to note that the velocity results are changed less than the length results and that the two sets of results are now in much better agreement with each other. The shape of both sets of results is now in good agreement with the experimental observation. The improvement is even more striking in the case of the asymmetry parameter β. The results of the length and velocity calculations agree to within 2% from near 0.5 eV and are both within the uncertainty of the measured value, 0.64 ± 0.02 (Kasdan and Lineberger, 1974), at 4880 Å. At higher energies the large discrepancy between the length and velocity results is removed, and the new results are in much better agreement with the old dipole velocity results, as expected.

Photodetachment of Rb^- and Cs^-

Patterson et al. (1974) have recently measured the photodetachment of the heavier alkali negative ions Rb^- and Cs^-. The first-structure splitting of the ^2P excited state of the neutral atoms is larger than for the lighter alkalies, and, as in K^-, structure is observed at both the $^2P_{1/2}$ and $^2P_{3/2}$ thresholds. In Rb^- and Cs^-, however, the effects are rather more spectacular. In this section we will confine our study to Cs^-, the results for Rb^- being qualitatively similar. Experimental results are shown in Figure 10. Within 3 meV of each threshold, a very sharp minimum is observed in the cross section. Below the $^2P_{1/2}$ threshold, the cross section drops by more then three orders of magnitude in a wavelength range of 10 Å and appears to be zero at the minimum. Below the $^2P_{3/2}$ threshold, however, although it drops one order of magnitude, it does not go to zero.

The minima can be explained if, as in the other alkalies, a resonance exists in the ^1P partial wave just below the threshold, but which leads to a window-type profile in the photodetachment cross section. According to the Fano (1961) theory, a zero minimum can be observed if there is only one accessible open channel. In this case, although LS coupling clearly does not apply, it is possible that the spin–orbit effect could be negligible for the ejected electron. If this were the case, there would indeed be only one channel open below the $^2P_{1/2}$ threshold, but two open at energies above this threshold and below the $^2P_{3/2}$ threshold. This would then lead to a nonzero minimum below this second threshold.

Although it is clear that some other coupling scheme is appropriate, the observed structure may be investigated by performing calculations in LS

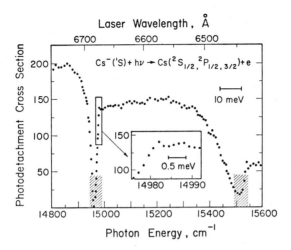

Figure 10. Photodetachment cross sections of Cs$^-$, taken from Patterson *et al.* (1974).

coupling. In these calculations, the $^2P_{1/2}$ and $^2P_{3/2}$ thresholds are replaced by a single threshold whose energy is their statistically weighted mean. In a three-state ($6s, 6p, 5d$) close-coupling calculation (omitting dielectric terms) Norcross (1974, private communication) does find a strong ^1P resonance whose effects encompass the threshold. The Wigner law holds only over an insignificant range, and the elastic cross section begins to rise, in contrast with the requirement of the threshold law, less than 0.015 meV above threshold.

Results of a photodetachment calculation, using a ground-state wave function calculated also in a three-state close-coupling closed-channel approximation, are shown in Figure 10. The picture is one of the threshold region dominated by a window-type resonance whose minimum is zero and which occurs at a few meV below threshold, the wings of which extend beyond the threshold region in both directions. If one tries to fit a Fano profile to the resonance, a resonance energy $E_r = 0.104$ Ry, a width $\Gamma = 0.06$ Ry, and a q factor of -0.015 are obtained. The opening up of the threshold appears as a tiny discontinuity, scarcely visible, in a rising cross section. On the whole, the important features of the calculated cross section are in accord with observation.

Conclusion

The main conclusion to be drawn from this work is that the Wigner threshold effect plays a minor role in determining the form of the photo-

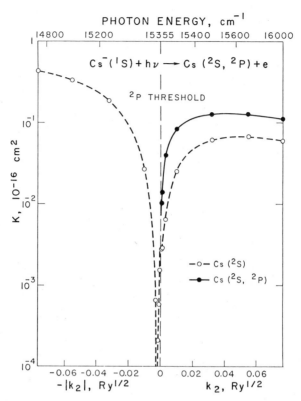

Figure 11. Photodetachment cross section of Cs⁻ calculated by Norcross; dashed curve, $K(6s)$; continuous curve, $K(6s) + K(6p)$.

detachment cross sections of the alkali negative ions at an ejected electron energy equal to the first excitation energy of the neutral atom. The structure observed is primarily caused by a resonance in the ^1P phase shift. In Li⁻ and Na⁻, for which the resonance is rising to a maximum just below threshold, the occurrence of the threshold leads to a sharp cusp. In K⁻, the resonance passes through a maximum below threshold, and the cross section is falling when the threshold intervenes, to give a step-like feature. In Rb⁻ and Cs⁻, the resonance takes the form of a narrow window-like feature below threshold. The threshold appears as an infinite derivative in the cross section just on the high-energy side, but the effect is rather insignificant.

It is clear that future calculations for the heavier alkalies should include the spin–orbit effect in the scattering problem, although this is not necessary to explain the main features of the photodetachment cross section in the threshold region.

References

Andrick, D., Eyb, M., and Hofmann, H. (1972). *J. Phys. B.*, **5**, L15.
Bardsley, J. N., and Nesbet, R. K. (1973). *Phys. Rev.*, A **8**, 203.
Burke, P. G., and Taylor, A. J. (1969). *J. Phys. B.*, **2**, 869.
Eyb, M., and Hofmann, H. (1974). *J. Phys. B.*, **8**, 1095.
Fano, U. (1961). *Phys. Rev.*, **124**, 1866.
Hotop, H., Patterson, T. A., and Lineberger, W. C. (1972). *Bull. Am. Phys. Soc.*, **17**, 1128.
Kasdan, A., and Lineberger, W. C. (1974). *Phys. Rev.*, A **10**, 1658.
Moores, D. L., and Norcross, D. W. (1972). *J. Phys. B.*, **5**, 1482.
Moores, D. L., and Norcross, D. W. (1974). *Phys. Rev.*, A **10**, 1646.
Norcross, D. W. (1974). *Phys. Rev. Letters*, **32**, 192.
Norcross, D. W., and Moores, D. L. (1973). In *Atomic Physics*, 3. Eds. S. J. Smith and G. K. Walters, p. 261. New York: Plenum.
Patterson, T. A., Hotop, H., Kasdan, A., Norcross, D. W., and Lineberger, W. C. (1974). *Phys. Rev. Letters*, **32**, 189.
Ross, M. H., and Shaw, G. L. (1961). *Ann. Phys.*, **13**, 147.
Weiss, A. W. (1968). *Phys. Rev.*, **166**, 70.
Wigner, E. P. (1949). *Phys. Rev.*, **73**, 1002.

10
Photodetachment Threshold Processes

W. C. LINEBERGER, H. HOTOP, AND T. A. PATTERSON

The development of tunable dye lasers has permitted the study of photodetachment threshold processes with a resolution of 10^{-4} eV or better. Several recent applications of this technique which relate to electron–atom scattering are briefly discussed, together with several proposed experiments which prove correlations of two electrons loosely bound in a Coulomb field.

Introduction

The energy resolution attainable in photodetachment experiments has recently been improved (Lineberger and Woodward, 1970; Hotop et al., 1973) to better than 1 meV by the use of continuously tunable laser light sources. These techniques now permit determination of atomic electron affinities with an accuracy of order 10^{-4} eV, as well as a thorough test of theoretical threshold laws. In this paper we give a brief summary of such measurements for the cases in which the neutral atom produced in the photodetachment process (a) is in its ground state, (b) is in an excited state, or (c) becomes ionized. The relationship of these processes to electron–atom scattering lies in the fact that the photodetachment process,

$$h\nu + A^- \xrightarrow{\sigma} A + e^-(k, L) \tag{1}$$

W. C. LINEBERGER • Department of Chemistry and Joint Institute for Laboratory Astrophysics, University of Colorado, and National Bureau of Standards, Boulder, Colorado 80302, U.S.A.   H. HOTOP and T. A. PATTERSON • Joint Institute for Laboratory Astrophysics, University of Colorado, and National Bureau of Standards, Boulder, Colorado 80302, U.S.A. W. C. Lineberger was an Alfred P. Sloan Foundation Fellow in 1972–75.

where k and L are the electron linear and orbital angular momenta, can in some senses be viewed as "half an electron–atom scattering process having only a restricted number of partial waves accessible." Within the constraints of this limitation, we can use the photodetachment to prepare electron–atom scattering systems in total energy states with an energy resolution that is essentially the optical resolution of the laser. Thus, in some cases we can investigate electron–atom scattering processes with an energy resolution that is several orders of magnitude higher than currently obtainable in high-resolution electron beam scattering experiments.

The basic experimental technique employed is to intersect a 2 keV mass-analyzed negative ion beam with the focused output of a pulsed, tunable dye laser. The fast neutral atoms produced in the photodetachment process are then detected by secondary electron emission on the first dynode of a windowless electron multiplier. The quantities measured are the ion beam current, the photon flux, the neutral atom signal, and the photon wavelength. The result of this measurement is a relative cross section for the production of neutrals as a function of photon wavelength. Although to date tunable laser photodetachment experiments have been performed only in the photon wavelength region 4400–9000 Å, present laser technology makes such experiments possible in the spectral region 2300 Å to 1 μm. Details of the experimental technique and procedures may be found elsewhere (Hotop et al., 1973).

Ground State Thresholds

The theoretically predicted behavior of the cross section σ_L for photodetachment of atomic negative ions near threshold $h\nu_{\text{thr}}$ by photons of energy $h\nu$ is given by (Wigner, 1948)

$$\sigma \propto h\nu(h\nu - h\nu_{\text{thr}})^{(L+1)/2} \propto h\nu(k)^{2L+1} \tag{2}$$

This functional dependence reflects the fact that in the photodetachment process the dominant long-range interaction between the separating electron–atom system is the centrifugal potential. If a p electron is detached, such as in Se^-, the outgoing electron can be in an s or d wave. The theoretical threshold behavior is given by the s-wave contribution $\sigma \propto k$, since the d-wave cross section is suppressed by the centrifugal barrier. In photodetachment threshold measurements one actually observes a series of thresholds corresponding to transitions from the various fine-structure substates of the negative ion to fine-structure states of the neutral atom. One of the more interesting questions which can be addressed by a high-resolution photodetachment experiment is: Over how large an energy range does Eq. (2) give a satisfactory

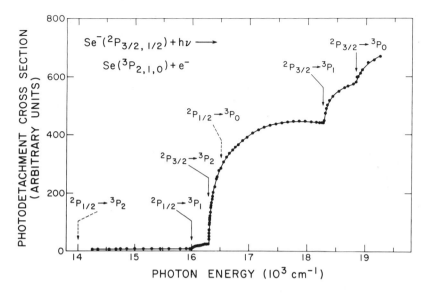

Figure 1. Se⁻ photodetachment cross section in the energy range 14,000–19,000 cm⁻¹. The individual fine-structure transition thresholds are labeled.

description of the cross section? In addition to the question of the threshold behavior of the cross section, one also can determine the relative transition strengths for the various fine-structure onsets; these strengths can differ substantially from simple statistical expectations.

A study of Se⁻ threshold photodetachment should show six s-wave thresholds corresponding to transitions from Se⁻ ($^2P_{1/2,3/2}$) to Se ($^3P_{2,1,0}$). From an investigation of an individual fine-structure threshold, one can obtain information on the range of validity of the Wigner threshold law. Figure 1 shows Se⁻ photodetachment cross section data obtained (Hotop et al., 1973) in the threshold region; the individual fine-structure transition thresholds labeled in the figure are obtained from knowledge of the neutral Se fine-structure energy separations. It is clear that the individual thresholds have qualitatively the shape predicted by the Wigner threshold law. In order to investigate this energy dependence further, one must look at the individual fine-structure thresholds in more detail; an example is the $^2P_{3/2} \rightarrow {}^3P_2$ partial cross section plotted in Figure 2 as a function of electron momentum k. In this representation the Wigner threshold law is a straight line passing through the origin; the Wigner law is seen to be a valid representation of the photodetachment cross section for only the first 5 meV (0.02 a.u.) above threshold. One can attempt to understand the departures from the threshold law in terms of the low-energy electron–atom elastic scattering cross section

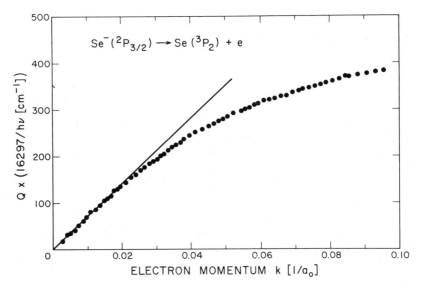

Figure 2. Se$^-$($^2P_{3/2}$)–Se($^3P_2$) partial photodetachment cross section plotted as a function of electron momentum. The straight line is the Wigner threshold law, which is seen to be valid for only the first 5 meV (0.02 a.u.) above threshold.

and the electron–atom polarization and quadrupole interactions, but only very limited success has been achieved (Hotop et al., 1973) in extracting such information from the photodetachment data to date.

Once the validity of the Wigner threshold law has been established, we can measure the relative strengths of the various fine-structure transitions and compare these results with the predictions of several threshold photodetachment models, a statistical model, and two complex models. The statistical model ignores the outgoing electron, assuming that the strengths of all transitions are inherently the same, except for multiplication by the level degeneracy ($2J + 1$) in the final state. This model fails to provide a good prediction of the transition strengths, particularly in the $^2P_{1/2} \to {}^3P_{2,1,0}$ branch. The complex model (Lineberger and Woodward, 1970) for Se$^-$ photodetachment is based on the idea that near threshold one should view the final state as a complex (e + Se) in LS coupling. The transition strengths are then determined by calculating the optical transition strengths going from the LS-coupled negative ion states to the intermediate LS-coupled state and further by assuming that each of these intermediates splits evenly in the alternative fine-structure levels of Se. This model provides rather good agreement with experiment for the $^2P_{1/2} \to {}^3P_{2,1,0}$ transitions but not for the $^2P_{3/2} \to {}^3P_{2,1,0}$ branch.

Rau and Fano (1971) have recently extended the idea of the complex model by considering in detail the dissociation of the intermediate complex into the various exit fine-structure levels. They use the method of frame transformation for projecting the LS-intermediate states onto the final jj-coupled states in a general treatment for negative ions of p^5 configuration. This model provides by far the best agreement with the Se^- data above. A virtue of their treatment is that their numbers for the various transition strengths are relatively insensitive to the physical parameters that describe the electron–atom interaction in the exit channel. This model has been further generalized by Rau (1975) in a paper that appears in this volume.

In addition to these studies of s-wave thresholds, the p-wave threshold has been studied in some detail (Hotop and Lineberger, 1973) in the case of Au^- photodetachment. As a result of some accidental cancellations, the Wigner threshold law was apparently valid for a much larger energy region (approximately 50 meV) than in the s-wave case. An interesting variant of these threshold measurements, which has not been attempted to date, is a study of the two-photon photodetachment threshold behavior. Using linearly polarized light, for example, one can detach a bound s electron with two photons and obtain an s-wave threshold. Alternatively, using circularly polarized light it should be possible to study higher-order partial waves that are not at present easily accessible experimentally.

Excited State Thresholds

In the vicinity of the threshold for photodetachment to an excited state of the neutral, Wigner (1947) has shown that the cross section may exhibit features arising from interference between the ground state channel and the excited state channel. In the case of an excited state channel involving an s-wave free electron, this interference manifests itself as a cusp in the ground state channel cross section having an energy dependence of the form $A + B|E - E_n|^{1/2}$, where E and E_n are the system energy and the threshold energy of the nth channel. The nature of these so-called Wigner cusps has been pursued in substantial detail in the study of scattering processes. Based on analogy with electron–atom scattering, it is reasonable to expect similar cusps in photodetachment cross sections near excited state thresholds. Moores and Norcross (1974) have examined theoretically the possibility of observing cusps in photodetachment and Moores (1975) has discussed these calculations in this volume. In fact, experiment demonstrates a cusp-like behavior in photodetachment of the light alkali negative ions at photon energies corresponding to the opening of the lowest np state of the neutral, and, as shown by Moores and Norcross, the experimental results (Patterson,

1974; Patterson et al., 1974) and theoretical calculations are in remarkably good agreement.

In contrast to the behavior in the light alkali ions, photodetachment of the heavy alkalis (Rb$^-$ and Cs$^-$) at photon energies corresponding to the opening of the lower np final state is dominated by doubly excited states of the alkali negation ion, which lie very close to the np threshold (Patterson et al., 1974). Figure 3 shows the Cs$^-$ photodetachment cross section in this energy region; the doubly excited state of Cs$^-$ gives rise to a Fano profile in the photodetachment cross section, but the smooth development of phase across the resonance is interrupted by the opening of the inelastic channel. This channel opening appears as a discontinuity in the resonance profile, as is seen in the inset in Figure 3. The Rb$^-$ cross section is qualitatively similar, but the resonances are significantly more narrow, the lowest-energy resonance in Rb$^-$ being only 150 μeV wide, by far the narrowest feature observed to date in electron–neutral atom scattering processes. These resonances and their interpretation in terms of configurations of the doubly excited state are discussed in more detail elsewhere (Patterson, 1974; Patterson et al., 1974).

Since, at each new photodetachment threshold, very-low-energy electrons associated with that threshold are produced, it is possible, if one

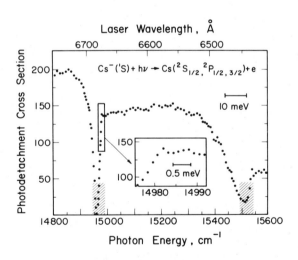

Figure 3. Cs$^-$ photodetachment cross section, 6800–6400 Å. The shaded regions indicate the confidence limits on the opening of the Cs($6^2P_{1/2}$) and ($6^2P_{3/2}$) exit channels, as determined by photoelectron spectroscopy (Kasdan and Lineberger, 1974).

can discriminate against the high-energy electrons, to look only at the low-energy electrons and see the new threshold in detail.

Two such experiments are currently being planned, and will certainly examine resonances associated with higher-lying excited states of the alkali neutrals. Using techniques similar to those described by Professor Read in this volume (Read, 1975), it should be possible to determine experimentally relative strengths for production of high Rydberg states in the alkalis and even extend the measurements into the region corresponding to ionization of the resulting alkali neutral. This latter process is the subject of the next section.

Two-Electron Photodetachment

In recent years a great deal of attention has been given, in both experimental and theoretical atomic physics, to correlation effects between electrons. An important prototype of such systems is the case of two highly excited electrons in the field of a nucleus. Such correlation effects, which should be most significant when the electrons have near-zero energy, are treated in the case of bound states by Lin (1975) and Fano (1975) in this volume. An experimental determination of the threshold law for electron–atom ionization is a particularly important test of current models for the correlation of two electrons in a Coulomb field. The theoretical predictions for threshold laws include linear (Geltman, 1969), the $E^{1.127}$ Wannier law (Wannier, 1953; Rau, 1971), and a more complicated oscillatory function (Temkin and Hahn, 1974; Temkin, 1974). The experimental state of such measurements (Read, 1975) has been such that one could not clearly distinguish between these several predictions. Recently, Cvejanović and Read (1974) have greatly improved the experimental situation in a study of the electron impact ionization of He. Their data lend strong support to the Wannier threshold law, provided that the distribution of excess energy between the two departing electrons is uniform. While this result is very suggestive, the limited resolution attainable with present electron beams (10 meV) precludes a completely definitive determination of the threshold behavior. Another way of experimentally probing this low-energy region is through two-electron photodetachment of negative ions using tunable dye lasers to probe the threshold region with optical resolution. Such an experiment now seems within the present state of the art and is currently being assembled by Slater, Read, and Lineberger at JILA.

These brief remarks have, we hope, indicated a few cases in which one can use the optical resolution obtainable in photodetachment of negative ions to obtain information on electron–atom scattering processes.

Acknowledgments

During the course of this work we have benefited from many discussions of threshold processes with our colleagues, especially U. Fano, A. R. P. Rau, S. Geltman, A. Temkin, F. H. Read, D. W. Norcross, and D. L. Moores. This work was supported by the National Science Foundation.

References

Cvejanović, C., and Read, F. H. (1974). *J. Phys. B.*, **7**, 1841.
Fano, U. (1975). Contribution to this volume.
Geltman, S. (1969). *Topics in Atomic Collision Theory*. New York: Academic Press.
Hotop, H., and Lineberger, W. C. (1973). *J. Chem. Phys.*, **58**, 2379.
Hotop, H., Patterson, T. A., and Lineberger, W. C. (1973). *Phys. Rev.*, A **8**, 762.
Kasdan, A., and Lineberger, W. C. (1974). *Phys. Rev.*, A **10**, 1658.
Lin, C. D. (1975). Contribution to this volume.
Lineberger, W. C., and Woodward, B. W. (1970). *Phys. Rev. Letts.*, **24**, 424.
Moores, D. L., and Norcross, D. W. (1974). *Phys. Rev.*, A **10**, 1646.
Moores, D. L. (1975). Contribution to this volume.
Patterson, T. A. (1974). Ph.D. Thesis, University of Colorado, Boulder.
Patterson, T. A., Hotop, H., Kasdan, A., Norcross, D. W., and Lineberger, W. C. (1974). *Phys. Rev. Letts.*, **32**, 189.
Rau, A. R. P. (1971). *Phys. Rev.*, A **4**, 207.
Rau, A. R. P. (1975). Contribution to this volume.
Rau, A. R. P., and Fano, U. (1971). *Phys. Rev.*, A **4**, 1751.
Read, F. H. (1975). Contribution to this volume.
Temkin, A. (1974). NASA Goddard Space Flight Center Report X-602-74-239.
Temkin, A., and Hahn, Y. (1974). *Phys. Rev.*, A **9**, 708.
Wannier, G. H. (1953). *Phys. Rev.*, **90**, 817.
Wigner, E. P. (1948). *Phys. Rev.*, **73**, 1002.

11
Multiple Photoionization of the Rare Gases

JAMES A. R. SAMSON AND G. N. HADDAD

The photoionization yield, or average charge produced per photon absorbed by Ne, Ar, Kr, and Xe, has been measured from the threshold of double photoionization to 107 eV. The probability of double ionization is zero at threshold for the rare gases. For Ar, Kr, and Xe double ionization increases rapidly, giving an abundance of 16–30% at 20 eV beyond the double ionization threshold. In this region double ionization is constant or increasing slowly. At the threshold for triple ionization the average charge per photon increases rapidly.

Introduction

There has been considerable interest recently in the process of multiple photoionization, both experimentally and theoretically. Detailed calculations on the double ionization of He have been made, taking into account ground state correlations between the electrons (Byron and Joachain, 1967; Brown, 1970). There is good agreement between theory and x-ray photoionization (Carlson, 1967). However, agreement is poor with electron impact experiments (Van der Wiel, 1969; El-Sherbini et al., 1970; Van der Wiel, 1970; Van der Wiel and Wiebes, 1971a,b). For atoms heavier than He there are no detailed calculations of double ionization. Calculations based on the sudden approximation theory using single electron wave functions fail to predict the magnitude of double ionization of the rare gases, especially near threshold. Chang et al. (1971) improved the agreement with theory for Ne ionized by photons of energy about 200 eV above the double ionization

---

JAMES A. R. SAMSON and G. N. HADDAD • Behlen Laboratory of Physics, University of Nebraska, Lincoln, Nebraska 68508, U.S.A.

threshold. The improvement was achieved by considering the contribution of several processes, namely, core rearrangement, virtual Auger transitions, and initial-state electron correlations. However, with the exception of He, no results have been published predicting the expected threshold behavior of double ionization in the rare gases. Amusia (1971) has pointed out the importance of knowing the probabilities of multiple-electron processes in understanding the interaction of the removed electron with those remaining. From the practical point of view, knowledge of the multiple photoionization yield of the rare gases is invaluable in extending the use of the rare gas ion chambers in measuring the absolute photon flux of wavelengths less than 300 Å (Samson and Haddad, 1974).

There are very few experimental results near the threshold of double ionization. The only detailed results are those of Cairns et al. (1969), who studied Xe from threshold to 50 eV beyond threshold. Van der Wiel and Wiebes (1971a,b) have studied Ar, Ne, and He by electron bombardment and have extracted optical oscillator strengths from their work from the double ionization threshold to several hundred electron volts beyond. In the region where their data overlap the x-ray photoionization results (Carlson, 1967) there is a major discrepancy between the two techniques. This is surprising, considering their excellent agreement with single photoionization data (Van der Wiel et al., 1969; El-Sherbini et al., 1970). We present here for the first time detailed experimental results of the threshold behavior of multiple photoionization of the rare gases Ne, Ar, Kr, and Xe.

Experimental

Previous work on multiple photoionization of gases has used either the technique of mass spectroscopy or photoelectron spectroscopy. The present work uses a new technique which simply measures the average charge produced per photon absorbed by the gas. This allows absolute determinations of the number of singly and doubly charged ions produced as a function of photon energy until the triple ionization threshold is reached. Beyond that point the abundances of the separate ions cannot be distinguished without the use of a mass spectrometer.

The apparatus consists of a single ion chamber (length L) terminated with a fluorescent screen of sodium salicylate. The fluorescent screen is used simply to monitor the photon flux emerging from a grazing-incidence vacuum-ultraviolet monochromator. The photoionization yield Y, defined as the number of charges produced divided by the number of photons absorbed in the length L, is given by

$$Y = \frac{i/e}{I_0[l - \exp(-\sigma n L)]} \quad (1)$$

where i is the ion current, I_0 is the number of photons incident per second, σ is the total absorption cross section, and n is the number density of the gas. Because σ, n, and L are constant and known for a given gas, the method consists simply in measuring the ion current and the photon flux as functions of wavelength. The cross sections have been measured previously (Samson, 1966). However, they were remeasured in the present work to a higher degree of accuracy ($\sim 3\%$) in the short-wavelength region. The technique for measuring the absolute photon flux at wavelengths shorter than 300 Å has been reported recently (Samson and Haddad, 1974). Briefly it requires the ion current from the gas under investigation to be compared to that of another rare gas whose photoionization yield is known. The yields of all the rare gases are unity from the first ionization potential to the threshold for double ionization (Samson, 1964). Helium was used to determine the yield of Ne for photon energies greater than 60 eV because the yield of He is unity for photon energies up to 79 eV. For energies up to 107 eV the yield of He was taken from the results of Carlson (1967). Neon was then used as the standard to determine the multiple ionization of the other rare gases.

To obtain maximum sensitivity the ion chamber was replaced by a gas jet and electron multiplier. The pressure around the multiplier was 10^{-5} torr. The ions were accelerated through 2.4 kV onto the first dynode of the multiplier. Xenon was studied with both systems. Similar results were obtained within experimental error. It was assumed, therefore, that the sensitivity of the multiplier under the present conditions was constant for all the rare gas ions.

The light source was a high-voltage condensed spark discharge in a low-pressure gas (Samson, 1967). The source produced a large number of discrete emission lines characteristic of the gas used in the discharge. The short-wavelength limit was imposed by the efficiency of the monochromator. In the present work a useful limit was 116 Å (107 eV).

Results

The photoionization yields or average charges produced per photon absorbed for Ne, Ar, Kr, and Xe are shown in Figures 1–4, respectively, as functions of photon energy. The overall accuracy of the data points is estimated to be $\pm 6\%$ for Ar, Kr, and Xe and $\pm 3\%$ for Ne.

The data of other authors have been expressed either as a ratio R^{n+} of the number of n-times charged ions to the number of singly charged ions, or as an abundance A^{n+} equal to the number of n-times charged ions divided by the total number of ions of all charges. These data have been converted to yields through the following relations:

$$Y = A^+ + 2A^{2+} + 3A^{3+} + \cdots$$

or, because $\sum_n A^{n+} = 1$,
$$Y = 1 + A^{2+} + 2A^{3+} + \cdots \tag{2}$$
and
$$Y = (1 + 2R^{2+} + 3R^{3+} + \cdots)/(1 + R^{2+} + R^{3+} \cdots) \tag{3}$$

Thus, in Figures 1–4 the abundance of doubly charged ions is simply equal to $(Y - 1)$ for photon energies less than the triple ionization threshold.

In Figure 1 the results for Ne are shown along with those of Carlson (1967) and of Lightner *et al.* (1971). In the region where the data overlap with those of Carlson's there is reasonable agreement. Excellent agreement is obtained with the electron impact work of Van der Wiel and Wiebes (1971b).

Multiple ionization of Ar, like Ne and He, shows a zero probability at threshold (Figure 2). The yield rises to a plateau giving a constant abundance of 16% of doubly ionized argon. This represents a ratio R^{2+} of doubly to singly charged ions of 19%. This is to be compared with 13–16% obtained by Carlson (1967). Although Carlson's data at 110 eV are plotted on our yield curve, they apply to double ionization only as there is no report of the abundance of Ar^{3+}. However, in light of the constancy of the double ionization abundance out to 110 eV, the rise in the curve at the triple ionization threshold is presumably caused by Ar^{3+}.

The double ionization threshold closely follows the electron impact data (Van der Wiel and Wiebes, 1971a) up to a photon energy of 65 eV.

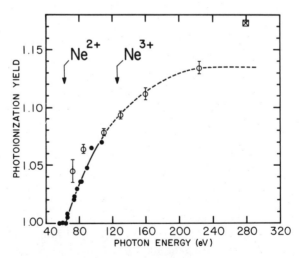

Figure 1. Photoionization yield of neon as a function of photon energy. Present data, ●; Carlson (1967), ⌀; Lightner *et al.* (1971), ⊠.

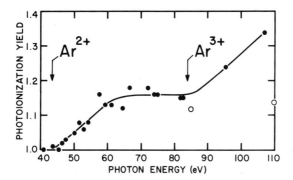

Figure 2. Photoionization yield of argon as a function of photon energy. Present data, ●; Carlson (1967), ○ (for double ionization only).

Beyond this energy the electron impact data continue to increase in disagreement with the present results.

The results for multiple ionization of Kr are shown in Figure 3. No other experimental data exist for Kr. At the threshold for double ionization krypton shows the characteristic rise from a yield of unity to a plateau giving an abundance of Kr^{2+} of about 14–15%. However, the abundance of doubly charged krypton continues to rise rapidly at higher photon energies just before the threshold for triple ionization. Presumably, the increase in double

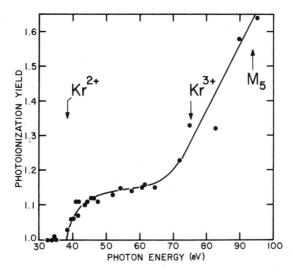

Figure 3. Photoionization yield of krypton as a function of photon energy.

ionization continues beyond the triple ionization threshold. Thus, no estimate of the abundance of triply ionized Kr can be made.

Figure 4 shows the results for multiple ionization of Xe. The solid data points represent the average of the yields obtained by both the ion chamber technique and the gas jet–electron multiplier combination. Included in the figure for comparison are the data points of Cairns *et al.* (1969). The characteristic plateau that shows in the Ar and Kr data is less distinct. However, the rapid rise in double ionization before the triple ionization threshold is more pronounced than in the Kr data. This increase may be caused by an interference between competing processes, one or more of which involve the d^{10} shell ($N_{5,4}$ level). In krypton the effect is less pronounced, and in argon, which has no d shell, the effect is absent.

The data of Cairns *et al.* (1969) are in good agreement with the present data. Beyond the $N_{5,4}$ edge Auger transitions are energetically possible, but according to their data the actual abundance (64%) of doubly charged ions peaks at 72 eV and then starts to decrease. At 83 eV they find an abundance of 59%.

The magnitude of multiple photoionization of the rare gases is surprisingly large. Further, double and triple ionization takes place at energies less than the threshold for Auger transitions. Those present a challenge for theoretical consideration. The recent successes of the random-phase approximation with exchange in describing single ionization in the rare gases would suggest that this technique should be developed further and applied to multiple ionization.

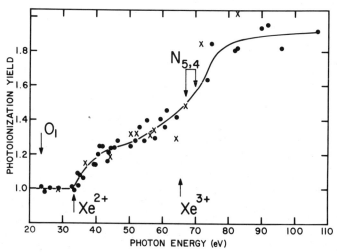

Figure 4. Photoionization yield of xenon as a function of photon energy. Present data, ●; Cairns *et al.* (1969), ×.

Acknowledgments

Research for this paper was supported by the Atmospheric Sciences Section, National Science Foundation, U.S.A.

References

Amusia, M. Ya. (1971). In *Atomic Physics 2*. Ed. P. G. H. Sanders, p. 249. New York: Plenum Press.
Amusia, M. Ya. (1973). Invited Lectures and Progress Reports, *VIII ICPEAC* abstracts, p. 171. Belgrade.
Amusia, M. Ya., and Kazachkov, M. P. (1968). *Phys. Letts.*, **28A**, 27.
Amusia, M. Ya., Kazachkov, M. P., Cherepkov, N. A., and Chernysheva, L. V. (1969). *VI ICPEAC* abstracts, p. 130. Cambridge: North-Holland.
Brown, R. L. (1970). *Phys Rev.*, A **1**, 587.
Byron, F. W. Jr, and Joachain, C. J. (1967). *Phys. Rev.*, **164**, 1.
Cairns, R. B., Harrison, H., and Schoen, R. I. (1969). *Phys. Rev.*, **183**, 52.
Carlson, T. A. (1967). *Phys. Rev.*, **156**, 142.
Carlson, T. A., Hunt, W. E., and Krause, M. O. (1966). *Phys. Rev.*, **151**, 41.
Chang, T. N., Ishihara, T., and Poe, R. T. (1971). *Phys. Rev. Letts.*, **27**, 838.
El-Sherbini, Th. M., Van der Wiel, M. J., and De Heer, F. J. (1970). *Physica*, **48**, 157.
Krause, M. O., Carlson, T. A., and Dismukes, R. D. (1968). *Phys. Rev.*, **170**, 37.
Lightner, G. S., Van Brunt, R. J., and Whitehead, W. D. (1971). *Phys. Rev.*, A **4**, 602.
Samson, J. A. R. (1964). *J. Opt. Soc. Am.*, **54**, 6.
Samson, J. A. R. (1966). In *Advances in Atomic and Molecular Physics*. Eds. D. R. Bates and I. Estermann, Vol. 2, p. 177. New York: Academic Press.
Samson, J. A. R. (1967). *Techniques of Vacuum Ultraviolet Spectroscopy*, p. 166. New York: Wiley.
Samson, J. A. R., and Haddad, G. N. (1974). *J. Opt. Soc. Am.*, **64**, 47.
Van der Wiel, M. J. (1970). *Physica*, **49**, 411.
Van der Wiel, M. J., El-Sherbini, Th. M., and Vriens, L. (1969). *Physica*, **42**, 411.
Van der Wiel, M. J., and Wiebes, G. (1971a). *Physica*, **53**, 225.
Van der Wiel, M. J., and Wiebes, G. (1971b). *Physica*, **54**, 411.

12
Nonstatistical Branching Ratios in Atomic Processes

A. R. P. Rau

In recent years, major departures from statistical branching ratios have been observed experimentally and explained theoretically for certain photodetachment and photoionization processes. The detailed theoretical analysis of the dynamics of these processes, while accounting for the nonstatistical ratios, has obscured a feature to be emphasized in this paper, namely, that the branching ratios are, in general, given by a geometrical factor. This factor, which is dynamics-independent and depends only on the configuration and quantum numbers of the system, is easily evaluated from angular-momentum algebra and should be useful in the analysis of experimental data. Any departures from such geometrical ratios will constitute direct measurements of the intermediate-coupling nature of the system under study.

1. Introduction

An initial expectation for the relative strengths of transitions originating from, or terminating in, alternative levels of angular momentum j is that they are in the ratio of the corresponding statistical weights $(2j + 1)$. In recent years, however, both experimentally (Samson and Cairns, 1968; Lineberger and Woodward, 1970; Hotop et al., 1973) and theoretically (Lu, 1971; Rau and Fano, 1971; Lee and Lu, 1973), several atomic processes have been studied which show significant departures from such statistical ratios. The occurrence of such departures and their significance has been emphasized by Fano (1972). It is, therefore, a privilege to dedicate to him this paper, which generalizes the concept of statistical ratios to a more general geometrical factor.

A. R. P. Rau • Department of Physics and Astronomy, Louisiana State University, Baton Rouge, Louisiana 70122, U.S.A.

2. Photodetachment of Negative Ions

The first really large departures from statistical ratios were observed in a high-resolution study of the total cross section for the photodetachment of S^- near threshold. The process, which involves the detachment of a p electron, is expected (Wigner, 1933) to be dominated near threshold by an s wave for the outgoing electron; on this basis, we can write

$$p^5(^2P_{3/2,1/2}) + h\nu \to p^4(^3P_{2,1,0})s$$

Six transitions are possible from the two fine-structure levels of S^- to the three fine-structure levels of the S atom, each marked by a characteristic threshold energy determined by the positions of the fine-structure levels. At each threshold, the cross section is expected to rise according to the Wigner law for s waves, $\sigma \propto E^{1/2}$, where E is the energy of the outgoing electron in that channel. The constants of proportionality in such a relation, called the step-heights at the various thresholds, could be expected to be given by the statistical weights; however, as shown in Table 1, the experimental results indicate significant departures from these values.

Lineberger and Woodward (1970) sketched a way of explaining the observation by considering the final state as an (e + S) complex created by the absorption of the photon by S^-. The complex can be regarded as being initially in LS coupling when the electron is close to the S atom and as dissociating subsequently into the final configuration, in jj coupling, of the electron receding to infinity. Rau and Fano (1971) carried out a detailed analysis along these lines for the probability amplitudes of the various transitions, which are matrix elements of the electric dipole operator **P**. For a particular light polarization, $P^{[1]}q$ (in tensorial notation), the matrix element involved is

$$(p^4(^3P)j_c m_j, ksj_e m_e | P_q^{[1]} | p^5(^2P)J'M') \tag{1}$$

where j_c and m_j are the quantum numbers of the residual S atom, $j_e = \tfrac{1}{2}$ and m_e those of the outgoing s electron with wave number k, and J' and M' those of the initial S^- ion. By introducing a complete set of LS-coupled states of the complex, Eq. (1) can be written as

$$\Sigma_{JM}(p^4(^3P)j_c m_j, ksj_e m_e | p^4(^3P)s(^2P)JM)(p^4(^3P)s(^2P)JM | P_q^{[1]} | p^5(^2P)J'M') \tag{2}$$

The factor on the right represents the absorption of the photon by the S^- ion, resulting in LS-coupled states of the complex with alternative values of J ($\tfrac{1}{2}$ and $\tfrac{3}{2}$) and M. This factor, the usual line-strength factor (Condon and Shortley, 1970; Shore and Menzel, 1968), can be easily worked out by standard procedures of spectroscopy. The factor on the left in Eq. (2) represents the dissociation of the LS-coupled states into the final channels in jj coupling.

It involves first of all a transformation of the angular wave functions from LS to jj coupling, which is given by a standard recoupling coefficient [Eq. (13.14) of Fano and Racah, 1959] and, secondly, it involves a consideration of the radial wave function of the outgoing electron in its passage from small r values to infinity. By expressing the wave function of the electron at small r in terms of the relevant scattering lengths, Rau and Fano (1971) developed the expression in Eq. (2) in terms of these scattering lengths, and presented detailed expressions for the cross section for each channel as a function of the energy or, alternatively, k.

The feature emphasized in this paper is that in squaring Eq. (2) the contributions from the alternative values of J add incoherently, if one is not looking for quantities such as angular distributions or polarizations of the different components involved. This is indeed what happens when we seek the total cross section for each channel. The step-heights at various thresholds derived in this way were found to be in good agreement with the experimental values as shown in Table 1. These step-heights were found to be simple numerical factors that do not involve the detailed dynamics of the outgoing electron expressed by the radial wave function and, therefore, by the scattering lengths. It was not emphasized at that time but seems quite clear now that, even on dimensional grounds alone, the scattering lengths a can only occur in the combination ka; therefore, dynamics can only influence, in a secondary manner, the behavior of the cross section away from threshold. The dominant behavior at threshold is just the behavior proportional to k given by the Wigner law with a purely numerical "geometrical" factor. Experimentally also, the clue to this dynamics-independent nature of the branching ratios is provided by the fact that the branching ratios are the same for the Se^- and S^- systems. Only the nature of the configuration and the quantum numbers seem important, not the specific atomic system one is looking at.

The main purpose of this paper is to emphasize this result—that branching ratios are given by a purely geometrical factor which is easily evaluated by Racah algebra—as a concept which should be very useful in the analysis of experimental data. The step-heights are not in simple statistical ratios but at the same time, for total cross section measurements, they do not involve the full dynamics of the process; they are simply geometrical factors, involving the line strengths of the LS-coupled states resulting from absorption of the photon multiplied by the projection of the angular part of the LS states onto the final jj states, and summed incoherently over unobserved quantum numbers. They may be written as

$$\zeta(J',j_c) = \Sigma_J |(p^4(^3P)j_c, sj_e|p^4(^3P)s(^2P)J)|^2 |(p^4(^3P)s(^2P)J|P^{[1]}|p^5(^2P)J'|)^2 \quad (3)$$

Both factors are given by angular-momentum algebra or may be drawn from standard tables (Condon and Shortley, 1970; Shore and Menzel, 1968) for

Table 1. Step-Heights at Different Thresholds for $p^5(^2P_{3/2,1/2}) \to p^4(^3P_{2,1,0})s$

The experimental results are from Lineberger and Woodward (1970) for S⁻ and Hotop et al. (1973) for Se⁻. The former are referred to $^2P_{1/2} \to {}^3P_2$ as a standard and the latter to $^2P_{3/2} \to {}^3P_2$. The geometrical ratios are from Rau and Fano (1971) or, alternatively, Eq. (4) of this paper, and t is a parameter characteristic of the experimental situation representing the population ratio of $^2P_{3/2}$ to $^2P_{1/2}$ in the negative ion targets.

Negative ion	Atom	Statistical ratio	Experiment S⁻	Experiment Se⁻	Geometrical ratios
$^2P_{1/2}$	$^3P_2$	2.5	2.5	2.5 ± 0.5	2.5
	$^3P_1$	1.5	5.75 ± 1.25	5.5 ± 0.5	4.5
	$^3P_0$	0.5	?	?	2
$^2P_{3/2}$	$^3P_2$	5	19 ± 3.75	100	$12.5t$
	$^3P_1$	3	10 ± 5	33 ± 3	$4.5t$
	$^3P_0$	1	?	15 ± 4	t

line strengths and LS → jj recoupling. Ignoring factors common to all the six transitions, Eq. (3) can be written in closed form as

$$\zeta(J',j_c) = (2J' + 1)(2j_c + 1) \sum_J (2J + 1) \begin{Bmatrix} \frac{1}{2} & J & 1 \\ 1 & 1 & J' \end{Bmatrix}^2 \begin{Bmatrix} J & j_c & \frac{1}{2} \\ 1 & \frac{1}{2} & 1 \end{Bmatrix}^2 \quad (4)$$

These are the values shown in the last column of Table 1 apart from the multiplying factor, t, which merely represents the population ratio of the initial $J' = \frac{3}{2}$ and $J' = \frac{1}{2}$ states of the S⁻ ion and is, therefore, characteristic of the experimental arrangement.

Equation (4) has been written in a form which demonstrates the factor (namely, the summation) which differentiates the geometrical branching ratios from the statistical ratio $(2J' + 1)(2j_c + 1)$. It is of interest to note that expression (4) collapses to the corresponding statistical factor if the fine-structure levels are not resolved in either the initial or the final state. This is easily verified because summation over either J' or j_c reduces Eq. (4), after two applications of the sum rule [Eq. (11.15) of Fano and Racah, 1959] for 6j coefficients, to $(2j_c + 1)$ or $(2J' + 1)$ respectively. This can also be confirmed by inspection of the entries in the last row of Table 1 (taking $t = 1$). That such a collapse under closure is a general characteristic is exhibited by the more general expressions developed in Section 4.

3. Photoionization

The analysis of the branching ratios that result from absorption of the photon in LS coupling with subsequent projection onto jj-coupled states is not restricted to the photodetachment of negative ions but may be expected to apply also to the photoionization of neutral atoms or positive ions. The presence of the long-range Coulomb force in such situations plays a role in the dynamics of the outgoing electron but is of no consequence to the purely geometrical factor that is involved in the branching ratio. The only difference introduced by the Coulomb force is that the threshold law is now independent of the orbital angular momentum of the outgoing electron; therefore, one should add the contributions of different possible partial waves, but these contributions add incoherently (once again, if no angular distribution is looked for). Consider, for instance, the photoionization of a rare gas atom

$$p^6(^1S_0) + h\nu \rightarrow p^5(^2P_{3/2,1/2})l$$

where $l = 0$ or 2 for the outgoing electron. The quantum numbers of the complex are, of necessity, $^1P_1$. The squared term, $|((^2P)l^1P_1|P^{[1]}|(^2P)p^1S_0)|^2$, is evaluated from Eqs. (17.10), (15.7), and (14.12) of Fano and Racah (1959) and is given, apart from common numerical factors, by $(2l+1)\begin{pmatrix} 1 & l & 1 \\ 0 & 0 & 0 \end{pmatrix}^2$. Multiplying this by the LS → jj recoupling coefficient, we have

$$\zeta(j_c) = \Sigma_{(lj_e)}(2l+1)\begin{pmatrix} 1 & l & 1 \\ 0 & 0 & 0 \end{pmatrix}^2 (2j_c+1)(2j_e+1)\begin{Bmatrix} \tfrac{1}{2} & \tfrac{1}{2} & 0 \\ 1 & l & 1 \\ j_c & j_e & 1 \end{Bmatrix}^2 \quad (5)$$

where $j_c = \tfrac{3}{2}, \tfrac{1}{2}$ characterizes the residual ion, j_e applies to the outgoing electron, and the summation runs over all possible combinations of (lj_e). The occurrence of the zero (singlet character) in the 9j symbol reduces the symbol to a 6j coefficient, $\begin{Bmatrix} 1 & l & 1 \\ j_e & j_c & \tfrac{1}{2} \end{Bmatrix}$, after which the summation over j_e can be carried out, followed by the summation over l, leaving behind, finally, $\zeta(j_c) = 2j_c + 1$. We note that the summation over l is not crucial to this reduction of the j_c dependence of $\zeta(j_c)$ to the statistical value, the reduction being essentially due to the zero in the 9j symbol; the occurrence of such isotropic components seems to lead generally, as considered in Section 4, to an averaging out of the geometrical ratio so that it collapses to the simpler statistical form.

Equation (5) would thus predict that the distribution of photoelectrons into $j_c = \tfrac{1}{2}$ and $\tfrac{3}{2}$ from the ionization of rare gases should be in the ratio 1:2. Experimentally, however, there is a slight departure, the numbers being 1.98 for neon, 1.87 ± 0.06, for argon, 1.64 for krypton, 1.44 for xenon (measurements of Samson and Cairns, 1968, as quoted in Lee and Lu, 1973). This is an indication of the progressive departure from LS coupling of the rare gas atoms, such a departure not being accounted for in our analysis, which assumed LS coupling for the initial state. The more detailed theoretical analysis in Xe, which accounted for the observed nonstatistical ratios, did in fact show that the departure from the value of 2 was due to the states of the Xe atom departing slightly from LS coupling (see Section VIC of Lu, 1971). Also, as discussed by Lee and Lu (1973), LS coupling becomes an increasingly better description as one goes to the lower rare gases, which correlates well with the branching ratio approaching 2 in such a sequence. It seems, therefore, that analysis of experimental data in terms of the geometrical branching ratios of this paper provides a handle for determining the nature of the coupling in the initial configuration, departures from the geometrical values serving as a measure of the departure from LS coupling.

A system very similar to the rare gases considered above [the same $p^6(^1S_0)$ configuration] but which would be a true LS-coupled initial state arises in the photodetachment of F^- or I^-. At threshold, the outgoing wave would be an s wave and it may be expected that $\zeta(\tfrac{3}{2})/\zeta(\tfrac{1}{2}) = 2.$

4. The General Process of Photoabsorption

We generalize in this section the results of the previous two sections to any atomic process which involves the absorption of a photon by a tightly-coupled LS state resulting in the excitation of an electron to a highly-excited state (which may be a continuum state) which is in fact described by jj coupling. Consider, for instance, a core with quantum numbers $(s_c l_c) j_c$ coupled to an electron $(\tfrac{1}{2}l')$ to form a state $(SL')J'$; after absorption of a photon, if the electron is described by $(\tfrac{1}{2}l)j_e$ and the combined final state by $(SL)J$, the transition strength $\zeta(j_c, L', J', l, L, J)$ is given by

$$\zeta(j_c, L', J', l, L, J) = \Sigma_{j_e}(2J' + 1)(2J + 1)\begin{Bmatrix} S & J & L \\ 1 & L' & J' \end{Bmatrix}^2 (2L + 1)(2L' + 1)$$

$$\times \begin{Bmatrix} l_c & L & l \\ 1 & l' & L' \end{Bmatrix}^2 (2l + 1)\begin{pmatrix} 1 & l' & l \\ 0 & 0 & 0 \end{pmatrix}^2 \begin{Bmatrix} s_c & \tfrac{1}{2} & S \\ l_c & l & L \\ j_c & j_e & J \end{Bmatrix}^2$$

$$\times (2L + 1)(2j_c + 1)(2j_e + 1) \qquad (6)$$

where the $9j$ symbol and factors to its right represent the LS $\to jj$ recoupling and the rest represent the line strengths of the complex formed by absorption of the photon [Eqs. (17.10), (15.7), and (14.12) of Fano and Racah, 1959].

When the photoabsorption leads to discrete states, alternative final states (L, J) will in general be nondegenerate in energy because of spin–orbit interactions between the core and the excited electron. If these states are not resolved or if they are degenerate, as in the situation when the electron is lifted to a continuum state, then one would further sum expression (6) over L and J. Finally, if the different l channels are degenerate, as at Coulomb thresholds, one would carry out a further summation over l and get

$$\zeta(j_c, L', J') = \Sigma_{lLJ}\, \zeta(j_c, L', J', l, L, J) \tag{7}$$

Equation (5) is a special case of this. Equation (7), combined with Eq. (6), shows the departure of the geometrical ratio, $\zeta(j_c, L', J')$, from the statistical $(2j_c + 1)(2L' + 1)(2J' + 1)$. Once again, one sees that the entire expression in Eqs. (7) and (6) collapses to the statistical result when one quantum number such as S or L is zero. Also if $\zeta(j_c, L', J')$ is summed over either j_c or J', closure relations again lead to the statistical result.

As a specific application of Eq. (7), consider the recent measurement (Hotop and Lineberger, 1973) of the photodetachment of Pt^-, $5d^9 6s^2(^2D_{5/2, 3/2})$. The detachment leads to a p wave, leaving behind Pt, $5d^9 6s(^3D_{3,2})$. A Wigner threshold law, $\sigma \propto k^3$, was seen clearly at the threshold corresponding to the ground state, $^3D_3$, but no appreciable rise was seen when the next threshold, that of $^3D_2$, opened up, although this threshold was still in the energy range of the experiment. On the basis of statistical weights, there should have been an appreciable rise amounting to $\frac{5}{7}$ of that observed for $^3D_3$. Applying Eq. (7), however, with $l = 1$, $L' = 2$, $J' = \frac{5}{2}$, we find $\zeta(j_c = 3)/\zeta(j_c = 2) = 4.2$ This branching ratio of 20% is, perhaps, consistent with the slight rise in cross section which is visible at the $^3D_2$ threshold in Figure 8 of Hotop and Lineberger (1973).

As a final remark, we note that very similar considerations could be useful even when more than one electron is excited. So long as the system can be thought of as a core plus a distinct collection of one or more electrons, excited by a photon to a combined highly-excited state, Eq. (6) still applies with the trivial replacement of the $\frac{1}{2}$ in the $9j$ symbol by the corresponding spin, s, of the system of electrons that is excited. In addition, of course, angular factors from matrix elements of the electron–electron interaction, $\langle (sl)|r_{12}^{-1}|(sl')\rangle$, are involved since it is only through such an interaction that multiple excitations can take place. With this simple extension, the considerations of this section may be of value in unraveling complicated multiple excitations, such as those to autoionizing states when several alternative channels may be present; a simple guideline for their relative strengths would be of invaluable help in the data analysis. Such an example will be considered elsewhere.

Acknowledgments

This work was supported in part by the U.S. Office of Naval Research through Contract No. N0014-69-A-0211-0004. I thank Professor Fano for a useful conversation, and the Aspen Center for Physics for its hospitality while this work was being completed.

References

Condon, E. U., and Shortley, A. H. (1970). *The Theory of Atomic Spectra*, Ch. IX. Cambridge: Cambridge University Press.
Fano, U. (1972). *Comm. Atom. Mol. Phys.*, **2**, 171.
Fano, U., and Racah, G. (1959). *Irreducible Tensorial Sets*. New York: Academic Press.
Hotop, H., and Lineberger, W. C. (1973). *J. Chem. Phys.*, **58**, 2379.
Hotop, H., Patterson, T. A., and Lineberger, W. C. (1973). *Phys. Rev.*. A **8**, 762.
Lee, C. M., and Lu, K. T. (1973). *Phys. Rev.*, A **8**, 1241.
Lineberger, W. C., and Woodward, B. W. (1970). *Phys. Rev. Letters*, **25**, 424.
Lu, K. T. (1971). *Phys. Rev.*, A **4**, 579.
Rau, A. R. P., and Fano, U. (1971). *Phys. Rev.*, A **4**, 1751.
Samson, J. A. R., and Cairns, R. B. (1968). *Phys. Rev.*, **173**, 80.
Shore, B. W., and Menzel, D. H. (1968). *Principles of Atomic Spectra*, p. 450. New York: Wiley.
Wigner, E. (1933). *Z. Physik*, **83**, 253.

13
Study of Atomic Structure by Means of (e, 2e) Impulsive Reactions

A. GIARDINI GUIDONI, G. MISSONI, R. CAMILLONI, AND G. STEFANI

> Impulsive (e, 2e) reactions on atoms are briefly introduced, and results obtained on Ne, Ar, and Kr are summarized and compared with similar photoelectron data. It is seen that the intensities of satellite peaks obtained at large or small momentum transfers are different. An attempt is made to give a unified treatment of all the satellite structure (shake-up and configuration-interaction peaks).

Introduction

Interactions of electron beams with atoms have been used for many years to study atomic structure. Among these interactions we will consider only those in which fast monoenergetic electrons collide with single atoms or molecules and two fast electrons emerge at large angles. These electrons are detected in coincidence and their energies and momenta are measured carefully. Usefulness of such impulsive or quasi-elastic (e, 2e) processes for the investigation of the atomic structure was suggested some time ago (Glassgold and Ialongo, 1968; Neudatchin et al., 1969; Amaldi, Ciofi degli Atti, 1970; Vriens, 1969, 1970, 1971; Levin et al., 1972) and more recently checked experimentally (Camilloni et al., 1972; Weigold et al., 1973; Hood et al., 1973).

A. GIARDINI GUIDONI • Laboratori Nazionali del CNEN, Frascati, Italy.   G. MISSONI • Laboratorio Ricerche di Base SNAM PROGETTI, Monterotondo, Italy.   R. CAMILLONI and G. STEFANI • Laboratorio Metodologie Avanzate Inorganiche del CNR, Rome, Italy.

These processes are carried out in such a way that

(a) the energy transfer is much larger than the energy necessary to separate the ejected electron from the atom;
(b) the momentum transfer is much larger than the average value $\langle q \rangle$ of the momentum of the bound electron.

In these conditions the wavelength and the time associated with the momentum and energy transfer are much smaller than the characteristic atomic sizes and times.

In this paper we report the results of (e, 2e) reactions obtained to date and compare these data with similar ones from photoelectron spectroscopy. An attempt is made to give a unified treatment of the satellite structures in which "shake-up" and "configuration-interaction" peaks appear as different cases of the same process.

1. The following physical assumptions are made in order to derive the cross section in impulse approximation:

(i) During the reaction, the binding forces of the system do not produce any appreciable change in its state. Their role is then merely to determine the eigenstate of the system.
(ii) The incident particle interacts only with one of the components of the system.
(iii) The waves representing the incoming and scattered electrons are not affected appreciably by the interaction with the rest of the atom.

Derivations of the cross section in the impulsive approximation have been given in several papers (Glassgold and Ialongo, 1968; Neudatchin et al., 1969; Amaldi and Ciofi degli Atti, 1970; Vriens, 1969, 1970, 1971; Levin et al., 1972; Weigold et al., 1973). In particular, L. Vriens (1969, 1970, 1971) has discussed the consequences of the long-range Coulomb interaction, while Hood et al. (1973) have given a formulation where distortion effects due to the interaction of the scattering electrons with the atomic field are accounted for. When all the assumptions are properly introduced in the calculation of the transition matrix element it can be factorized as

$$T = T_{ee} F_\lambda(\mathbf{q}) \tag{1a}$$

$$F_\lambda(\mathbf{q}) = \langle \Phi_\lambda^I \mathbf{q} | \Phi^A \rangle = \int e^{-i\mathbf{q}\mathbf{r}_i} d\mathbf{r}_i \int \Phi_\lambda^I(\mathbf{r}_1 \cdots \mathbf{r}_{i-1} \cdot \mathbf{r}_{i+1} \cdots \mathbf{r}_z)$$
$$\times \Phi^A(\mathbf{r}_1 \cdots \mathbf{r}_z) \cdot d\mathbf{r}_1 \cdots d\mathbf{r}_{i-1} \cdot d\mathbf{r}_{i+1} \cdots d\mathbf{r}_z \tag{1b}$$

where $\mathbf{q}$ is the momentum of the bound electron determined by the momentum conservation

$$\mathbf{P}_0 - \mathbf{q} = \mathbf{P}_e + \mathbf{P}_s \quad (2a)$$

$\mathbf{P}_0$, $\mathbf{P}_s$, and $\mathbf{P}_e$ are the observed momenta of the incoming and outgoing electrons; T_{ee} is the Mott matrix element for the Coulomb scattering of the two free electrons, calculated under the condition

$$E_0 = \varepsilon_\lambda + E_s + E_e \quad (2b)$$

i.e., off the energy shell. ε_λ is the energy separation of the ionic state, and E_0, E_s, and E_e are the kinetic energies of the free electrons.

We will discuss in the following the meaning of the term $F_\lambda(\mathbf{q})$, since it is directly connected with the structural properties of the atomic system. First we note that the atomic and ionic wave functions are not orthogonal, fulfilling two different Schrödinger equations,

$$H(z)\Phi^A = E_A \Phi^A \qquad H(z-1)\Phi^I_\lambda = E_I \Phi^I_\lambda \quad (3)$$

so that more than one ionic final state is allowed in the integral (1b). In the framework of the HFSCF approximation the wave functions are written in terms of Slater's determinants built with a suitable set of orthonormal spin orbitals.

This approach describes well the ground states of closed shells or of open shells with maximum multiplicity. Otherwise the method does not account adequately for electron correlations. This failure is overcome by expanding the wave functions according to the method of configuration interaction (CI) as

$$\langle \Phi | = C_0 \langle \Phi_{HF} | + \Sigma_j C_j \langle \Phi_j | \quad (4)$$

The summation is over Slater's determinants with different configurations. Ionization modifies the atomic potential in different ways, depending on where the hole is created. For inner-shell ionizations the main effect is a sudden change in the screening, while the removal of one of the external electrons affects the correlation terms.

Outer-Shell Ionization

(a) Consider first the case in which the atom is represented by the ground HF configuration, while the ionic state is a CI expansion (4); for instance, this is the case of $3^{-1}s$ hole state of Ar^{II} (Luyken, 1973). $F_\lambda(\mathbf{q})$, for a given final state λ of the ion, becomes

$$F_\lambda(\mathbf{q}) = C_0^\lambda \langle \Phi^I_{\lambda,HF} \mathbf{q} | \Phi^A \rangle + \sum_{j=1}^{k-1} C_j^\lambda \langle \Phi^I_j \mathbf{q} | \Phi^A \rangle \quad (5)$$

where the configuration in $\langle \Phi^I_{\lambda,\mathrm{HF}}|$ is obtained by suppressing the orbital nl from the atomic ground configuration. The other terms, according to the Brillouin theorem, involve only two or more excited electrons providing the same multiplicity and parity as the ionic λ state. Then, expressing $\langle \Phi^A|$ as a product of the nl orbital times the remainder minor, we have

$$F_\lambda(\mathbf{q}) = C_0^\lambda \langle \mathbf{q}|\varphi_{nl}\rangle \langle \Phi^I_{\lambda\mathrm{HF}}|\Phi^R(nl)\rangle = C_0^\lambda k_\lambda f_{nl}(\mathbf{q}) \tag{6}$$

Other products $\langle \Phi^I_j|\Phi^R(nl)\rangle$ are zero because of the orthogonality of the angular momentum wave functions. The coefficient k_λ is different from 1 because of the difference in the radial wave functions contained in $\langle \Phi^I_\lambda|$ and $|\Phi^R(nl)\rangle$. The factor $f_{nl}(\mathbf{q})$, which gives rise to the angular correlation, is the Fourier transform of the bound-electron wave function. Thus the same factor $f_{nl}(\mathbf{q})$ is expected for the ionization of the same electron state φ_{nl}.

The probability of populating the ionic state λ (configuration-interaction peaks) is proportional to $|C_0^\lambda k_\lambda|^2$, which is called the spectroscopic factor.

(b) A different behavior is expected when the correlation effects are strong in the initial state of the atom so as to perturb at least the single-particle character of the outer electron states. By applying the expansion (4) to the atomic ground state, $F_\lambda(\mathbf{q})$ becomes

$$F_\lambda(\mathbf{q}) = C_0 \langle \Phi^I_\lambda \mathbf{q}|\Phi^A_{\mathrm{HF}}\rangle + \Sigma_j C_j \langle \Phi^I_\lambda \mathbf{q}|\Phi^A_j\rangle \tag{7}$$

When the λ state of the ion is approximated by means of a single configuration, the previous treatment implies that in (7) only the μth atomic configuration which contains as a minor the λth ionic one contributes:

$$F_\lambda(\mathbf{q}) = C_\mu \langle \mathbf{q}|\varphi_{nl}\rangle \langle \Phi^I_\lambda|\Phi^R_\mu(nl)\rangle = C_\mu k_{\mu\lambda} f_{nl}(\mathbf{q}) \tag{8}$$

The angular correlation of the outgoing electrons may now be different for the various ionic states, depending on the character of the φ_{nl} orbital present in the μth atomic configuration Φ^A_μ. Calculation on C and Be were done by Levin (1972). In both cases (a) and (b) we expect at least as many peaks in the energy spectrum as interacting configurations.

Inner-Shell Ionization

Correlation effects are not important for the core electrons. Hence removal of an inner electron is related with states of the ion having besides a deep hole a possible excitation of one outer electron. It is unimportant to consider the possibility of CI in the initial state since this would complicate calculations unduly. Considering a single configuration for the atomic wave function, we have for each state of the ion,

$$F_\lambda(\mathbf{q}) = \langle \Phi_\lambda^I | \mathbf{q} | \Phi^A \rangle = \langle \mathbf{q} | \varphi_{nl} \rangle \langle \Phi_\lambda^I | \Phi^R(nl) \rangle = f_{nl}(\mathbf{q}) k_\lambda \qquad (9)$$

The ionic configuration $\langle \Phi_\lambda^I |$ which gives a nonzero value to (9) must contain the same ordered set of quantum numbers l and m, but possibly different principal quantum numbers n. Each of these allowed final states ("shake-up") is associated with the same Fourier transform $f_{nl}(\mathbf{q})$ of the single-particle initial state.

We conclude that (e, 2e) reactions provide different kinds of information for the atom or the ion.

For the ion, information is relative to the separation energies of the various excited states. The presence of a satellite structure has been explained as a direct consequence of the change in the atomic Hamiltonian. The difference between shake-up and configuration-interaction peaks is ascribable to the different effects of deep and external ionization on the atomic dynamics.

Information on atoms comes from the angular distributions and is relative to the single-particle state of the bound electrons. This results from the assumption that there is only one electron which exchanges its momentum in the interaction.

2. The first (e, 2e) measurements were carried out on thin carbon films (~ 100 Å). Energy spectra and angular distributions were taken at 15 keV (Amaldi et al., 1969) and then at 9 keV (Camilloni et al., 1972). Data indicate that the angular distribution is sensitive to the details of the 1s electron wave function. The first data on noble gases were produced by the group at Flinders (Hood et al., 1973), who studied external shells in He and Ar. From a comparison of these data with different theoretical predictions, it was possible to discriminate between hydrogenic and HF wave functions in the case of Ar. The experiment was carried out at various energies (from 400 eV to 1200 eV) in order to study the distortion of the incoming and outgoing electrons by the atomic potential. In this energy range the distortion is shown to be negligible, in that the shape of the angular correlation does not depend on the energies and is uniquely related to bound-electron wave function. The excited states of the Ar ion were also observed at Flinders. In the energy spectrum a broad peak 24 eV higher than the $3p$ peak, whose angular correlation indicated a typical s shape, was recorded. On the basis of optical data and Luyken's calculations this peak was attributed to the unresolved excited states of the Ar ion $3d'(^2S)$ and $4d'(^2S)$. In a more recent experiment carried

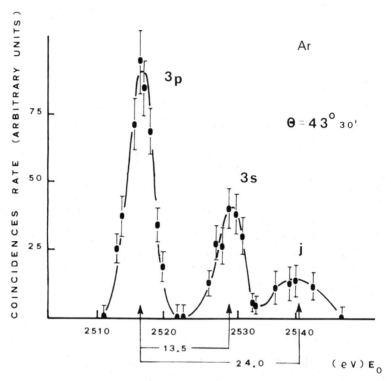

Figure 1a. Energy spectrum measured under symmetrical coplanar conditions. Angular resolution $\Delta\Omega = 3 \times 10^{-4}$ ster. Ar spectrum for angles $\Theta_s = \Theta_e = 43°\,30'$.

out at Belfast (J. Williams, private communication), at an incident energy of 250 eV and resolution of 1 eV, this peak was resolved into three peaks of different intensities.

In our experiment, outer shells of Ne, Ar, and Kr have been investigated at higher incident energies (2500 eV) and resolution of 4 eV. The results on Ar, reported in Figure 1a, fairly match the previous ones. Energy spectra are also shown in Figures 1b and 2a. In the Ne spectrum two peaks, 27.0 ± 0.5 eV apart, are observed which can be assigned to the 2p and 2s subshell ionization respectively, in agreement with the optical value of 26.9 eV. In the Kr spectrum three peaks have been observed: The two larger ones are attributed to 4p and 4s ionization respectively; the third, located at 20 eV from the 4p peak, is broader than expected from the experimental resolution, suggesting contributions from more than one final excited state. Angular distributions have been preliminarily measured (Figure 2b). The 4p peak exhibits a p-like

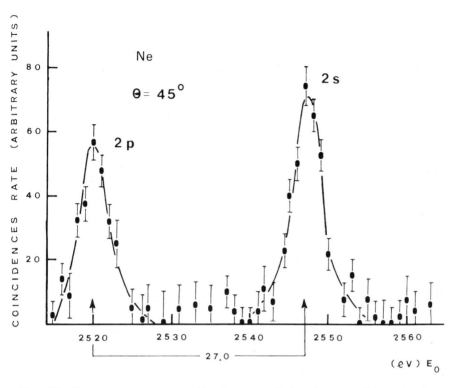

Figure 1b. Energy spectrum measured under symmetrical coplanar conditions. Angular resolution $\Delta\Omega = 3 \times 10^{-4}$ ster. Ne spectrum for angles $\Theta_s = \Theta_e = 45°$.

momentum distribution with a minimum at the value $\mathbf{q} = 0$ of the momentum of the bound electron. The 4s peak, as well as the third peak, gives rise to a momentum distribution typical of an s state, i.e., with a maximum at $\mathbf{q} = 0$.

3. In Table 1 data relative to Ne, Ar, and Kr are reported and compared with similar data obtained from photoelectron spectroscopy (Wuilleumier and Krause, 1974; Carlson *et al.*, 1970; Spears *et al.*, 1974) and radiative decay (Tan and McConkey, 1974). The prominent features are:

(i) the intensities of the satellite peaks increase as a function of the atomic number Z;
(ii) the increment is markedly different, depending on whether the ionization is performed with high momentum transfer [i.e., (e, 2e)] or with low momentum transfer (i.e., photoionization);

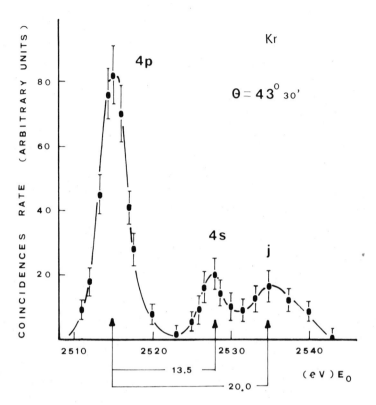

Figure 2a. Energy spectrum of Kr measured under symmetrical coplanar conditions. Angular resolution $\Delta\Omega = 3 \times 10^{-4}$ ster. $\Theta_s = \Theta_e = 43° 30'$.

(iii) within the limits of the errors quoted, there is an indication that the intensities of the satellite 3s peak in Ar depend on the incident energy.

The dependence in (i) is predicted by theoretical calculations (Luyken, 1972; Minnhagen *et al.*, 1969) and can be understood as due to the increase of correlation effects in high-Z atoms. The difference in (ii) may be attributed to the fact that factorization of the transition amplitude, as in (1), is possible only for large momentum and energy transfer.

In these respects it is remarkable that the relative intensities of the satellite peaks of the Ar $3s^{-1}$ computed by Luyken are in fair agreement with the (e, 2e) data and quite different from data obtained with photoelectron spectroscopy.

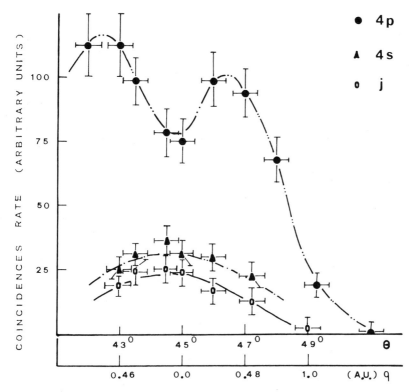

Figure 2b. Energy spectrum of Kr measured under symmetrical coplanar conditions. Angular resolution $\Delta\Omega = 3 \times 10^{-4}$ ster. Angular distributions observed for the $4p$, $4s$, and j peaks measured at 20 eV from the $4p$. In the lower scale the q values corresponding to the scattering angles are reported.

Conclusion

From the above consideration it follows that more extended and precise measurements are needed on outer-shell ionizations, to study the configuration mixing in the initial state, and on inner-shell ionizations, for which (e, 2e) data are lacking completely. Also, a study of the dependence of the satellite intensities on the incident energy is hoped for in the optics of a more extensive comparison of (e, 2e) and photoelectron spectroscopy data.

Acknowledgments

We thank Professors U. Fano and U. Amaldi, Jr. for useful discussions.

Table 1

Ionic hole state	Ionic state designation	Excitation intensity[a]			Relative intensity[b]						Shape of the angular distribution
		Optical[a]	Photoionization	(e, 2e)	Photoionization				(e, 2e)	Radiative decay (1500 eV) (Tan and McConkey, 1974)	Calculated Leyken (1972)	
					E_0 (1151.4 eV)	E_0 (1253.6 eV)	E_0 (1486 eV)	Flinders (400 eV)	Rome (2500 eV)			
Ne $2p^{-1}$	$2p^5(^2P^0)$	0	0^c	0^R								p^c
$2s^{-1}$	$2p^6(^2S)$	26.91^a	26.89^c	27.0^R								s^c
$2p^{-1}$	$3p'(^2P^0)$	24.27			2.61	6.7	6	3			1.3	p^c
$2p^{-1}$	$3p'(^2D^0)$	34.38	34.31									
$2s^{-1}$	$3s''(^2S)$	34.30										
$2s^{-1}$	$3d'(^2S)$	38.31			0.8							p^c
$2s^{-1}2p^{-1}$	$3p''(^2P^0)$	37.86	37.97									
$2s^{-1}2p^{-1}$	$3d'(^2D^0)$	37.99										
$2s^{-1}2p^{-1}$	$3d'(^2P)$	37.88			1.7	2.1	2.0					s^c
$2s^{-1}$	$3p(^2S)$		57.1^d		2.6	3.9	3.3					s^c
$2s^{-1}$	$3p(^2S)$		66.3^d									

Atomic Structure Study by (e, 2e) Reactions

Ar 3p^{-1}	3p^5($^2P^o$)	0	0						
3s^{-1}	3p^6(^2S)	13.49	13.5^e	13.5^{RF}	15	4	3	}72	
3s^{-1}	4s''(^2S)	20.74			15	15	19			0.15	p^F s^F
3p^{-1}	4p'($^2P^o$)	21.39	21.6^e	24.0^{RF}	8	7	6	}52 ± 10	26	49.8	s^F
3s^{-1}	3d'(^2S)	22.82	22.9^e						14	11.3	
3s^{-1}	4d'(^2S)	25.44	25.4^e								
Kr 4p^{-1}	4p^5($^2P^o$)	0	0	0							p^R s^R
4s^{-1}	4p^6(^2S)	13.51	13.5^e	13.5^R		9	7				
4p^{-1}	5p'($^2P^o$)	18.61	18.4								s^R
4s^{-1}	5s''(^2S)	18.08									
4s^{-1}	4d'(^2S)	19.94	20.0^e	20.0^R		25	31	}170 ± 40			
4s^{-1}	6s''(^2S)	20.39									

[a] Excitation energies as obtained from optical data from Moore (1949, 1952, 1958); in case of Kr 4d'(^2S) from Minnhagen et al. (1969).
[b] Photoionization and (e, 2e) intensities are given relative to the 100 value for the lowest-energy peak having the same hole state.
[c] Photoionization data as obtained from Wuilleumier and Krause (1974).
[d] Photoionization data for Ne from Carlson et al. (1970).
[e] Photoionization data for Ar and Kr from Spears et al. (1974).

References

Amaldi, U., Jr., and Ciofi degli Atti, C. (1970). *Nuovo Cimento*, **66**, 129.
Amaldi, U., Jr., Egidi, A., Marconero, R., and Pizzella, G. (1969). *Rev. Sci. Instr.*, **40**, 1001.
Camilloni, R., Giardini Guidoni, A., Tiribelli, R., and Stefani, G. (1972). *Phys. Rev. Letters*, **29**, 618, and references cited therein.
Carlson, T. A., Krause, M. O., and Moddeman, W. E. (1970). 'Les Processus Electroniques Simples et Multiples du domaine X et X-V,' Paris, Sept. 21–25, 1970, 'Colloques Internationaux.'
Glassgold, A. E., and Ialongo, G. (1968). *Phys. Rev.*, **175**, 151.
Hood, S. T., McCarthy, I. E., Teubner, P. J. O., and Weigold, E. (1973). *Phys. Rev.*, A **8**, 475.
Levin, V. G. (1972). *Phys. Letters*, **39A**, 125.
Levin, V. G., Neudatchin, V. G., and Smirnov, Yu. F. (1972). *Phys. Status Solidi* (6), **49**, 489.
Luyken, B. F. J. (1972). *Physica*, **60**, 432.
Minnhagen, L., Strihed, H., and Petersson, B. (1969). *Arkiv Fysik*, **39**, 471.
Moore, C. E. (1949/1952, 1958). Atomic Energy Levels, NBS No. 467, Vols. I–III. Washington, D.C.: U.S. GPO.
Neudatchin, V. G., Novoskol'tseva, G. A., and Smirnov, Yu. F. (1969). *JETP*, **28**, 540.
Spears, D. P., Fischbeck, H. J., and Carlson, T. A. (1974). *Phys. Rev.*, A **9**, 1603.
Tan, K.-H., and McConkey, J. W. (1974). *J. Phys. B*, **7**, No. 6.
Vriens, L. (1969). *Physica*, **45**, 400.
Vriens, L. (1970). *Physica*, **49**, 602.
Vriens, L. (1971). *Phys. Rev.*, B **4**, 3008.
Weigold, E., Hood, S. T., and Teubner, P. J. O. (1973). *Phys. Rev. Letters*, **30**, 475.
Wuilleumier, F., and Krause, M. O. (1974). To be published in *Phys. Rev.*, A **10**, 242.

14
Phase Shift Analysis and Dispersion Relations

B. H. Bransden

> The problem of determining partial wave amplitudes from measured cross sections is discussed for the case of electron scattering by neutral atoms, and the recent analyses of experimental data are reviewed.

1. Introduction

Most theoretical calculations of cross sections for the scattering of low-energy electrons by atoms make use of the principle of angular-momentum conservation to simplify the problem, so that the scattering amplitude for each angular-momentum eigenstate, f_l, is determined separately. If the target is of spin zero, such as helium, and in the absence of interactions depending on electron spin, the elastic scattering amplitude $f(k, \theta)$ is related to the partial wave amplitudes f_l, by the partial wave expansion

$$f(k, \theta) = \sum_{l=0}^{\infty} \frac{1}{2ik} f_l(k) P_l(\cos \theta) \qquad (1)$$

where θ is the angle of scattering and k the momentum of the incident beam. For neutral targets, the expansion converges and only a finite number of amplitudes f_l, from $l = 0$ to $l = L$ are of significance.*

---

*When the target is charged, the long-range Coulomb interaction defines the Coulomb amplitude $f_c(k, \theta)$, and the difference $f(k, \theta) - f_c(k, \theta)$ can be expanded in a convergent series. For an introduction to this and other aspects of scattering theory, Bransden (1970) can be consulted.

B. H. Bransden • Department of Physics, University of Durham, England.

The amplitudes f_l can be written in the form

$$f_l = S_l - 1 \quad \text{where } S_l = \eta_l \exp(2i\delta_l) \tag{2}$$

where the phase shifts δ_l and inelasticity parameters η_l are real, with $0 \leq n_l \leq 1$. These conditions ensure the unitarity of the scattering matrix required by conservation of probability. In general $|S_l| \leq 1$, but below the first inelastic threshold $n_l = 1$ (all l) and $|S_l| = 1$.

The only measurable quantities relating to elastic scattering of spinless particles by a spinless target are the differential cross section $I(k, \theta) = |f(k, \theta)|^2$ and the total cross section $Q(k) = 4\pi \operatorname{Im} f(k, 0)/k$. If the energy is below the first inelastic threshold, then the measurement of Q is not theoretically independent of I, since

$$Q = \int_{(4\pi)} I(\theta) \, d\Omega \tag{3}$$

and the momentum transfer cross section,

$$Q_D = \int_{(4\pi)} I(\theta)(1 - \cos\theta) \, d\Omega \tag{4}$$

is also determined by I. In practice, $I(\theta)$ may only be measured over a limited angular range, in which case Q and Q_D represent additional information.

The objective of a phase shift analysis is to determine the amplitudes f_l from the measured values of I, Q, and Q_D. The success of this parameterization of the data in terms of the f_l can, as we shall see, provide information about the accuracy and consistency of the data, in addition to providing numerical values of the amplitudes to be compared with theoretical predictions.

2. Phase Shift Analysis

The technique of phase shift analysis has been developed to a large extent in nuclear physics (see, for example, Bransden and Moorhouse, 1973, pp. 54, 195; Arndt and MacGregor, 1967). The conventional procedure is to minimize the expression

$$F[x_1, x_2, \ldots, x_\lambda] \equiv \sum_{i=1}^{M} \left[\frac{I_i^E - I_i^P(x)}{\varepsilon_i} \right]^2 \tag{5}$$

with respect to parameters $x_1, \ldots, x_\lambda$, where the quantities I_i^E are a set of measured cross sections at a given momentum, each with statistical error ε_i, and $I_i^P(x)$ are the same quantities parameterized in terms of the x's. The x's can either be taken to be the phase parameters δ_l, η_l or can be functions of these parameters, and the minimization is restricted to the field $-\pi \leqslant \delta_l \leqslant +\pi$; $0 \leqslant \eta_l \leqslant 1$. Several computer programs* have been written to perform minimizations in multidimensional space, of which a typical and widely used example is that of Powell and Fletcher (1964).

In general, there will be several sets of parameters $[x]$ which minimize F, but some of these can be eliminated, since if the errors ε_i are independent and normally distributed, the expected value of F at the required minimum, $F_{\min}$, will equal the number of degrees of freedom (the difference between the number of pieces of data M and the number of parameters λ). If F cannot be reduced to this value, this may indicate inconsistencies among the measurements and the presence of systematic errors. Further, if a major part of $F_{\min}$ arises from perhaps a single data point, it can be concluded that this point is in error and reminimization can be attempted excluding the suspect point.

Near a minimum at $[\bar{\mathbf{x}}]$, F is of the form

$$F = F_{\min} + \Sigma_{ij} G_{ij}(x_i - \bar{x}_i)(x_j - \bar{x}_j) \qquad (6)$$

$\mathbf{G}$ is a real symmetric matrix and its inverse, $\mathbf{G}^{-1}$, is known as the error matrix. The diagonal elements of the error matrix $(\mathbf{G}^{-1})_{ii}$ are equal to the standard deviation of each parameter x_i, and the off-diagonal elements of $\mathbf{G}^{-1}$ are associated with correlations between the parameters.

At a particular energy, several data sets, perhaps from different laboratories, may exist with different normalization errors. This situation can be met by introducing parameters λ_n, so that λ_n takes the value unity if the normalization of the nth data set is correct. F is now expressed as

$$F = \sum_{i,n} \left[\frac{\lambda_n I_{i,n}^E - I_{i,n}^P}{\varepsilon_i} \right]^2 + \sum_n \left[\frac{\lambda_{n-1}}{\Delta_n} \right]^2 \qquad (7)$$

where Δ_n is the normalization error on the nth data set. F is now minimized with respect to both $[\mathbf{x}]$ and $[\lambda]$.

3. The Problem of Uniqueness

Because of incomplete or erroneous data, sets of phase parameters can sometimes be found which correspond to different minima of F, but which give a similar goodness of fit. Which, if any, of these sets represents the correct phase shifts can then only be decided by using theoretical ideas, the most useful of which is that δ_l and η_l should be continuous functions of k. A more fundamental problem is to decide whether, if we know the cross sections exactly and if the corresponding phase parameters are also known exactly,

---
* The problems of minimization in many dimensions are discussed in Fletcher (1969) and Künzi et al. (1968).

there are other sets of phase parameters that also fit the cross section exactly.

In discussing the uniqueness problem, we first notice the trivial ambiguity that $|f(\theta)|$ and Q are invariant under the substitution $\delta_l \to -\delta_l$ (all l). Sufficient conditions for the phase shifts to be unique in certain cases have been given by Martin (1969), Tortorella (1972), and others, but the first explicit discussion of a nontrivial ambiguity for the scattering of spinless particles below the inelastic threshold ($\eta_l = 1$) was given by Crichton (1966) for the case $L = 2$. He found numerically that, starting from given values of $\delta_0, \delta_1, \delta_2$ defining the amplitude $f(k, \theta)$, a different amplitude $f'(k, \theta)$ existed with $|f'| = |f|$ and $\delta'_2 = \delta_2, \delta'_1 \neq \delta_1, \delta'_0 \neq \delta_0$, provided δ_2 lay in the interval $12.5° \leqslant \delta_2 \leqslant 24.2°$.

The problem has received a detailed discussion by Berends and Ruijsenaars* (1973a). Since $f(k, \theta)$ is a polynomial in $\cos\theta$ it can be written in terms of its zeros F_i as

$$f(k, \theta) = f(k, 0) \prod_{i=1}^{L} \frac{\cos\theta - F_i}{1 - F_i} \tag{8}$$

The partial wave amplitudes can be expressed in terms of $f(k, 0)$ and F_i. For example, if $L = 2$,

$$f_0 = 2ikf(k, 0)(F_1 F_2 + \tfrac{1}{3})/\{(1 - F_1)(1 - F_2)\}$$
$$f_1 = -2kif(k, 0)(F_1 + F_2)/\{3(1 - F_1)(1 - F_2)\} \tag{9}$$
$$f_2 = 4ikf(k, 0)/\{15(1 - F_1)(1 - F_2)\}$$

All possible amplitudes $f'(k, \theta)$ satisfying $|f'| = |f|$ and $\operatorname{Im} f'(k, 0) = \operatorname{Im} f(k, 0)$ can be obtained from $f(k, \theta)$ by applying combinations of the transformations (a) $\operatorname{Re} f(k, 0) \to -\operatorname{Re} f(k, 0)$, (b) $F_i \to F_i^*$. In general, disregarding the trivial ambiguity $\delta_l \to -\delta_l$, we see that there are 2^{L-1} different sets of transformations (a) and (b). Not all of these sets are physical, since we require $|f_l + 1| < 1$ (for each l), and it is this requirement that restricts the Crichton ambiguity to a certain range of δ_2. A complete analysis of the cases $L = 2$ and $L = 3$ has been given by Berends and Ruijsenaars (1973a). For the case $L = 2$, it is, in fact, possible to display δ_2 ($= \delta'_2$), $\delta_1, \delta'_1, \delta_0, \delta'_0$ as functions of the parameter $|F_2|^2$ explicitly. The number of ambiguities increases with L, and for $L = \infty$ a continuum of ambiguities can occur in the inelastic region (see, for example, Atkinson et al., 1973a,b). To avoid excessive ambiguity it is clear that the higher partial waves must be taken as known; but it is often not sufficient just to set $f_l = 0$, above some small value of L. Since large l values correspond to peripheral scattering, which can be treated theoretically with some confidence, the higher partial waves can be fixed at

---

* See also Atkinson et al. (1973d).

the calculated values, leaving the large amplitudes to be determined by the data, through minimization.

The important case of electron scattering by a spinless target but allowing for spin-dependent interactions has also been discussed by Berends and Ruijsenaars (1973b). For each value of l, there are two amplitudes $f_{l\pm}$, for which the total angular momentum is $(J \pm \tfrac{1}{2})$. The differential cross section for elastic scattering is

$$I(k, \theta) = |f(k, \theta)|^2 + |g(k, \theta)|^2 \tag{10a}$$

where

$$f(k, \theta) = \frac{1}{k} \sum_{l=0}^{L} [(l + 1)f_{l+} + lf_{l-}]P_l \cos \theta \tag{10b}$$

and

$$g(k, \theta) = \frac{i}{k} \sum_{l=1}^{L} (f_{l+} - f_{l-}) \sin \theta P_l^1 (\cos \theta) \tag{10c}$$

The famous Fermi–Yang ambiguity,* which caused difficulty in the early discussion of π–p scattering, is removed by a measurement of the polarization of the electron in the final state. If the initial state is unpolarized, the polarization is given by

$$P(k, \theta) = 2 \operatorname{Re}[f^*(k, \theta)g(k, \theta)]/I(k, \theta) \tag{11}$$

There is also the generalized Minami ambiguity, for which $I(k, \theta)$, $P(k, \theta)$, and $Q(k)$ are unaltered by the replacements (for all l)

$$f'_{(l+1)-} = -f^*_{l+}, f'_{(l+1)+} = -f^*_{l-} \tag{12}$$

A measurement of the rotational coefficients A and R (see Bransden and Moorhouse, 1973) that determine the change of polarization after scattering with an initially polarized beam resolves this ambiguity. Above the inelastic threshold, further ambiguities exist for $L > 1$, if only I, P, and Q are measured.

A problem related to the uniqueness problem is that of the stability of the fit of a differential cross section by a given set of phase parameters. To investigate this problem it is assumed that we know $I(k, \theta)$ and the corresponding phase shifts, and we seek the change in phase shifts corresponding to a small increment in $I(k, \theta)$. This allows the systematic examination of the uncertainty in the phase shifts generated by the experimental error. In the spinless case with $L = 2$, Atkinson et al. (1973c) have shown that the Crichton

---

* The Fermi–Yang ambiguity occurs for $L = 1$. It is found that $|f'| = |f|$, if $\delta_{1+} - \delta_{1-} = \delta'_{1+} - \delta'_{1-}$ and $2 \exp(i\delta_{1+}) + \exp(i\delta_{1-}) = 2 \exp(i\delta'_{1+}) + \exp(i\delta'_{1-})$.

ambiguity is not isolated but a continuation away from both of the possible amplitudes is possible.

4. The Practical Removal of Ambiguities

(a) Dispersion Relations

The sets of phase parameters which give rise to the same differential and total cross sections can be divided into two which differ in the sign of the corresponding values $\mathrm{Re}\, f(k, 0)$. A knowledge of this sign will remove the Crichton ambiguity completely and, in general, will diminish the number of ambiguous solutions. Further, a knowledge of the magnitude of $\mathrm{Re}\, f(k, 0)$ helps to resolve difficulties associated with inadequate experimental data. Provided the total cross section Q is known with sufficient accuracy, it is possible to calculate $\mathrm{Re}\, f(k, 0)$ from the forward dispersion relation (Gerjuoy and Krall, 1960, 1962):

$$\mathrm{Re}\, f(k, 0) = f_B(k, 0) + \frac{1}{2\pi} \int_0^\infty \frac{k'^2 Q(k')\, dk'}{(k'^2 - k^2)} + \sum_j \frac{R_j}{k^2 + |\varepsilon_j|} \qquad (13)$$

where f_B is the first-order perturbation expression for f. The last term only arises when stable bound states with energies ε_j of the negative ion formed from the incident electron and the target exist. The residues R_j can be calculated from the wave function of the negative ions.*

To calculate the integral over the total cross section, measurements of Q can be employed, but these necessarily exist only over a finite momentum interval, say, $0 \leqslant k \leqslant k_m$. For $k > k_m$, a theoretical approximation to Q must be used, and the Bethe–Born approximation is a possibility if k_m can be chosen to be sufficiently large. For this purpose, Inokuti and McDowell (1974) have systematically evaluated the integrated elastic cross section in the first Born approximation for electron scattering by neutral atoms from helium to neon. By adding these cross sections to the total inelastic cross sections previously computed (Dehmer et al., 1974), the Born approximation to the total cross sections can be obtained. There is some uncertainty as to how large the momentum k_m must be before the Born cross sections can be considered to represent the true cross section within some specified degree of accuracy. In this connection, de Heer et al. (1974) have shown that over the energy range 400 to 3000 eV the measured elastic cross section in helium is compatible with the Born approximation, but for neon and argon the Born cross sections are too large. This can be explained by remembering

---

* For electron scattering by hydrogen, the residue has been calculated by Spruch and Rosenberg (1960).

that the effective interaction depends on Z, and the convergence of the Born series depends on the parameter (Z/k) being small. It has been shown by de Heer *et al.* that the elastic scattering for neon and argon can be parameterized by an effective interaction, from which cross sections can be computed classically over this energy range.

A test on the adopted values of the total cross section can be obtained by evaluating the dispersion relation at zero momentum. When there is no stable bound state of the whole system, we have a sum rule

$$-A = f_B(0,0) + \frac{1}{2\pi} \int_0^\infty Q(k')\, dk' \qquad (14)$$

where A is the scattering length.* A can be obtained from measurements of the total cross section near zero energy. A recent example of the utility of this technique is given by the case of positron scattering by helium. Bransden *et al.* (1974) were able to show that the measured total cross sections (Canter *et al.*, 1973) could not be reconciled with the sum rule, the measured cross section appearing to be too small at high energies. By assuming the accuracy of the Born approximation at high energies, Bransden *et al.* were able to predict the excepted shape of the total cross sections as a function of energy. This stimulated a remeasurement of the parameters of the positron beam, providing a corrected set of cross sections, which are shown in Figure 1 (Coleman *et al.*, 1974). The remeasured cross sections are in good agreement with the sum rule. In fact,

$$-(A + f_B) = 1.26 \pm 0.06 \qquad \frac{1}{2\pi}\int_0^\infty Q(k)\, dk = 1.21 \pm 0.05$$

(b) Continuity in Momentum

Ambiguities among the phase parameters can be removed not only by using dispersion relations, but also by requiring the partial wave amplitudes to have the behavior expected on general physical grounds. For example, near zero momentum we expect only the s wave to be of importance; as the momentum increases, first the p and then the d wave will become appreciable, although care must be taken not to miss a low-energy resonance in a partial wave otherwise expected to be small. The general continuity of the phase parameters with momentum can be taken into account in several ways. In "energy-independent" analyses, the phase parameters are found at a number of energies over the interval of interest. From the various sets of partial wave amplitudes, the set or sets that are continuous in momentum are picked out. This can be done "by eye," or the process can be automated

---
* $A = -\lim_{k\to 0}(\tan \delta_0/k)$.

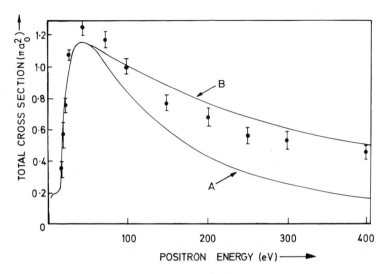

Figure 1. The total cross section for positron scattering by helium. Curve (A): measurements of Canter *et al.* (1973); curve (B): shape of the cross section suggested by Bransden *et al.* (1974) based on (a) the measurement below 50 eV, (b) the Born approximation above 400 eV, (c) an interpolation satisfying the sum rule (14). Vertical bars: remeasured cross section of Coleman *et al.* (1974).

using a computer program to search for the path of least distance (suitably defined) in phase parameter space (Johnson and Steiner, 1967). To avoid inadvertently moving from one Crichton ambiguity to another, the zeros of the amplitudes F_i can be plotted as functions of energy, to ensure that the transformations $F_i \to F_i^*$ have not occurred.

This procedure is satisfactory if the data are closely spaced in energy, otherwise an "energy-dependent" technique is often more valuable. In this case, continuity is imposed on the phase parameters at the start by parameterizing these quantities as functions of momentum. For example, in electron–helium scattering below the inelastic threshold, the phase shifts can be represented by effective range expansions of the type*

$$k \cot \delta_0 = -A + a_{10}k - \frac{\pi\alpha^2}{3}\log k + \sum_{n=2}^{m} a_{n0}k^n \tag{15}$$

$$\delta_l = k^2 \left(b_l + \sum_{n=1}^{m'} a_{nl}k^n \right) \qquad l \neq 0 \tag{16}$$

where $b_l = \pi\alpha/\{(2l-1)(2l+1)(2l+3)\}$ and α is the dipole polarizability, which together with the scattering length A is assumed to be known.

---

* The modification of these expressions near resonances is given in the papers cited.

The coefficients a_{nl} are identified with the minimization parameters $[x]$, and all the data over the energy range of interest are fitted simultaneously. The advantage of this technique is that the continuity of the phase shifts is automatic, but there can be a danger of underparameterizing the data if the energy interval over which the fit is attempted is too large. This approach has been used with considerable success in a program of analyses started by Bransden and McDowell (1969, 1970) at Durham and since continued at Durham and Royal Holloway College, London.* In practice, it is found that for the cases so far investigated and by varying the phase shifts for $L \leq 2$ there is only one solution that reduces F to the expected value.

5. Recent Applications

In selecting systems which can be successfully analyzed, one of the chief considerations is the existence of reasonably extensive data sets. For this reason, Bransden and McDowell (1969, 1970) discussed electron–helium scattering in their first attack on the problem. This initial work was extended to neon and argon (McDowell, 1971b,c) and more recently to the interesting case of electron scattering by mercury, which exhibits strong polarization effects (Hutt and Bransden, 1974). As new measurements are made it is intended to refine the original analyses, and this has been carried out for electron scattering by helium and neon (Naccache and McDowell, 1974).

(a) Electron–Helium Scattering

Because of the existence of a fairly complete set of experimental data, Bransden and McDowell (1969, 1970) chose the electron–helium system for detailed analysis at energies up to 20 eV. The phase shifts δ_0, δ_1, and δ_2 were parameterized in terms of eleven parameters, while the higher phase shifts were obtained by applying the first Born approximation to the peripheral interaction $V_p = -\alpha/r^4$. The detailed fits obtained showed that the data at that time were not completely consistent, but nevertheless the data could be fitted within 1% by the predicted phase shifts.

The application of the dispersion relations enabled $\operatorname{Re} f(k, 0)$ to be predicted to within $\pm 15\%$. A greater accuracy could not be achieved because of the lack of accurate measurements of the total cross section at the higher energies.† In the calculation of the first-order amplitude f_B, the exchange contribution is sensitive to the choice of helium wave function, and

---
*A computer program for this analysis has been published by Naccache and McDowell (1973).
† The only measurements available are those of Golden and Bandel (1965) for the range 3 to 19.2 eV and Normand (1930) from 20 to 400 eV.

care must be taken to use a sufficiently accurate wave function, but this factor is not a serious limitation for helium.

New and more accurate differential cross section data for electron–helium scattering (Andrick and Bitsch, 1971; McConkey and Preston, 1973) have allowed a more precise determination of the low-energy phase shifts and the extension of the energy range above the inelastic threshold up to 100 eV (Naccache and McDowell, 1974). In the inelastic region the parameterization (15) and (16) was unsatisfactory, and for the real part of the phase shifts the following forms were adopted:

$$\delta_0(k^2) = \pi - \tan^{-1}[ka_1(1 + a_4k^3)/\{1 - 1.45661a_1^{-1}k$$
$$- 0.925k^2 \ln k^2 + a_2k^2 + a_3k^3\}] \quad (17a)$$

$$\delta_1(k^2) = b_1k^2(1 + a_5k + a_6k^2 + a_7k^3)/(1 + a_8k^3) \quad (17b)$$

$$\delta_2(k^2) = b_2k^2[(1 + a_9k + a_{10}k^2)/(1 + a_{11}^2k^2)]^2 \quad (17c)$$

where a_i are the variable parameters.

The corresponding inelasticity parameters were expressed as $n_l = \exp(-\mu_l)$, and the imaginary part of the phase shift μ_l was approximated to first order as

$$\mu_l = -\pi \int_0^\infty V_1(r)[J_{l+1/2}(kr)]^2 \, dr \quad (18)$$

Here V_1 is the imaginary part of the optical potential, taken to be of a general form suggested by the work of Furness and McCarthy (1973):

$$V_1(r) = \theta(k^2 - k_T^2)[A + B(k^2 - k_T^2)^\gamma]e^{-\lambda r}/r \quad (19)$$

where k_T is the threshold momentum and A, B, γ, and λ are variable parameters, determined through the minimization procedure.

It is of interest that the values of the parameters obtained ($\lambda = 0.425$, $\gamma = 3.8$, $A = 0$, $B = 5.22 \times 10^{-4}$) are consistent with what would be expected from the theoretical work of Furness and McCarthy on the e^-–He interaction ($\lambda \simeq 0.5$, $\gamma \simeq 4.0$), but the strength of the interaction is remarkably small. A good fit to the data can be obtained up to 100 eV by varying the phase shifts for $l \leq 2$ and taking the higher-order phase shifts as calculated from the peripheral interaction. The s wave is stable to alterations of the data set and is believed to be determined to within $\pm 2\%$ below 13 eV and within $\pm 5\%$ up to 100 eV, while the p wave is expected to be accurate to within $\pm 5\%$ below 13 eV and $\pm 15\%$ overall. The best phase shifts are shown in Table 1. These phase shifts agree within the errors with the polarized orbital calculations of Duxler et al. (1971) and the variational calculations of Sin Fai Lam and Nesbit (1972). Calculations based on many-body theory (Knowles and McDowell, 1973; Yarlagadda et al., 1973) are also consistent

Table 1. Phase Shifts for e^-–He Scattering
(Real Parts)

(Naccache and McDowell, 1974—Data Set H)

E	$l = 0$	$l = 1$	$l = 2$
1.5	2.73	0.054	–
5.1	2.36	0.154	–
11.1	2.05	0.237	0.035
15.1	1.94	0.274	0.052
19.1	1.87	0.306	0.066
31.1	1.77	0.384	0.084
50.0	1.71	0.482	0.070
100.0	1.66	0.688	0.005

with the $l = 0$ and $l = 1$ phase shifts. The d-wave phase shifts, which are small, are less well determined and further measurements, particularly for $\theta > 90°$, are required. The experimental phase shifts are of similar size to those predicted by theory, but appear to increase too rapidly with energy, at energies below 20 eV.

The comparison of the $\operatorname{Re} f(k^2, 0)$ calculated from the experimental phase shifts with that calculated from the dispersion relation is shown in Figure 2. The uncertainty in the value of $\operatorname{Re} f(k^2, 0)$ predicted from the phase shifts is estimated to be less than $\pm 8\%$, for $k^2 \leqslant 1.25$, while the error associated with the dispersion relation prediction was estimated as $\pm 10\%$. The two sets of results are compatible within the errors for energies below 12 eV but at higher energies agreement is less good. The source of the discrepancy appears to be in the measured total cross section values between 20 and 100 eV, which could be up to 20% too large, and further total cross section measurements are required above 20 eV.

(b) Neon and Argon

A fresh analysis of electron–neon scattering by Naccache and McDowell (1974) now makes use of the new measurement of the differential cross section up to 100 eV by McConkey and Preston (1973) and by Andrick and Bitsch (1971), and these have been fitted together with the diffusion cross section of Robertson (1972) and the total cross section of Salop and Nakano (1970) up to 60 eV. It is clear that the data sets of McConkey and Preston and of Andrick and Bitsch are inconsistent in normalization, but, if this is allowed for, a fit can be obtained to all the data up to 60 eV. The quality of the fit obtained is illustrated in Figure 3, which shows the data at 14 eV. It will be seen that there is some appreciable difference in shape between the

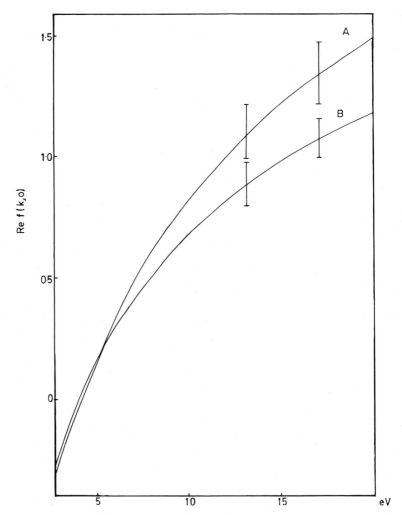

Figure 2. Re $f(k, 0)$ for electron–helium scattering: curve (A) from the forward dispersion relation; curve (B) from the phase shift analysis of Naccache and McDowell (1974).

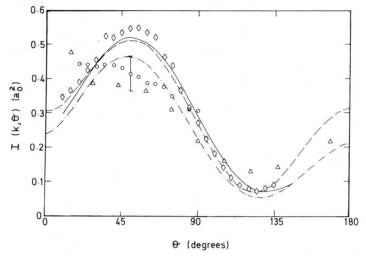

Figure 3. Electron–neon elastic differential cross sections at 14 eV. (———), measurements of Bitsch (1972) at 13.6 eV; (— — —), cross section at 14 eV constructed from the phase shifts found by Bitsch (1972) on analyzing his data at 13.6 eV; ◯, measurements of McConkey and Preston (1973) at 14 eV with $\pm 12\%$ error bars; ◇, measurements of Andrick (1969) at 13.6 eV; △, measurements of Ramsauer and Kollath (1932) at 13.8 eV; (— · —), fit to the data in the phase shift analysis of Naccache and McDowell (1974).

measurements, particularly in the forward direction. The phase shifts obtained by Naccache and McDowell are shown for certain energies in Table 2.

The application of the dispersion relation to the electron–neon system is frustrated because the Born exchange amplitude has not yet been calculated with sufficient accuracy,* and, on the other hand, the total cross

Table 2. Phase Shifts for e^-–Ne Scattering (Real Parts)

(Naccache and McDowell, 1974)

E	$l = 0$	$l = 1$	$l = 2$
1.5	−0.232	0.0051	0.0060
5.0	0.536	0.0679	0.0239
11.0	0.812	0.232	0.0739
21.0	0.938	0.414	0.177
30.0	1.086	0.492	0.290
40.0	1.348	0.500	0.356
60.0	−1 532 ± 0.15	−0.563 ± 0.06	0.186 ± 0.02

* Unpublished work by K. H. Winters and M. Knowles has shown the inadequacy of Hartree–Fock functions for this calculation.

section measurements (Normand, 1930) above 20 eV are difficult to accept without question. In particular, the measured cross section at 30 Ry is about a factor of two below an estimate based on the Bethe–Born approximation (Inokuti and McDowell, 1974; McDowell, 1974), and although the Born approximation may well not be accurate at this energy, the discrepancy is remarkably large, and new measurements of the total cross section are urgently required.

For argon, the original phase shift analysis (McDowell, 1971b,c) did not produce a good fit, but a revised analysis with new data is in progress, and an accurate fit has been obtained up to 40 eV (Naccache and McDowell, 1975).

(c) The Electron–Mercury System

This system is of particular importance because of the use of mercury as a polarizer or analyzer. The analysis of Hutt and Bransden (1974) has concentrated on the data below the inelastic threshold in the range 0.5 to 4.5 eV, although an extension above the inelastic threshold is in progress.

In addition to differential cross section data (Glatzel, 1973), the polarization of an initially unpolarized beam has been measured at 0.5 and 2.0 eV (Glatzel, 1973) and at 3.5 eV (Deichsel et al., 1966) and also the total cross section (Brode, 1929; Jones, 1928).† The method employed is quite similar to that for the rare gases, the phase shifts for $l \leqslant 3$ being varied, theoretical estimates of Walker (1973) for the small partial waves with $l > 3$ being used.

It is possible to fit the differential cross section well at all energies and angles. The polarization at 0.5 and 2 eV can also be fitted but that at 3.5 eV less well. Figure 4 illustrates the fit at 2 eV. The total cross section data cannot be fitted either in magnitude or shape and appear to be inconsistent with the other data (see Figure 5), but Bederson and Kieffer (1971) have argued that the method used by Brode is unreliable.

The large $l = 0$ and $l = 1^{\pm}$ phase shifts are shown in Figure 6. At 3.5 eV, comparison can be made with Walker's (1971) calculation, which makes allowance for the distortion of the atom during the collision. The agreement is not unreasonable considering the uncertainties in the data. Not surprisingly, the phase shifts are in complete disagreement with earlier calculations in which distortion is neglected (Walker, 1970). The small phase shifts

† New measurements by Duweke et al. (1974) of both the differential cross section and polarization in this energy range are now being analyzed.

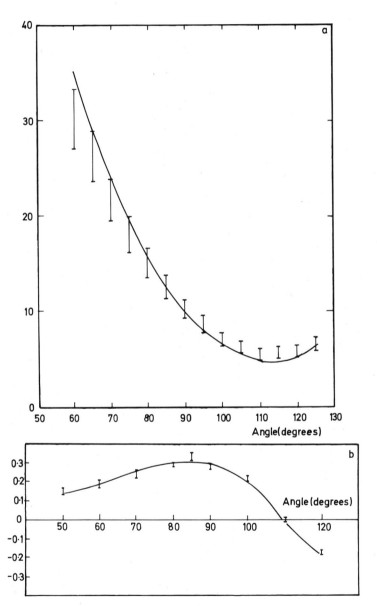

Figure 4. Electron–mercury elastic scattering at 2 eV. (a) Differential cross section measurements; (b) polarization measurements. Solid curve: fit from the phase shift analysis of Hutt and Bransden (1974); data from Glatzel (1973).

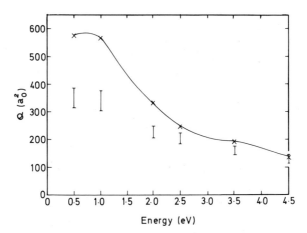

Figure 5. Total cross section from electron–mercury scattering. Solid curve: fit from the phase shift analysis of Hutt and Bransden (1974); Vertical bars, measurements of Brode (1929).

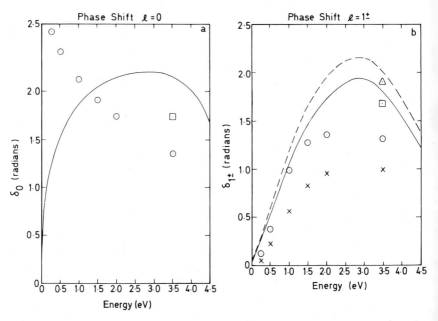

Figure 6. The $l = 0$ (a) and $l = 1\pm$ (b) phase shifts for electron–mercury scattering from the analysis of Hutt and Bransden (1974). (———), $l = 0$ and $l = 1^+$ phase shifts from the analysis; (— — —), $l = 1^-$ phase shift from the analysis; $\square$, the $l = 0$ and $l = 1^+$ and $\triangle$, the $l = 1^-$ Walker (1971) calculated allowing for distortion; $\odot$, the $l = 0$ and $l = 1^+$ and $\times$, the $l = 1^-$ phase shifts of Walker (1969) calculated without allowance for distortion.

for $l = 2^\pm$ and $3^\pm$ are not well determined but are of the correct sign and order of magnitude.

The accuracy of even the large phase shifts will remain doubtful until a better data set is available, including a remeasurement of the total cross section.

An extension of this analysis above the inelastic threshold is in progress, using for the inelasticity parameters an optical model potential, similar in concept to that used for the rare gases.

6. Conclusions and Outlook

There is no doubt that where an adequate data set exists, the phase shift analysis supplemented by forward dispersion relations can be used not only to provide a convenient and physical description of the data, but also to provide a severe and detailed test of theoretical predictions. Of the systems so far analyzed only that for electron scattering by helium has a really adequate data set and this is still not beyond reproach. In particular the total cross section for e^-–He scattering requires remeasurement, although enough has been done to establish the validity of the dispersion relation and its use as a working tool.

Of the other systems so far analyzed, that of e^-–Ne has the best data set, but here there are obvious inconsistencies in normalization to be resolved and the total cross section is in considerable doubt at higher energies. There is good hope here and in the interesting case of e^-–Hg that better data sets will be available shortly. It is clearly desirable that the analysis should be extended to a number of systems that typify electron–atom interactions. In particular e^-–H and electron–alkali metal systems will be important to examine.

The concepts of phase shift analysis can, of course, be extended in various ways. It is possible to attempt an analysis of several open channels simultaneously to find the elements of the reaction matrix. Such studies carried out at energies just above the lowest inelastic thresholds could yield very detailed information to confront our theoretical ideas.

References

Andrick, D. (1969). Dissertation, Freiburg.
Andrick, D., and Bitsch, A. (1971). *Abstracts 7th IPEC Conference*, 1, 7.
Arndt, R. A., and MacGregor, I. (1967). In *Methods of Computational Physics*, Vol. 6. New York: Academic Press.
Atkinson, D., Johnson, P. W., and Warnock, R. L. (1973a). *Comm. Math. Phys.*, 33, 221.

Atkinson, D., Maekebeke, M., and de Roo, M. (1973c). Groningen Preprint.
Atkinson, D., Mahoux, G., and Yndurain, F. J. (1973b). *Nucl. Phys.*, **B54**, 263.
Atkinson, D., Johnson, P. W., Mehta, N., and de Roo, M. (1973d). *Nucl. Phys.*, **55**, 125.
Bederson, B., and Kieffer, L. J. (1972). *Rev. Mod. Phys.*, **43**, 601.
Berends, F. A., and Ruijsenaars, S. N. M. (1973a). *Nucl. Phys.*, **B56**, 507.
Berends, F. A., and Ruijsenaars, S. N. M. (1973b). *Nucl. Phys.*, **B56**, 525.
Bitsch, A. (1972). Thesis, Trier–Kaiserslӓutern.
Bransden, B. H. (1970). *Atomic Collision Theory*. New York: Benjamin.
Bransden, B. H., Hutt, P. K., and Winters, K. H. (1974). *J. Phys. B*, **7**, L129.
Bransden, B. H., and McDowell, M. R. C. (1969). *J. Phys. B.*, **2**, 1187.
Bransden, B. H., and McDowell, M. R. C. (1970). *J. Phys. B.*, **3**, 29.
Bransden, B. H., and Moorhouse, R. G. (1973). *The Pion–Nucleon System*. Princeton: Princeton University Press.
Brode, R. B. (1929). *Proc. Phys. Soc.*, **A125**, 134.
Canter, K. F., Coleman, P. G., Griffith, T. V., and Heyland, G. R. (1973). *J. Phys. B.*, **6**, L201.
Coleman, P. G., Griffiths, T. C., Heyland, G. R., and Killeen, T. L. (1974). Contribution to this volume.
Crichton, J. H. (1966). *Nuovo Cimento*, **45A**, 256.
Deichsel, H., Reichert, E., and Steidl, H. (1966). *J. Phys. B.*, **6**, 2280.
de Heer, F. J., Wingerden, B. V., Jansen, R. H. J., and Los, J. (1974). Contributions to this volume.
Dehmer, J. L., Inokuti, M., and Saxon, R. P. (1974). *Phys. Rev.*, in press.
Duweke, M., Kirchner, N., and Reichert, E. (1974). To be published.
Duxler, W. M., Poe, R. T., and LaBahn, R. W. (1971). *Phys. Rev.*, A **4**, 1935.
Fletcher, R. (1969). *Optimization*. New York: Academic Press.
Furness, J. B., and McCarthy, I. E. (1973). *J. Phys. B.*, **6**, 2280.
Gerjuoy, E., and Krall, N. A. (1960). *Phys. Rev.*, **119**, 705.
Gerjuoy, E., and Krall, N. A. (1962). *Phys. Rev.*, **127**, 2105.
Glatzel, M. (1973). Thesis, Physics Institute, Mainz.
Golden, D. E., and Bandel, H. W. (1965). *Phys. Rev.*, **138**, A14.
Hutt, P. K. and Bransden, B. H. (1974). *J. Phys. B.*, **7**, 2223.
Inokuti, M., and McDowell, M. R. C. (1974). *J. Phys. B.*, **7**, 2382.
Johnson, C. H., and Steiner, H. M. (1967). Radiation Lab. Reports 18001, University of California.
Jones, T. J. (1928). *Phys. Rev.*, **32**, 459.
Knowles, M., and McDowell, M. R. C. (1973). *J. Phys. B.*, **6**, 300.
Künzi, H. P., Tzschach, H. G., and Zender, C. A. (1968). *Numerical Methods of Mathematical Optimization*. New York and London: Academic Press.
Martin, A. (1969). *Nuovo Cim.*, **45A**, 131.
McConkey, T. W., and Preston, J. A. (1973). *Abstracts VIII IPEC Conference*, **1**, 273.
McDowell, M. R. C. (1971a). *J. Phys. B.*, **4**, 1649.
McDowell, M. R. C. (1971b). In *Atomic Physics 2*, Ed. P. G. H. Sandars. New York: Plenum Press.
McDowell, M. R. C. (1971c). *J. Phys. B.*, **4**, 1649.
McDowell, M. R. C. (1974). In *Comments on Atomic and Molecular Physics*, in press.
Naccache, P. F., and McDowell, M. R. C. (1973). *Compt. Phys. Com.*, **6**, 77.
Naccache, P. F., and McDowell, M. R. C. (1974). *J. Phys. B.*, **7**, 2203.
Naccache, P. F., and McDowell, M. R. C. (1975). *J. Phys. B.*, to be submitted.
Normand, C. E. (1930). *Phys. Rev.*, **35**, 1217.

Powell, M. J. A., and Fletcher, R. (1964). *Compt. J.*, **6**, 163.
Ramsauer, A., and Kollath, R. (1932). *Ann. Phys. L.*, **12**, 529.
Robertson, A. G. (1972). *J. Phys. B.*, **5**, 648.
Sin Fai Lam, A. L., and Nesbit, R. K. (1972). *Phys. Rev.*, A **6**, 2118.
Spruch, L., and Rosenberg, L. (1960). *J. Appl. Phys.*, **31**, 2104.
Salop, A., and Nakano, H. J. (1970). *Phys. Rev.*, A **2**, 127.
Tortorella, M. (1972). *J. Math. Phys.*, **13**, 1764.
Walker, D. W. (1969). *J. Phys. B.*, **2**, 356.
Walker, D. W. (1970). *J. Phys. B.*, **3**, 788.
Walker, D. W. (1971). *Adv. Phys.*, **20**, 251.
Yarlagadda, B. S., Csanak, G., Taylor, H. S., Schneider, B., and Yaris, R. (1963). *Phys. Rev.*, A **7**, 146.

15
Corrections for Forward Scattering to Positron–Helium Total Cross Section Measurements

P. G. Coleman, T. C. Griffith, G. R. Heyland, and T. L. Killeen

> In the cross section data for positron collisions with atoms and molecules published by Canter et al. (1972, 1973, 1974) no attempt was made to allow for forward scattering and forward inelastic collisions. At energies above 50 eV the measured cross sections were, therefore, too low. A method of deducing the true cross sections over the entire energy range 2–400 eV has now been developed and the corrected cross sections for positron–helium scattering are given.

The development of an efficient low-energy positron source in recent years has made possible the measurement of total scattering cross sections for positrons in gases. In the authors' time-of-flight system (Coleman et al., 1973) positrons emerging from such a source, with an energy distribution peaked at 1 eV with f.w.h.m. of 1 eV, are accelerated to any required energy and timed along a 0.9-m flight path. A resulting time spectrum exhibits a peak corresponding to the positrons, and measurement of its attenuation on leaking gas into the flight tube yields a value for the total cross section (Canter et al., 1972, 1973, 1974; Coleman et al., 1974).

The time width of such peaks is determined not only by the positrons' energy distribution and angular spread, but also by an inherent time delay in the emission of the slow positrons from the source moderator of up to 8 ns; this is the width of all the peaks above 100 eV, where the energy and angle factors become unimportant.

P. G. Coleman, T. C. Griffith, G. R. Heyland, and T. L. Killeen • Department of Physics and Astronomy, University College London, Gower St., London WC1E 6BT, England.

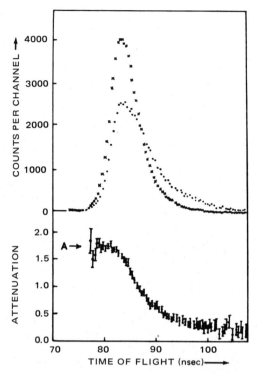

Figure 1. *Top*: Time Spectrum for 300 eV positrons in vacuum and in 5×10^{-3} torr He. *Bottom*: Attenuation plotted channel by channel.

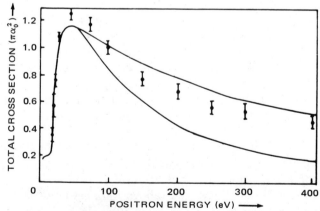

Figure 2. Results of measurements by Canter *et al.* (bottom curve) and theoretical values (Bransden *et al.*, 1974; top curve). Experimental points are corrected values.

Positrons scattered into the forward hemisphere, including those which have suffered inelastic energy loss, are kept to a helical path by a high magnetic field. Their measured time of flight may therefore be only a little longer than that of an unscattered positron, and may lie within the peak limits chosen for cross section evaluation. Consequently such forward scattering is not measured, and the resulting cross section value is too low, as already discussed by the authors (Canter et al., 1973, 1974).

It has recently been suggested (Bransden et al., 1974) that the theoretically expected total positron–helium cross sections should begin to diverge from the published experimental results above 50 eV, until at 400 eV they should approach those given by the Born approximation—almost three times greater than the experimental results. In response to this suggestion the attenuation of the positron beam over the entire energy range has been remeasured. The peaks were recorded in many more channels than previously, so that any distortion could be observed in detail.

Figure 1 shows a time spectrum for 300 eV positrons, after background subtraction, in vacuum and in 5×10^{-3} torr of helium. The additional counts at longer times in the gas peak correspond to positrons scattered, either elastically or inelastically, into the forward hemisphere. In the treatment of the data it is assumed that as one moves across the peak from right to left, i.e., toward the shorter time intervals, progressively fewer positrons can be scattered into each channel. Thus the attenuation should approach the true value near the left-hand edge of the peak. This is illustrated in the lower graph in Figure 1, where the attenuation is plotted channel by channel. The corrected value for the cross section is calculated from the asymptotic value A, using a simple fitting program incorporating a χ^2 test on A.

The results of a series of such measurements are shown in Figure 2. At positron energies below 50 eV no distortion of the peak was observed, and therefore no corrections to the previously published values are necessary. At 50 eV, however, the correction was found to be 15%, rising to 200% at 400 eV. This suggests that the dominant scattering processes may become progressively more forward-peaked with increasing positron energy.

Acknowledgments

The authors are grateful to Professor B. H. Bransden for valuable discussions. This work was supported by the Science Research Council.

References

Bransden, B. H., Hutt, P. K., and Winters, K. H. (1974). *J. Phys. B.*, **7**, L129.
Canter, K. F., Coleman, P. G., Griffith, T. C., and Heyland, G. R. (1972). *J. Phys. B.*, **5**, L167.

Canter, K. F., Coleman, P. G., Griffith, T. C., and Heyland, G. R. (1973). *J. Phys. B.*, **6**, L201.
Canter, K. F., Coleman, P. G., Griffith, T. C., and Heyland, G. R. (1974). *Appl. Phys.*, **3**, 249.
Coleman, P. G., Griffith, T. C., and Heyland, G. R. (1973). *Proc. Roy. Soc. Lond.*, A. **331**, 561.
Coleman, P. G., Griffith, T. C., and Heyland, G. R. (1974). *Appl. Phys.*, **4**, 89.

16

Testing of Classical and Quantum-Mechanical Criteria for Elastic Scattering of Electrons by Noble Gases

B. VAN WINGERDEN, F. J. DE HEER,
R. H. J. JANSEN, AND J. LOS

> Our experimental data for differential elastic scattering of 100–3000 eV electrons by He, Ne, and Ar are analyzed on the basis of classical and quantum-mechanical criteria. For He, the differential cross section $\sigma(\theta)$ is a universal function of momentum transfer K in a large angular and energy region, showing that there the Born approximation is valid. In contrast, such a behavior is not found for Ne and Ar. However, for these targets a universal relation is found between $\sigma(\theta)\sin(\theta)/E$ and $E\theta$ which corresponds to a classical approximation. The ranges of E and θ for which the criteria predict the universal relations to be valid are consistent with our experiment.

There exist several ways of analyzing data from elastic scattering experiments. In this paper we apply, in the region where the Born approximation is not valid, a classical model to the scattering of particles by a center of force using an inverse power potential. The experimental data are remarkably consistent with the simple criteria that apply to this classical analysis.

Jansen et al. (1975) have measured differential cross sections for elastic scattering of electrons by He, Ne, and Ar at energies between 100 and 3000 eV and angles between 5° and 55° with an angular resolution of about 0.5° and a reproducibility of 10′. The usual way of analyzing experimental

---

B. VAN WINGERDEN, F. J. DE HEER, R. H. J. JANSEN, and J. LOS • FOM—Instituut voor Atoom- en Molecuulfysica, Amsterdam-Wgm, Netherlands.

scattering results is by plotting the differential cross section $d\sigma/d\Omega$ versus momentum transfer K. When the Born approximation is valid, this method should yield a universal curve for each gas, i.e., all experimental points should line on the same curve, independent of the impact energy. Jansen et al. (1975) indeed find such a universal curve for He except for the low-energy and small-angle results, where polarization and exchange are important. In the case of Ne and Ar, however, $d\sigma/d\Omega$ cannot be represented by such a curve. For all except very small K values $d\sigma/d\Omega$ increases with higher impact energy E in contrast to He. Calculated Born cross sections for Ne and Ar fall above all their experimental data.

To understand this behavior we applied the criterion for (quantum-mechanical) Born scattering. Using a screened Coulomb potential $V(r) = (Ze^2/r)\exp(-r/a)$, where $a = a_0/Z^{1/3}$ is the screening parameter and Z the atomic number, the criterion for the validity of the Born approximation is given by Merzbacher (1970):

$$I \equiv (\mu Z e^2/k\hbar^2)\sqrt{\{\ln(1 + 4k^2a^2)^{1/2}\}^2 + \{\arctan 2ka\}^2} \ll 1 \quad (1)$$

where μ is the reduced mass (here the electron mass) and k the momentum of the incident electron. In Table 1 we given I_{min} for the experimental range mentioned above. Criterion (1) directly explains the disagreement between the Born approximation and the experimental results on Ne and Ar.

The criteria for classical scattering directly follow from the Heisenberg uncertainty principle (Mott and Massey, 1965), applied to the wavelength λ of the incoming electron [Eq. (2)] and the scattering angle θ [Eq. (3)]:

$$\lambda \ll a \quad (2)$$

$$\theta \gg \lambda/a \quad (3)$$

Approximating θ by $V(a)/E$, where $V(r)$ is the interaction potential with effective range a and E the impact energy, and substituting $V(a) = Ze^2/a$, relation (3) leads to

$$\mu Z e^2/k\hbar^2 \gg 1 \quad (4)$$

From Eqs. (2) and (4) we calculated the minimum and maximum energies

Table 1

Gas	I_{min}	E_{min}(eV)	E_{max}(eV)
He	0.47	22	54
Ne	2.04	63	1360
Ar	3.45	93	4400

Table 2

E (eV)	θ_{min}(rad)		
	He	Ne	Ar
3000	0.08	0.15	0.18
1000	0.15	0.25	0.31
500	0.21	0.36	0.43
100	0.46	0.80	0.97

(see Table 1) and from Eq. (3) the minimum angles (see Table 2) for classical scattering.

Smith *et al.* (1966) have shown that in a classical approach, using a forward impact expansion, there exists a universal relation between $\sigma(\theta) \sin(\theta)/E$ and $E\theta$. In this way the results of Jansen *et al.* (1975) for He, Ne, and Ar are given in a double logarithmic plot (see Figures 1, 2, and 3). From these plots we see that the existence and limitations of this universal relation are in good agreement with the E_{min} and E_{max} from Table 1. A comparison of Figures 2 and 3 with Table 2 shows that only in the angular range where the classical theory is valid do the experimental data fit to the universal relation of Smith *et al.* (1966). From Figures 2 and 3, one sees that in a large range of $E\theta$ the universal relation is linear. This can be described by an inverse power potential $V(r) = Ar^{-s}$, where A is a constant. The slope of the straight line is correlated to the parameter s. For both Ne and Ar we

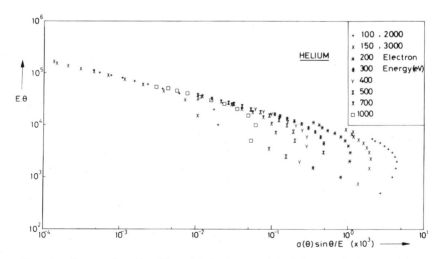

Figure 1. Cross sections for differential elastic scattering of electrons by He, plotted as $\sigma(\theta) \sin(\theta)/E$ versus $E\theta$.

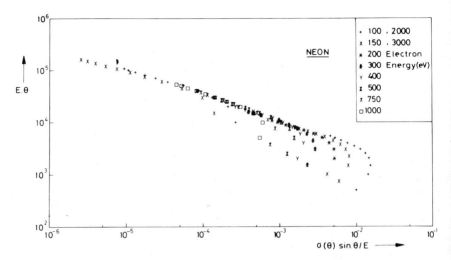

Figure 2. Cross sections for differential elastic scattering of electrons by Ne, plotted as $\sigma(\theta)\sin(\theta)/E$ versus $E\theta$.

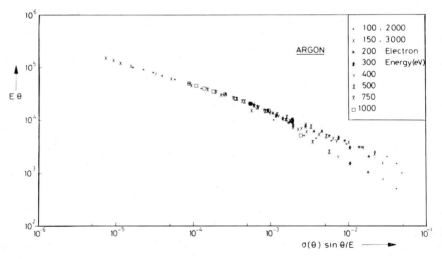

Figure 3. Cross sections for differential elastic scattering of electrons by Ar, plotted as $\sigma(\theta)\sin(\theta)/E$ versus $E\theta$.

find $s = 1.3$. The experimental fact that $d\sigma/d\Omega$ vs. K increases with the impact energy corresponds, in this classical picture, with an inverse power potential with $s > 1$.

Applying the inverse power potential, the distance of closest approach r_{min} can be calculated for the universal curve in the experimental angular range. Using Eqs. (5) and (6),

$$1 - V(r_{min})/E - b^2/r_{min}^2 = 0 \tag{5}$$

where b is the impact parameter, and

$$\sigma(\theta) = b^2/s\theta^2 \tag{6}$$

which is valid for sufficiently small angles, we find that r_{min} varies from $0.3a_0$ to $2.6a_0$ for Ne and from $0.25a_0$ to $3.2a_0$ for Ar. This indicates that, especially for high-energy and large-angle scattering, the screening effect of the atomic electrons is small.

Acknowledgment

This work was sponsored by F.O.M. with financial support by Z.W.O.

References

Jansen, R. H. J., de Heer, F. J., Luyken, H. J., van Wingerden, B., and Blaauw, H. J. (1975). *J. Phys. B.*, in press.

Merzbacher, E. (1970). *Quantum Mechanics*, 2nd ed. New York: Wiley.

Mott, N. F., and Massey, H. S. W. (1965). *The Theory of Atomic Collisions*. Oxford: Clarendon Press.

Smith, F. T., Marchi, R. P., and Dedrick, K. G. (1966). *Phys. Rev.*, **150**, 79.

17
Spin Polarization in Electron–Atom Scattering

BENJAMIN BEDERSON AND THOMAS M. MILLER

> A summary of recent work at New York University on the scattering of low-energy electrons by alkali atoms with spin analysis is presented. In particular, new measurements of total cross sections for the scattering of low-energy electrons by lithium in the energy range of 0.6 to 10 eV and differential elastic spin exchange cross sections for sodium at 0.6 eV are presented and compared to recent calculations. In addition, the problem of superelastic scattering from laser-excited hyperfine states for the $3(P)$–$3(S)$ transitions in sodium with spin analysis are discussed. It is shown how such measurements, performed using the recoil technique, are equivalent to the time-reversed double-coincidence experiment with polarized beams, and can be used to obtain magnitudes and phases of the direct and exchange scattering amplitudes.

The first part of this paper will be an updating of the review presented at the Third International Conference on Atomic Physics (Bederson, 1972). There we discussed the then current state of affairs in low-energy electron scattering by the alkalies, including total, momentum-transfer, differential elastic, differential and total elastic spin exchange, and some differential n^2S–n^2P cross sections, with and without spin analysis. At that time the situation, taking experiment and theory together, was beginning to shape up very nicely. In fact, with the exception of some spin-exchange data at thermal energies from optical pumping experiments (Balling, 1966), theory and experiment appeared to be in excellent harmony, at least for potassium. Examples of this agreement are shown in Figure 1, which is a summary of the elastic differential spin-exchange cross sections at several energies below the first excitation threshold for potassium, from the work of Collins et al. (1971).

BENJAMIN BEDERSON and THOMAS M. MILLER • New York University, 4 Washington Place, New York, New York 10003, U.S.A. T. M. Miller's present address is Stanford Research Institute, Menlo Park, California 94025, U.S.A.

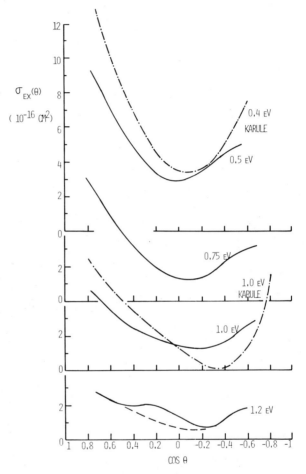

Figure 1. A comparison of experimental and theoretical differential spin-exchange cross sections for electron–potassium scattering. The two-state close-coupling results of Karule (1972) (dot-dashed curve) are given for 0.4 and 1.0 eV. The solid curves are the measurements of Collins et al. (1971). At 1.2 eV the experimental cross section contains some inelastic contributions, and the dashed line is an estimate of the elastic part.

These data are actually obtained as ratios of the differential exchange to "full" differential cross sections, which are measured independently. The curves show the resulting spin-exchange cross sections, obtained by comparing the measured ratio to the measured differential cross section at each energy and angle. These spin-exchange data were taken using the "recoil" technique (e.g., Bederson and Fite, 1968). Briefly, a spin-polarized and velocity-selected alkali beam is cross-fired at right angles by a low-energy electron beam. The recoil-scattered atoms are collected by a "hot-wire" detector which moves in an arc about the scattering center. In line with the rotatable detector is an "$E-H$" gradient balance magnet (Bederson et al., 1960), which possesses the property of being able to selectively pass a single hyperfine state without distortion of the atomic trajectories, while rejecting others, provided that it possesses a unique negative effective magnetic moment. In the alkali ground states, in "high" magnetic fields, this is equivalent to transmitting the $M_J = -\frac{1}{2}$ spin state, while rejecting the $M_J = +\frac{1}{2}$ state. In our experiment the polarizer and analyzer magnets are usually set for extinction, that is, they transmit opposite states, so that with the analyzer operative, only those atoms which have undergone a change of spin state while being scattered will be detected. Accordingly, we measure the "full" differential cross section with the analyzer inoperative, and the differential exchange cross section with the analyzer operative (Figure 2).

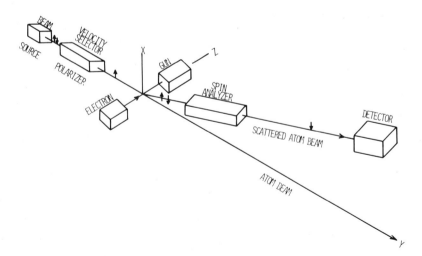

Figure 2. A block diagram of the atomic beam recoil apparatus (Collins et al., 1971; Goldstein et al., 1972; Rubin et al., 1969) used to study electron–alkali collisions with spin analysis of the scattered atom beam. The spin analyzer passes only those atoms which have changed their spin in the interaction region.

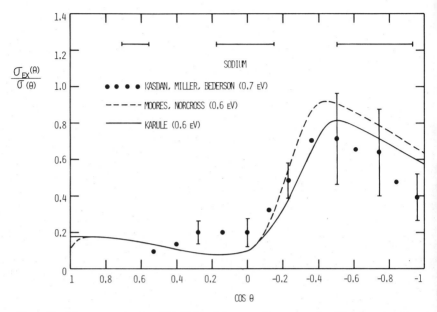

Figure 3. Measurements of $\sigma_{ex}(\theta)/\sigma(\theta)$ for electron–sodium scattering. The close-coupling results are two-state (Karule) and four-state (Moores, Norcross).

Figure 3 shows more recent data taken using sodium. Here the ratio $\sigma_{ex}(\theta)/\sigma(\theta)$ is plotted vs. $\cos\theta$ and compared to close-coupling calculations (Moores and Norcross, 1972).* Again the agreement with theory is seen to be very good. The overall agreement between close-coupling theory and experiment is even better shown in the total cross section for sodium (Figure 4), where the theory curve includes the elastic ($2S$–$3S$) and inelastic ($3S$–$3P$), ($3S$–$4S$), and ($3S$–$3D$) excitations. Other recent experiments, particularly the differential measurements of Andrick *et al.* (1972) and Gehenn and Wilmers (1971) show similarly excellent agreement.

Recently we have been trying to extend our beam measurements to lithium. Lithium presents an interesting challenge to the experimentalist. This is another way of saying that lithium is a beast, more difficult to form in a beam and to detect than the other alkalies. Thus we have not yet been as successful with lithium as with sodium and potassium, although our experimental difficulties have been mostly overcome, and better data are now being obtained. We do have what we believe is a reliable total cross section curve

---

* Sinfailam and Nesbet (1973) have recently performed an elaborate set of calculations for the light alkalies using the Bethe–Goldstone approximation. For sodium and lithium there is excellent agreement with the close-coupling calculations of Moores and Norcross.

Spin Polarization in Electron–Atom Scattering

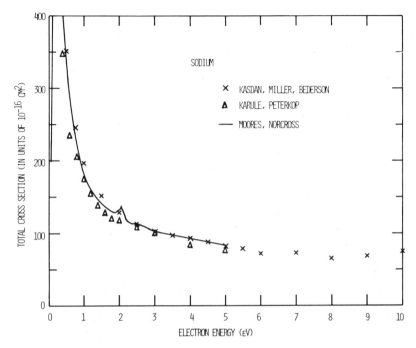

Figure 4. Measurements of total cross sections for electron scattering by sodium (see Kasdan *et al.*, 1973).

for low-energy electrons on lithium (Figure 5), and preliminary data (this is one of several runs) have been obtained for $\sigma_{ex}(\theta)/\sigma(\theta)$ (Figure 6). Although we have exhaustively investigated possible sources of systematic errors, the measured total cross sections stubbornly persist at about 20% above close-coupling theory. The measured values agree well with a semiempirical calculation by Inokuti and McDowell (1975), however. The main source of possible systematic error is uncertainty in knowledge of the average velocity, but we are confident that this has been determined experimentally for our lithium beam to better than 5%. The preliminary spin-exchange data for lithium show a similar disagreement with close-coupling theory. These data, it should be noted, are relative, and, accordingly, most systematic errors attributable to uncertainties in beam and apparatus parameters cancel in the determination. However, we have not completely eliminated the possibility of a systematic error in determining the velocity distribution of the lithium beam.

No beam data exist for rubidium and cesium elastic spin exchange. The total cross sections for these alkalies, measured by Rubin and co-workers (Visconti *et al.*, 1971), agree in general shape with the recent close-coupling

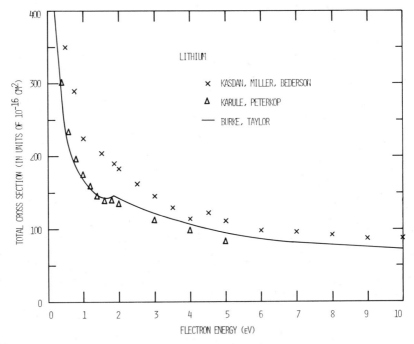

Figure 5. Measurements of total cross sections for electron scattering by lithium. The measurements extend to higher energies (50 eV) where they are considerably above the close-coupling results of Burke and Taylor (1969) but in agreement with semi-empirical calculations of Inokuti and McDowell (1974); inelastic effects dominate at 50 eV.

calculations of Burke and Mitchell (1973, 1974), although there is a remaining difference of about 15% in the absolute values, with the experimental values lying below the theoretical values. The cusp structure at the first excitation threshold, as calculated by Burke and Mitchell, is confirmed by some recent unpublished work by Robeson and Rubin.*

Thus, questions are now being raised concerning the differences between theory and experiment, ranging through the alkalies from lithium to cesium. Are nonrelativistic, few-state close-coupling approximations equally accurate for all of the alkalies? In the heavy alkali case, there could be problems

---

* It should be noted that there is a conversion error in the Visconti data as plotted in by Burke and Mitchell (1973, 1974). Making this correction, and also including an estimated correction for angular resolution not included in the published curve, gives an experimental curve which lies about 15% below the Burke and Mitchell computation. We thank Dr. Rubin for calling these corrections to our attention. We also thank Dr. Rubin for informing us of the results of his recent unpublished work with cesium.

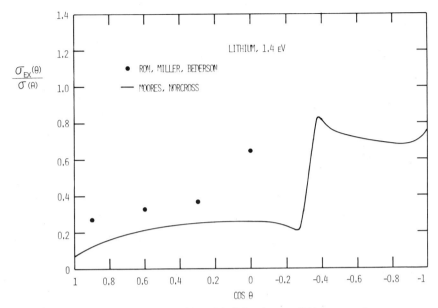

Figure 6. Measurements of $\sigma_{ex}(\theta)/\sigma(\theta)$ for electron–lithium scattering.

associated with obtaining good target wave functions (although for cesium the static electric-dipole polarizability obtained from Norcross's semi-empirical wave function is in excellent agreement with experiment; these wave functions were also used by Burke and Mitchell). On the other hand, for lithium there may be some convergence problems in the close-coupling expansion because of the lack of a sufficiently well established core.

The validity of nonrelativistic computations for the alkalies is currently being questioned. Evidence in the lighter alkalies indicates that these effects are small, although not necessarily completely negligible. For example, Wilmers (1971) has studied the spin–orbit interaction at 5 eV in potassium by determining the polarization produced by elastic scattering of unpolarized electrons as a function of scattering angle. On average the polarization effect is very small, although an effect of about 8% is seen at two angles at which the differential cross section goes through minima (40° and 140°). This effect is not large enough to upset the overall agreements between theory and experiment in potassium. A paper by Walker (1975) contributed to this volume (see also Carse and Walker, 1973) considers a complete relativistic treatment of the electron–cesium problem, including spin–orbit, spin–spin, and relativistic (Mott scattering) interactions. He concludes that these contribute substantially to, for example, spin exchange, and obtains substantially different values for these cross sections than those obtained in the recent nonrelativistic close-coupling calculation of Burke and Mitchell (1973, 1974). Clearly, more work remains to be performed on elastic spin exchange in the heavier alkalies.

With respect to polarized beam experiments in inelastic collisions, no new data have been published for the alkalies since our last report. At that time, the differential "spin-flip" cross section was reported for potassium (Goldstein *et al.*, 1972). In this work a polarized atomic beam is excited by unpolarized electrons, and the final spin state of the scattered atom after decay to the ground state is measured. The spin-flip cross section refers to the cross section for which this process results in a change of spin state. Differential spin-flip data were obtained over a range of energies and angles. These data agreed only qualitatively with the two-state close-coupling calculations. And, in fact, one would be surprised, were such computations based upon a few-state expansion to give good results for spin-flip cross sections, which are more sensitive to small variations in the scattering amplitudes than is the full differential cross section.

It should be noted that Hanne and Kessler (1975) have been successful in measuring depolarization of polarized electrons used in the triplet excitation of mercury. Assuming pure LS coupling, such an excitation can only proceed by spin exchange; but the data indicate some deviation from pure exchange excitation of the triplet. We believe that this is the first experiment in low-energy atomic physics which actually uses a beam of polarized electrons to perform a scattering experiment.

Having presented this brief summary, we would now like to offer a more general discussion of excited-state collision experiments. Hertel and Stoll (1973, 1974) have recently demonstrated the feasibility of using a laser to selectively populate a given excited state (the $3^2P_{1/2,3/2}$ states) in sodium and to observe the superelastic cross section $3P \to 3S$, as well as several of the inelastic cross sections, e.g., $3P \to 4D$. Hertel and Stoll (1973) and Macek and Hertel (1974) have also published analyses of these experiments, from the point of view of general angular-momentum theory. The latter paper is presented in the language of state multipoles and the former in terms of more conventional vector-coupling language. In these analyses, of course, one first obtains an expression for the excitation or deexcitation, starting from a specific quantum state and ending in another specific quantum state; one must then sum over all quantum numbers which are not observables in the experiment. Moores *et al.* (1974) have already extended close-coupling calculations to elastic and superelastic (including spin-flip) collisions with the first excited state of sodium.

As Macek and Hertel point out, such experiments are fully equivalent to appropriate time-reversed coincident experiments, time-independent scattering theory, of course, always describing a complementary pair of collision problems, each being the time and space reverse of the other (Shakeshaft et al., 1973). Thus, even without spin preparation or analysis, a superelastic collision experiment prepared by optically pumping an alkali with a single-mode tunable dye laser will yield information not ordinarily obtainable in a scattering experiment. Specifically, the phase difference Δ between f^1, f^0 can be obtained, where f^1, f^0 are the excitation amplitudes with $\Delta M_l = \pm 1, 0$ respectively, provided exchange, spin–orbit, and relativistic interactions can be neglected (of course, for very low energies in sodium, the latter two effects are small, although exchange is not). Thus, such experiments are closely related to the coincidence measurements of Kleinpoppen and co-workers (Eminyan et al., 1973).

Here we would like to refer back to a related analysis performed by Rubin et al. (1969) in connection with the $4S \rightarrow 4P$ excitation in potassium with spin analysis. We showed how to relate the various direct and exchange scattering amplitudes to the possible observables of a scattering experiment. Integrating over photon angles and polarization and electron spin polarization we obtained an expression for $R(\theta) = \sigma_{\text{spin-flip}}(\theta)/\sigma(\theta)$, where $\sigma_{\text{spin-flip}}(\theta)$ and $\sigma(\theta)$ are the spin-flip and "full" differential cross sections. At the time, a more complete experiment involving double coincidence with spin analysis seemed a long distance in the future. However, the arrival of the tunable dye laser has changed that, and such a "perfect" scattering experiment is now almost completely feasible (with the exception of the use of polarized electrons).

Rubin et al. (1969) showed that the relative probability P that an electron is inelastically scattered into (θ, ϕ) while a delayed photon of linear polarization $\hat{e}$ is emitted into angles (A, B) and the atom, originally in the ground state with spin quantum number m_s, ends up with spin quantum number M_s is given by

$$P \sim \sum_{j,\mu} C_a^2 C_b^2 |f^\mu|^2 [(\delta_{a,\pm 1}\tfrac{1}{2}\cos^2 A + \delta_{a,0}\sin^2 A)_\| + (\tfrac{1}{2}\delta_{a,\pm 1})_\perp]$$

$$+ \sum_{\substack{jj' \\ \mu \neq \mu'}} C_a C_{a'} C_b C_{b'} |f^\mu\|f^{\mu'}|\,\text{Re}\left\{\frac{1-i\eta}{1+\eta^2} e^{-i\Delta}[(e^{2i(B-\varphi)}\delta_{a,\pm 1}\delta_{b,\pm 2}\tfrac{1}{2}\cos^2 A \right.$$

$$\left. - e^{\pm i(B-\varphi)}\delta_{a,\pm 1}\delta_{b,\pm 1}\sin A \cos A)_\| - (\tfrac{1}{2}e^{2i(B-\varphi)}\delta_{c,1}\delta_{b,2})_\perp]\right\} \quad (1)$$

In Eq. (1): (i) We have ignored exchange; the inclusion of exchange introduces the exchange amplitudes $g^{1,0}$, for which Eq. (1) is easily generalized (see Rubin et al., 1969). This is not necessary here, for our present purposes. (ii) We have ignored hyperfine structure (hfs). Nuclear coupling certainly messes things up a bit. The best that can be said about hfs is that Eq. (1) can be generalized to include it, although no new information is thereby added to the problem from the viewpoint of collision theory. Working at "high" fields, where nuclear and atomic spins are decoupled in the ground state, effectively eliminates the second term in Eq. (1), as will be seen below. (iii) We have written the Clebsch–Gordan coefficients and Kronecker deltas in abbreviated form for the sake of clarity. These are defined as follows:

$$C_a = C(1, \tfrac{1}{2}, j; \mu, m_s, \mu + m_s)$$

$$C_b = (1, \tfrac{1}{2}, j; \mu + m_s - M_s, M_s, \mu + m_s)$$

$$\delta_{a,i} = \delta_{\mu + m_s - M_s, \pm i}$$

$$\delta_{b,i} = \delta_{\mu - \mu', i}$$

$$\delta_{c,i} = \delta_{\mu + m_s, i}$$

$\mu = z$ component of orbital angular momentum

$\parallel, \perp$ = parallel and perpendicular components of polarization vector

a', b' = same as a, b except with j', μ' rather than j, μ

θ, φ and A, B are the polar and azimuthal scattered electron and photon angles respectively (see Figure 7), and Δ is the phase difference between the scattering amplitudes f^0 and f^1; the axis of quantization lies along the applied magnetic field and the incident (or inverse scattered) electron. The second set of terms are interference terms, not normally observable in a scattering experiment, since φ and/or B are usually averaged out to zero. They represent the interference effects due to the fact that the atom is simul-

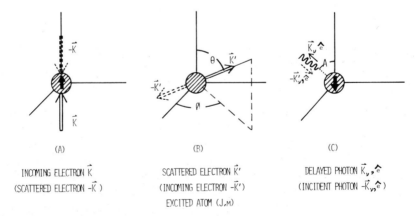

Figure 7. Scattering diagrams showing double differential electron-delayed photon excitation experiment (solid lines) and equivalent time-inversed superelastic experiment (dashed lines, parenthetic descriptions), both with spin analysis.

taneously decaying from two different states. These are the same terms appearing in level-crossing experiments. They are significant when the interference factor, η, is small ($\sim$ unity or less). η is defined as

$$\eta(j, m; j', m') = [E(j, m) - E(j'm')]\Gamma/\hbar$$

where $E(j, m)$ is the energy of the (j, m) state and Γ is the radiative lifetime ($\sim 10^{-8}$ s). η is large if $j \neq j'$, so that interference can only occur among fine-structure states of the same j. For a magnetic field of about 10 gauss $\eta \approx 1$ (hfs is still being neglected!); hence, interference can only be seen in "weak" fields. Thus, one can obtain $|f^1|$ and $|f^0|$ at "high" fields, and Δ in "weak" fields.

The recoil experiment in the time reverse mode, that is, using laser excitation, can be employed to obtain f^0 and f^1, including their phase difference, neglecting exchange. Inclusion of exchange yields five of the seven parameters characterizing $f^{0,1}$, $g^{0,1}$ (since polarized electrons are not being used). A complete analysis of this experiment, including effects of hfs, will be published elsewhere. Figure 7 shows the generalized coincidence experiment and its time-reverse equivalent. The laser excitation is in the X–Z plane ($\varphi = 0$). The electron polar scattering angles (θ, φ) are obtained from the equivalent atomic recoil angles (ψ, χ) (Shakeshaft et al., 1973). The spin states, before and after the time-inverted problem, are m_s, M_s; these are easily measured in the recoil experiment.

The superelastic collisions are readily distinguishable, since these result in recoil atoms scattered backwards, that is, opposite to the direction of the incident electrons. Thus, we see that the combination of the dye laser and the recoil experiment now permits experiments involving determination of combinations of the phases of direct and exchange scattering amplitudes. At high fields, where $\eta \gg 1$, or by integrating over all azimuth angles, it is easy to show that one can obtain $|g^0|^2, |g^1|^2, |f^0|^2, |f^1|^2$.

An experiment of this kind is now being performed by Drs. N. Bhaskar and B. Jadusliwer at New York University.

Acknowledgments

Work done at New York University described in this report has been supported by the U.S. Army Research Office, Durham, the Air Force Office of Scientific Research, and The National Science Foundation.

We thank the Center for Theoretical Studies, University of Miami, Coral Gables, Fla., under support by Army Research Grant DAHCO 4-71-C-0050 for the hospitality offered one of us (B.B.) while some of the ideas discussed in this paper were being developed.

References

Andrick, D., Eyb, M., and Hofmann, H. (1972). *J. Phys. B.*, **5**, L15.
Balling, L. C. (1966). *Phys. Rev.*, **151**, 1.
Bederson, B. (1972). In *Atomic Physics*, *3* (Proceedings of the Third International Conference on Atomic Physics), p. 401. New York: Plenum Press.
Bederson, B., Eisinger, J., Rubin, K., and Salop, A. (1960). *Rev. Sci. Inst.*, **31**, 852.
Bederson, B., and Fite, W. L. (Eds.) (1968). *Methods in Experimental Physics.* Vol. 7A, p. 89. New York: Academic Press.
Burke, P. G., and Mitchell, J. F. B. (1973). *J. Phys. B.*, **6**, L161.
Burke, P. G., and Mitchell, J. F. B. (1974). *J. Phys. B.*, **7**, 214.
Burke, P. G., and Taylor, A. J. (1969). *J. Phys. B.*, **2**, 869.
Carse, G. D., and Walker, D. W. (1973). *J. Phys. B.*, **6**, 2529.
Collins, R. E., Bederson, B., and Goldstein, M. (1971). *Phys. Rev.*, A **3**, 1976.
Eminyan, N., MacAdam, K., Slevin, J., and Kleinpoppen, H. (1973). In *Abstracts of 8th International Conference on Physics of Electronic and Atomic Collisions*, p. 318. Belgrade, Yugoslavia.
Gehenn, W., and Wilmers, M. (1971). *Z. Physik*, **244**, 395.
Goldstein, M., Kasdan, A., and Bederson, B. (1972). *Phys. Rev.*, A **5**, 660.
Hanne, G. F., and Kessler, J. (1975). Contribution to this volume.
Hertel, I. V., and Stoll, W. (1973). In *Abstracts of 8th International Conference on Physics of Electronic and Atomic Collisions*, p. 321. Belgrade, Yugoslavia.
Hertel, I. V., and Stoll, W. (1974). *J. Phys. B.*, **7**, 570.
Inokuti, M., and McDowell, M. R. C. (1974). *J. Phys. B.*, **7**, 238.
Karule, F. M. (1972). *J. Phys. B.*, **5**, 2051.
Kasdan, A., Miller, T. M., and Bederson, B. (1973). *Phys. Rev.*, A **8**, 1562.
Macek, J., and Hertel, I. V. (1974). *J. Phys. B.*, **7**, 2173.
Moores, D. L., and Norcross, D. W. (1972). *J. Phys. B.*, **4**, 1458.
Moores, D. L., Norcross, D. W., and Sheory, V. B. (1974). *J. Phys. B.*, **7**, 371.
Rubin, K., Bederson, B., Goldstein, M., and Collins, R. E. (1969). *Phys. Rev.*, **182**, 201.
Shakeshaft, R., Macek, J., and Gerjuoy, E. (1973). *J. Phys. B.*, **6**, 794.
Sinfailam, A. L., and Nesbet, R. K. (1973). *Phys. Rev.*, A **7**, 1987.
Visconti, P. J., Slevin, J. A., and Rubin, K. (1971). *Phys. Rev.*, A **3**, 1310.
Walker, D. W. (1975). Contribution to this volume.
Wilmers, M. (1971). *Messung der Spinpolarisation bei der elastischen Streuung langsamer Electronen an Kalium*, Ph.D. Dissertation, University of Mainz.

18

Conservation of Total Spin in Electron–Atom Collisions

D. W. Walker

> The interactions which can cause changes in total spin in electron–atom collisions (*spin–orbit, spin–spin, and spin–other-orbit interactions*) are investigated for hydrogenic systems and for cesium by solution of the Dirac equation in a distorted wave approximation which takes the interaction between the incident and valence electron as either the Breit or the Møller interaction. The results obtained indicate that spin–spin and spin–other-orbit interactions are important only at incident electron energies of 10 keV or more, but that spin–orbit interactions are important for heavy atoms even at very low energies.

Experiments of the "spin-exchange" type (Bederson, 1969) rely on the assumption that the total spin of a system consisting of an alkali atom and a colliding electron is conserved during the collision process, in order to deduce separate "direct" and "exchange" scattering amplitudes. This is a good approximation for light atoms, but for heavier systems, where spin–orbit coupling effects are known to be important, its validity is more doubtful.

A useful picture in which to discuss the dynamics of processes in which the total spin can change is the Breit–Pauli approximation (Bethe and Salpeter, 1957; Jones, 1975). This approximation can be used to consider a system of two interacting electrons in an electrostatic potential due to the nucleus (and also core electrons, if required). Relativistic effects are represented as a perturbation expansion in the fine-structure constant $\alpha = 1/137.04$ about the nonrelativistic solutions to the problem under discussion, and

---

D. W. Walker • Physics Department, University of Edinburgh, Edinburgh EH9 3JZ, Scotland. Present address: Research School of Chemistry, Australian National University, Canberra, A.C.T. 2600, Australia.

the lowest-order terms (in α^2) retained. Four of these terms (the "fine-structure" interactions) do not commute with the total spin S of the system:

(i) spin–orbit interaction (which, for clarity, we will refer to as core spin–orbit interaction), which is the coupling of the spin and orbital angular momenta of the same electron in the electrostatic field of the nucleus;
(ii) mutual spin–orbit interaction, which is similar to (i) but due to the electrostatic field of the other electron;
(iii) spin–other-orbit interaction, which is the coupling of the spin angular momentum of one electron with the orbital angular momentum of the other;
(iv) spin–spin interaction, which is the coupling of the spins of the two electrons.

All four of these interactions are essentially magnetic effects, but the first two arise from a field which an observer in the rest frame of the nucleus (which is the usual reference frame in electron–atom collisions) regards as electrostatic, but which the moving electrons regard as having magnetic components, while the third and fourth require explicit inclusion of magnetic interactions between the two electrons. In principle, all these effects are of equal importance, since they all appear to the same order in α in the perturbation expansion, but in practice, for low-energy electron–atom collisions, the core spin–orbit effect is likely to be much larger than the other three. This is because all four terms are dominated by their behavior at small separations of the particles, and since the electron–electron interaction is repulsive, small separations are unlikely except at very high energies, whereas the electron–nucleus interaction is attractive and becomes stronger with increasing atomic number.

To calculate the effect of these interactions in specific cases, it is more convenient to use the Dirac equation, with the interaction between the electrons given by either the Breit or the Møller interaction. (Both these interactions are first-order approximations to the retarded electromagnetic interaction, the Breit interaction containing the additional assumption of zero energy transfer. If the energy difference between the valence and continuum electrons is much smaller than the rest energy of the electron, 511 keV, then these two interactions may be regarded as equivalent.) In this treatment, the "one-body" relativistic effects, which include the core spin–orbit interaction, are included to all orders in α, while the "two-body" effects, including spin–other-orbit and spin–spin interactions, are included to order α^2. The mutual spin–orbit term is technically a two-body term, and hence included to order α^2, but part or all of it is often included to higher order, e.g., by including a screening effect due to the bound electron when calculating the continuum electron wave function. The necessary theory for this approach is given by Carse and Walker (1973) and the results of calculations using it by Walker (1974, 1975). Transition matrix elements may be calculated, in this case using a distorted wave approximation, and then used to find a set of scattering amplitudes $B_\tau^m(A \to A')$. m and τ refer to spin states of the incident and scattered beams, $m = +\frac{1}{2}$ indicating an incident beam polarized in the positive z direction, $m = -\frac{1}{2}$ in the $-z$ direction (the polarizations referring to the rest frame of the electron), while $\tau = \pm\frac{1}{2}$ refers to the same states in the scattered electron. The electrons are assumed to be incident in the positive z direction. The target states A are specified by the principal quantum number n, the orbital angular momentum l, the total angular momentum j, and its projection μ_A. If we restrict ourselves to s states in the target, then the quantum numbers μ_A reduce to spin projection quantum numbers m_s. In this case, the amplitude $B_\tau^m(\mu_A \to \mu'_A)$ is the same as the transfer matrix element $M_{\mu'_A \tau, \mu_A m}$ discussed by Wilmers (1972, 1973) and by Burke and Mitchell (1974).*

Nonrelativistically, there are three paradigm cases in the scattering of polarized electrons by polarized s-state atoms. These may be characterized in the following way:

(i) singlet initial state, $m = \frac{1}{2}, \mu_A = -\frac{1}{2}$;
   (a) direct scattering will give a final state $\tau = \frac{1}{2}, \mu_{A'} = -\frac{1}{2}$;
   (b) exchange scattering will give $\tau = -\frac{1}{2}, \mu_{A'} = \frac{1}{2}$;
(ii) triplet initial state, $m = \mu_A = \frac{1}{2}$, leads to a final state $\tau = \mu_{A'} = \frac{1}{2}$.

Hence, singlet scattering can be used to separate direct and exchange scattering processes if either the final target or the scattered electron polarizations are observed. This also means that, in nonrelativistic calculations, it is always possible to separate direct and exchange processes (in practice by taking appropriate combinations of singlet and triplet amplitudes), even if the formulation of the problem apparently does not permit this (e.g., as in a close-coupling calculation). In relativistic calculations, conservation of total spin is not assumed, so processes to final states other than those given above can occur. This means that direct and exchange amplitudes can no longer be separated. In the results given below, we do in fact give separate direct and exchange results, but this is only possible because the dynamic approximation used (the distorted wave approximation) involves calculation of direct and exchange matrix elements separately. The fact that we cannot separate direct and exchange processes also means that the triplet process [labeled (ii) above] is the most crucial test of total spin conservation, since

---
* M is Burke and Mitchell's notation; Wilmers uses S.

any final state with $\tau = -\frac{1}{2}$ or $\mu_{A'} = -\frac{1}{2}$ involves a change in S_z, irrespective of whether a direct or exchange process has occurred.

As our first example, we discuss the 1s–2s transition in hydrogenic ions. Calculations have been made (Walker, 1975) for nuclear charges $Z = 2$, 25, 50, and 100, at incident electron energies of $0.8E_I$, E_I, and $4E_I$, where E_I is the ionization energy of the 1s state.* The 1s–2s transition was chosen rather than the elastic 1s–1s scattering in order to avoid the interaction between the two electrons being swamped by the nuclear Coulomb scattering, which is many orders of magnitude larger for large Z. Since the energies involved are large for large Z ($E_I = 162$ keV for $Z = 100$), we should be able to see the effects of all four fine-structure interactions.

Earlier (Walker, 1974) we gave total cross sections for the processes in which the total spin S changed which were many orders of magnitude smaller than the cross sections for the nonrelativistically allowed processes. This is also true if the Coulomb interaction (which allows for core and mutual spin–orbit interactions only) is replaced by the Breit or Møller interaction, although, for $Z = 100$ at $E = 4E_I$ (0.65 MeV), the forbidden processes are as large as 0.03 of the permitted ones. The reason for this is given by Jones (1975), who shows that, provided one averages over any fine-structure states in the target, the lowest-order (α^2) contribution of all four fine-structure interactions to the total scattering cross section is zero in the Breit–Pauli approximation. Hence we are looking at higher-order (α^4 and above) contributions to the total cross sections, not all of which are included. For this reason, we concentrate on angular distributions.

In Figure 1 we plot the final target polarization

$$P_T = \frac{I(\frac{1}{2}) - I(-\frac{1}{2})}{I(\frac{1}{2}) + I(-\frac{1}{2})}$$

where $I(\mu_{A'})$ is the differential cross section for exciting a final state $\mu_{A'}$ from the specified initial state, and the z component (P_z) of the polarization of the scattered electron, for a system initially in a triplet state. In the forward and backward directions, $P_z = P_T = 1$. This occurs because in 0° and 180° scattering, the component of the orbital angular momentum along the incident beam (L_z) cannot change, and since the total angular momentum **J** is conserved (and hence J_z), S_z is also conserved. At other angles, P_T and P_z are not equal to one. The maximum deviation occurs for large Z (at $Z = 100$, P_T differs from 1 by up to 11%, while P_z goes negative, compared with $Z = 2$, where the maximum deviation is about 1 part in 10^7). Results from the Møller and Breit interactions are similar in magnitude, but if the Coulomb interac-

---

* Some preliminary results, in which the interaction between the two electrons was taken to be the Coulomb interaction, are given in an earlier publication (Walker, 1974).

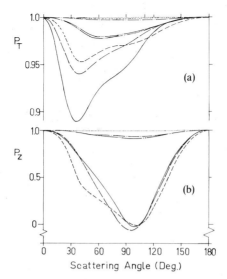

Figure 1. (a) Target polarization P_T, and (b) z component of the scattered electron polarization (P_z), as a function of electron scattering angle θ for triplet scattering from hydrogenic ions (1s–2s transition). $Z = 100$: (- - - -), Coulomb interaction; (— — —), Breit interaction; (———), Møller interaction; $Z = 50$: (— · —), Coulomb interaction; (— ·· —), Breit interaction (indistinguishable from Møller for P_z); (— ·· —), Møller interaction; $Z = 25$: (······), Coulomb interaction; (— —·— —), Breit and Møller interaction (indistinguishable from Coulomb for P_z).

tion alone is used, the deviations are much smaller, which indicates that spin–spin and spin–other-orbit interactions are important here. The deviations appear to increase slightly with increasing electron energy for a given Z. Values of P_T and P_z at the point of maximum deviation from 1 are given in Table 1 for the three energies considered.

Table 1. Minimum Values of Target Polarization (P_T) and Scattered Electron Polarization (P_z) for Triplet Scattering from Hydrogenic Ions after Excitation of the 1s–2s Transition, for Incident Electron Energies (E) of 0.8, 1, and 4 times the Ionization Energy (Møller Interaction)

	$Z = 25$			$Z = 50$			$Z = 100$	
E (keV)	P_T	P_z	E (keV)	P_T	P_z	E (keV)	P_T	P_z
6.9	0.9976	0.9930	28.2	0.9742	0.8719	129	0.9330	−0.4687
8.5	0.9979	0.9947	35.2	0.9779	0.9156	162	0.8893	−0.0261
34.3	0.9958	0.9870	141	0.9635	0.8407	642	0.8368	0.0110

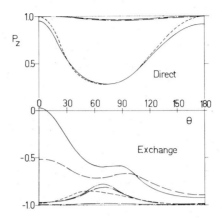

Figure 2. Scattered electron polarization component P_z for singlet scattering from hydrogenic ions. Other details as for Figure 1. If no curve is plotted for the Breit interaction, the curve is indistinguishable from the Møller interaction curve.

In Figure 2 we give values of P_z for singlet scattering, the direct processes being shown in the top half of the diagram and the exchange in the bottom half. The restriction on forward and backward angles is not so stringent in this case, the requirement (since $S_z = 0$) being $P_T = -P_z$. At most other angles, the target polarization is smaller than P_z, although this is not necessarily the case. This can be explained if we view the scattering process as an initial scattering in the potential field of the atom, followed by an interaction with the valence electron and then a further potential scattering. The spin changes arising from core spin–orbit coupling occur in the first and third stages of this process, but those which affect the target polarization are those which occur in the first stage only (there being no spin–orbit coupling in the target s states); hence, the target polarizations can be expected to be smaller than the scattered electron polarization.

The difference between the results obtained here and those expected nonrelativistically are larger in the singlet than in the triplet case. This may in part reflect the necessity of making additional approximations in order to separate direct and exchange processes. Even so, at $Z = 25$, P_T differs from the nonrelativistic values by at most 1%, and P_z by 2%, at $E = E_I$ (8.5 keV), and by 4% and 12% respectively at $E = 4E_I$ (34 keV).

Another way of looking at the breakdown of total spin conservation is to consider the ratio of the exchange to the direct cross sections for the "direct" singlet process leading to a final state $\mu_{A'} = -\frac{1}{2}$ (scattered electron spin not observed), and the direct/exchange ratio for the "exchange" process leading to $\mu_{A'} = \frac{1}{2}$. For $Z = 100$ these ratios are large, the maximum values (at $E = E_I$) being 0.52 (at 0°) for the first ratio and 0.31 (at 95°) for the second, but the proportion drops rapidly with decreasing energy and atomic number, the maxima at $E = E_I$ being 2.6×10^{-4} and 1.1×10^{-2} respectively for $Z = 25$, and 5×10^{-8} and 3×10^{-7} respectively for $Z = 2$.

As a second example, we consider elastic scattering from the ground states of hydrogen and cesium. The hydrogen calculation followed the same lines as those on the hydrogenic ions, but for cesium allowance had to be made for the core electrons. These were represented by a model potential, including provision for the polarizability of the core, which was adjusted to give a fit to the lowest energy levels of cesium. Magnetic interactions with the core electrons were ignored.

The calculations for cesium were made at 1.427, 5, 13.6, 35, and 100 eV, and those for hydrogen at 1.427 and 13.6 eV. The Breit interaction was used in all cases. The choice of 1.427 eV requires a comment. This energy is above the threshold for the $6s$–$6p_{1/2}$ transition for cesium, but below that for the $6s$–$6p_{3/2}$ transition. Since we have not considered coupling to either channel, this does not concern us. More important, it is also one of the energies chosen by Burke and Mitchell (1974) in their calculation of the polarization of the scattered electrons due to the splitting of the $6p$ excitation threshold due to spin–orbit coupling in the target. This calculation used nonrelativistic R-matrix elements, which were recoupled into j–j coupling, fitted as a function of energy, and then extrapolated to the correct threshold energies. In this way, spin–orbit coupling effects in the target were to some degree included, but the effect of spin–orbit coupling on the continuum electron completely neglected. This does not matter if total cross sections are calculated, since, as we have already pointed out, the fine-structure interactions give zero in this case, but Burke and Mitchell also gave a plot of electron spin polarization versus scattering angle, the peaks in which were 3% or less. In Figure 3 we give the final electron spin polarization (which is perpendicular to the scattering plane) for the scattering of an unpolarized beam by an unpolarized target atom, using our model. The dotted line is scattering by the atom regarded as a central potential (i.e., Mott scattering) and is virtually identical to the direct scattering obtained from the distorted wave approximation, while the solid curve is the full distorted wave result, including both direct and exchange matrix elements. The differential cross section for this process is included in the bottom half of the diagram. It is clear from these results that the exchange effect is large, which, combined with the neglect of coupling to the two $6p$ states, makes the actual numbers obtained here of doubtful reliability, but it is clear that Mott scattering (i.e., the core spin–orbit interaction of the continuum electron) will have a dominant effect on any polarization processes observed. In contrast, the maximum polarization obtained for hydrogen at this energy (on a 1°

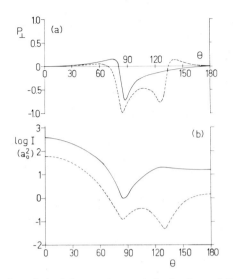

Figure 3. Scattering of unpolarized electrons by unpolarized cesium at 1.427 eV. (a) Polarization $P_\perp$ perpendicular to the scattering plane. (b) Differential cross section I (log scale). Solid line: full distorted wave approximation; dashed line: potential scattering.

tabulation interval) was 4×10^{-5}, which is also consistent with our knowledge of Mott scattering. The effect of spin–other-orbit and spin–spin interactions is negligible for both atoms, both at this energy and the others considered. For cesium at 1.427 eV, the matrix elements of the magnetic interactions are about 10^{-4} of the corresponding Coulomb matrix elements, and the effect on the cross sections and polarizations is of a similar size.

In Figure 4, we plot P_T and P_z for cesium at 1.427 eV. The sharp peaks, coinciding with the minima in the cross section, which are such a feature of Mott scattering, are much in evidence. It is also significant that the polarizations in the singlet case are very small at small angles and at 180°: it is characteristic of Mott scattering that there are no spin changes in 0° and 180° scattering. For hydrogen, the deviations in P_T and P_z from the nonrelativistic values are small, 1 part in 10^8 or less. In a similar way, the proportion of exchange scattering in the "direct" processes and direct scattering in the "exchange" processes is very small, except near the minima in the cross section, where the exchange/direct ratio for "direct" scattering from cesium approaches 1, compared with 10^{-5} at 0°.

The results for cesium at higher energies show an interesting effect. The scattered electron polarizations are large at all the energies investigated, but the target polarization P_T tends rapidly to the nonrelativistic value with increasing energy; e.g., for triplet scattering at 13.6 eV, the minimum values

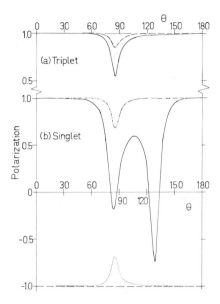

Figure 4. (a) Triplet scattering from cesium at 1.427 eV. (— — —), P_T; (———), P_z; (b) singlet scattering from cesium at 1.427 eV. Direct: (— — —), P_T; (———), P_z; exchange: (— · —), P_T; (· · · · · · ·), P_z.

of P_T are 0.9986 and 0.9966, corresponding to the two minima in the cross section, and at 35 eV and 100 eV, $P_T = 1$ to 1 part in 10^4 or better at all angles. This occurs because, in elastic scattering, interaction with the valence electron is no longer a necessary part of the process, and with increasing electron energy, and hence deeper penetration of the incident electron into the atom, it becomes much less important than the scattering by the nucleus and core electrons. This is exemplified by the reduction in the importance of exchange of the continuum electron with the valence electron: at 1.427 eV the exchange cross section (for an unpolarized beam on an unpolarized target) is ten times the direct one, while by 100 eV it is down to 0.3% of the direct cross section.

It seems clear from the results given here that the processes which cause changes in the total spin of a system consisting of an electron and an alkali-like atom or ion fall into two distinct groups:

(1) A set of processes which are important only at high energies (of the order of a few keV or more), and which can be identified with the magnetic (spin–other-orbit and spin–spin) interactions, and possibly also with the mutual spin–orbit interaction.
(2) An effect which is important in heavy atoms even at very low incident electron energies, which we can identify as being due to

the coupling of the spin and orbital angular momenta of the continuum electron in the field of the nucleus and core electrons.

The second effect is the only one which is likely to be of significance in spin-exchange experiments on alkali systems. This process has not been investigated in detail for the alkalis, and would require more sophisticated calculations than those made here in order to obtain accurate cross section and polarization data. However, the qualitative features can be deduced from the very extensive experimental (Kessler, 1969) and theoretical (Walker, 1971) data on the inert gases, since the most important single factor is the atomic number Z. The results for the inert gases give the polarization of the scattered electron perpendicular to the scattering plane $(P_\perp)$ for the scattering of an unpolarized beam by an unpolarized target, but for scattering of polarized beams, the other polarization components are closely related to $P_\perp$ and exhibit similar behavior. The target polarization P_T can be expected to be comparable to the electron polarization at low energies (less than 5 eV, say) but will decrease with increasing energy. For the light alkalis (lithium and sodium) one would not expect to notice any polarization effects (the maximum calculated polarization on a 1° tabulation for neon is 0.03%), although there is a theoretical possibility of very high, narrow peaks of polarization. For argon, these peaks are about 0.1° wide and have an energy spread of 0.1 eV (at half height, i.e., 50% polarization), and so would still not be observable experimentally, but polarizations of 1–2% occur over quite large angular and energy regions, and 10% polarizations have been measured (Schackert, 1968). This is consistent with the measurement of Wilmers (1972), who obtained a polarization of 9% in scattering from potassium at 5 eV. For the heavier inert gases, and for mercury, significant polarizations are found at most angles except in the forward direction (below about 20–30°) and at 180°, 5% being common in krypton, 10% in xenon, and 20% in mercury, and the peaks (up to 100% polarization) increase in number and in width (both energy and angular) with increasing atomic number, until in xenon and mercury they are the dominant feature of the polarization curves. The behavior of rubidium should be similar to that of krypton, while cesium should resemble xenon (a deduction which the calculations here seem to confirm).

To summarize: (1) The dominant process causing changes in the total spin in low-energy scattering of electrons off alkali atoms is likely to be spin–orbit coupling in the nuclear (and core electron) potential. (2) This process will not affect small-angle scattering, nor scattering at 180°. (3) It increases in importance with increasing atomic number, so that it is not likely to be significant for lithium or sodium, but could affect scattering from potassium by a few percent, will be important for rubidium, and will be a dominant feature in scattering from cesium.

References

Bederson, B. (1969). *Comments Atom. Molec. Phys.*, **1**, 41.
Bethe, H. A., and Salpeter, E. E. (1957). *Quantum Mechanics of One and Two Electron Systems*. Berlin: Springer.
Burke, P. G., and Mitchell, J. F. B. (1974). *J. Phys. B.*, **7**, 214.
Carse, G. D., and Walker, D. W. (1973). *J. Phys. B.*, **6**, 2529.
Jones, M. (1975), *Phil. Trans. Roy. Soc. A*, **277**, 587.
Kessler, J. (1969). *Rev. Mod. Phys.*, **41**, 3.
Norcross, D. W. (1973). *Phys. Rev.*, A **7**, 606.
Schackert, K. (1968). *Z. Phys.*, **213**, 316.
Walker, D. W. (1971). *Adv. Phys.*, **20**, 257.
Walker, D. W. (1974). *J. Phys. B.*, **7**, 97.
Walker, D. W. (1975). *J. Phys. B.*, **8**, 760.
Wilmers, M. (1972). Dissertation, University of Mainz.
Wilmers, M. (1973). *Rev. Colombiana Fisica*, **9**, 12.

19
Spin Polarization of Electrons by Resonance Scattering

E. REICHERT

In formal description, as well as in experimental observations, there are many similarities between atomic and nuclear resonance processes. One aspect here is the spin polarization of spin-1/2 particles scattered via the formation of compound states. The experimental observations of electron spin polarization by resonance scattering on neon are in agreement with formal descriptions developed originally in the theory of nuclear resonance processes. At the moment there are difficulties in understanding quantitatively the experimental results found in resonance scattering on mercury. It is not believed that these difficulties arise because of inapplicability of the general theory. They will probably be removed if correct phase shifts of the potential scattering are known and the inelastic decay of the resonance states lying above the first threshold is properly taken into account.

In 1946 J. Schwinger proposed to produce polarized nucleons via the reactions

$$n + {}^4He \rightarrow {}^5He^* \rightarrow {}^4He + n$$
$$p + {}^4He \rightarrow {}^5Li^* \rightarrow {}^4He + p$$

At collision energies of some MeV the colliding particles form compound states with spectroscopic classification ${}^2P_{3/2,1/2}$ that are widely split by spin–orbit interaction. At the resonance energies p-wave scattering is largely different in the two spin states of the impinging nucleons. Therefore high values of spin polarization of the scattered neutrons or protons should be observable. Detailed theoretical investigations by Wolfenstein (1949) and experiments done by Critchfield and Dodder (1949) and Heusinkveld and Freier (1952) have confirmed the arguments of Schwinger.

E. REICHERT • Institut für Physik, Johannes Gutenberg Universität, Mainz, W. Germany.

In electron–atom collisions similar resonance processes occur. The first observation of a "split" resonance was made by Simpson and Fano in 1963 in the scattering of electrons on neon. Figure 1 shows a transmission curve of the neon resonances published by Kuyatt et al. in 1965. The structure seen was classified by Simpson and Fano as being a $^2P_{3/2,1/2}$ resonance due to the formation of a compound Ne$^-$ state with electron configuration $(\ldots 2p^5 3s^2)^2 P_{3/2,1/2}$.

Many other resonance doublets have been observed since that time. Some examples are shown in the succeeding figures. The lowest resonances seen in electron scattering on the heavier noble gases (Figures 2, 3, and 4) are again classified to be $^2P_{3/2,1/2}$ resonances (Kuyatt et al., 1965). The two most prominent structures seen in the cross section of mercury (Figure 5) are $D_{3/2,5/2}$ resonances, probably due to formation of Hg$^-$ states with classification $(\ldots 6s6p^2)^4 P_{3/2,5/2}$ (Fano and Cooper, 1965).

In those cases mentioned where only elastic resonances come into play (neon, Figure 1; argon, Figure 2), the general formalism describing the elastic scattering of spin-$\frac{1}{2}$ particles at spin-0 targets is applicable (Franzen and Gupta, 1965; Suzuki and Tanaka, 1973; Heindorff et al., 1973). The scattering process is described by a 2×2 matrix M, which can be expressed in terms of the Pauli spin matrices $\sigma_x, \sigma_y, \sigma_z$ and the unit matrix $\mathbb{1}$:

$$M = f(E, \theta)\mathbb{1} + g(E, \theta)\boldsymbol{\sigma} \cdot \mathbf{n}$$

where $\mathbf{n}$ is a unit vector normal to the plane of scattering and $f(E, \theta)$ and $g(E, \theta)$ are complex numbers that depend on collision energy E and scattering angle θ (see, for example, Mott and Massey, 1965).

The cross section I^+ of electrons whose spins are oriented in direction $\mathbf{n}$ ("spin up") is

$$I^+ = |f + g|^2$$

The cross section I^- of electrons with spin orientation antiparallel to $\mathbf{n}$ ("spin down") is

$$I^- = |f - g|^2$$

If the primary electron beam is unpolarized, the cross section will be

$$I = \tfrac{1}{2}(I^+ + I^-) = |f|^2 + |g|^2$$

The spin polarization $\mathbf{P}$ after scattering in this case is

$$\mathbf{P} = \frac{I^+ - I^-}{I^+ + I^-}\mathbf{n} = \frac{2\,\mathrm{Re}\,f^*g}{I}\mathbf{n} \quad \text{(Mott polarization)}$$

The point is that, even for a light atom like neon, the "spin-flip" amplitude $g(E, \theta)$ is not negligible at the energies of the $^2P_{3/2,1/2}$ resonance states.

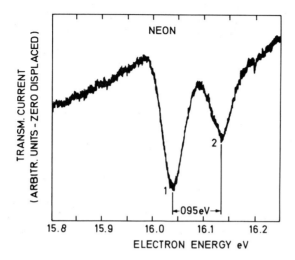

Figure 1. Resonance scattering of electrons on neon. Transmitted current as a function of electron energy (from Kuyatt et al. 1965).

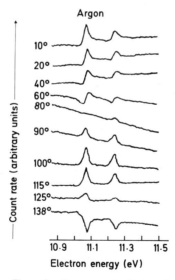

Figure 2. Resonance scattering of electrons on argon. Angular dependence of the resonance doublet $^2P_{3/2,1/2}$ below the first excited state (from Weingartshofer et al., 1974).

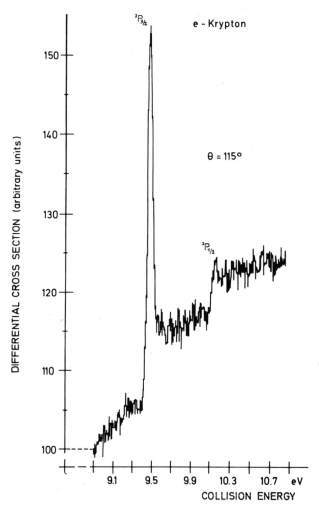

Figure 3. Resonance scattering of electrons on krypton. $^2P_{3/2,1/2}$ resonance at a scattering angle $\theta = 115°$ (from Heindorff et al., 1974).

This can be seen by decomposing the amplitudes into partial waves:

$$f = \frac{1}{2ik} \sum_e [(e+1)\{\exp(2i\delta_e^+) - 1)\} + e\{\exp(2i\delta_e^-) - 1\}]P_e(\cos\theta)$$

$$g = \frac{1}{2k} \sum_e [\exp(2i\delta_e^+) - \exp(2i\delta_e^-)]P'_e(\cos\theta)$$

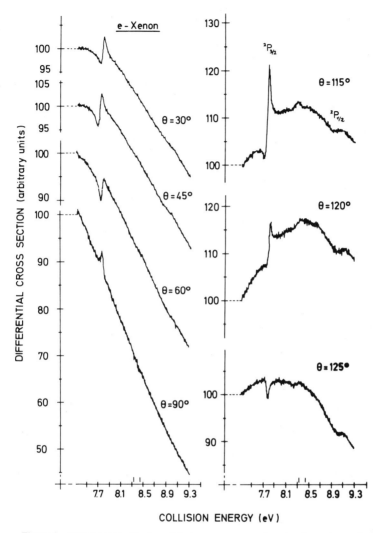

Figure 4. Resonance scattering of electrons on xenon. Angular dependence of the $^2P_{3/2,1/2}$ resonance doublet (from Heindorff et al., 1974).

in standard notation. Far from resonance, there is practically no difference between δ_e^+ and δ_e^- in the case of a light target atom. Here $g(E, \theta)$ vanishes and $f(E, \theta)$ reduces to the well-known formula describing potential scattering without spin.

At the resonance energies, however, the phases of those partial waves in which the resonance takes place (p waves in the case of neon) change

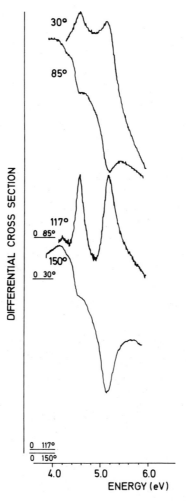

Figure 5. Resonance scattering of electrons on mercury. Angular dependence of the $^2D_{3/2,5/2}$ resonances (from Düweke et al., 1973).

rapidly, with collision energy obeying a Breit–Wigner energy dependence (Fano, 1961):

$$\delta_1^\pm = \delta_{1\,\text{pot}} + \delta_{\text{res}}^\pm; \qquad \delta_{\text{res}}^\pm = \cot^{-1}\left\{\frac{E_{\text{res}}^\pm - E}{\Gamma^\pm/2}\right\}$$

where $\delta_{1\,\text{pot}}$ is the potential phase shift of p waves, $\delta_{\text{res}}^\pm$ the resonance phase shift, E_{res}^+ the energy of $P_{3/2}$ resonance, E_{res}^- the energy of $P_{1/2}$ resonance, and $\Gamma^\pm$ the natural half-width of $P_{3/2}$ and $P_{1/2}$ resonance, respectively.

E_{res}^+ and E_{res}^- have different values in case of "split" resonances. In consequence, $g(E, \theta)$ is not negligible at energies around the resonance energies, but may influence the shape of the resonance cross section to a large extent and produce high values of spin polarization of the scattered electrons. In Figures 2 and 4, for example, the resonance peaks in 90° scattering are solely due to $|g|^2$. In electron scattering on neon, polarization values up to 100% should be attainable in principle, as has been shown by Franzen and Gupta (1965) and Suzuki et al. (this volume).

Experimental studies of spin polarization by resonance scattering have been published for scattering on neon only (Reichert and Deichsel, 1967; Heindorff et al., 1973). Figure 6 shows the apparatus used by Heindorff et al. Three types of measurements are made:

(1) Angular distribution $I(\theta)$ of the scattered current at fixed collision energy just outside resonance. The scattering angle is varied by moving the monochromator in chamber I. The scattered current is detected by M2. The purpose of this measurement is to get information about the potential phase shifts in the partial wave expansion. These phase shifts must be known if a quantitative analysis of the resonance scattering is to be made. Figure 7 shows an example of measured values of $I(\theta)$ compared with theoretical data derived from

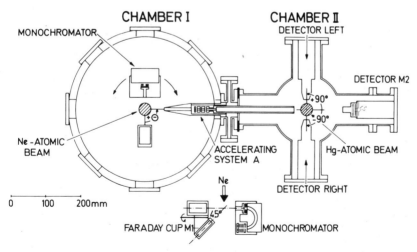

Figure 6. Resonance scattering of electrons on neon. Apparatus used by Heindorff et al., (1973). *Top*: horizontal cross section of apparatus. *Bottom*: vertical section through scattering center I.

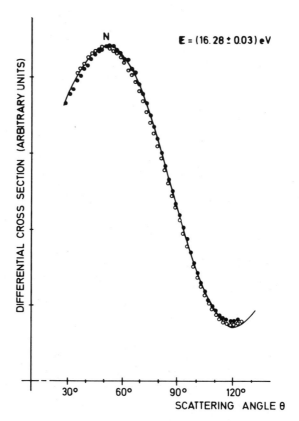

Figure 7. Scattering of electrons on neon. Relative differential cross section for $E = (16.28 \pm 0.03)$ eV; (●), experimental points for $(+\theta)$ scattering; (○), experimental points for $(-\theta)$ scattering; (———), theory (D. G. Thompson, 1971) (from Heindorff et al., 1973).

phases given by Thompson (1971). Because of the good agreement between experiment and theory, the phases of Thompson are used for further analysis by Heindorff et al.

(2) Energy dependence $I(E)$ of scattered current at fixed scattering angles. Examples are shown in the upper parts of Figures 8, 9, and 10. Out of the area below the resonance peaks in the 90° curve (Figure 8) the natural half-width of the neon resonances is evaluated to be $\Gamma^\pm = 1.4$ meV, in agreement with Ehrhardt et al. (1968). The instrumental energy profile of the primary electron beam can be read off the same curve.

(3) Energy dependence $P(E)$ of spin polarization of scattered electrons at fixed scattering angle. In this measurement a mercury atomic beam is turned on in chamber II. Out of the asymmetry of currents detected in detector "left" and "right," the polarization of electrons hitting the mercury atoms is evaluated in a manner described by Deichsel *et al.* (1965, 1966). Examples are shown in the lower parts of Figures 8, 9, and 10.

The dashed curves are calculated, taking into account the potential phases given by Thompson, a natural half-width of the neon resonances

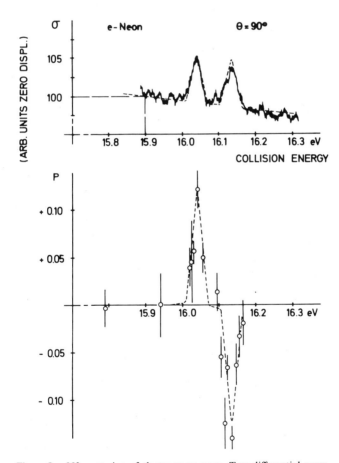

Figure 8. 90° scattering of electrons on neon. *Top*: differential cross section. *Bottom*: spin polarization; (— — —), calculated shape (from Heindorff *et al.*, 1973).

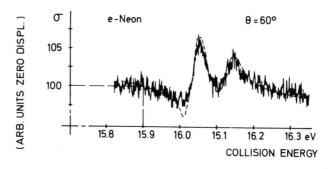

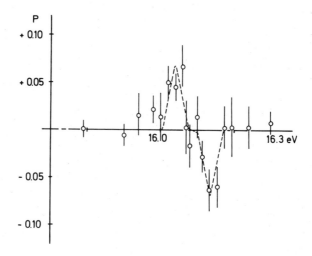

Figure 9. 60° scattering of electrons on neon. *Top*: differential cross section; *Bottom*: spin polarization, (---) calculated shape (from Heindorff et al., 1973).

of 1.4 meV, and an instrumental energy profile taken from Figure 8. The agreement with the experimental data is seen to be good.

Experimental studies of spin polarization by resonance scattering on mercury are now underway in Mainz. The analysis of the scattering process is somewhat more complicated than in the case of neon.

Splitting of potential phases $\delta_e^{\pm}$ outside resonance is not negligible here. This can be seen from Figure 11. High values of spin polarization of electrons scattered by mercury at energies below the resonance region are observed (Düweke et al., 1974).

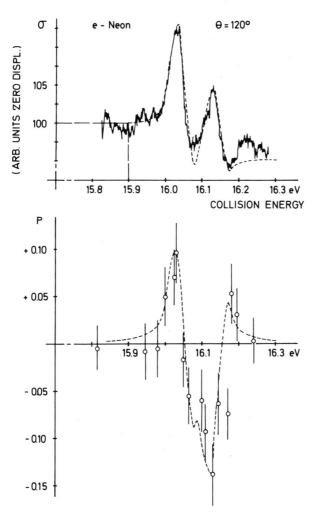

Figure 10. 120° scattering of electrons on neon. *Top*: differential cross section; *Bottom*: spin polarization; (---) calculated shape (from Heindorff et al., 1973).

Preliminary results of spin polarization at energies in the range of the $D_{3/2, 5/2}$ resonances measured by Düweke et al. (1975) are shown in Figure 12. The observed behavior of polarization is not quite understood yet.

A rough calculation, using potential phases given by Walker (1969, 1970) but not taking into account inelastic processes, predicts a crossing of the polarization curve to high positive values at energies between the resonances.

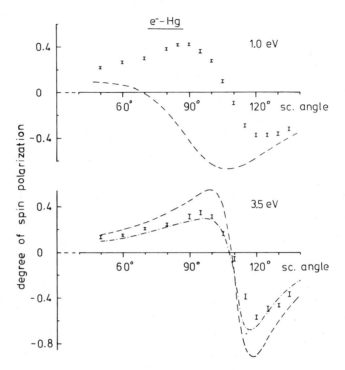

Figure 11. Electron scattering on mercury for electron energies below the first excitation threshold. Angular dependence of spin polarization: (———), with exchange; (—·—), with exchange and distortion theory (Walker, 1969, 1970); I, experimental values (from Düweke et al., 1974).

This is not found in the measured data shown in Figure 12. It is hoped that inclusion of the possibility of inelastic decay of the $D_{5/2}$ resonance in the calculations will remove this discrepancy.

NOTE: P. D. Burrow and J. A. Michejda (1975) presented a paper at the Stirling Symposium in which a new energy calibration of the mercury resonances is reported. According to this measurement, the energies of the $^4P_{3/2,5/2}$ states in Hg$^-$ are 4.9 eV and 5.49 eV respectively. These values are roughly 350 meV above those given in Figure 5 and Figure 12 of this report. With the resonance energies observed by Burrow and Michejda, not only is the $D_{5/2}$ resonance above the first inelastic threshold of Hg, but so is the $D_{3/2}$ resonance. In this case, both Hg$^-$ states may decay into inelastic channels, which further complicates the analysis of the polarization data shown in Figure 12.

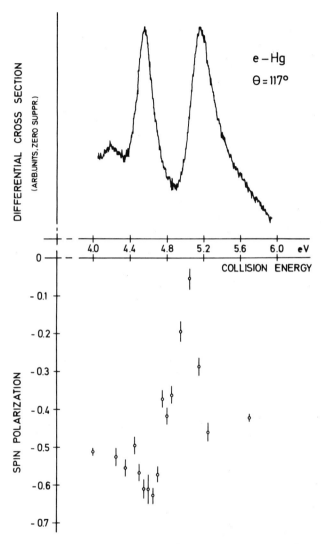

Figure 12. Resonance scattering of electrons on mercury at $\theta = 117°$. *Top*: differential cross section; *Bottom*: spin polarization (from Düweke *et al.*, 1975, preliminary results).

References

Burrow, P. D., and Michejda, J. A. (1975). Stirling Symposium, contributed papers, p. 50.
Critchfield, C. L., and Dodder, D. C. (1949). *Phys. Rev.*, **76**, 602.
Deichsel, H., and Reichert, E. (1965). *Z. Phys.*, **185**, 169.
Deichsel, H., Reichert, E., and Steidl, H. (1966). *Z. Phys.*, **189**, 212.
Düweke, M., Kirchner, N., Reichert, E., and Staudt, E. (1973). *J. Phys. B.*, **6**, L208.
Düweke, M., Kirchner, N., Reichert, E., and Schön, S. (1975). To be published.
Ehrhardt, H., Langhans, E., Linder, F., and Taylor, H. S. (1968). *Phys. Rev.*, **173**, 222.
Fano, U. (1961). *Phys. Rev.*, **124**, 1866.
Fano, U., and Cooper, J. W. (1965). *Phys. Rev.*, **138A**, 400.
Franzen, W., and Gupta, R. (1965). *Phys. Rev. Letts.*, **15**, 819.
Heindorff, T., Höfft, J., and Dabkiewicz, P. (1974). To be published.
Heindorff, T., Höfft, J., and Reichert, E. (1973). *J. Phys. B.*, **6**, 477.
Heusinkveld, M., and Freier, G. (1952). *Phys. Rev.*, **85**, 80.
Kuyatt, C. E., Simpson, J. A., and Mielczarek, S. R. (1965). *Phys. Rev.*, **138A**, 385.
Mott, N. F., and Massey, H. S. W. (1965). *The Theory of Atomic Collisions*. Oxford: Clarendon Press.
Reichert, E., and Deichsel, H. (1967). *Phys. Letters*, **25A**, 560.
Schwinger, J. (1946). *Phys. Rev.*, **69**, 681.
Schwinger, J. (1947). *Phys. Rev.*, **73**, 406.
Simpson, J. A., and Fano, U. (1963). *Phys. Rev. Letters*, **11**, 158.
Suzuki, T., and Tanaka, H. (1973). *J. Phys. Soc. Japan*, **34**, 566.
Thompson, D. G. (1971). *J. Phys. B.*, **4**, 468.
Walker, D. W. (1969). *J. Phys. B.*, **2**, 356.
Walker, D. W. (1970). *J. Phys. B.*, **3**, 788.
Weingartshofer, A., Willmann, K., and Clarke, E. M. (1974). *J. Phys. B.*, **7**, 79.
Wolfenstein, L. (1949). *Phys. Rev.*, **75**, 1664.

20

Influence of Spin Polarization on Resonance Scattering by Neon

T. Suzuki, H. Tanaka, M. Saito, and H. Igawa

> The influence of the spin polarization of incident electrons on the differential cross section of elastic resonance scattering by neon is calculated numerically. The differential cross section of the resonance scattering is found to be strongly dependent on the degree of spin polarization of the incident electrons. These results suggest that the resonance scattering can be used as a spin-polarization analyzer of incident electrons in place of Mott scattering.

1. Introduction

It is well known that the fine-structure splitting of the peak in the resonance scattering is an indication of strong spin–orbit coupling in the compound state (Farago, 1971). In the recorder traces of the transmission of electrons through neon, Simpson and Fano (1963) and then Kuyatt *et al.* (1965) found a doublet structure of 0.095 eV separation, 0.6 eV below the first excited state. It was suggested by Simpson and Fano (1963) that the resonance scattering processes may take place by the formation of a compound state Ne$^-$ through the addition of a 3s electron to the lowest excited state of the neon atom resulting in a configuration of $[(1s^2 2s^2 2p^5 3s^2)P_{3/2}P_{1/2}]$. Andrick and Ehrhardt (1966) and Andrick *et al.* (1968) measured the differential cross section of resonance scattering from neon at various scattering angles. Finally, the possibility of spin polarization of the resonance-scattered electron was predicted by Franzen and Gupta (1965), and their prediction was examined experimentally for neon by Reichert and Deichsel (1967).

T. Suzuki, H. Tanaka, M. Saito, and H. Igawa • Department of Physics, Sophia University, Tokyo, Japan 102.

In all the above-mentioned investigations, however, only the case of nonpolarization for the incident electron spin was assumed. It seems to be most probable that there should be some influence of the incident electron spin polarization on the resonance scattering, because the electron spin polarization generally has an important role in that process. The present paper treats the influence of the spin polarization of the incident electrons on the differential cross section in resonance scattering, with numerical calculation for the case of neon.

2. Theoretical

In general, the differential cross section $\sigma(\theta)$ for electron scattering by a spherically symmetric atom is characterized by the two well-known scattering amplitudes $f(\theta)$ and $g(\theta)$, taking into account the degree of the spin polarization for the incident electrons as follows:

$$\sigma(\theta) = I_0[1 + P_0 P(\theta)] I(\theta)$$

where

$$I(\theta) = |f(\theta)|^2 + |g(\theta)|^2$$

$$P(\theta) = 2 \operatorname{Re} [f(\theta) g^*(\theta)/I(\theta)]$$

Here I_0 and P_0 are, respectively, the intensity and the degree of spin polarization for the incident electron beam. As is well known, f and g contain the relativistic phase shifts.

In the resonance scattering, the presence of a configuration between a continuum and a discrete level produces additional phase shifts as given by Fano (1961). In the case of neon, the additional phase shifts $\delta^\pm$ appear only in p-wave scattering, which is caused by a resonance split with energies $E_{3/2}$ and $E_{1/2}$. The phase shifts $\delta^\pm$ at an energy E can be expressed as

$$\delta^\pm = -\cot^{-1}[\varepsilon \pm x]$$

where

$$\varepsilon = [(E_{3/2} + E_{1/2})/2 - E](2/\Gamma)$$

$$x = (E_{3/2} - E_{1/2})/\Gamma$$

Here δ_1 is the relativistic phase shift in the absence of resonance and Γ is a width for both levels. In this way, the differential cross section depends on the interference between the resonance and the mean potential scattering. The detailed expression for $P(\theta)$ and $I(\theta)$ including $\delta^\pm$ is omitted here. (It will be published in another place.) Thus, in principle, $\sigma(\theta)$ can be computed. In practice, however, empirical values for $E_{3/2}$, $E_{1/2}$, and Γ are utilized.

The values for $E_{3/2}$ and $E_{1/2}$ were determined with a fairly good accuracy by Kuyatt *et al.* (1965). On the other hand, the value of Γ is different with different authors. Haselton (1973) found a value of 8.9 meV based upon the total cross section measurement, whereas Heindorff *et al.* (1973) obtained a value of 1.4 meV, deduced from the measurement of differential cross section and spin polarization. The experimental half-width for both levels, $E_{3/2}$ and $E_{1/2}$, is a convolution value of Γ itself, with a definite energy resolution ΔE. In practice, neither is ΔE zero, nor is Γ known *a priori*. One therefore has to try to find a proper "fitting combination" of Γ and ΔE. Among various reasonable combinations, we found that a combination of $\Gamma = 10$ meV and $\Delta E = 40$ meV gives an excellent "fit" of the calculated curves to the experimental results of the differential cross section for neon obtained by Andrick *et al.* (1968). The fit is surprisingly good for all scattering angles. (The result of the comparison will be published elsewhere.) With these values for Γ and ΔE, we calculated the differential cross section for resonance scattering at various scattering angles between 5° and 145° for the three cases $P_0 = 0.0, 0.5$, and 1.0.

3. Results and Discussion

The computed results show that the differential cross section of the resonance scattering from neon is strongly dependent on the degree of spin polarization of the incident electron as was expected, except for small scattering angles (Figure 1). Therefore, we can estimate the degree of spin polarization of the incident electrons by measuring the differential cross section at a suitable scattering angle. For example, at 90°, as can be seen from Figure 1, the variation of the cross section with P_0 is quite large; there are two peaks with the same height for $P_0 = 0$, but one peak and one dip for $P_0 = 1.0$. Therefore, we can quantitatively determine P_0, the degree of spin polarization of the incident electrons, by measuring the ratio $\sigma(E_{3/2})/\sigma(E_{1/2})$ at a particular angle.

We propose with our results to use the resonance scattering as a spin polarization analyzer in place of Mott scattering. For Mott scattering, it is necessary to accelerate the electrons up to about 50–200 keV. The use of these high voltages inevitably introduces insulation problems. Moreover, the asymmetry factor in Mott scattering (the so-called Sherman function) is 0.3–0.5 at its maximum value. In our resonance scattering method, not only is the influence of the degree of spin polarization quite remarkable, but also high acceleration of the electrons is not necessary. Only the energy

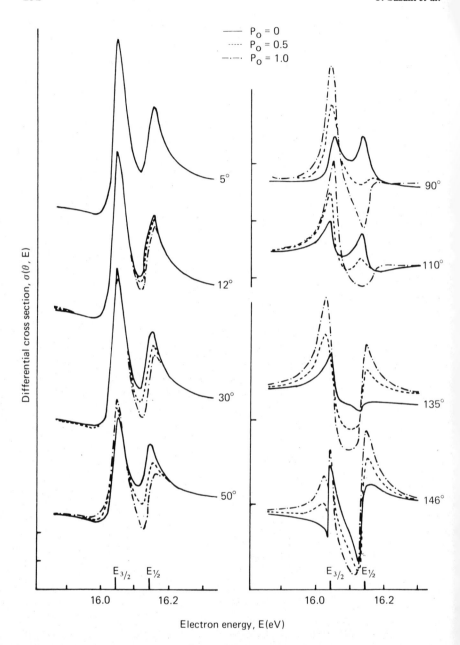

Figure 1. Theoretical differential cross section for resonance scattering from neon with different degrees of spin polarization of the incident electrons. $\Gamma = 10$ meV and $\Delta E = 40$ meV were assumed.

of the electrons, the degree of polarization of which is to be examined, should be adjusted to between 5 and 20 eV depending on the first excitation energy of the inert gas target atoms. The resonance scattering method, however, demands a comparatively good energy resolution, at least 20 meV.

We have only calculated the case for neon; however, it is quite easy to carry out a similar procedure for other inert gas atoms choosing suitable fitting parameters. Further numerical calculations for the other inert gases, Ar and Kr, and also an experiment to test our theoretical proposals are now underway.

Acknowledgments

The authors sincerely thank Professors U. Fano and T. Yamanouchi for their kind encouragement. One of us (T.S.) acknowledges the support of the Royal Society and the Japan Society for Promotion of Science in providing the opportunity to sum up the present work in Britain. We also thank Professor P. S. Farago and Dr. D. M. Campbell for their kind reading of the manuscript.

References

Andrick, D., and Ehrhardt, H. (1966). *Z. Phys.*, **192**, 99.
Andrick, D., Ehrhardt, H., and Eyb, M. (1968). Abstract of contributed papers: *Int. Conf. Atom. Phys.*, p. 110 (New York University).
Fano, U. (1961). *Phys. Rev.*, **124**, 1866.
Farago, P. S. (1971). *Rep. Prog. Phys.*, **34**, 1055.
Franzen, W., and Gupta, R. (1965). *Phys. Rev. Letts.*, **15**, 819.
Haselton, H. H. (1973). *Bull. Am. Phys. Soc.*, **18**, 710.
Heindorff, T., Höfft, J., and Reichert, E. (1973). *J. Phys. B.*, **6**, 477.
Kuyatt, C. E., Simpson, J. A., and Mielczarek, S. R. (1965). *Phys. Rev.*, **138**, 385.
Reichert, E., and Deichsel, H. (1967). *Phys. Letts.*, **25A**, 560.
Simpson, J. A., and Fano, U. (1963). *Phys. Rev. Letts.*, **17**, 31.

21
Asymmetry in the Single Scattering of Electrons from One-Electron Atoms

P. S. FARAGO

> It is argued that the measurement of the asymmetry in the single scattering of unpolarized electrons from a spin-polarized assembly of heavy alkali atoms is eminently suitable for the study of the interference between spin–orbit coupling and spin exchange effects. The reasons are: (a) the observable asymmetry is a unique consequence of such an interference, and (b) the experiment is a single-scattering experiment and hence more efficient than one involving electron polarization analysis.

The spin dependence of electron scattering from atomic targets arises from two processes of fundamentally different nature, namely, spin–orbit interaction and spin exchange. While the effects of both processes can be detected through the analysis of the polarization of the scattered electrons, the pattern of single scattering retains its azimuthal symmetry, irrespective of the polarization of the target, provided that, firstly, the incident electron beam is unpolarized and, secondly, only one of the two processes is operative. The purpose of this note is to reiterate (Farago, 1974) that if the two effects occur concurrently their interference can lead to a change in the scattering pattern. Such an interference yields an asymmetry in the single scattering of unpolarized electrons if the target atoms are polarized at right angles to the scattering plane, and the detection of such an asymmetry is a unique evidence of the simultaneous presence of and interference between spin–orbit interaction and spin exchange effects.

Let us consider an assembly of spin-$\frac{1}{2}$ target atoms of polarization **P** and an initially unpolarized electron beam. The statistical operator of the

P. S. FARAGO • Department of Physics, University of Edinburgh, Scotland.

system is given by

$$\hat{d}_i \propto (\hat{\sigma}_0 + \mathbf{P} \cdot \hat{\boldsymbol{\sigma}}) \otimes \hat{\sigma}_0$$

where $\hat{\sigma}_0$ is the 2×2 unit matrix, $\hat{\boldsymbol{\sigma}}$ is the Pauli spin vector operator and $\otimes$ symbolizes direct product. Denoting the scattering matrix connecting the initial and final spin states by $\hat{M}$, the differential cross section is given by

$$\sigma(\theta) = \frac{\text{Tr}\,(\hat{M}\hat{d}_i\hat{M}^\dagger)}{\text{Tr}\,\hat{d}_i} \qquad (1)$$

This quantity depends on the target polarization $\mathbf{P}$, because, in general, the only restriction on the form of the scattering matrix $\hat{M}$ is determined by postulating its invariance under rotation and reflection of coordinates and under time reversal. The resulting expression (Mott and Massey, 1965; Wilmers, 1972; Drukarev, 1973; Burke and Mitchell, 1974) can be written in the form

$$\hat{M} = a_1\hat{I} + a_2(\hat{\boldsymbol{\sigma}}_1 \cdot \mathbf{n}) + a_3(\hat{\boldsymbol{\sigma}}_2 \cdot \mathbf{n}) + a_4(\hat{\boldsymbol{\sigma}}_1 \cdot \mathbf{n})(\hat{\boldsymbol{\sigma}}_2 \cdot \mathbf{n}) + a_5(\hat{\boldsymbol{\sigma}}_1 \cdot \mathbf{p})(\hat{\boldsymbol{\sigma}}_2 \cdot \mathbf{p})$$
$$+ a_6(\hat{\boldsymbol{\sigma}}_1 \cdot \mathbf{q})(\hat{\boldsymbol{\sigma}}_2 \cdot \mathbf{q}) \qquad (2)$$

In this expression a_j $(j = 1, \ldots, 6)$ are mutually independent complex amplitudes, $\hat{I}$ is the 4×4 unit matrix, $\boldsymbol{\sigma}_1 = \sigma_0 \otimes \boldsymbol{\sigma}$ and $\boldsymbol{\sigma}_2 = \boldsymbol{\sigma} \otimes \sigma_0$ are the target and projectile spin operators respectively, and $\mathbf{n}$, $\mathbf{p}$, $\mathbf{q}$ are unit vectors defining the reference frame linked to the scattering geometry:

$$\mathbf{n} = \frac{\mathbf{k}_i \times \mathbf{k}_f}{|\mathbf{k}_i \times \mathbf{k}_f|} \qquad \mathbf{p} = \frac{\mathbf{k}_i + \mathbf{k}_f}{|\mathbf{k}_i + \mathbf{k}_f|} \qquad \mathbf{q} = \frac{\mathbf{k}_i - \mathbf{k}_f}{|\mathbf{k}_i - \mathbf{k}_f|}$$

with $\mathbf{k}_i$ and $\mathbf{k}_f$ denoting the momenta of the incident and scattered electron respectively.

Substitution of equation (2) and (1) gives

$$\sigma(\theta) = \sum_{j=1}^{6} |a_j|^2 + 2(\mathbf{P} \cdot \mathbf{n})\,\text{Re}\,(a_1 a_2^* + a_3 a_4^*)$$

and hence a scattering asymmetry

$$A \equiv \frac{\sigma(\theta) - \sigma(-\theta)}{\sigma(\theta) + \sigma(-\theta)} = 2(\mathbf{P} \cdot \mathbf{n})\,\text{Re}\,(a_1 a_2^* + a_3 a_4^*)\{\sum|aj|^2\}^{-1} \qquad (3)$$

This scattering asymmetry vanishes irrespective of the target polarization, i.e., even if $\mathbf{P} \cdot \mathbf{n} \neq 0$, if only spin–orbit interaction or only spin exchange effects are present, for the following reasons. In the absence of spin exchange the scattering matrix $\hat{M}$ must be independent of the target spin operator $\hat{\boldsymbol{\sigma}}_1$, and, therefore, in this case $a_2 = a_4 = a_5 = a_6 = 0$. In the absence of spin–orbit coupling the total spin $\hat{\boldsymbol{\sigma}}_1 + \hat{\boldsymbol{\sigma}}_2$ is conserved, and, as a consequence,

$\hat{M}$ is subject to more stringent symmetry conditions than in the general case leading to the following constraints: $a_2 = a_3 = 0$ and $a_4 = a_5 = a_6$.

It should be stressed that although the polarization of initially unpolarized electrons scattered by an unpolarized target depends on the simultaneous presence of spin exchange and spin–orbit coupling effects according to the relation

$$\mathbf{P}' = 2 \operatorname{Re}(a_1 a_3^* + a_2 a_4^*)\{|a_j|^2\}^{-1}\mathbf{n} \qquad (4)$$

and vanishes in the absence of spin–orbit coupling effects by virtue of the condition $a_2 = a_3 = 0$, it does, however, remain finite if spin–orbit interaction operates in the absence of spin exchange, since in this case both a_1 and a_3 remain finite.

It has been pointed out before (Farago, 1974) that the detection of the asymmetry under the conditions outlined above is a very convenient approach for the study of spin–orbit coupling effects in the presence of spin exchange because it is a single-scattering experiment, while any experiment involving electron spin-polarization analysis requires a second scattering process. For this reason, in terms of obtainable accuracy in a given period of observation, an asymmetry measurement is more efficient than a polarization analysis. From the point of view of studying the interference between spin–orbit coupling and spin exchange effects, the asymmetry measurements in question have the further advantage that the mere existence of the scattering asymmetry hinges on the interference between the two simultaneously operative mechanisms, while this is not true in the case of the polarization of initially unpolarized electrons scattered from unpolarized targets.

Burke and Mitchell (1974) calculated the polarization, after elastic scattering, of initially unpolarized electrons from unpolarized cesium atoms. By considering the consequence of the fine-structure splitting of the $6p_{1/2}$–$6p_{3/2}$ states they predict a pronounced polarization effect at energies which lie between the excitation thresholds of these levels, and it was conjectured (Farago, 1974) that at these energies the scattering asymmetry discussed above would be comparable. In a paper appearing in this volume, Walker (1975) considers spin–orbit coupling effects of other origin and argues that the coupling between the spin and orbital angular momentum of the projectile electron moving in the field of the scattering center plays a dominant part masking the polarization effect due to fine-structure splitting.

Recent preliminary calculations of the asymmetry in single scattering from polarized targets performed by Walker (1974) confirm the views expressed on the basis of algebraic arguments. Indications are that under the conditions where spin exchange effects and spin–orbit coupling effects are of comparable importance their interference will manifest itself in a significant asymmetry. This will be the case if the target consists of heavy spin-

Table 1

	E(eV)	θ	σ/a_0^2	$A(\%)$	$P(\%)$
(1)	1.427	88°	1.05	−59	−72
(2)	5.0	123°	0.073	30	60
(3)	13.6	70°	0.060	7	39
(4)	35.0	143°	0.012	1	31
(5)	100.0	143°	0.003	0.87	−37

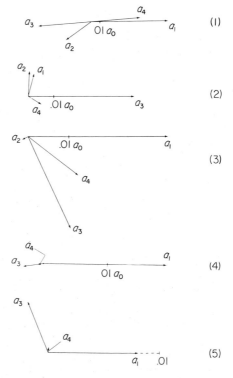

Figure 1. Vector diagrams illustrating the relative magnitudes and phases of scattering amplitudes $a_1, \ldots, a_4$ in the elastic scattering of electrons from cesium atoms. Relevant parameter values are listed in Table 1.

polarized alkali atoms, such as cesium, at low-electron energies (below 15 eV, say) in the neighborhood of the minima of the differential cross section. If, however, one of the two effects loses its significance in the scattering process, the peak values of the asymmetry decrease drastically while the peak values of the spin polarization of the scattered electrons remain high. This is borne out by the calculations for higher incident electron energies, where the effect on the scattering by the valence electrons, and hence that of spin exchange, is rapidly diminishing.

The relative behavior of the polarization of the scattered electrons from unpolarized targets and of the asymmetry of single scattering from polarized targets can be readily visualized with the aid of diagrams in which the amplitudes $a_1, \ldots, a_4$ are represented by vectors in the complex plane. Five such diagrams are shown in Figure 1, based on calculations by Walker (1974) for different electron energies. The moduli of the amplitudes are drawn in units of the Bohr radius, a_0, as marked; the orientation of the diagrams (1)–(5) relative to one another is arbitrary. At each energy the scattering angle is chosen to yield maximum asymmetry A. The relevant parameters and the calculated values of the spin-independent differential cross section, $\sigma(E, \theta)/a_0^2$, asymmetry, A, and polarization, P', are listed in Table 1. Note that under the conditions corresponding to diagrams (4) and (5), $|a_2| \ll |a_4|$ and a_2 can no longer be displayed on the scale shown.

Acknowledgment

The author is grateful to Dr. D. W. Walker for helpful discussions and for making results available prior to publication.

References

Burke, P. G., and Mitchell, J. F. B. (1974). *J. Phys. B.*, **7**, 214.
Drukarev, G. F. (1973). Invited Lectures and Progress Repts., *8th Int. Conf. Phys. Electronic Atom. Collisions, Belgrade*.
Farago, P. S. (1974). *J. Phys. B*, **7**, L28.
Mott, N. F., and Massey, H. S. W. (1965). *The Theory of Atomic Collisions*, Chap. X.6. 3rd edn. Oxford: Oxford University Press.
Walker, D. W. (1974). *J. Phys. B.*, **7**, L489.
Walker, D. W. (1975). Contribution to this volume.
Wilmers, M. (1972). Dissertation, Gutenberg University, Mainz.

22

Intense Source for Highly Polarized Electrons Using the Fano Effect

W. von Drachenfels, U. T. Koch, T. M. Müller, and W. Paul

A source of polarized electrons using the Fano effect on cesium is described. It produces 3×10^9 electrons at a polarization of 90%. Referring to the more detailed description of the apparatus given in von Drachenfels et al. (1974), this article reports an improvement of the light source which increases the repetition rate to 1 Hz.

A source of polarized electron pulses has been built at the University of Bonn (von Drachenfels *et al.*, 1974). It uses the Fano effect (Fano, 1969) on cesium, i.e., unpolarized cesium is photoionized by circularly polarized light ($\lambda \sim 300$ nm) to give highly polarized photoelectrons.

The electron source consists of two main parts: the light source (laser system) and the oven system, in which a beam of cesium atoms interacts with the light. The resulting photoelectrons are then extracted and accelerated to allow polarization analysis by Mott scattering.

The laser system used so far consists of a laser oscillator and amplifier stage (Figure 1). The oscillator produces 70 mJ at 610 nm with a pulse length of 1 μs and 0.05 nm bandwidth. The amplifier gives a gain of 4, i.e., 280 mJ final output. The radiation is then frequency-doubled by using a KDP (potassium dihydrogen phosphate) crystal. The UV output was routinely 1.5 mJ, the maximum 3 mJ with a pulse length of 0.5 μs.

The oven system is shown in Figure 2. It has an array of 20 atomic beams in parallel which are all crossed by the UV light beam. In order to increase the running time while keeping the amount of cesium low, the cesium

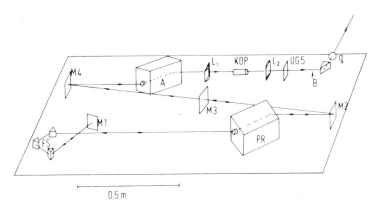

Figure 1. Setup of laser system. PR, pumping reflector; M1, M2, M4, 100% reflecting mirrors; M3, mirror with 30% reflection; FS, frequency selector (dispersing prisms); A, amplifier; L_1, L_2, cylindrical lenses; B, position of bolometer; q, quarter-wave plate.

from the atomic beams is condensed and recycled within the oven system. Thus, an increase in running time by a factor of 30 has been achieved. The density in the beams is 10^{12} atoms/cm^3 and the interaction length is 80 mm. This yields 2×10^9 photoelectrons for every millijoule of UV input. The photoelectron polarization was measured to be $90 \pm 7\%$ for all intensities. This is consistent with Fano's theory and other experiments (Heinzmann et al., 1970; Baum et al., 1972) The maximum intensity was 3×10^9 electrons per pulse. Further details of the apparatus and the measurements have been given by von Drachenfels et al. (1974).

These measurements were performed with a repetition rate of only 3/min because of the insufficient cooling of the amplifier lamp of the laser system. The oscillator, however, is capable of running at a repetition rate of 1 pps. At this rate it delivers about 30 mJ at 610 nm. In order to get a higher repetition rate of our source, we explored the possibility of frequency doubling the output of the oscillator alone. The output of our KDP crystal, however, was insufficient (~ 0.3 mJ). The power density in the crystal can be increased by stronger focusing of the incident light, but this in turn increases the beam divergence, which again reduces the conversion efficiency. A way to avoid this can be found by using a special frequency-doubling crystal with *uncritical phase matching* (Adhav and Wallace, 1973). Here, the sensitivity to divergence is reduced by an order of magnitude or more.

For our wavelength region (~ 600 nm), a KDA (potassium dihydrogen arsenate) crystal can be used. We focused the output of our oscillator (31 mJ) onto a KDA crystal with a strongly focusing lens. The output reliably achieved was 1.5 mJ UV. Under these conditions the oscillator runs with a

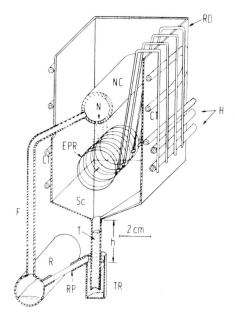

Figure 2. Cross section of recirculating oven system and interaction region. NC, nozzle chamber; N, nozzles; R, reservoir; Sc, shield; F, feeder pipe; RP, return pipe; TR, trap; H, ceramic holders; RD, rods holding extraction rings; CT, copper tubes; EPR, extraction potential rings; T, descending tube; h, ascending height for Cs.

repetition rate of 1 Hz. With this UV input, our polarized electron source produces 3×10^9 electrons per pulse at 1 Hz.

References

Adhav, R. S., and Wallace, R. W. (1973). *IEEE J. Quant. Electronics*, **QE, 9**, 855.
Baum, G., Lubell, M. S., and Raith, W. (1972). *Phys. Rev.*, **A5**, 1073.
Drachenfels, W. von, Koch, U. T., Lepper, R. D., Müller, T. M., and Paul, W. (1974). *Z. Physik*, **269**, 387.
Fano, U. (1969). *Phys. Rev.*, **178**, 131.
Heinzmann, U., Kessler, J., and Lorenz, J. (1970). *Z. Physik*, **240**, 42.

23
Electron Impact Excitation of Light Atoms at Intermediate Energies

M. R. C. McDowell

Perturbation (Born) series, Coulomb-projected Born, eigenfunction expansion, distorted wave, eikonal, and many-body (Green's function) methods for the theoretical study of electron impact excitation of hydrogen and helium at energies from the ionization threshold to the Born region are reviewed. The results of recent calculations are compared, and analyzed in the light of absolute measurements of total and differential cross sections, polarization of the emitted photons, and electron–photon coincidence measurements.

1. Introduction

Theoretical studies of electron impact excitation of light atoms at intermediate energies have advanced appreciably in the last five years. They have been much aided by the parallel development of techniques for experimental work, which yielded absolute measurements of differential cross sections in several cases.

It has long been accepted that the first Born approximation (FBA) gives an accurate account of total cross sections for particular inelastic processes at some sufficiently high energy, though the grounds for this belief vary from physicist to physicist. It remains unclear how high this energy must be for FBA results to be reliable within some preassigned error.

It is equally clear, both on experimental and theoretical grounds, that the FBA can never describe inelastic differential cross sections at large angles, but, again, there is no way of specifying in advance the angular range over

M. R. C. McDowell • Mathematics Department, Royal Holloway College, Englefield Green, Surrey, England.

which one can expect FBA results to be accurate within a preassigned error at a specified energy.

It is therefore necessary to develop more complete theoretical treatments, and, by applying them to a wide variety of cases, both extend our knowledge of the limits of validity of the FBA and attempt to obtain results consistently valid over wider ranges of angle and energy.

At very low energies—essentially below the first ionization threshold— eigenfunction expansion techniques have shown their power. Close-coupling calculations, including correlation effects (Burke et al., 1967a; Taylor and Burke, 1967; Burke and Taylor, 1969), give total and differential cross sections in close agreement with absolute experimental measurements (Williams, 1974; Williams and Willis, 1974). Burke, Smith, Seaton, and others have extensively developed these methods and made detailed predictions of resonance phenomena in hydrogen, helium, and other atoms (see, for example, the paper by Burke in this volume). These methods cannot be applied in their original form in energy regions where many channels are open and closely spaced in energy.

I will attempt in the remainder of this paper to review briefly several of the alternative methods that have been proposed for the treatment of inelastic processes at energies between the ionization threshold and the range of validity of the FBA (i.e., at "intermediate energies"). For reasons of space I will confine myself to specific transitions in hydrogen and helium, namely,

$$e + H(1s) \to e + H(2s, 2p) \tag{1}$$

$$e + He(1^1S) \to e + H(2^1S, 2^1P, 2^3S, 2^3P) \tag{2}$$

where absolute experimental measurements are available for comparison. This has the additional advantage that exact target eigenfunctions are available for hydrogen, and those for helium are of high accuracy; hence, differences between theoretical and experimental results for the cross sections must arise from either inadequate approximations in our scattering theory or experimental error, or both.

The measurable parameters common to all the transitions under study are the total cross section $Q_{if}(k^2)$ and the inelastic differential cross section $\sigma_{if}(\theta, k^2)$. In addition, the excited state may decay through photon emission [H(2p), He $(2^1P, 2^3P)$], and measurements of the optical polarization of the emitted photon can be made at, say, 90°, $P(90)$.

The scattered electron and emitted photon may be measured in coincidence (Macek and Jaecks, 1971; Eminyan et al., 1974). In simple cases [H(2p), He(2^1P)], the coincidence rate for fixed scattering angle and specified angle of emission of the photon may be expressed in terms of the orientation parameters of the excited state (Fano and Macek, 1973), which for our purpose are the moduli of the relative phase χ of the components of that state, and the ratio of the differential cross sections for excitation of that level:

$$\lambda = \sigma_0/(\sigma_0 + 2\sigma_1) \tag{3}$$

(see, however, Morgan and McDowell, 1975).

We shall assume that spin–orbit and other relativistic effects are unimportant for the cases considered. Measurements with spin-polarized atoms and electrons are not yet available and will not be considered.

The theoretical models are outlined in Section 2 and compared with absolute experimental data in Section 3, our conclusions being briefly summarized in tabular form on page 271.

2. Theoretical Models

The total cross section for excitation of an atom from an initial state i to a final state f by incident electrons of energy k_i^2 Ry may be written

$$Q_{if}(k_i^2) = \frac{1}{2\pi^2} \frac{k_f}{k_i} \int_{-1}^{+1} |T_{if}|^2 \, d(\cos\theta)\pi a_0^2 \tag{4}$$

where the T-matrix element is

$$T_{if} = \langle \psi_f, V_f, \Psi_i^+ \rangle \tag{5}$$

The total wave function in the initial channel Ψ_i^+ satisfies

$$(H - E)\Psi_i^+ = 0 \tag{6}$$

with appropriate boundary conditions, where ψ_f is the final unperturbed wave function defined by

$$H = H_f + V_f, \qquad H_f \psi_f = E\psi_f \tag{7}$$

and V_f is the interaction in the final channel.

The first Born approximation is obtained by noting that a formal expansion of Ψ_i^+,

$$\Psi_i^+ = \sum_{p=0}^{\infty} (G_i^+ V_i)^p \psi_i \tag{8}$$

may be made, with

$$H = H_i + V_i, \quad G_i^+ = \lim_{\varepsilon \to 0_+} \left\{ \frac{1}{E - H_i + i\varepsilon} \right\} \tag{9}$$

and retaining only the leading term so that

$$T_{if}^{(\text{FBA})} = \langle \psi_f, V_f, \psi_i \rangle \tag{10}$$

An obvious improvement is to retain the next-highest-order term to obtain the second Born approximation:

$$T_{if}^{(SBA)} = \langle \psi_f, V_f(\psi_i + G_i^+ V_i \psi_i) \rangle \tag{11}$$

The second-order term cannot be computed exactly, even in the simplest case. Further subsidiary approximations must be employed. They normally involve taking explicit account of the lowest N states and adopting a mean excitation energy, together with closure, to complete the sum. Different authors (Holt et al., 1971; Woollings and McDowell, 1972, 1973) make different choices of this mean excitation energy $\bar{\varepsilon}$.

The results for inelastic transitions show a marked improvement on the FBA differential cross sections at large angles. This is attributable to inclusion of the initial state as an intermediate state, elastic scattering being the dominant intermediate process. At smaller angles Berrington et al. (1973) conclude that the version adopted by Woollings and McDowell leads to a substantially improved account of the differential cross sections for He($2^1S, 2^1P$) at $E_i \geqslant 200$ eV.

The labor involved in using even a simplified version of the SBA suggests that its application to more complex atoms would be prohibitively expensive. The Eikonal–Born series (EBS) method of Byron and Joachain discussed in (iii) below is preferable, and a second-order diagonalization procedure developed by Baye and Heenan (1974) shows promise.

(i) Eigenfunction Expansion Methods

Where the close-coupling method is based on an expansion of Ψ_i^+ in terms of target eigenfunctions,

$$\Psi_i^+ = \sum_{q=1}^{N} \mathscr{A} \psi_{i_q}(\mathbf{r}) F_q(\mathbf{r}') + \chi(\mathbf{r}, \mathbf{r}') \tag{12}$$

necessarily truncated at some finite N, but with a correlation function χ representing some of the effects of the missing terms ($\mathscr{A}$ being an appropriate antisymmetrizing operator), Burke and Webb (1970) chose to expand in terms of pseudostates. The hope is that a suitable choice of expansion functions (of the appropriate rotational symmetries) will adequately represent the complete set of discrete and continuous states which should be present in (12). Burke and Webb studied the 2s and 2p excitations in atomic hydrogen, choosing a basis of the 1s, 2s, and 2p states of the target, together with modified $\overline{3s}$ and $\overline{3p}$ pseudostates,

$$|\overline{3l}\mu\rangle = \tilde{R}_{3l}(r) Y_{l\mu}(\hat{\mathbf{r}}) \tag{13}$$

with
$$\tilde{R}_{3p}(r) = (a + br)e^{-\varepsilon r}$$
$$\tilde{R}_{3s}(r) = (c + dr + er^2)e^{-\zeta r} \qquad (14)$$

The linear parameters were adjusted to ensure orthogonality to lower states of the same angular symmetry and to normalize the wave functions. Then the exponents were chosen to give a joint dummy threshold at $k^2 = 1.0$ and to provide considerable correlation contributions in the final state. About 25% of the pseudostates arise from target s and p continuum states. This pseudostate approach has also been adopted by Callaway and his collaborators (Callaway and Wooten, 1974; Callaway et al., 1975).

An alternative approach has been taken by Bransden and Coleman (1972 and later papers). For simplicity of exposition, neglecting exchange, they again expand in target eigenstates, so that the scattering functions $F_q(\mathbf{r}')$ satisfy

$$(\nabla^2 + k_n^2)F_n(\mathbf{r}') = 2\sum_m V_{nm}F_m(\mathbf{r}') \qquad (15)$$

with

$$V_{nm} = \int \phi_n^*(\mathbf{r})V_i\phi_m(\mathbf{r})\,\mathbf{dr} \qquad (16)$$

The lowest N states are treated exactly, but for $m > N$ only coupling terms with states $p \in \{1, \ldots, N\}$ are retained, so that

$$F_m(\mathbf{r}') = 2\sum_{q=0}^{N}\int G_0^+(k_m^2, \mathbf{r}, \mathbf{r}')V_{mq}F_q(\mathbf{r})\,\mathbf{dr} \qquad (17)$$

and G_0^+ is the free particle Green's function at $E = k_m^2 + i\varepsilon$.

The equations for the scattering functions in the retained states are then (Mittleman and Pu, 1962)

$$(\nabla^2 + k_n^2)F_n(\mathbf{r}') = 2\sum_{m=0}^{N} V_{nm}(\mathbf{r}')F_m(\mathbf{r}') + 4\sum_{m=0}^{N}\int \mathbf{dr}\, K_{nm}(\mathbf{r}', \mathbf{r})F_m(\mathbf{r}) \qquad (18)$$

with kernels K_{nm} defined in Bransden and Coleman (1972).

These are still too complicated to be solved exactly, and so a mean excitation energy $\bar{\varepsilon}_m$ for $m > N$ is introduced which together with closure enables the optical potential $K_{nm}(\mathbf{r}, \mathbf{r}')$ to be written

$$K_{nm}(\mathbf{r}, \mathbf{r}') = G_0^+(\bar{\varepsilon}_m, \mathbf{r}, \mathbf{r}')\left[\tilde{V}_{nm} - \sum_{q=0}^{N}V_{nq}V_{qm}\right] \qquad (19)$$

with

$$\tilde{V}_{nm} = \int \mathbf{dr}''\phi_n^+(\mathbf{r}'')V_i(\mathbf{r}'', \mathbf{r})V_i(\mathbf{r}'', \mathbf{r}')\phi_m(\mathbf{r}'') \qquad (20)$$

and the ϕ_n being target eigenstates.

The parameter $\bar{\varepsilon}_m$ is then chosen to ensure that the diagonal elements of $\tilde{V}_{nm}$ behave correctly as $-\alpha_n/r^4$ asymptotically, the resulting equations being solved initially in an impact parameter model (though work on a partial wave treatment and on the inclusion of exchange is in progress (Winters et al., 1973, 1974). We will refer to this method as the second-order optical potential, or SOP method.

(ii) Distorted-Wave Methods

Madison and Shelton (1971, 1973) have carefully formulated a distorted-wave model for the problem. They write the total Hamiltonian as

$$H = H_{atom} + T_0 + V_i = H_i + V_i \qquad (21)$$

in the initial channel, T_0 being the kinetic energy operator for the incident electron, and choose in addition a model Hamiltonian,

$$H_D = H_{atom} + T_0 + U_f = H_i + U_f \qquad (22)$$

where U_f is an arbitrary interaction. The unperturbed states ψ_q can be written as products of target eigenfunctions ϕ_q and plane waves. The U-potential distorted wave in channel f is written as W_f^- and satisfies

$$|W_f^- \phi_f\rangle = |W_f^- \phi_f\rangle + \frac{1}{E - i\varepsilon - H_f} U_f |W_f^- \phi_f\rangle \qquad (23)$$

so that after some algebra the T-matrix element becomes, for an n-electron atom,

$$T_{if} = (n+1)\langle W_f^- \phi_f | V_f - U_f | \mathscr{A} \Psi_i^+ \rangle \qquad (24)$$

which is still exact, but unhelpful, as Ψ_i^+ is unknown.

They assume that with a suitable choice of U_f an initial channel interaction U_i can be found, so that a good approximation to Ψ_i^+ is

$$\Psi_i^+ \simeq \phi_i W_i^+ \qquad (25)$$

with $|W_i^+ \phi_i\rangle$ satisfying (23) with U_f replaced by U_i.

They then choose

$$U_q = \frac{2Z}{r_{n+1}} + \langle q|V_i|q\rangle_{sp} \qquad (26)$$

the last term being the spherical average of the static potential in channel q. Thus, finally, their distorted-wave approximation is

$$T_{if}^{DW} = \langle W_f^- \phi_f | V_f - U_f | \phi_i W_i^+ \rangle$$
$$\quad - n\langle W_f^- \phi_f | V_f - U_f | \phi_i(1, \ldots, n-1; n+1) W_i^+(n)\rangle$$
$$= T_{if}^{DW}(\text{direct}) - T_{if}^{DW}(\text{exchange}) \qquad (27)$$

It is important to note that, unlike the other models discussed, the potentials ($V_f - U_f$) employed here are short-range and do not have a polarization behavior asymptotically. Further, the distorted waves, W_f^- and W_i^+, which appear in both channels are elastic scattering solutions without exchange in the channel static potential.

An alternative treatment has been given by McDowell, Myerscough, and Morgan (McDowell et al., 1973, 1975a,b). They choose

$$\Psi_i^+ = \mathscr{A}[\phi_i(\mathbf{r}) + \phi_{\text{pol}}(\mathbf{r}, \mathbf{r}')]F^q(\mathbf{r}') \qquad (28)$$

where $\phi_{\text{pol}}(\mathbf{r}, \mathbf{r}')$ is the dipole part of the first-order perturbed target wave function, provided $r > r'$, and zero otherwise. The distorted waves F^q are expanded in partial waves,

$$F^q = \sum_{l=0}^{\infty} A_l U_l^{(q)}(r) P_l(\cos \theta) \qquad (29)$$

the radial functions $U_l^q(r)$ being spin dependent and satisfying the adiabatic exchange equations (Temkin and Lamkin, 1961; Duxler et al., 1971):

$$\left[\frac{d^2}{dr^2} + k_i^2 - \frac{l(l+1)}{r^2} - 2V_{00}(r) - 2V_{\text{pol}}(r)\right] U_l^{(q)}(r) = X^q R_i(r) \qquad (30)$$

where

$$V_{\text{pol}}(r) \underset{r \to \infty}{\to} -\frac{\alpha_i}{r^4} \qquad (31)$$

and $X^q(r)$ is a nonlocal zero-order exchange interaction.

The T-matrix element then becomes (for spin q)

$$T_{if}^{\text{DWPO}} = \langle \phi_f \varkappa_f | V_f | [\phi_i + \phi_{\text{pol}}] F^q(\mathbf{r}') \rangle$$
$$- \langle \phi_f \varkappa_f | V_f | \phi_i(\mathbf{r}') F^q(\mathbf{r}) \rangle \qquad (32)$$

the function $\varkappa_f$ being an outgoing plane wave in the e–H case and the polarization function ϕ_{pol} being omitted in the exchange term, consistently with neglect of exchange polarization in obtaining adiabatic exchange distorted waves. We shall refer to this method as the distorted-wave-polarized-orbital (DWPO) method.

(iii) Eikonal Methods

Various forms of Glauber's eikonal approximation (Glauber, 1953) have been applied to the excitation problems considered here. A detailed review has recently been published (Gerjuoy and Thomas, 1974) so that only a brief outline is required in our paper. Again, neglecting exchange, the problem may be treated by analogy with potential scattering. We take

$$\Psi_i^+ = B e^{iS} \tag{33}$$

with B and S real, but B slowly varying with wavelength,

$$(\nabla^2 B/B) \ll k_i^2 \tag{34}$$

so that, assuming incident plane waves and straight-line trajectories,

$$S = k_i z - \frac{1}{v_i} \int_{-\infty}^{z} V_i(x, y, z') \, dz' \tag{35}$$

and

$$\Psi_i^+ \simeq e^{i\mathbf{k}_i \cdot \mathbf{r}_{n+1}} \exp\left\{-\frac{i}{v_i}\phi(r)\right\} \tag{36}$$

with

$$\phi(r) = \int_{-\infty}^{z} V_i(x, y, z') \, dz'$$

The Glauber T matrix is then (for inelastic events)

$$T_{if}^{(G)} = \langle \psi_f | V_f | e^{-(i/v)\phi(\mathbf{r}_{n+1})} \psi_i \rangle \tag{37}$$

In this form it is clearly a high-energy, small-angle approximation and there is no further loss of generality in assuming that the component of momentum transfer parallel to $\mathbf{k}_i$ can be neglected. Take $\hat{\mathbf{k}}_i$ along Z, and let $\mathbf{b} = (b, \theta)$ be the trace of the trajectory in a plane perpendicular to $\hat{\mathbf{k}}_i$, so that after some algebra,

$$T_{if}^{(G)} = -ik_i \int d\mathbf{b} \, e^{i\mathbf{k} \cdot \mathbf{b}} \langle \phi_f | 1 - e^{-(i/v_i)\phi(\mathbf{r}_{n+1})} | \phi_i \rangle \tag{38}$$

Note that if $|\phi/v_i| \ll 1$, then the expansion of the exponential gives

$$T_{if}^{(G)} \xrightarrow[v_i \to \infty]{} T_{if}^{(FBA)} \tag{39}$$

Various modifications have been suggested to allow (38) to be extended to larger scattering angles (cf. Gerjuoy and Thomas, 1974), but all are inconsistent with the step from (37) to (38). For spherically symmetric potentials, integration over θ gives

$$T_{if}^{(G)} = -2\pi i k_i \int_0^{\infty} b \, db \, J_0(Kb) \langle f | e^{2i\chi(b)} - 1 | i \rangle \tag{40}$$

where

$$\chi(b) = -\frac{1}{v_i} \int_0^{\infty} V_i(\sqrt{z^2 + b^2}) \, dz$$

Now, at high energies, provided $k_i \simeq k_f$, $K \simeq 2k_i \sin(\theta/2)$ and $J_0[2k_i b \sin(\theta/2)]$ is the small angle limit of a Legendre function, and a first-order correction term may be derived (Wallace, 1973).

An alternative derivation of the many-body form of the Glauber approximation due to Gerjuoy and Thomas (1974) shows that it is closely related to the second-order optical potential method of Bransden and Coleman (1972), by using a mean excitation energy $\bar{\varepsilon}_m = k_i^2$ and closure to reduce the Lippman–Schwinger equation to potential scattering form. However, this particular choice of $\bar{\varepsilon}_m$ implies that, although the Glauber approximation takes account of the long-range polarization interaction, it does so for the incorrect polarizability. This also follows from the fact that the second-order term in the expansion of (40) in powers of $\chi(V_i)$ is purely real; it is the missing term that gives the polarizability.

A more elaborate eikonal approximation has recently been suggested by Flannery and McCann (1974, and this volume), and allows for different trajectories in each channel. They replace (33) by a truncated sum over target eigenstates,

$$\Psi_i^+(\mathbf{r}, \mathbf{r}_{n+1}) = \sum_{q=1}^{N} A_q(\mathbf{b}, z) \psi_q e^{iS_q(\mathbf{b}, z)} \qquad (41)$$

and specify the eikonal phase S_q in channel q as the solution of the "static" (V_{qq}) potential scattering problem in the classical ($\hbar \to 0$) limit. Provided V_{qq} is slowly varying with channel wavelength, the eikonal coefficients A_q satisfy

$$iK_p \frac{\partial A_p}{\partial z} = \sum_{q=1}^{N}{}' A_q V_{pq} e^{i(S_q - S_p)} \qquad p = 1, \ldots, N \qquad (42)$$

where K_q is the local channel wave number

$$K_q = [k_q^2 - 2V_{qq}(\mathbf{r}_{n+1})]^{1/2}$$

Factoring out the azimuthal dependence of the potentials (which for a transition in which the orbital magnetic quantum number changes by Δ is $e^{-i\Delta\phi}$), the formal solution to (42) is the many-state eikonal (ME) T matrix,

$$T_{if,\Delta}^{ME} = -2\pi i^{\Delta+1} \int_0^\infty J_\Delta(bk_f \sin\theta)\{I_1(b,\theta) - I_2(b,\theta)\} b\, db \qquad (43)$$

where

$$I_1(b, \theta, \alpha) = \int_{-\infty}^{\infty} K_f(b, z) \frac{C_f}{\partial z} e^{i\alpha z}\, dz$$

$$I_2(b, \theta, \alpha) = i \int_{-\infty}^{\infty} [K_f(K_f - k_f) + V_{ff}] C_f e^{i\alpha z}\, dz$$

and α is the difference between the Z component of $\mathbf{K}$ and $\mathbf{K}_{\min}$

$$\alpha = 2k_f \sin^2\left(\frac{\theta}{2}\right)$$

while

$$C_q = A_q e^{-i\Delta\phi} \exp\left\{i \int_{-\infty}^{z} (K_q - k_q)\, dz\right\}$$

All channels are now coupled through $\partial C_f/\partial z$, and the integrals over z are taken over a classical trajectory in the appropriate channel (usually assumed rectilinear). Neglecting couplings except those connecting initial and final channels yields the FBA, and with obvious simplifications the method reduces to the Glauber approximation.

A different approach to improving the Glauber and Born models has been investigated by Byron and Joachain (1973), and a detailed review is in preparation (Byron and Joachain, 1975). They start from an examination of the energy dependence of each term of expansions of Born and Glauber series in powers of V_i. Each term of the Born series (7) apart from the FBA contains both real and imaginary parts, whereas the Glauber series (38)

$$T_{if}^{(G)} = -ik_i \int d\mathbf{b}\, e^{i\mathbf{K}\cdot\mathbf{b}} \left\langle \phi_f \bigg| \sum_{q=1}^{\infty} \frac{1}{q!}\left(\frac{i}{v}\phi\right)^q \bigg| \phi_i \right\rangle \qquad (44)$$

may be written

$$T_{if}^{(G)} = \frac{k_i}{i} \sum_{q=1}^{\infty} \frac{i^q}{q!} f_q^{(G)} \qquad f_q^{(G)} = \langle \phi_f | \chi(b)^q | \phi_i \rangle$$

(the z axis being chosen perpendicular to k_i). Thus the Glauber terms are alternatively real or pure imaginary. The first Glauber term is identical to the FBA, to which the dominant correction at large k_i^2 is Re $f_{B2} \propto k_i^{-1}$ and Im f_{B2} which behaves as $k_i^{-1} \ln k_i$ at small angles. However, the higher Born terms are dominated by Re f_{B3}, and this may be replaced by $f_3^{(G)}$. Byron and Joachain (1973) then take

$$T_{if}^{(EBS)} = T_{if}^{FBA} + T_{if}^{SBA} + T_{if}^{(G3)}$$

which is equivalent to correcting the usual (simplified) second Born approximation by adding the third-order Glauber term. In addition, if the resulting expression for $T_{if}^{(EBS)}$ is to be correct to order k_i^{-2} at large k_i^2, the leading k_i^{-2} term of the exchange amplitude must be included. It can be shown that this is, in fact, the usual Ochkur approximation.

Unpublished results (F. W. Byron, private communication, 1974) for $e + H(1s) \rightarrow e + H(2s)$ differential cross sections at 100 eV are in very close agreement with those obtained in the DWPO approximation (McDowell et al., 1975b).

The eikonal and distorted-wave approaches have been combined in an eikonal distorted-wave theory by Chen et al. (1972) and applied to e + H and e − He scattering by them and by Joachain and Vanderporten (1973, 1974). The small-angle results are rather unsatisfactory compared with those of the DWPO or EBS models. At large angles, the model, which allows for the elastic intermediate channel, necessarily gives the correct shape. A more detailed comparison of the EDW results with those of other models will be given elsewhere (Bransden and McDowell, 1976).

(iv) Green's Function Methods

Taylor and his colleagues (Csanak et al. 1973) have used the elegant Green's function method of Martin and Schwinger (1959) and extended it to inelastic transitions in atoms. A detailed review of the theory has been given elsewhere (Csanak et al. 1971); a progress report appears in this volume. The T-matrix element T_{if} is constructed by starting with the probability distribution $\phi_{\mathbf{k}_i}(\mathbf{r}_1, t_1)$ of having a wave packet $\psi^+(\mathbf{r}_1 t_1)|0\rangle$ present in the position $\mathbf{r}$, with momentum $\mathbf{k}_1$ at time t_1 and ending up with a probability distribution $\phi_{\mathbf{k}_1}(\mathbf{r}_2, t_2)$ of that wave packet propagating to $\psi(\mathbf{r}_2 t_2)|0\rangle$, allowing for all time orderings and intermediate states. That is,

$$T_{if} = \lim_{\substack{t_1 \to -\infty \\ t_2 \to +\infty}} \int d\mathbf{K}_1 \int d\mathbf{K}_2 \, \phi^*_{\mathbf{k}_f}(2) \langle f | \tau(\psi(\mathbf{r}_2, t_2)\psi^+(\mathbf{r}_1 t_1)|i\rangle \phi^\Omega_{\mathbf{k}_i}(1))$$

where τ is the time-ordering operator, and the bracket is the off-diagonal matrix element χ_{if} satisfying

$$\chi_{if}(1, 2) = \int G(1, 1')G(2, 2')\Gamma(1'2'34)\chi_{if}(34) \, dK'_1 \, dK'_2 \, dK_3 \, dK_4$$

the G's being one-particle propagators and Γ the derivative with respect to G of the self-energy (or optical potential) Σ,

$$\Gamma = \frac{\partial \Sigma}{\partial G} \simeq \frac{\partial \Sigma_{\text{HF}}}{\partial G}$$

approximated by taking the Hartree–Fock self-energy.

Inserting $\chi_{if}(1, 2)$ in the T-matrix element, the integrals over t_1 and t_2 combine the probability distributions $\phi_{\mathbf{k}_f}$ with $G(2, 2')$ and $\phi_{\mathbf{k}_i}$ with $G(1, 1')$ to yield distorted waves $f^-_{\mathbf{k}_i}(2)$ and $f^+_{\mathbf{k}_i}(1)$ both in the field of the ground state, all other channel effects, including those of the final channel being absorbed in the off-diagonal pseudopotential V_{0n}. Thus, the many-body Green's function (GF) matrix element is

$$T_{if}^{(GF)} = \frac{1}{i} \int d\mathbf{r}_1 \int d\mathbf{r}_2 \, f_{\mathbf{k}_i}^+(1) V_{0n}(1,2) f_{\mathbf{k}_f}^-(2)$$

and the $f_{\mathbf{k}}^\pm$ are elastic scattering solutions of the fixed core Hartree–Fock equations in the initial channel, but with momenta $\mathbf{k}_i$ and $\mathbf{k}_f$. Taylor's method then depicts the incident and scattered electrons as traveling in the field of the ground state, any photon emission identifying the final state taking place when the "scattered" electron is at asymptotic distances.

Note that exchange effects, but not polarization, are included for both distorted waves and that polarization is the dominant correction to the static potentials in $V_{0n}(1, 2)$.

(v) *The Coulomb-Projected Born Approximation*

The choice of unperturbed final channel Hamiltonian in (6) is arbitrary. Commonly it is chosen, as in deriving the Born series, to consist of the target Hamiltonian, together with the kinetic energy operator for the scattered electron. An alternative choice is to include the interaction of that electron with the target nucleus in H_f, so that at large separations in the final channel the total wave function becomes the product of a target eigenfunction and a Coulomb wave.

This approach has been adopted by Geltman and Hidalgo (1971; Hidalgo and Geltman, 1972) in their Coulomb-projected Born approximation (CPB). Thus, the CPB T-matrix element becomes

$$T_{if}^{CPB} = \left\langle \phi_f(\mathbf{r}) \varkappa_{\mathbf{k}_f}(z, \mathbf{r}') \frac{1}{\mathbf{r}-\mathbf{r}'} \phi_i(\mathbf{r}) e^{i\mathbf{k}_i \cdot \mathbf{r}'} \right\rangle$$

for a hydrogenic target.

The distorted wave $\varkappa_{\mathbf{k}_f}(z, \mathbf{r}')$ is pure Coulomb and does not take account of long-range polarization interactions. Exchange is not included in this model, either in the T matrix or in obtaining the radial functions, but an extension of the model to include the exchange matrix element has been carried out by Stauffer and Morgan (1975), and applied to $1s \to 2s$ and $2p$ transitions in hydrogen (CPBE).

3. Predictions of the Models

(i) *Total Cross Section for Excitation of H(2s)*

Absolute cross sections for

$$e + H(1s) \to e + H(2s)$$

have been measured by Kaupilla et al. (1970) and normalized via the 2p measurements of Long et al. (1969) to the Born approximation for 2p at 200 eV. The earlier measurements of Hils et al. (1966) were extended by Koschmieder et al. (1973) at low energies; we have normalized these results to those of Kaupilla et al. at 0.8 Ry.

We suggest elsewhere (McDowell et al., 1975a) that the Born values for 2p may be 10% too high at 200 eV compared with the predictions of the theories discussed in Section 2. This would also affect the 2s normalization.

For the present we retain the Kaupilla normalization, and in Figure 1 compare the experimental results from threshold to 200 eV with the predictions of several theoretical models. Kaupilla et al. present their results graphically, including the cascade contribution of $0.23Q(3p)$. We have used our DWPO values of $Q(3p)$ to obtain the 2s cross section (McDowell et al. 1975a). It is clear that the 4-state impact parameter version of the SOP model is inadequate in this case at energies below 150 eV. The pseudostate calculation of Burke and Webb (1970) improves on this but gives results which are appreciably too low. The experimental results for $E \geqslant 50$ eV are bracketed above by the 4-state ME of Flannery and McCann (1974) and

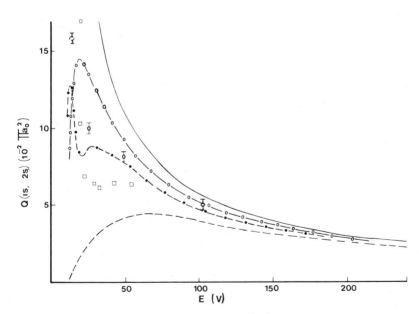

Figure 1. Total cross section (in units of $10^{-2}\pi a_0^2$) for $e + H(1s) \to e + H(2s)$. Theoretical results: (— — —), SOP 4-channel (Sullivan et al., 1972); (O — O), ME 4-channel (Flannery and McCann, 1974); (● - - ●), DWPO (McDowell et al., 1975a); (□), pseudostates (Burke and Webb, 1972); (———), Born (FBA). The experimental results (⌀) are those of Kaupilla et al. (1970), corrected for cascade.

below by the DWPO II results of McDowell *et al.* (1975a). All these models tend rapidly to FBA for $E_i \geqslant 200$ eV.

(ii) Total Cross Section for Excitation of H(2p)

Absolute total cross sections for

$$e + H(1s) \rightarrow e + H(2p)$$

have been measured by Williams and Willis (1974) at low energies. They are in good agreement with the relative measurements of Long *et al.* (1969), which are normalized to the FBA at 200 eV. These latter results are shown in Figure 2 and compared with a number of theoretical calculations. They are compatible at all energies $20 \leqslant E \leqslant 200$ eV with experiment within $\pm 10\%$ except for the 4-state SOP results and the FBA, both of which are too large at energies below 150 eV. All the theoretical models (except the FBA) give an accurate account of the long-range polarization, but whereas the DWPO and pseudostate calculations allow for exchange in calculating the distorted radial functions, the other two calculations do not.

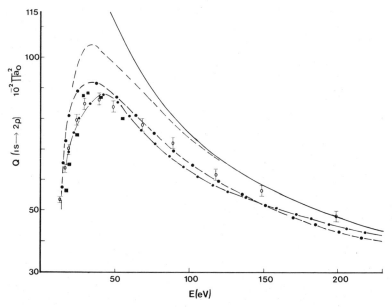

Figure 2. As Figure 1, but for $e + H(1s) \rightarrow e + H(2p)$. The experimental results (ϕ) shown are those of Long *et al.* (1969), normalized to the Born approximation at 200 eV.

This suggests that the long-range polarization potential is not important in determining the total cross section, the large difference between the 4-state SOP results and the other non-FBA calculations being attributed to the use of an impact parameter approximation to the solution of the SOP equations.

(iii) Lyman α Polarization

Results for the linear optical polarization of Ly_α at 90° to the incident beams

$$P(90) = \frac{3(1-x)}{7+11x} \qquad x = \sigma_1/\sigma_0$$

have been reported in the FBA, Glauber (Gerjuoy and Thomas, 1974), and DWPO (McDowell et al., 1975a) models, and are compared with the experimental measurements of Ott et al. (1970) in Figure 3. The DWPO results are in essentially perfect agreement with experiment for $E \geqslant 20$ eV, though the Glauber results show a substantial improvement on the FBA values.

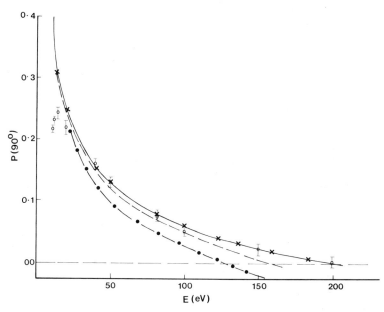

Figure 3. Optical polarization of Lyα at 90° to the incident beam. The experimental data ($\bar{\phi}$) of Ott et al. (1970) are compared with the FBA (● — ●), Glauber (— — —), and DWPO (× — ×) results.

However, the Glauber model predicts zero polarization at 150 eV, DWPO agreeing with experiment in predicting it at 200 eV. Below 20 eV all three models tend to the threshold value of 43% (Percival and Seaton, 1958), whereas the measured value near threshold is 14%, in good agreement with close-coupling predictions (Taylor and Burke, 1967; Callaway *et al.*, 1975).

(iv) *Differential Cross Sections for the n = 2 Levels*

Williams and Willis (1975) have obtained absolute measurements of the large-angle $n = 2$ differential cross section at a number of energies. These are in good agreement with a recent absolute normalization of the earlier work of K. G. Williams (1969) at 20° by Bohm (1974).

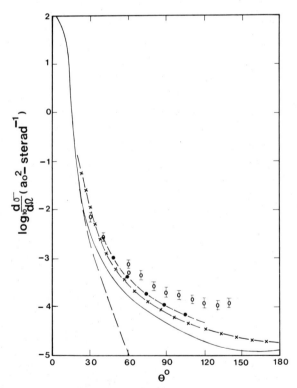

Figure 4. Differential cross sections for $e + H(1) \to H(2) + e$ at 200 eV (McDowell *et al.*, 1975b). The measurements (○) are those of Williams and Willis (1975), the theoretical results shown being FBA (— — —), DWPO (———), CPB (× — ×), and 4-state SOP (● — ●).

A detailed comparison of theory and experiment at 50(50)200 eV will be given elsewhere (McDowell et al., 1975b), the results at 200 eV being shown in Figure 4.

The FBA fails dramatically at angles greater than 30°, the difference with the DWPO results at 180° being a factor of 10^6! However this enormous difference is largely accounted for by departures of the low-angular-momentum ($l \leqslant 10$) partial waves from their Born form. It follows that any partial-wave treatment will have great difficulty in obtaining accurate large-angle differential cross sections.

In the present case, all the non-FBA models whose results are shown (SOP, ME, CPB, DWPO) give differential cross sections of the correct shape, the best absolute values being obtained by the 4-state SOP calculation (Sullivan et al., 1972). Nevertheless, the remaining discrepancy with experiment is a factor of two at 120° and 200 eV.

The distorting potentials in all these calculations are static or adiabatic and it would be interesting to examine the importance of nonadiabatic effects on large-angle scattering. Such calculations are now in progress.

Glauber calculations of $I_{1 \to 2}(\theta)$ at large angles have been reported by Tai et al. (1970) and Gosh and Sil (1970), and are in good agreement with the other theoretical models discussed. However, the difficulty of justifying the Glauber model at large angles remains, and its success there must, for the present, be regarded as fortuitous.

(v) Coincidence Measurements of Ly_α Photons and Scattered Electrons

It would be of interest, in principle, to extend the coincidence measurements of Eminyan et al. (1974) to Ly_α. Expressions for the coincidence rate have been given by Macek and Jaecks (1971), and it can be shown to depend on essentially two parameters. These are the relative phase χ of the two components of the $2p$ state,

$$|2p\rangle = a_0|10\rangle + a_1\{|1,1\rangle - |1,-1\rangle\}$$

with

$$a_0 = a_{00}\,e^{i\chi_0} \qquad a_1 = a_{10}\,e^{i\chi_1}$$

and

$$\chi = \chi_0 - \chi_1$$

together with the ratio of differential cross sections,

$$\lambda = \sigma_0/(\sigma_0 + 2\sigma_1)$$

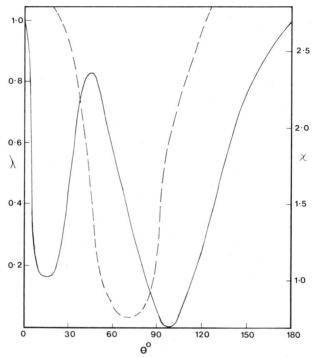

Figure 5. Orientation parameters χ (— — —) and λ (———) for the $2p$ state of H, calculated for $E = 100$ eV in the DWPO approximation (McDowell et al., 1975b).

These two parameters determine the components of the alignment tensor and orientation vector (Fano and Macek, 1973). Measurements of χ and λ provide a more sensitive test of theory than do total or differential cross section measurements.

McDowell et al. (1975b)* have carried out extensive calculations for the Ly$_\alpha$ case in the DWPO model. At 100 eV their results for λ are in agreement with Born approximation for $\theta \leq 10°$ (whereas the differential cross section in FBA seemed good out to 30° at this energy), but FBA predicts $\chi \equiv 0$!

The corresponding DWPO values are shown in Figure 5, χ decreasing smoothly from a value near π at small angles to a broad flat minimum of 0·78 between 60° and 80°, while λ shows an oscillatory behavior with minima at 15° and 95°, and maximum at 45°. The corresponding differential cross section is quite smooth in this angular range (15° $\leq \theta \leq$ 45°) and well

---

*Note Added in Proof. It was later realized that the simple model outlined here does not apply to hydrogen (Morgan and McDowell, 1975). The results for λ are unaffected.

represented by the FBA. However, the FBA values of λ,

$$\lambda_{FBA} = \frac{(1 - x\cos\theta)^2}{1 + x^2 - 2x\cos\theta} \qquad x = k_f/k_i$$

increase smoothly from a minimum at 20° to their maximum at 180°.

(vi) Total Cross Section for Excitation of $He(2^1S)$

There do not appear to be many absolute measurements for the excitation cross sections, results at some energies being given by Rice *et al.* (1972), Vriens *et al.* (1968), and Lassettre (1965). Of the theoretical models considered here, there are published results for the CPB, DW, SOP, Glauber, and GF methods. A graphical comparison for 2^1S is given in Figure 6.

As with all calculations of transitions in complex atoms, some of the differences between the results obtained in different models are attributable to different choices of approximate target wave functions, but it is not in general possible to separate out this effect.

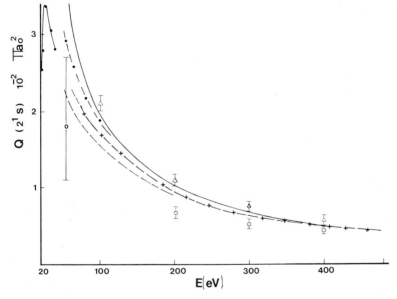

Figure 6. Total cross sections for excitation of $He(2^1S)$. Theoretical results are FBA (———), CPB (× -- ×), SOP (— — —), and 3-state close coupling (— — — —) (Marriott, 1964). The experimental points are due to Rice *et al.* (1972), (ϕ), Vriens *et al.* (1968), renormalized by Chamberlain *et al.* (1970), ($\square$), and Lassettre (1965), ($\triangle$). Also shown are the DW–SOP results of Bransden and Winters (1974) (● -- ●).

The theoretical models all give results which converge rapidly to the Born approximation for $E \geqslant 400$ eV, with respect to total cross sections. At lower energies the most consistent interpretation is to prefer the CPB and SOP results to the FBA, since they are consistent both with the experimental result of Rice *et al.* (1972), of 55.4 eV and with the 3-state 1^1S–2^1S–2^3S close-coupling calculation of Marriott (1964). This interpretation rules out the experimental results of Lassettre (1965), which are usually of high reliability, and for that reason it must be treated with caution. Further calculations in the DWPO and GF models are in progress, and additional experimental results would be helpful.

Winters (1974) has recently obtained distorted wave solutions of the SOP model, rather than impact parameter results, for this transition at 50, 81.6, and 100 eV. These results for the total cross section are above the earlier SOP values, and close to the Born values. A more detailed discussion is given elsewhere (Scott and McDowell, 1975).

(vii) *Differential Cross Sections for $He(2^1S)$, $He(2^3S)$*

Absolute differential cross section measurements have been reported by Trajmar (1973) and Truhlar *et al.* (1970, and private communication to E. C. Beaty), both normalized via 2^1P, and by Opal and Beaty (1972). The most interesting feature is the very strong minimum in the observed cross section at angles near 50° at energies below 100 eV.

As usual, the FBA model is unsatisfactory at large angles ($\theta > 40°$), whereas the other models do show the strong backward enhancement we noted in e–H collisions, and which is also characteristic of the observations in helium. The CPB calculations fail to show the minimum in the observed cross sections, these results at 81.6 eV being shown in Figure 7. The new DW–SOP calculations of Winters (1974) again give a reasonable account of the large-angle scattering and, though failing near the observed minimum, they do produce a clear shoulder in the differential cross section near 60°, as does the Glauber approximation (Gerjuoy and Thomas, 1974; Yates and Tenney, 1972). However, the Glauber results are quite accurate near the minimum ($\theta = 50°$), and at smaller angles.

At lower energies (Figure 8), for example, at 40.1 eV, the Glauber model fails completely, the minimum value obtained being an order of magnitude higher than experiment and shifted from 50 to 60°.

Calculations in the DWPO model are in progress (Scott and McDowell, 1975), and preliminary many-body results have been obtained by Taylor and his colleagues; they obtain minima, but the theoretical result is too deep at 29.6 eV, correct at 34 eV, and too shallow at 40.1 eV. Ormonde and Golden (1973) have carried out a 5-state (1^1S, 2^3S, 2^1S, 2^1P, 2^3P) close-coupling

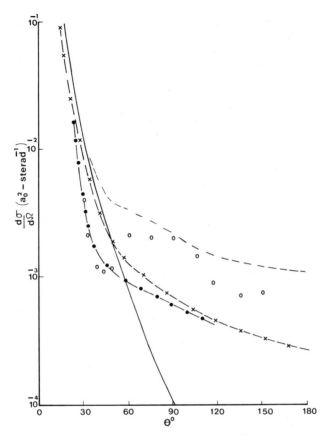

Figure 7. Differential cross sections for He(2^1S) at 81.6 eV. The experimental data shown as open circles are those of Opal and Beaty (1972). Theoretical models are FBA (———), DW (— — —), CPB (× - - ×), and Glauber (● — ●).

calculation at 30, 40, and 50 eV which gives results in good general accord with experiment at 40 and 50 eV.

At higher energies the DW–SOP calculations of Winters (1974) are in excellent agreement with the measurements of Crooks *et al.* (1972) at all angles, both lying an order of magnitude above the Glauber values at 100 eV for $\theta \geqslant 90°$, the CPB results of Hidalgo and Geltman (1972) being lower by a factor of 5 than the DW–SOP values at 100 eV and 180°.

Thus it appears that distorted-wave calculations can account successfully for the details of the small-angle 2^1S inelastic differential scattering in helium for impact energies $E \geqslant 100$ eV, without including nonadiabatic

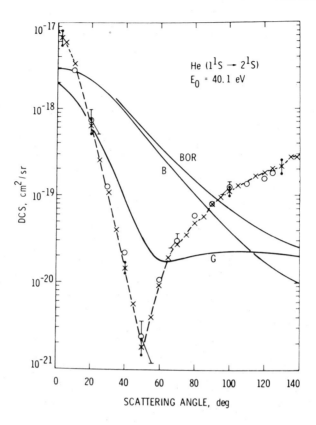

Figure 8. As Figure 7, but at 40.1 eV (Trajmar (1973) (× — ×)), together with the data of Hall *et al.* (1972a) at 39.2 eV (⌽). The Glauber (G), FBA (B) and Born–Ochkur–Rudge predictions (BOR) are shown for comparison. (By permission of the American Institute of Physics from S. Trajmar, 1973.) RMS error bars are given at some angles.

effects, though both exchange and polarization effects are important. At larger angles many-body theory succeeds in reproducing the deep minimum observed, though simple DW calculations probably cannot; Taylor's model includes nonadiabatic effects.

The position for 2^3S is quite similar, straightforward DW theories giving an accurate model at large angles for $E \geqslant 100$ eV, though tending to miss the detailed structure at moderate angles; at lower energies (50 eV) the deep flat minimum observed in the 2^3S differential cross section by Hall *et al.* (1972, 1973) is not obtained in the DW models so far published, though a shallow minimum is predicted by the many-body theory calculation. However, such models do reproduce the overall shape of the cross section correctly, unlike the simpler Born–Oppenheimer or Born–Ochkur–Rudge treatments, whose predictions (Trajmar, 1973) bear no relation to experiment. On the other hand, a recent four-channel variational calculation by Thomas and Nesbet (1974) at 29·6 eV does predict the main features of the differential cross section correctly.

(viii) *Total Cross Sections for He (2^1P, 2^3P)*

There are numerous experimental measurements of the 2^1P excitation. We give the values of Donaldson *et al.* (1972), normalized via the oscillator strength, which are in good agreement with the measurements of van Eck and de Jongh (1970), but 20% higher at 100 eV than those of the NBS group (Vriens *et al.*, 1968; Chamberlain *et al.*, 1970). They are compared with the theoretical results in Figure 9. The CPB and SOP models give results which are larger than the FBA cross section at energies below 200 eV, whereas the experimental results, which are in good agreement with the FBA for $E \geqslant 400$ eV, fall below it at lower energies.

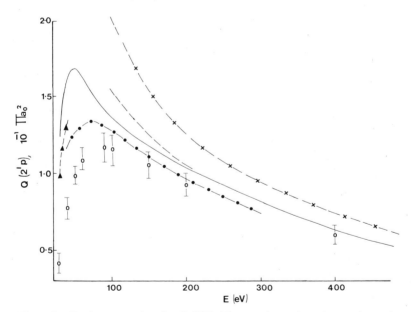

Figure 9. Total cross sections for He(2^1P). The experimental results are those of Donaldson *et al.* (1972), ($\bar{\text{Q}}$). Theoretical models are SOP (— — —), FBA (———), CPB (× — ×), and DW (● — ●); 2-state close coupling, (▲ — ▲) (Truhlar *et al.*, 1973).

The DW calculations by Madison and Shelton (1973) look to be a substantial improvement and are compatible with experiment for $E \geqslant 150\,\text{eV}$, though again tending to be high at lower energies. The $1^1S\text{–}2^1P$ close-coupling results calculated at those energies (29·0, 34·0, and 39·0 eV) by Truhlar et al. (1973) appear to be a factor of three too high compared with the measurements of Donaldson et al. (1972) or of de Jongh and van Eck (1971), and their slope is too large compared with the DW results. However, one would also expect coupling to 2^1S to be important in this case.

(ix) Differential Cross Sections for $He(2^1P)$

There are recent absolute experimental measurements by Truhlar et al. (1973), Opal and Beaty (1972), Hall et al. (1973), and Crooks (1972). The experimental differential cross section is well-behaved, showing the forward enhancement, FBA-type behavior out to 60°, and appreciable flattening at large angles (cf. Figure 10).

The first reported non-plane-wave calculations were in the Glauber model (Yates and Tenney, 1972), and were found to be in good agreement with experiment for $26.5 \leqslant E \leqslant 82\,\text{eV}$ and $\theta \leqslant 60°$. The 2-state close-coupling calculations (Truhlar et al., 1973) give a poor account of the experimental shape at 30 eV, having a deep minimum and being much too large at large angles, but they appear quite accurate at 40 eV for $\theta \leqslant 120°$, the oscillations in the theoretical values being due to a failure to include a sufficient number partial waves [cf. McDowell et al., (1975b) for a discussion].

At higher energies (200 eV) there is substantial qualitative agreement between the measurements of Opal and Beaty (1972), which agree with those of Williams (1969), and the CPB calculations of Hidalgo and Geltman (1972). However, the distorted-wave calculations of Madison and Shelton (1973) is again in very close agreement with the experimental data (at 200 eV) for $\theta < 120°$, though experimental values tend to be lower at larger angles.

This agreement between the DW results and experiment remains evident at energies as low as 48.2 eV, where the results of Hall et al. (1973) and Crooks (1972) are consistent, but large divergences appear for $E \leqslant 40\,\text{eV}$.

Again, results in the many-body theory (Green's function) approach by Taylor and his colleagues (Thomas et al., 1974) show considerable promise and appear superior to the DW results of Madison and Shelton (1973) at energies below 80 eV. This would not be suprising, since the GF model includes polarization correctly, and takes full acount of exchange in calculating the distorted waves. Both these effects will undoubtedly become more important at lower energies. DWPO calculations now in progress may help to clarify these points.

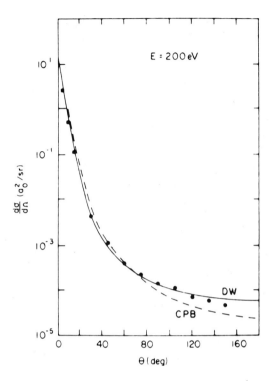

Figure 10. Differential cross sections for He(2^1P) at 200 eV from Madison and Shelton (1973), showing the experimental data of Opal and Beaty (1973) (solid circles) compared with the CPB and DW calculations. (By permission of the American Institute of Physics, from Madison and Shelton, 1973.)

Apart from the work by Taylor, there are no published results in the models considered here for 2^3P, or for excitation of the $n = 3$ levels. Work on these transitions is now required.

(x) *Electron–Photon Correlation Measurements*

In presenting a paper in honor of Ugo Fano, at Stirling, it would be improper to leave the topic of inelastic electron–atom collisions at intermediate energies without briefly mentioning the important experiment by Eminyan *et al.* (1973, 1974), even though it will be described in detail elsewhere in this volume. This is an electron–phonon coincidence experiment in helium, in which a λ5800 photon is collected in coincidence with a scattered electron which has lost energy corresponding to excitation of 2^1P, and the measurement is carried out in the scattering $(\hat{\mathbf{k}}_i, \hat{\mathbf{k}}_f)$ plane. The expected number of coincidences is a function of photon emission angle θ_v,

$$N_v \propto \{\lambda \sin^2 \theta_v + (1 - \lambda) \cos^2 \theta_v - [(1 - \lambda)]^{1/2} \sin 2\theta_v \cos \chi\}$$

and the measurements, carried out for $40 \leq E \leq 200$ eV and

$$\theta = \cos^{-1}(\hat{\mathbf{k}}_1, \hat{\mathbf{k}}_f)$$

in the range $16 \leq \theta \leq 40°$ for $20 \leq \theta_v \leq 140°$, allow a determination of λ and $|\chi|$.

Theoretical calculations have been carried out by Madison and Shelton, and by Taylor and his colleagues in the DW and GF models respectively, (private communications to H. Kleinpoppen), and their values for λ are compared with experiment and the FBA prediction in Figure 11. The DW results are superior in this case to the GF calculations, the Born model failing for $\theta > 20°$. The agreement of DW theory with experiment at 80 eV is excellent, but it should be noted that the agreement with FBA at 50 eV (not shown) is also extremely good. The FBA theory implies $\chi = 0$ (mod π), whereas the observed values at, say, 80 eV of $|\chi|$ increase smoothly with θ ($15 \leq \theta \leq 40°$). The only published theoretical predictions are the

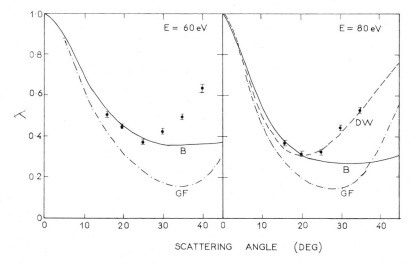

Figure 11. Excitation of He(2^1P). Measurements of $\lambda = \sigma_0(2\sigma_1 + \sigma_0)$ at 60 eV and 80 eV by Eminyan *et al.* (1974) compared with FBA (———), DW (— — —), and GF (— · —) calculations. (By permission of the authors.)

Table 1. Summary of Model Properties

	Model						
	PS	SOP	DW	DWPO	GF	ME	CPB
Direct and exchange	Yes	No?	Yes	Yes	Yes	No	No?
Distortion in both channels	No	No	Yes	No	Yes	Yes	No
Distorted wave with exchange	Yes	No?	No	Yes	Yes	No	No
Distorted wave with polarization	Yes	Yes	No	Yes	No	No	No
Potential with polarization	No	Yes	No	No	Yes	No	No
Full wave treatment	Yes	No	Yes	Yes	Yes	No	Yes
Impact parameter	No	Yes	No	No	No	Yes	No
Partial-wave analysis	Yes	No	Yes	Yes	Yes	No	No
Nonadiabatic effects	Yes	No	No	No?	No?	No	No

A query (?) indicates that work is in progress to extend the model.

DW results, which are too large but give an accurate account of the angular variation. Tai et al. (1970) have noted that the Glauber assumption that $\hat{K} \perp \hat{k}_i$ implies $\sigma_0 = 0$; hence in this (as in the FBA) approximation $\chi_G \equiv 0$ while $\lambda_G \equiv 0$ also; however, many-channel eikonal methods (Flannery and McCann, 1974) may prove useful.

These results are extremely encouraging, and there is no doubt that other theorists will quickly join in the fun.

References

Baye, D., and Heenan, P. H. (1974). *J. Phys. B.*, **7**, 928 and 938.
Berrington, K. A., Bransden, B. H., and Coleman, J. P. (1973). *J. Phys. B.*, **6**, 436.
Bohm, K. (1974). Private communications via J. A. Williams.
Bransden, B. H., and Coleman, J. P. (1972). *J. Phys. B.*, **5**, 537.
Bransden, B. H., and Winters, K. H. (1974). *J. Phys. B.*, **8**, 1236.
Bransden, B. H., and McDowell, M. R. C. (1976). In preparation for *Physics Reports*.
Burke, P. G., Ormonde, S., and Whitaker, W. (1967a). *Proc. Phys. Soc.*, **92**, 319.
Burke, P. G., Taylor, A. J., and Ormonde, S. (1967b). *Proc. Phys. Soc.*, **92**, 345.
Burke, P. G., and Taylor, A. J. (1969). *J. Phys. B.*, **2**, 44.
Burke, P. G., and Webb, T. G. (1970). *J. Phys. B.*, **3**, L.133.
Byron, F. W. Jr. and Joachain, C. J. (1973). *Phys. Rev.*, **A8**, 1267.
Byron, F. W. Jr., and Joachain, C. J. (1975). In preparation for *Physics Reports*.
Callaway, J., and Wooten, J. W. (1974). *Phys. Rev.*, **A9**, 1924.
Callaway, J., McDowell, M. R. C., and Morgan, L. A. (1975). *J. Phys. B.*, **8**, 2181.
Chamberlain, G. E., Mielczarek, S. R., and Kuyatt, C. E. (1970). *Phys. Rev.*, **A2**, 1905.
Chen, J. C. Y., Joachain, C. J., and Watson, K. M. (1972). *Phys. Rev.*, **A5**, 2460.
Crooks, G. B. (1972). Ph.D. Thesis, University of Nebraska.
Crooks, G. B., DuBois, R. D., Golden, D. E., and Rudd, M. E. (1972). *Phys. Rev. Lett.*, **29**, 327.

Csanak, G., Taylor, H. A., and Yaris, R. (1971). *Adv. Atom. Molec. Phys.*, **7**, 287.
Csanak, G., and Taylor, H. S. and Tripathy, D. N. (1973). *J. Phys. B.*, **6**, 2040.
Donaldson, E. G., Hender, M. A. and McConkey, J. W. (1972). *J. Phys. B.*, **5**, 1192.
Duxler, W., LaBahn, R. A. and Pöe, R. T. (1971). *Phys. Rev.*, **A4**, 1935.
Eminyan, M., MacAdam, K. B., Slevin, J., and Kleinpoppen, H. (1972). *Phys. Rev. Letters*, **31**, 576.
Eminyan, M., MacAdam, K. B., Slevin, J., and Kleinpoppen, H. (1974). *J. Phys. B.*, **7**, 1519.
Fano, U., and Macek, J. (1973). *Rev. Mod. Phys.*, **45**, 553.
Flannery, M. R., and McCann, K. J. (1974). *J. Phys. B.*, **7**, L223.
Geltman, S., and Hidalgo, M. B. (1971). *J. Phys. B.*, **4**, 1299.
Gerjuoy, E., and Thomas, B. K. (1974). *Rep. Prog. Phys.*, **37**.
Glauber, R. J. (1953). *Phys. Rev.*, **91**, 459.
Gosh, A. S., and Sil, N. C. (1970). *Indian J. Phys.*, **44**, 153.
Hall, R. I., Joyez, G., Mazeau, J., Reinhardt, J., and Schermann, C. (1972). *Compts Rendus*, **272**, 743.
Hall, R. I., Joyez, G., Mazeau, J., Reinhardt, J., and Schermann, C. (1973). *J. de Physique*, **34**, 827.
Hidalgo, M. B., and Geltman, S. (1972). *J. Phys. B.*, **5**, 617.
Hils, D., Kleinpoppen, H., and Koschmieder, H. (1966). *Proc. Phys. Soc.*, **89**, 35.
Holt, A. R., Hunt, J., and Moiseiwitsch, B. L. (1971). *J. Phys. B.*, **4**, 1318.
Joachain, C. J., and Vanderporten, R. (1973). *J. Phys. B.*, **6**, 622.
Joachain, C. J., and Vanderporten, R. (1974). *J. Phys. B.*, **7**, 817.
Kaupilla, W. E., Ott, W. R., and Fite, W. L. (1970). *Phys. Rev.*, **A1**, 1099.
Kleinpoppen, H. (1974). Invited Papers, 4th International Conference on Atomic Physics, Heidelberg.
Koschmieder, H., Raible, V., and Kleinpoppen, H. (1973). *Phys. Rev.*, **A8**, 1365.
Lassettre, E. N. (1965). *J. Chem. Phys.*, **43**, 4479.
Long, R. L., Cox, D. M., and Smith, S. J. (1969). *J. Res. N.B.S.*, **72A**, 521.
Macek, J., and Jaecks, D. A. (1971). *Phys. Rev.*, **A4**, 2288.
Madison, D. H., and Shelton, W. N. (1971). Tech. Rep. No. 9, Physics, Florida State University.
Madison, D. H., and Shelton, W. N. (1973). *Phys. Rev.*, **A7**, 499.
Marriott, R. (1964). In *Atomic Collision Processes*. Ed. M. R. C. McDowell, p. 114. Amsterdam: North Holland.
Martin, P. C., and Schwinger, J. (1959). *Phys. Rev.*, **115**, 1342.
McDowell, M. R. C., Myerscough, V. P., and Morgan, L. A. (1973). *J. Phys. B.*, **6**, 1435.
McDowell, M. R. C., Myerscough, V. P., and Narain, U. (1974). *J. Phys. B.*, **7**, L.195.
McDowell, M. R. C., Myerscough, V. P., and Morgan, L. A. (1975a). *J. Phys. B.*, **8**, 1053.
McDowell, M. R. C., Myerscough, V. P., and Morgan, L. A. (1975b). *J. Phys. B.*, **8**, 1838.
Mittleman, M. H. and Pu (Pöe) R. T. (1962). *Phys. Rev.*, **126**, 370.
Morgan, L. A., and McDowell, M. R. C. (1975). *J. Phys. B.*, **8**, 1073.
Opal, C. B., and Beaty, E. C. (1972). *J. Phys. B.*, **5**, 627.
Ormonde, S., and Golden, D. E. (1973). *Phys. Rev. Letters*, **31**, 1161.
Ott, W. R., Kaupilla, W. E., and Fite, W. L. (1970). *Phys. Rev.*, **A6**, 1089.
Percival, I. C., and Seaton, M. J. (1958). *Phil. Trans. Roy. Soc.*, **A751**, 113.
Rice, J. K., Truhlar, D. G., Cartwright, D. C., and Trajmar, S. (1972). *Phys. Rev.*, **A5**, 762.
Scott, T., and McDowell, M. R. C. (1975). *J. Phys. B.*, **8**, 1851.
Stauffer, A. D., and Morgan, L. A. (1975). *J. Phys. B.*, **8**, 2172.
Sullivan, J., Coleman, J. P., and Bransden, B. H. (1972). *J. Phys. B.*, **5**, 2061.
Tai, H., Bossel, R. H., Gerjuoy, E., and Franco, V. (1970). *Phys. Rev.*, **A6**, 1819.
Taylor, A. J., and Burke, P. G. (1967). *Proc. Phys. Soc.*, **92**, 336.

Temkin, A., and Lamkin, J. C. (1961). *Phys. Rev.*, **121**, 788.
Thomas, L. D., and Nesbet, R. K. (1974). Abstract, *IVth Int. Conf. Atom. Phys., Heidelberg*.
Thomas, L. D., Csanak, G., Taylor, H. S., and Yarlagadda, B. S. (1974). *J. Phys. B.*, **7**, 1719.
Trajmar, S. (1973). *Phys. Rev.*, **A8**, 191.
Truhlar, D. G., Rice, J. K., Kupperman, A., Trajmar, S., and Cartwright, D. C. (1970). *Phys. Rev.*, **A1**, 1819.
Truhlar, D. G., Trajmar, S., Williams, W., Ormonde, S., and Torres, B. (1973). *Phys. Rev.*, **A8**, 2475.
van Eck, J., and de Jongh, J. P. (1970). *Physica*, **47**, 141.
Vriens, L., Simpson, J. A., and Mielszarek, S. R. (1968). *Phys. Rev.*, **165**, 7; **170**, 163.
Wallace, S. J. (1973). *Phys. Rev.*, **D8**, 1846.
Williams, J. F. (1974). *J. Phys. B.*, **7**, 56.
Williams, J. F., and Willis, B. A. (1974). *J. Phys. B.*, **7**, L61.
Williams, J. F., and Willis, B. A. (1975). *J. Phys. B.*, **8**, 1641.
Williams, K. G. (1969). Abstracts *VI ICPEAC*, p. 731. Cambridge, Mass.: M.I.T. Press.
Winters, K. H. (1974). Ph.D. Thesis, University of Durham.
Winters, K. H., Clark, C. D., Bransden, B. H., and Coleman, J. P. (1973). *J. Phys. B.*, **6**, L247.
Winters, K. H., Clark, C. O., Bransden, B. H., and Coleman, J. P. (1974). *J. Phys. B.*, **7**, 788.
Woollings, M. J., and McDowell, M. R. C. (1972). *J. Phys. B.*, **5**, 1320.
Woollings, M. J., and McDowell, M. R. C. (1973). *J. Phys. B.*, **6**, 450.
Yates, A. C., and Tenney, A. (1972). *Phys. Rev.*, **A5**, 247; **6**, 1451.

24
The Multichannel Eikonal Treatment of Electron–Atom Collisions

M. R. FLANNERY AND K. J. MCCANN

A multichannel treatment of atomic collisions is presented and applied to the excitation of atomic hydrogen and helium by electrons with incident energy above the ionization threshold. The calculated cross sections compare very favourably with other refined theoretical procedures and with various experiments.

Introduction

A variety of methods have been proposed for the theoretical description of electron–atom collisions at low and intermediate energies. The close-coupling expansion with its pseudostate modifications (Burke and Webb, 1970) and the polarized-orbital distorted-wave model of McDowell et al. (1973) are among those that follow from the full wave treatment of the collision. Other methods, termed semiclassical—the eikonal approximations of Bates and Holt (1966), Callaway (1968), Byron (1971), and of Chen et al. (1972), the impact parameter approach (Bransden and Coleman, 1972), and the Glauber approximation (cf. Tai et al., 1972)—all essentially separate the relative motion of the incident electron (described by an eikonal-type or Born wave function for the electron in a static field) from the internal electronic motions of the atomic system which is described by a multistate expansion. In this paper, a new generalization of the eikonal approximation to multichannel scattering is presented.

M. R. FLANNERY and K. J. MCCANN • School of Physics, Georgia Institute of Technology Atlanta, Georgia 30332, U.S.A.

Theory

Consider the collision of a particle B of mass M_B and incident velocity $\mathbf{v}_i$ along the Z axis with a one-electron atomic system (A + e) of mass $(M_A + m)$. The subsequent analysis can be immediately generalized so as to cover multielectron systems. Let $\mathbf{R}$, $\mathbf{R}_B$, $\mathbf{r}$, and $\mathbf{r}_a$ denote the A–B, B–(Ae) center of mass, e–(AB) center of mass, and e–A separations, respectively. In the (ABe) center-of-mass reference frame, the scattering amplitude for a direct transition between an initial state i and a final state f of the collision system, of reduced mass μ, is

$$f_{if}(\theta, \varphi) = -\frac{1}{4\pi}\frac{2\mu}{\hbar^2}\langle \Psi_f(\mathbf{k}_f; \mathbf{r}, \mathbf{R})|V(\mathbf{r}, \mathbf{R})|\Psi_i^+(\mathbf{k}_i; \mathbf{r}, \mathbf{R})\rangle_{\mathbf{r}, \mathbf{R}} \quad (1)$$

in which $V(\mathbf{r}, \mathbf{R})$ is the instantaneous electrostatic interaction between the collision species, and where the scattering is directed along the final relative momentum $\hbar \mathbf{k}_f$ $(1, \theta, \varphi)$. The final stationary state of the isolated atoms in channel f is Ψ_f, and Ψ_i^+ is the solution of the time-independent Schrödinger equation,

$$\left[-\frac{\hbar^2}{2\mu}\nabla_R^2 + H_e(\mathbf{r}) + V(\mathbf{r}, \mathbf{R})\right]\Psi_i^+(\mathbf{r}, \mathbf{R}) = E_i \Psi_i^+(\mathbf{r}, \mathbf{R}) \quad (2)$$

solved subject to the asymptotic boundary condition

$$\Psi_i^+(\mathbf{r}, \mathbf{R}) \xrightarrow{\text{large } R} \sum_n \left[e^{i\mathbf{k}_n \cdot \mathbf{R}_B}\delta_{ni} + f_{in}(\theta, \varphi)\frac{e^{ik_n R_B}}{R_B}\right]\varphi_n(\mathbf{r}_a) \quad (3)$$

in which $\varphi_n(\mathbf{r}_a)$ are eigenfunctions of the Hamiltonian $H_e(\mathbf{r}) \approx H_e(\mathbf{r}_a)$ for the isolated atomic system (A + e) with internal electronic energy ε_n such that the total energy E_i in channel i is $\varepsilon_i + \hbar^2 k_i^2/2\mu$, which is conserved throughout the collision.

The Multichannel Eikonal Approximation

The eikonal approximation to (2) sets

$$\Psi_i^+(\mathbf{r}, \mathbf{R}) = \sum_n A_n(\boldsymbol{\rho}, Z) \exp i\, S_n(\boldsymbol{\rho}, Z)\varphi_n(\mathbf{r}_a) \exp\left\{i\left(\frac{m}{M_A + m}\mathbf{k}_n \cdot \mathbf{r}_a\right)\right\} \quad (4)$$

where the nuclear separation $\mathbf{R} \equiv (R, \Theta, \Phi) \equiv (\rho, \Phi, Z)$ in spherical and cylindrical coordinate frames respectively. The eikonal S_n in (4) is the characteristic-function solution of the classical Hamiltonian–Jacobi equation (i.e., the Schrödinger equation in the $\hbar \to 0$ limit) for the A–B relative motion under the static interaction $V_{nn}(\mathbf{R}) = \langle \varphi_n | V | \varphi_n \rangle$, and is therefore given by

$$S_n(\boldsymbol{\rho}, Z) = k_n Z + \int_{-\infty}^{Z} [\kappa_n(\mathbf{R}) - k_n] dZ \tag{5}$$

in which the local wave number $\kappa_n(\mathbf{R})$ of relative motion at $\mathbf{R}$ is

$$[k_n^2 - (2\mu/\hbar^2) V_{nn}(\mathbf{R})]^{1/2}$$

and where dZ is an element of path length along the trajectory, which, at present, is taken as a straight line. For electron–atom collisions, κ_n is always real. The use of the actual classical trajectory with its "built-in" turning point is therefore not as essential as, for example, in positron–atom collisions, where κ_n becomes imaginary for sufficiently close rectilinear encounters. The general problems associated with the choice of classical trajectory within a multichannel framework are at present unresolved, although the force-common-turning-point, two-state procedure of Bates and Crothers (1970) is attractive.

On assuming that the main variation of Ψ_i^+ on $\boldsymbol{\rho}$ is contained in S_n, i.e., provided $V_{nn}(\mathbf{R})$ varies slowly over many wavelengths $2\pi/\kappa(\mathbf{R})$ of relative motion, and the coefficients $A_n(\boldsymbol{\rho}, Z)$ therefore vary primarily along Z, substitution of (4) in (2) yields the set of coupled differential equations

$$\frac{i\hbar^2}{\mu} \kappa_f \frac{\partial B_f(\boldsymbol{\rho}, Z)}{\partial Z} + \left[\frac{\hbar^2}{\mu} \kappa_f(\kappa_f - k_f) + V_{ff}(\mathbf{R}) \right] B_f(\boldsymbol{\rho}, Z)$$

$$= \sum_{n=1}^{N} B_n(\boldsymbol{\rho}, Z) V_{fn}(\mathbf{R}) \exp\{i(k_n - k_f)Z\}, \quad f = 1, 2, \ldots, N \tag{6}$$

a set of N coupled equations to be solved for

$$B_f = A_f \exp\left\{ i \int_{-\infty}^{Z} (\kappa_f - k_f) dZ \right\},$$

subject to the asymptotic condition $B_f(\boldsymbol{\rho}, -\infty) = \delta_{fi}$, which ensures that $\Psi_i \sim \varphi_i(\mathbf{r}_a) \exp\{ik_i Z\}$ as $Z \to -\infty$. The scattering amplitude (1), with the undistorted final wave $\Psi_f = \varphi_f(\mathbf{r}_a) \exp\{i\mathbf{k}_f \cdot \mathbf{R}\}$ inserted, is therefore

$$f_{if}(\theta, \varphi) = -\frac{1}{4\pi} \frac{2\mu}{\hbar^2} \int \exp\{i\mathbf{K} \cdot \mathbf{R}\} d\mathbf{R} \sum_{n=1}^{N} B_n(\boldsymbol{\rho}, Z) V_{fn}(\mathbf{R}) \exp\{i(k_n - k_i)Z\} \tag{7}$$

where $\mathbf{K}$ is the momentum change $\mathbf{k}_i - \mathbf{k}_f$ caused by the collision. Since the electrostatic interaction $V(\mathbf{r}, \mathbf{R})$ is composed of central potentials,

$$V_{fi} = \langle \varphi_f | V | \varphi_i \rangle = V_{fi}(R, \Theta) \exp\{i\Delta\Phi\},$$

where $\Delta = M_i - M_f$ is the change in the azimuthal quantum number of the atom. Hence, with the substitution

$$C_f(\rho, Z) = B_f(\rho, Z) \exp\{-i\Delta\Phi\},$$

the set of phase Φ-independent equations

$$\frac{i\hbar^2}{\mu} \kappa_f(\rho, Z) \frac{\partial C_f(\rho, Z)}{\partial Z} + \left[\frac{\hbar^2}{\mu} \kappa_f(\kappa_f - k_f) + V_{ff}(\rho, Z)\right] C_f(\rho, Z)$$

$$= \sum_{n=1}^{N} C_n(\rho, Z) V_{fn}(\rho, Z) \exp\{i(k_n - k_f)Z\} \tag{8}$$

is obtained, solved subject to the boundary condition $C_f(\rho, -\infty) = \delta_{if}$. On completion of the Φ-integration in (7), the scattering amplitude reduces to

$$f_{if}(\theta, \varphi) = -i^{\Delta+1} \int_0^\infty J_\Delta(K'\rho)[I_1(\rho, \theta) - iI_2(\rho, \theta)]\rho \, d\rho \tag{9}$$

where K' is the XY component $k_f \sin\theta$ of $\mathbf{K}$ and where J_Δ are Bessel functions of integral order. Both the functions

$$I_1(\rho, \theta; \alpha) = \int_{-\infty}^\infty \kappa_f(\rho, Z) \left[\frac{\partial C_f(\rho, Z)}{\partial Z}\right] \exp\{i\alpha Z\} \, dZ \tag{10}$$

and

$$I_2(\rho, \theta; \alpha) = \int_{-\infty}^\infty \left[\kappa_f(\kappa_f - k_f) + \frac{\mu}{\hbar^2} V_{ff}\right] C_f(\rho, Z) \exp\{i\alpha Z\} \, dZ \tag{11}$$

contain a dependence on the scattering angle θ via

$$\alpha = k_f(1 - \cos\theta) = 2k_f \sin^2\frac{\theta}{2} \tag{12}$$

the difference between the Z component of the momentum change $\mathbf{K}$ and the minimum momentum change $k_i - k_f$ in the collisions. Equations (8)–(12) are the basic formulas given by the present multichannel eikonal description for the scattering amplitude, and they can be easily generalized so as to cover collisions involving multielectron systems. It is apparent that a variety of approximations readily follow. Note that in the absence of all couplings except that connecting the initial and final channels, i.e., $C_n = \delta_{ni}$, either (7) or (9) directly yields

$$f_{if}(\theta, \theta) = -\frac{1}{4\pi} \frac{2\mu}{\hbar^2} \int V_{fi}(\mathbf{R}) \exp\{i\mathbf{K}\cdot\mathbf{R}\} \, d\mathbf{R} \tag{13}$$

which is the Born-wave scattering amplitude.

Moreover, it can be shown (Flannery and McCann, 1974) by successive approximation that the above equations can reproduce (a) the customary one-channel eikonal expression (cf. Bransden, 1970) for elastic scattering and (b) the distorted-Born-wave expression of Chen et al. (1972). Also, in the heavy-particle or high-energy limit the usual impact-parameter and Glauber formulas are recovered.

In summary, the present method (i) has defined a scattering amplitude rather than an excitation probability, the key quantity occurring in time-dependent impact-parameter treatments; (ii) has acknowledged different local momenta of relative motion in various channels; and (iii) has automatically included an infinite number of partial waves, via the eikonal in (4), which are distorted by the static interactions associated with the various channels and which in turn are coupled to the internal electronic motions via A_n in (4).

Results and Discussion

As examples of the preceding analysis, the full eikonal equations (8)–(12) are now applied to the examination of the excitation processes

$$e + H(1s) \rightarrow e + H(2s \text{ or } 2p) \tag{14}$$

and

$$e + He(1\,^1S) \rightarrow e + He(n\,^1S \text{ or } n\,^1P), \quad n = 2, 3 \tag{15}$$

in which the initial and all final channels with the same n are closely coupled. Note that the resulting set of coupled equations in which exchange is neglected are *not* the semiclassical analogues or even approximations to the actual coupled differential equations obtained from the full close-coupling method (cf. Burke and Webb, 1970).

In Figures 1–3 are displayed the total cross sections, labeled FE,

$$\sigma_f(E_i) = 2\pi \frac{k_f}{k_i} \int_0^\pi |f_{if}(\theta, \varphi)|^2 \sin \theta \, d\theta \tag{16}$$

computed as functions of electron-impact energy E_i, together with various theoretical results and experimental measurements, as referenced in the captions. For process (14), curve EB is an approximation to FE in which $\kappa_n = k_n$ and I_2 in (9) and the term in square brackets in (8) are ignored. For (15), two sets of orthogonal wave functions for He ($n = 2$) are used. The cross sections labeled FE1 and FE2 refer to calculations performed with the analytical wave functions respectively given by Flannery (1970) and the multiparameter Hartree–Fock frozen-core set of McEachran and Cohen (1969) and Crothers and McEachran (1970).

The figures provide an indication of the overall ability of the present method, and very little need be said. The situation appears rather encouraging.

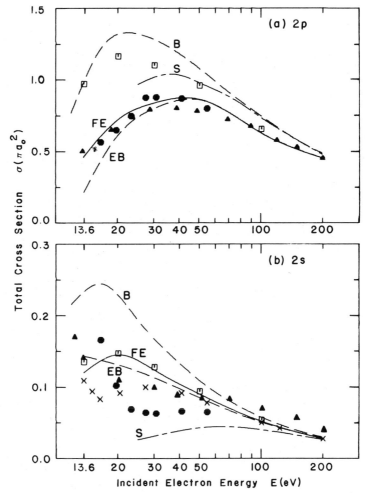

Figure 1. Cross sections for (a) the $2p$ and (b) the $2s$ excitations of H(1s) by electron impact. *Theory:* FE, full eikonal treatment; EB, eikonal approximation ($\kappa_n = k_n$); S, second-order potential method (Sullivan *et al.*, 1972); B, Born approximation; ●, pseudostate method (Burke and Webb, 1970); ×, polarized-orbital distorted-wave model (McDowell *et al.*, 1974); ▫, four-state impact-parameter method. *Experiment:* ▲ (a) Long *et al.* (1968); (b) Kauppila *et al.* (1970).

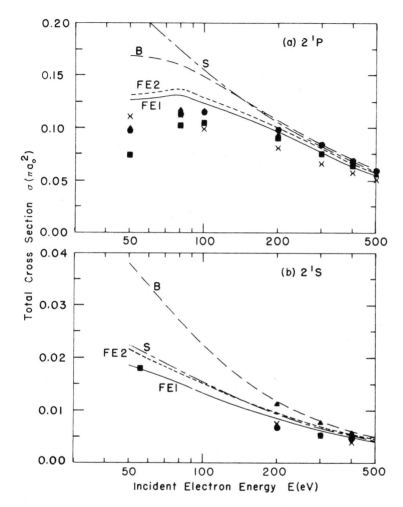

Figure 2. Total cross sections for (a) the 2^1P and (b) the 2^1S excitations of $He(1^1S)$ by electron impact. *Theory:* FE1: Four-channel eikonal treatment with first set of atomic wave functions; FE2: Four-channel eikonal treatment with second set of atomic wave functions; S: Second-order potential method with first set of atomic wave functions (Berrington et al., 1973); B: Born approximations (Bell et al., 1969). *Experiment:* [$2(^1P)$] ▲: Donaldson et al. (1972); ■: Jobe and St. John (1967); ×: Moustafa-Moussa et al. (1969); ●: van Eck and de Jongh (1970); (2^1S), ▲: Lassettre et al. (1970); ×: Miller et al. (1968); ●: Vriens et al. (1968).

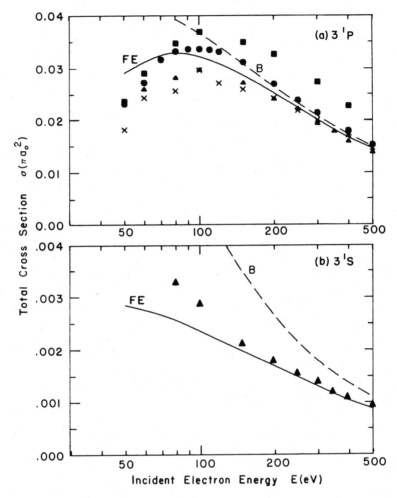

Figure 3. Total cross sections for (a) the 3^1P and (b) the 3^1S excitations of $He(1^1S)$ by electron impact. *Theory:* FE: Four-channel eikonal treatment with Hartree–Fock frozen core wave functions; B: Born approximation (Bell *et al.*, 1969). *Experiment:* ●: Donaldson *et al.* (1972); ▲: Moustafa-Moussa *et al.* (1969); ×: van Eck and de Jongh (1970); ■: St. John *et al.* (1964).

Acknowledgment

This research was sponsored by the Air Force Aerospace Research Laboratories, Air Force Systems Command, United States Air Force, Contract F 33615-74-C-4003.

Note Added in Proof

We have recently presented in *J. Phys. B.*, **8**, 1716 (1975) a more elaborate ten-channel treatment of differential and integral cross sections for the excitation processes

$$e + He(1^1S) \rightarrow e + He(2^1S, 2^1P, 3^1S, 3^1P, 3^1D)$$

This work also includes comparison with experiment and theoretical predictions of the λ and χ parameters which provide, as functions of θ and E, the orientation and alignment vectors together with the circular polarization fractions of the radiation emitted from the n^1P levels. We have also recently reported [*Phys. Rev. A.*, **12**, 846 (1975)] a similar treatment of the $n = 3$ excitation and $n = 1$ de-excitation arising in $e - He(2^{1,3}S)$ collisions. Finally, a pseudostate and seven-state eikonal treatment of the $2s$, $2p$, $3s$, and $3p$ excitations in $e - H(1s)$ collisions is presented in *J. Phys. B.*, **7**, L522 (1974).

References

Bates, D. R., and Crothers, D. S. F. (1970). *Proc. Roy. Soc. A*, **315**, 465.
Bates, D. R., and Holt, A. R. (1966). *Proc. Roy. Soc. A.*, **292**, 168.
Bell, K. L., Kennedy, D. J., and Kingston, A. E. (1969). *J. Phys. B.*, **2**, 26.
Berrington, K. A., Bransden, B. H., and Coleman, J. P. (1973). *J. Phys. B.*, **6**, 436.
Bransden, B. H. (1970). *Atomic Collisions Theory*, New York: Benjamin, Inc., p. 82.
Bransden, B. H., and Coleman, J. P. (1972). *J. Phys. B.*, **5**, 537.
Burke, P. G., and Webb, T. G. (1970). *J. Phys. B.*, **3**, L1313.
Byron, F. W. (1971). *Phys. Rev.* **4**, 1907.
Callaway, J. (1968). In *Lectures in Theoretical Physics*. Eds. S. Geltman, K. T. Mahanthappa, and W. E. Brittin, **11C**, p. 119. New York: Gordon and Breach.
Chen, J. C. Y., Joachain, C. J., and Watson, K. M. (1972). *Phys. Rev.*, **5**, 2460.
Crothers, D. S. F., and McEachran, R. P. (1970). *J. Phys. B.*, **3**, 976.
Donaldson, F. G., Hender, M. A., and McConkey, J. W. (1972). *J. Phys. B.*, **5**, 1192.
Flannery, M. R. (1970). *J. Phys. B.*, **3**, 306.
Flannery, M. R., and McCann, K. J. (1974). *J. Phys. B.*, **7**, 2518.
Jobe, J. E., and St. John, R. M. (1967). *Phys. Rev.*, **164**, 117.
Kauppila, W. E., Ott, W. R., and Fite, W. L. (1970). *Phys. Rev.*, **141**, 1099.
Lassettre, E. N., Skerbele, A. and Dillon, M. A. (1970). *J. Chem. Phys.*, **52**, 2797.
Long, R. L., Cox, D. M., and Smith, S. J. (1968). *J. Res. Nat. Bur. Standards*, **72A**.
Miller, K. J., Mielczarek, S. R., and Krauss, M. (1968). *J. Chem. Phys.*, **51**, 945.
McDowell, M. R. C., Morgan, L. A., and Myerscough, V. P. (1973). *J. Phys. B*, **6**, 1435.
McDowell, M. R. C., Myerscough, V. P., and Narain, U. (1974). *J. Phys. B.*, **7**, L195.
McEachran, R. P., and Cohen, M. (1969). *J. Phys. B.*, **3**, 976.
Moustafa-Moussa, H. R., de Heer, F. J., and Schulten, J. (1969). *Physics*, **40**, 517.
St. John, R. M., Miller, F. L., and Lin, C. C. (1964). *Phys. Rev.*, **A134**, 888.
Sullivan, J., Coleman, J. P., and Bransden, B. H. (1972). *J. Phys. B.*, **5**, 2061.
Tai, H., Bassel, R. H., Gerjuoy, E., and Franco, V. (1970). *Phys. Rev.*, **A1**, 1819.
Van Eck, J., and de Jongh, J. P. (1970). *Physics*, **47**, 141.
Vriens, L., Kuyatt, C. E., and Mielczarek, S. K. (1968). *Phys. Rev.*, **165**, 7.

25
Recent Progress in the Application of Eikonal–Born Series Methods in Atomic Physics

F. W. BYRON, JR.

An analysis is made of intermediate energy electron–atom scattering via a comparison of the Born series and the Glauber series. The importance of a consistent treatment of the leading corrections to the first Born approximation is stressed, and the Glauber approximation is used to obtain the most complicated of the necessary corrections, namely, the real part of the third-order term. The fact that the Glauber series, at least for s–s transitions, has a range of angular validity encompassing all angles enables us to give a complete picture of angular distributions in elastic e^- + H and e^- + He scattering, as well as in the excitation of atomic hydrogen to the 2s state. Moreover, it is shown that for 1s–2s excitation at large angles, more terms of perturbation theory are necessary for a satisfactory description than are needed at large angles in elastic scattering. For 1s–2p excitation, we show that a simple Glauber calculation fails seriously at large angles but Born series methods can be used to obtain the correct asymptotic form of the 1s–2p differential cross section in this region.

The eikonal-Born series (EBS) method recently introduced in potential scattering (Byron *et al.*, 1972, 1973) and in electron–atom scattering (Byron and Joachain, 1973a,b, 1974) consists simply of an attempt to analyze the momentum ($\mathbf{k}_i$) and momentum transfer ($\mathbf{K}$) dependence of the terms in the Born series with the aim of understanding what is meant by a consistent leading correction to the first Born approximation. This is particularly appropriate for atomic physics, where the coupling constant measuring the strength of the potentials is of order unity, at least for simple atoms.

F. W. BYRON, JR. • Department of Physics, University of Massachusetts, Amherst, Mass. 01002, U.S.A.

In the case of potential scattering, the basic results obtained in recent years (Byron et al., 1972, 1973) may be summarized as follows. We recall that the eikonal approximation has the extremely simple form

$$f_E = \frac{k_i}{2\pi i}\int e^{i\mathbf{K}\cdot\mathbf{b}}[e^{i\chi(\mathbf{b})} - 1]\,d\mathbf{b} \tag{1a}$$

with χ given by

$$\chi = -\frac{1}{k_i}\int_{-\infty}^{\infty} V(\mathbf{b}, z)\,dz \tag{1b}$$

the z direction being taken to be perpendicular to the momentum transfer. The eikonal scattering amplitude, f_E, when expanded as a power series in the potential strength, has a one-to-one correspondence with the terms in the Born series and gives in each order the leading term (in powers of k_i^{-1}) of the Born series for all scattering angles. This relationship, which holds for potentials of the "Yukawa type," is remarkable in light of the traditional derivations of the eikonal approximation, which have made it appear to be uniquely a small-angle approximation. The results just described do *not* hold for potentials whose analytic properties are different from Yukawa type potentials. For example, for long-range potentials such as the polarization or Van der Waals potential, the eikonal approximation is useful only for angles $\theta \lesssim k_i^{-1/2}$.

Despite the remarkable all-angles property of the eikonal approximation, the method was found to suffer from certain deficiencies in the weak and intermediate coupling situations, the intermediate coupling case being of particular interest for atomic physics. The key point can be seen by writing the Born and eikonal series schematically as follows:

$$\begin{array}{cccccc}
f_B = \bar{f}_{B1} & + \operatorname{Re}\bar{f}_{B2} & + i\operatorname{Im}\bar{f}_{B2} & + \operatorname{Re}\bar{f}_{B3} & + i\operatorname{Im}\bar{f}_{B3} & + \cdots \\
\| & \wr\! & \wr\! & \wr\! & \wr & \\
A(K) & B(K)/k_i^2 & iC(K)/k_i & D(K)/k_i^2 & O(k_i^{-3}) & \\
\| & & \| & \| & & \\
f_E = \bar{f}_{E1} & + \quad 0 & + i\operatorname{Im}\bar{f}_{E2} & + \operatorname{Re}\bar{f}_{E3} & + \quad 0 & + \cdots
\end{array} \tag{2}$$

In the middle line of this equation we give the leading term (in an expansion in k_i^{-1}) of each order in the Born series. It is clear from the above equation that if A, B, C, and D are all of the same order (coupling constant of order unity), then the terms in B, C, and D are all equally important in correcting the first Born differential cross section given by

$$\frac{d\sigma_{B1}}{d\Omega} = |\bar{f}_{B1}|^2 = |\bar{f}_{E1}|^2$$

each term giving a correction of order k_i^{-2}. This means that the eikonal series, which has no real part in second order, misses one of the three pieces which correct the first Born result. However, to put things in a more positive light, one may also say that this remarkable approximation gives two of the leading correction terms, including the most difficult one to obtain (i.e., Re $\bar{f}_{B3}$), accurately at all angles.

The many-body generalization of the eikonal approximation, that is, the Glauber approximation (Glauber, 1959), has also been studied in the above spirit for electron–hydrogen and electron–helium elastic scattering (Byron and Joachain, 1973a, b, 1974). The Glauber approximation results when one makes a special form of the closure approximation in the many-body Green's function. The closure approximation replaces the energy difference between the ground and excited states by an average excitation energy, Δ_i. If, further, one uses an average excitation energy $\Delta_i = 0$, then the Green's function takes the form

$$G(\mathbf{r}, \mathbf{r}', \xi, \xi') = -\frac{1}{4\pi} \frac{e^{ik_i|\mathbf{r}-\mathbf{r}'|}}{|\mathbf{r}-\mathbf{r}'|} \delta(\xi - \xi') \qquad (3a)$$

and leads, according to Eq. (1), to

$$f_G = \frac{k_i}{2\pi i} \int e^{i\mathbf{K}\cdot\mathbf{b}} \, f|e^{i\chi(b,\xi)} - 1|i\rangle \, d\mathbf{b} \qquad (3b)$$

Here $V(\mathbf{r}, \xi)$ is the electron–atom potential and ξ denotes the ensemble of atomic coordinates. We should note here that χ can be evaluated analytically for electron scattering by any atom. As in the potential theoretic eikonal method, we follow Glauber (Glauber, 1959) by taking the z axis to be perpendicular to the momentum transfer. Equations (3a) and (3b) have been used extensively to analyze electron–atom scattering in recent years (Gerjuoy and Thomas, 1974).

In the many-body situation the possibility of detailed analysis is reduced due to the increased complexities of the Born series. However, when one evaluates the second Born term for elastic scattering in the closure approximation, using a nonzero value of Δ_i, one finds that the leading part of the second Born term (in powers of k_i^{-1}) agrees with the corresponding term in the Glauber series at all angles, except at very small ones ($\theta \lesssim k_i^{-2}$), where the Glauber approximation begins to show the effects of a spurious logarithmic divergence at $\theta = 0$. Note that just as in potential scattering, the second term in the Glauber series for elastic scattering is purely imaginary. Figure 1 shows the comparison between Im $\bar{f}_{B2}^{el}$ and Im $\bar{f}_{G2}^{el}$. The agreement between these two quantities, shown for electron–hydrogen elastic scattering at 100 eV, is striking. Also shown for comparison is the contribution from just the 1s intermediate state. At small angles it does very poorly, while at

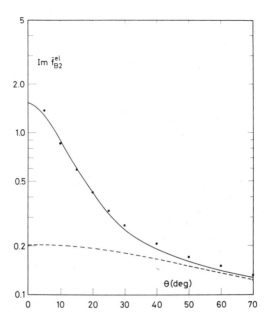

Figure 1. The quantity $\text{Im}\,\bar{f}_{B2}^{el}$ for the elastic scattering of electrons by atomic hydrogen at 100 eV is shown in the solid curve. The dots represent $\text{Im}\,\bar{f}_{G2}^{el}$, and the dashed curve shows what would be obtained for the second Born term if only the 1s state of the target were included as an intermediate state.

larger angles it rapidly tends to $\text{Im}\,\bar{f}_{B2}$, as we would expect. This tells us that an optical model calculation using only the lowest-order static potential would not give a very good representation of elastic scattering at small angles.

At first glance, the agreement between $\text{Im}\,\bar{f}_{B2}$ and $\text{Im}\,\bar{f}_{G2}$ may seem surprising since one knows that there are long-range forces present in electron–atom scattering, and we have just stated that for such potentials the eikonal approximation fails at large angles. However, one must remember that the long-range potentials are important only for rather small momentum transfers, while for large momentum transfers the short-range potentials (of the Yukawa type) are expected to dominate. These are precisely the type of potentials for which the eikonal approximation has its remarkable all-angles property.

By analogy with potential scattering we expect the real part of the second Born term to be important for elastic scattering, and this is indeed the case. This term is of order k_i^{-2}, except at small angles ($\theta \lesssim k_i^{-2}$). Here polarization effects play a critical role, and the real part of the second Born term is of order k_i^{-1} and thus gives the leading correction to $\bar{f}_{B1}$. Since the real part of the third Born term will also be of order k_i^{-2}, Byron and Joachain (1973a), arguing by analogy with the potential theory case, took the real part of the third Glauber term to approximate the real part of the third Born term, thus obtaining a direct amplitude which is correct to order k_i^{-2}, that is,

$$f_{EBS} = \bar{f}_{B1} + \operatorname{Re} \bar{f}_{B2} + i \operatorname{Im} \bar{f}_{B2} + \operatorname{Re} \bar{f}_{G3} \tag{4}$$

Interestingly enough, in order to obtain the leading correction to the first Born result one must also use the leading term of the exchange amplitude, $\bar{g}_{B1}$, in the small-angle region. Here, Byron and Joachain (1973a) used the first term of $\bar{g}_{B1}$ when expanded in powers of k_i^{-1}, i.e. the Ochkur term, which comes from the interaction of the incident electron with the ejected electron. This gives, e.g., for hydrogen,

$$g \simeq \bar{g}_{B1} \simeq g_{Och} = -\frac{2}{k_i^2} \int e^{i\mathbf{k}\cdot\mathbf{r}} [\phi_{1s}(\mathbf{r})]^2 \, d\mathbf{r} \tag{5}$$

with a similar expression for helium.

Putting f and g together in the familiar manner for hydrogen and helium, one obtains the differential cross sections for elastic scattering. Figure 2 shows the situation for electron–helium scattering at 500 eV, where the very precise experimental values of Bromberg (1969) are available. The agreement between experiment and the EBS method is seen to be excellent.

Recently progress has been made in extending the EBS method to inelastic processes (Byron and Latour, 1976). Here we shall see that interesting light is shed on these problems, but that in order to obtain the same kind of experimental agreement found in elastic scattering at large angles, one would have to carry the EBS method to higher orders of perturbation theory and to come to grips with the problem of obtaining a fuller understanding of the exchange amplitude.

The analysis of the $1s$–$2s$ excitation process in atomic hydrogen is very similar to that of elastic scattering. Evaluation of the second Born term in the closure approximation is straightforward, in a manner virtually identical to that used by Byron and Joachain (1973a). One finds that the imaginary part of the second Born term agrees at all angles (in the region $k_i \gg 1$) with the second Glauber term. The solid curve in Figure 3 shows the second Born term (with the $1s$ and $2s$ states included exactly) for $E = 100$ eV. The dots indicate the Glauber approximation, $\bar{f}_{G2}^{2s}$. The agreement is seen to be excellent at all angles. The dashed curve shows what results when one uses just the $1s$ and $2s$ states in computing the second Born term. This is, for example, the result that would be obtained for $\bar{f}_{B2}^{2s}$ by using the distorted-wave Born approximation or the $1s$–$2s$ close-coupling equations. Clearly, at

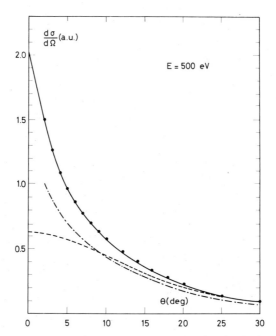

Figure 2. The differential cross section for the elastic scattering of electrons by helium at 500 eV. The solid curve is the result of the EBS method, the dash-dotted curve shows the Glauber approximation, and the dashed curve gives the result of the first Born approximation. The dots are the experimental values of Bromberg (1969).

small angles ($\theta \lesssim 30°$) this is not a good approximation to the full result, while at larger angles it is very good. The large-angle agreement is not surprising since wide-angle scattering will be dominated asymptotically by processes going through elastic intermediate states, because in these states the incident electron is exposed to the central Coulomb singularity.

These large-angle contributions will vary like $k_i^{-1} K^{-2}$ for large K. It is interesting to note that since $\bar{f}_{B1}^{2s}$ falls off like K^{-6}, $\operatorname{Im} \bar{f}_{B2}^{2s}(\simeq \operatorname{Im} \bar{f}_{G2}^{2s})$ will dominate over the first Born term at large angles and will in fact give the *leading contribution* to the differential cross section. At smaller angles, where $\bar{f}_{B1}^{2s} \simeq 1$, $\operatorname{Im} \bar{f}_{B2}^{2s}$ gives only a piece of the leading *correction* to $\bar{f}_{B1}^{2s}$. As in the case of elastic scattering, $\operatorname{Re} \bar{f}_{B2}^{2s}$ and $\operatorname{Re} \bar{f}_{B3}^{2s}$ are needed. If we use $\operatorname{Re} \bar{f}_{G3}^{2s}$ to approximate $\operatorname{Re} \bar{f}_{B3}^{2s}$, then we can form

$$f_{EBS}^{2s} \simeq \bar{f}_{B1}^{2s} + \operatorname{Re} \bar{f}_{B2}^{2s} + i \operatorname{Im} \bar{f}_{B2}^{2s} + \operatorname{Re} \bar{f}_{G3}^{2s} \tag{6}$$

for the direct amplitude.

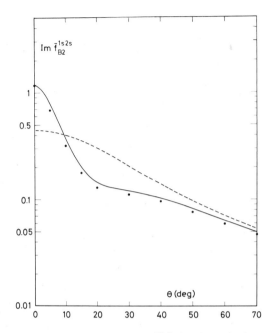

Figure 3. The quantity $\mathrm{Im}\,\bar{f}_{B2}^{1s2s}$ for the excitation of atomic hydrogen to the $2s$ state is shown in the solid curve for an electron energy of 100 eV. The dots represent $\mathrm{Im}\,\bar{f}_{G2}^{1s2s}$, and the dashed curve shows what would be obtained if only the $1s$ and $2s$ states of the target were included as intermediate states.

In elastic scattering we found that $\mathrm{Re}\,\bar{f}_{B2}$ was a very important term, being of order k_i^{-1} at small angles and of order k_i^{-2} at other angles. The same dependence is found here, but with two very important differences. First, at small angles, $\mathrm{Re}\,\bar{f}_{B2}^{2s}$ behaves like

$$\mathrm{Re}\,\bar{f}_{B2}^{2s} \simeq -\frac{512\sqrt{2}}{729}\frac{\pi}{k_i}\left[1 - \frac{k_i K}{(k_i^2 K^2 + \Delta_i \Delta_f)^{1/2}}\right] + O(k_i^{-2}) \qquad (7\mathrm{a})$$

where $\Delta_f = \Delta_i - 3/4$ is the energy difference between the average excitation energy and the final state energy (all measured in rydbergs). This is to be compared with the elastic result

$$\mathrm{Re}\,\bar{f}_{B2} \simeq \frac{\pi}{k_i}\left[1 - \frac{k_i K}{(k_i^2 K^2 + \Delta_i^2)^{1/2}}\right] + O(k_i^{-2}) \qquad (7\mathrm{b})$$

In the elastic case, the value of θ inside which this k_i^{-1} contribution is important is $\theta_c^{el} \leqslant \Delta_i/k_i^2$, whereas for $1s$–$2s$ excitation we have $\theta_c^{2s} \leqslant (\Delta_i \Delta_f)^{1/2}/k_i^2$.

Since average excitation energies are likely to be in the vicinity of 1 Ry, it is clear that the angular range in which the k_i^{-1} term will play an important role is less than that for elastic scattering.

The second difference is one which is not well understood at present. It arises from the fact that although each term in the k_i^{-2} part of $\mathrm{Re}\,\bar{f}_{B2}^{2s}$ is of the same size as $\mathrm{Re}\,\bar{f}_{G3}^{2s}$, when they are added together there is a tremendous amount of cancellation for values of Δ_i near 1 Ry. Thus, $\mathrm{Re}\,\bar{f}_{B2}^{2s} \ll \mathrm{Re}\,\bar{f}_{G3}^{2s}$, and in addition $\mathrm{Re}\,\bar{f}_{B2}^{2s}$ is very sensitive to Δ_i. Of course, since it is small compared to $\mathrm{Re}\,\bar{f}_{G3}^{2s}$, the sensitivity does not extend to the full EBS amplitude, which in this case will resemble rather closely the full Glauber amplitude, in marked contrast with the elastic scattering case. Figure 4 shows some of the details of $\mathrm{Re}\,\bar{f}_{B2}^{2s}$. At large angles, sensitivity to Δ_i is removed by adding in the 1s and 2s intermediate states exactly, but at smaller angles the results continue to show great sensitivity to Δ_i.

Putting all these results together and using the lowest-order exchange amplitude as before, we obtain the differential cross section shown in Figure 5. The differences between the Glauber cross section and the EBS cross section

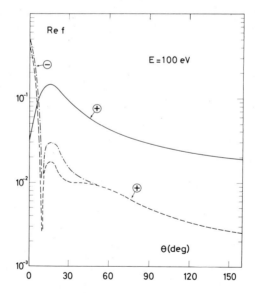

Figure 4. The upper curve (solid) shows the quantity $\mathrm{Re}\,\bar{f}_{G3}^{1s2s}$ for the excitation of atomic hydrogen to the 2s state by electron impact at 100 eV. The lower curves show $\mathrm{Re}\,\bar{f}_{B2}^{1s2s}$ for the same process with two different values of the excitation energy (1.0 and 2.0 Ry).

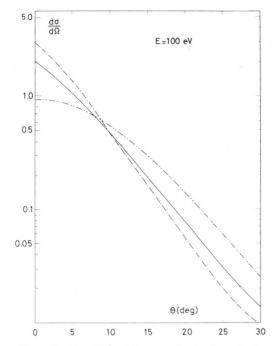

Figure 5. The differential cross section for the excitation of the 2s state of atomic hydrogen by electron impact at 100 eV. The solid curve is the Glauber approximation, the dash-double-dotted curve is the first Born approximation, and the dashed-dot curve is the EBS result using $\Delta_i = 1.0$ Ry.

are evident, but they are by no means as striking as in the elastic case. At large angles ($\theta \geqslant 30°$) the cross section becomes dominated by the second and higher Born terms so that the EBS method gives only the *leading* contribution. The first correction to this leading term will come from $\operatorname{Im} \bar{f}_{B3}^{2s}$, $\operatorname{Im} \bar{f}_{B4}^{2s}$, and $\operatorname{Im} \bar{g}_{B2}^{2s}$ in addition to the terms already present. One could presumably approximate $\operatorname{Im} \bar{f}_{B4}^{2s}$ by $\operatorname{Im} \bar{f}_{G4}^{2s}$, but the remaining terms are far from being simple to obtain.

It is interesting that in the first correction to the leading term for wide-angle scattering the second Born exchange amplitude already makes its appearance. In fact, $\operatorname{Im} \bar{g}_{B2}^{2s}$ (which is of order $k_i^{-3} K^{-2}$) plays a larger role, in absolute magnitude, in a process like 2s excitation when it interferes with the direct amplitude ($\operatorname{Im} \bar{f}_{B2}^{2s}$ is of order $k_2^{-1} K^{-2}$) than in, say, the excitation of the 2^3S state of helium, which is a "pure" rearrangement collision. Thus, it may well be simpler to get information on wide-angle exchange scattering from 2^1S excitation than from the pure rearrangement process.

Turning now to $2p$ excitation we find results rather different from those just discussed for $2s$ excitation. We may summarize as follows. Firstly, the piece of the second Born term which corresponds to the second Glauber term is now $\text{Re } \mathbf{f}_{B2}^{2p}$, the first Born term being pure imaginary. Here we use a vector scattering amplitude to summarize the three magnetic sublevels. For momentum transfers of the order of unity or less, the real part of the second Born amplitude takes the form

$$\text{Re } \mathbf{f}_{B2}^{2p} = \frac{1}{k_i} T_1(K)\mathbf{K} + \frac{1}{k_i^2} T_2(K)\hat{k}_i \tag{8}$$

Since the Glauber amplitude is simply proportional to $\mathbf{K}$ with no term in $\hat{k}_i$, it is clear that the $m = 0$ amplitudes will differ considerably in the region $K \simeq 1$, where $K_z \sim k_i^{-1}$. We should point out that it is precisely the presence of terms proportional to $\mathbf{k}_i$ which give rise to the function $\chi(k_i, \theta)$ measured by Eminyan et al. (1973). In Glauber theory, since the amplitude $\mathbf{f}_G^{2p}$ is proportional to $\mathbf{K}$, the function χ vanishes identically.

Secondly, at large angles, although $\text{Re } \mathbf{f}_{B2}^{2p}$ and $\text{Re } \mathbf{f}_{G2}^{2p}$ are of the same order of magnitude, they are not equal. Even more significant from the practical point of view is the fact that the non-Glauber second-order term, $\text{Im } \mathbf{f}_{B2}^{2p}$, is larger at wide angles than the term $\text{Re } \mathbf{f}_{B2}^{2p}$, so that in the large-$K$ region where $\mathbf{f}_{B1}^{2p}$ becomes negligible ($K \gtrsim 1$) the Glauber approximation does not give the leading term in the scattering amplitude. One can show that

$$\text{Re } \mathbf{f}_{B2}^{2p} \sim \mathbf{K}/k_i K^4 \sim \text{Re } \mathbf{f}_{G2}^{2p} \tag{9a}$$

$$\text{Im } \mathbf{f}_{B2}^{2p} \sim \mathbf{K}/k_i^2 K^2 \tag{9b}$$

whereas

$$\mathbf{f}_{B1}^{2p} \sim \vec{\mathbf{K}}/K^8 \sim \mathbf{f}_{G1}^{2p} \tag{9c}$$

Hence, the EBS result will give a large-angle cross section

$$\frac{d\sigma_{EBS}^{2p}}{d\Omega} \sim k_i^{-4} K^{-2} \tag{10a}$$

which is the same order of magnitude as the cross section for $2s$ excitation at wide angles,

$$\frac{d\sigma_{EBS}^{2s}}{d\Omega} \sim k_i^{-2} K^{-4} \sim \frac{d\sigma_G^{2s}}{d\Omega} \tag{10b}$$

For the Glauber cross section for $2p$ excitation, one has

$$\frac{d\sigma_G^{2p}}{d\Omega} \sim k_i^{-2} K^{-6} \tag{10c}$$

which is smaller than the EBS result by a factor of k_i^{-2} (since $K \sim k_i$ for large angles).

The best experimental evidence on this point is in the excitation of helium to the 2^1P state. At 200 eV Suzuki and Takayanagi (1973) find that the cross sections for 2^1S and 2^1P excitation in helium are very nearly equal for angles greater than about 60°. Figures 6 and 7 show the cross sections $d\sigma^+/d\Omega$ and $d\sigma^\circ/d\Omega$ at $E = 100$ eV obtained by following the EBS procedure outlined above. Note that already at angles of 20°–30° the differences between Born, Glauber, and EBS results are striking. It should be pointed out, however, that the use of Im f_{G3}^{2p} to replace Im f_{B3}^{2p} is rather suspect at small angles in light of Eq. (8), particularly for the $m = 0$ amplitude.

Finally, we should emphasize that as far as integrated cross sections are concerned, only angles less than about 20° play any role, so that the differences between Born, Glauber, and EBS are greatly reduced. The Glauber result,

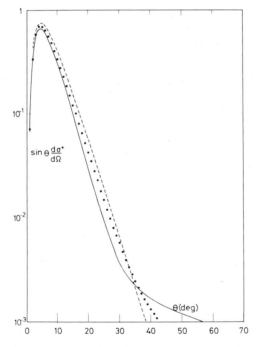

Figure 6. The differential cross section for the excitation of atomic hydrogen to the $2p^+$ state by 100 eV electron impact. The solid curve represents the EBS results, the dotted curve gives the Glauber approximation, and the dashed curve shows the first Born approximation.

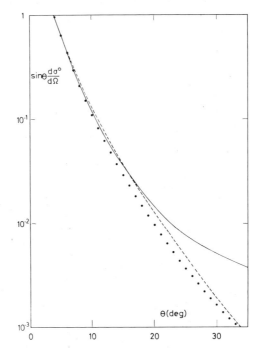

Figure 7. Same as Figure 8 but for the $2p^\circ$ state.

σ_G^{2p}, is less than the Born result, σ_B^{2p}, by about 9%, while σ_{EBS}^{2p} is less than σ_B^{2p} by about 12%. It is, however, interesting that these modifications are achieved in rather different ways. The cross section $\sigma_G^{2p^\circ}$ is less than $\sigma_B^{2p^\circ}$ by 5%, while $\sigma_{EBS}^{2p^\circ}$ is smaller by only 1%. On the other hand, $\sigma_G^{2p^+}$ is less than $\sigma_B^{2p^+}$ by about 12%, while $\sigma_{EBS}^{2p^+}$ falls 20% below $\sigma_B^{2p^+}$.

Acknowledgments

Much of the work reported above was done while the author was a Fulbright Research Scholar at the Université Libre de Bruxelles. He is grateful to the Fulbright Commission for making his year in Brussels possible and to Professor C. J. Joachain for his hospitality at the Université Libre de Bruxelles during the past academic year.

References

Bromberg, J. P. (1969). *J. Chem. Phys.*, **50**, 3906.
Byron, F. W. Jr., and Joachain, C. J. (1973a). *Phys. Rev.*, **A8**, 1267.

Byron, F. W. Jr., and Joachain, C. J. (1973b). *Phys. Rev.*, **A8**, 3266.
Byron, F. W. Jr., and Joachain, C. J. (1974). *J. Phys. B.*, **7**, L212.
Byron, F. W. Jr., Joachain, C. J., and Mund, E. H. (1972). *Bull. Am. Phys. Soc.*, **17**, 542.
Byron, F. W. Jr., Joachain, C. J., and Mund, E. H. (1973). *Phys. Rev.*, **D8**, 2622.
Byron, F. W. Jr., and Latour, L. J. Jr. (1976). *Phys. Rev. A.* (to be published).
Eminyan, M., MacAdam, K. B., Slevin, J., and Kleinpoppen, H. (1973). *Phys. Rev. Lett.*, **31**, 576.
Gerjuoy, E., and Thomas, B. K. (1974). SRCC Report No. 194, University of Pittsburgh.
Glauber, R. J. (1959). In *Lectures in Theoretical Physics*, Vol. I, pp. 314–414. New York: Interscience.
Suzuki, H., and Takayanagi, T. (1973). *Abstr. VIIIth ICPEAC.* Eds., B. C. Cobič and M. V. Kurepa, pp. 286–287. Belgrad Institute of Physics.

26
Ab Initio Optical Model Theory of Elastic Electron–Atom Scattering

F. W. Byron, Jr. and Charles J. Joachain

> Elastic electron–atom scattering at intermediate and high energies is analyzed by using an optical potential obtained from first principles. In addition to the static interaction, this optical potential accounts for long-range forces (arising from the polarization of the target) and absorption effects. Exchange effects are also included by using a pseudopotential. The theory is illustrated by a detailed treatment of electron–helium and electron–neon collisions. Our results are in excellent agreement with recent absolute experimental data.

1. Introduction

The nonrelativistic elastic scattering of electrons by atoms is one of the most fundamental problems in atomic collision theory. In this paper we present a theoretical treatment of such processes within the framework of the optical model formalism (Mittleman and Watson, 1959, 1960; Mittleman, 1961; Goldberger and Watson, 1964; Fetter and Watson, 1965; Joachain and Mittleman, 1971a,b; Byron and Joachain, 1974; Joachain 1974). Our approach, which is based on a multiple-scattering expansion of the optical potential in terms of the full electron–atom interaction (see, e.g., Goldberger and Watson, 1964; Joachain and Quigg, 1974), applies essentially to the region of intermediate and high energies. It consists in the construction of a local, central, complex pseudopotential which is derived from first

---

F. W. Byron, Jr. • Department of Physics and Astronomy, University of Massachusetts, Amherst, Mass. 01002, U.S.A.   Charles J. Joachain • Physique Théorique, Faculté des Sciences Université Libre de Bruxelles, Brussels, Belgium, and Institut de Louvain, Louvain-la-Neuve, Belgium.

principles. In addition to the static (charge cloud) interaction, this potential also accounts for polarization, absorption, and exchange effects. The only parameter in this theory is an average excitation energy, which we have fixed by requiring that the long-range form of the polarization potential be $-\bar{\alpha}/(2r^4)$, where $\bar{\alpha}$ is the experimentally determined dipole polarizability of the target atom. In contrast to our recent eikonal optical model computations (Byron and Joachain, 1974), we have here used the optical potential to perform a full-wave calculation of the scattering amplitude, thus avoiding the difficulties associated with the use of the eikonal approximation.

After recalling some basic equations, we present in Section 2 our approach to the determine of the optical potential. Section 3 is devoted to the application of our method to the analysis of the elastic scattering of fast electrons by helium and neon atoms. Our results are compared with recent absolute experimental data (Crooks and Rudd, 1972; Bromberg, 1974; Jansen et al., 1974; Gupta and Rees, 1974). We find that our theoretical values agree very well with experiment.

2. Theory

2.1. Basic Equations

Let us consider the nonrelativistic elastic scattering of an electron by a neutral atom having Z electrons. Since we are interested in the intermediate- and high-energy region, we shall first consider direct scattering; exchange effects will be studied later. The initial and final wave vectors of the electron will be denoted by $\mathbf{k}_i$ and $\mathbf{k}_f$, respectively, with $k = |\mathbf{k}_i| = |\mathbf{k}_f|$. The nucleus of the atom being the origin of our coordinate system, we label by $\mathbf{r}$ and $\mathbf{r}_i$ ($i = 1, 2, \ldots, Z$), respectively, the positions of the incident and atomic electrons. We shall also use the symbol X to denote all the target coordinates. All quantities will be expressed in atomic units.

The free motion of the two colliding particles is described by the Hamiltonian

$$H_0 = K + h \qquad (1)$$

where $K = -\frac{1}{2}\nabla_r^2$ is the kinetic energy operator of the projectile electron and h is the internal target Hamiltonian with eigenkets $|n\rangle$ and eigenenergies w_n. We shall use the symbols $|0\rangle$ and w_0 to denote, respectively, the ground-state eigenket and energy of the target atom. The full Hamiltonian of the system is such that

$$H = H_0 + V \qquad (2)$$

where V, the interaction potential between the projectile electron and the target atom, is given by

$$V = \sum_{i=1}^{Z} \frac{1}{|\mathbf{r}_i - \mathbf{r}|} - \frac{Z}{r} \tag{3}$$

Let us now write the equivalent one-body Schrödinger equation for elastic scattering, namely,

$$[K + V_{opt} - \tfrac{1}{2}k^2]\psi_i^{(+)} = 0 \tag{4}$$

where $\psi_i^{(+)}$ is the elastic-scattering wave function describing the motion of the projectile in the optical potential V_{opt}. Neglecting for the moment exchange effects between the incident and target electrons, we write the direct part V_{opt}^d of the optical potential to second order in a multiple scattering expansion (in terms of the full electron–atom interaction V) as

$$V_{opt}^d = V^{(1)} + V^{(2)} \tag{5}$$

Here the first-order or *static* potential $V^{(1)}$ is given by

$$V^{(1)} = V_S = \langle 0|V|0\rangle \tag{6}$$

while the second-order part $V^{(2)}$ is such that

$$V^{(2)} = \sum_{n \neq 0} \frac{\langle 0|V|n\rangle \langle n|V|0\rangle}{\tfrac{1}{2}k^2 - K - (w_n - w_0) + i\varepsilon}, \quad \varepsilon \to 0^+ \tag{7}$$

The static potential $V^{(1)}$ is readily evaluated for simple target atoms or when an independent particle model (e.g., the Hartree–Fock method) is used to describe the ground-state wave function of the target. We note that since this potential is real and of short range, it does not account for absorption or polarization effects which play an important role in the energy range considered here, particularly at small momentum transfers. We remark, however, that the static interaction (6) correctly reduces for small values of r to the Coulomb interaction $-Z/r$ acting between the incident particle and the nucleus of the target atom. We therefore expect that when $k \gg 1$ the static interaction will describe correctly the large-angle direct elastic scattering, which corresponds to small impact parameters. That this is indeed the case has been shown recently for the elastic scattering of fast electrons by helium (Byron and Joachain, 1973b) and will also be illustrated below for electron–neon elastic scattering.

We now examine the second-order piece $V^{(2)}$ of the direct optical potential. Since we are considering the scattering of fast electrons, we shall follow the method of Joachain and Mittleman (1971a,b) and replace the differences $w_n - w_0$ in Eq. (7) by an average excitation energy Δ. Performing the summation on n in Eq. (7) by closure, we then obtain for $V^{(2)}$ (in the coordinate representation) the nonlocal, complex expression

$$\langle \mathbf{r}|V^{(2)}|\mathbf{r}'\rangle = G_0^{(+)}(k', \mathbf{r}, \mathbf{r}')A(\mathbf{r}, \mathbf{r}') \tag{8}$$

where

$$G_0^{(+)}(k', \mathbf{r}, \mathbf{r}') = -(2\pi)^{-3} \int d\mathbf{p}\, \frac{e^{i\mathbf{p}\cdot(\mathbf{r}-\mathbf{r}')}}{p^2/2 - k'^2/2 - i\varepsilon}, \quad \varepsilon \to 0^+ \tag{9}$$

is the single-particle free propagator (with $k'^2 = k^2 - 2\Delta$) and

$$A(\mathbf{r}, \mathbf{r}') = \langle 0|V(\mathbf{r}, X)V(\mathbf{r}', X)|0\rangle - \langle 0|V(\mathbf{r}, X)|0\rangle \langle 0|V(\mathbf{r}', X)|0\rangle \tag{10}$$

The Schrödinger equation (4) satisfied by the scattering wave function $\psi_i^{(+)}$ then becomes

$$[-\tfrac{1}{2}\nabla_\mathbf{r}^2 + V^{(1)}(\mathbf{r}) - \tfrac{1}{2}k^2]\psi_i^{(+)}(\mathbf{r}) + \int G_0^{(+)}(k', \mathbf{r}, \mathbf{r}')A(\mathbf{r}, \mathbf{r}')\psi_i^{(+)}(\mathbf{r}')\, d\mathbf{r}' = 0 \tag{11}$$

2.2 Determination of a Local Equivalent Potential

The above equation is still difficult to solve because of the nonlocal character of the second-order pseudopotential. However, a careful analysis based on the optical eikonal theory (Joachain and Mittleman, 1971a,b; Byron and Joachain, 1974) and the all-angles properties of the Glauber approximation (Byron and Joachain, 1973a; Byron et al., 1973) shows that $V^{(2)}$ can be approximated by a complex, *local* potential containing two terms. That is,

$$V^{(2)} = V_{\text{pol}} + iV_{\text{abs}} \tag{12}$$

The first term V_{pol} takes into account the polarization of the target by the incident particle. It is given by (Byron and Joachain, 1974)

$$V_{\text{pol}}(r) = -\frac{\pi}{2ka^3\rho}\langle 0|\mathscr{Z}^2|0\rangle\left([I_0(\rho) - L_0(\rho)] - \frac{1}{\rho}[I_1(\rho) - L_1(\rho)]\right) \tag{13}$$

where $\rho = r/a$, $a = k/2\Delta$, $\mathscr{Z}$ denotes the sum of all the z coordinates of the bound electrons, I_n is a modified Bessel function, and L_n is a modified Struve function. The term V_{pol} is dominant at large distances where it has the form

$$V_{\text{pol}}(r) = -\frac{\bar{\alpha}}{2r^4}\left(1 + \frac{6a^2}{r^2} + \frac{135a^4}{r^4} + \cdots\right) \tag{14}$$

where $\bar{\alpha}$ is the dipole polarizability of the target atom obtained in the closure approximation. That is,

$$\bar{\alpha} = \frac{2\langle 0|\mathscr{Z}^2|0\rangle}{\Delta}. \tag{15}$$

It is worth noting that at large r the function $V_{pol}(r)$ lies below its asymptotic form $-\bar{\alpha}/(2r^4)$. This is just the opposite of what would be obtained from a Buckingham polarization potential having the form $-\bar{\alpha}/2(r^2 + a^2)^2$.

Returning to Eq. (12), we now determine the absorption potential V_{abs} in the following way. We first write the optical eikonal scattering amplitude as

$$f_{OE} = \frac{k}{2\pi i} \int d^2\mathbf{b}\, e^{i\mathbf{K}\cdot\mathbf{b}} \{e^{-(i/k)[\chi_{st}(\mathbf{b}) + \chi_{pol}(\mathbf{b}) + i\chi_{abs}(\mathbf{b})]} - 1\} \tag{16}$$

where $\mathbf{K} = \mathbf{k}_i - \mathbf{k}_f$ and

$$\chi_{st}(\mathbf{b}) = \int_{-\infty}^{+\infty} V^{(1)}(\mathbf{b}, z)\, dz \tag{17a}$$

$$\chi_{pol}(\mathbf{b}) = \int_{-\infty}^{+\infty} V_{pol}(\mathbf{b}, z)\, dz \tag{17b}$$

$$\chi_{abs}(\mathbf{b}) = \int_{-\infty}^{+\infty} V_{abs}(\mathbf{b}, z)\, dz \tag{17c}$$

Using the work of Joachain and Mittleman (1971a,b) to obtain the absorption part of the eikonal phase shift function, χ_{abs}, we then solve the Abel integral equation

$$\chi_{abs}(b) = 2 \int_b^\infty \frac{V_{abs}(r)}{\sqrt{r^2 - b^2}} r\, dr \tag{18}$$

to obtain

$$V_{abs}(r) = -\frac{1}{\pi} \int_r^\infty \frac{\chi'_{abs}(b)}{\sqrt{b^2 - r^2}}\, db \tag{19}$$

We remark that, because of the all-angles properties of the eikonal approximation (Byron and Joachain, 1973a,b,c; Byron et al., 1973), the above procedure is not restricted to small momentum transfers.

Finally, let us consider exchange effects. These may be accounted for by using a pseudopotential V_{opt}^{ex} (Mittleman and Watson, 1960; Furness and McCarthy, 1973). We have used here the exchange pseudopotential of Mittleman and Watson, namely,

$$V_{opt}^{ex} = -\frac{1}{\pi k}\left\{kp_F - \tfrac{1}{2}[k^2 - p_F^2]\ln\left(\frac{k + p_F}{|k - p_F|}\right)\right\} \tag{20}$$

where $p_F(\mathbf{r})$ is the Fermi momentum at $\mathbf{r}$. The full optical potential then reads

$$V_{opt} = V_{opt}^d + V_{opt}^{ex} = V_S + V_{pol} + iV_{abs} + V_{opt}^{ex} \tag{21}$$

We emphasize that in this theory the only free parameter is the average excitation energy Δ. We determine this parameter by the requirement that the quantity $\bar{\alpha}$, as given by Eq. (15), should be the experimental dipole polarizability of the target atom.

Having obtained *a priori* a local, complex optical potential in the way described above, it is now a straightforward matter to perform a full-wave calculation of the scattering amplitude by using the partial-wave method to solve the Schrödinger equation for the scattering wave function.

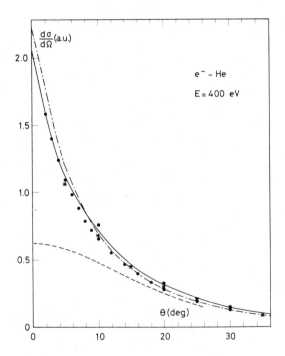

Figure 1. Differential cross section (in atomic units a_0^2/sr) for elastic electron–helium scattering at an incident electron energy of 400 eV. *Solid curve*: present optical model theory, with $\Delta = 1.16$ a.u.; *dash-dotted curve*: eikonal-Born series results of Byron and Joachain (1973a,b); *dashed curve*: first Born approximation. The squares are the experimental data of Crooks and Rudd (1972), the dots are those of Bromberg (1974), and the asterisks refer to those of Jansen *et al.* (1974).

3. Elastic Scattering of Fast Electrons by Helium and Neon Atoms

As an illustration of the theory presented in Section 2, we now analyze the elastic scattering of electrons by helium and neon atoms at intermediate energies. The ground-state wave functions used in performing our calculations are the Hartree–Fock wave functions of Clementi (1965).

3.1. Electron–Helium Elastic Scattering

We display in Figure 1 the differential cross section (for scattering angles such that $0° \leqslant \theta \leqslant 35°$) at an incident electron energy of 400 eV, while Table I gives a picture of the scattering over the full angular range at an incident energy of 200 eV. In both cases we compare our present optical model results (obtained by using an average excitation energy $\Delta = 1.16$ a.u.) with our eikonal-Born series (EBS) calculations (Byron and Joachain, 1973a,b) and with the recent absolute experimental data of Crooks and Rudd

Table 1. Differential Cross Sections (in a_0^2/sr) for Elastic Electron–Helium Scattering at an Incident Electron Energy of 200 eV

θ (deg)	Experimental values			Theoretical values	
	Crooks and Rudd (1972)	Bromberg (1974)	Jansen et al. (1974)	EBS method (Byron and Joachain, 1973a,b)	Present optical model theory
0	–	–	–	3.16	2.88
5	–	1.73	1.85	2.12	1.86
10	1.93	1.08	1.23	1.34	1.21
15	–	7.47(−1)	8.38(−1)	8.65(−1)	8.23(−1)
20	7.13(−1)	5.27(−1)	5.92(−1)	5.84(−1)	5.77(−1)
25	–	3.79(−1)	4.32(−1)	4.06(−1)	4.11(−1)
30	3.25(−1)	2.76(−1)	3.20(−1)	2.88(−1)	2.96(−1)
40	–		1.68(−1)	1.54(−1)	1.59(−1)
50	1.03(−1)		1.11(−1)	8.80(−2)	9.14(−2)
60	–			5.43(−2)	5.63(−2)
70	4.23(−2)			3.62(−2)	3.72(−2)
80	–			2.59(−2)	2.61(−2)
90	2.33(−2)			1.96(−2)	1.93(−2)
110	1.41(−2)			1.31(−2)	1.20(−2)
130	1.05(−2)			1.00(−2)	8.66(−3)
150	8.43(−3)			8.53(−3)	7.04(−3)
180	–			7.84(−3)	6.28(−3)

(1972), Bromberg (1974), and Jansen et al. (1974). Also shown in Figure 1 are the results obtained by using the first Born approximation. Apart from one experimental point of Crooks and Rudd (at 200 eV and $\theta = 10°$) which is almost certainly in error, we see that our optical model results agree within about 15% with the three sets of experimental data. In fact the agreement is often much better, particularly at high energies.

We also note that the present optical model results agree well with our eikonal–Born series calculations. As in the case of our eikonal optical model calculations (Byron and Joachain, 1974), the optical model results for the real part of the scattering amplitude lie consistently below the corresponding EBS values at small scattering angles. This is due to the fact that a second-order optical model misses all terms containing two virtual excitations of the target. Among such terms are those due to elastic scattering in virtual

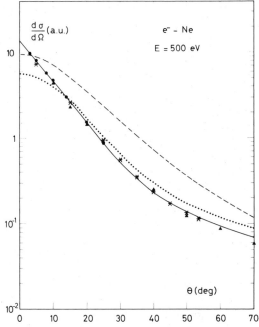

Figure 2. Differential cross section (in atomic units a_0^2/sr) for elastic electron–neon scattering at an incident electron energy of 500 eV. *Solid curve:* present optical model theory, with $\Delta = 1.5$ a.u.; *dotted curve:* static potential only; *dashed curve:* first Born approximation. The dots are the experimental results of Bromberg (1974), the asterisks those of Jansen et al. (1974), and the triangles refer to those of Gupta and Rees (1974).

excited states, which induce long-range forces in higher orders. For example, no second-order optical model can produce long-range forces in third order. It is also worth noting that the EBS method includes exchange effects through the Ochkur amplitude (which is of order k^{-2} at fixed momentum transfer), while the present optical model theory uses an exchange pseudopotential V_{opt}^{ex}. The differences between these two ways of including exchange effects are negligible only at small angles and high energies.

3.2. Electron–Neon Elastic Scattering

We now turn to our electron–neon optical model calculations, performed with an average excitation energy $\Delta = 1.5$ a.u. We show in Figure 2 the differential cross section at 500 eV in the angular range $0° \leq \theta \leq 70°$, while Table 2 displays our results over the entire angular range at 300 eV. In both cases we compare our results with the experimental data of Bromberg (1974), Jansen et al. (1974), and Gupta and Rees (1974). The agreement with the experimental values of Bromberg and of Jansen et al. is very good. The results of Gupta and Rees also agree well with our calculations at small angles, but there is room for improving the agreement at large angles.

Table 2. Differential Cross Sections (in a_0^2/sr) for Elastic Electron–Neon Scattering at an Incident Electron Energy of 300 eV

θ (deg)	Experimental values			Present optical model theory
	Bromberg (1974)	Jansen et al. (1974)	Gupta and Rees (1974)	
0	–	–	–	15.0
5	9.03	8.14	–	9.52
10	5.56	5.18	5.65	5.72
15	–	3.26	3.40	3.44
20	2.14	2.05	2.08	2.05
25	1.31	1.28	1.26	1.22
30		8.12(−1)	8.65(−1)	7.29(−1)
35		5.25(−1)	5.59(−1)	4.50(−1)
40		3.54(−1)	3.86(−1)	2.94(−1)
45		2.52(−1)	3.04(−1)	2.05(−1)
50		1.87(−1)	2.29(−1)	1.53(−1)
70			1.02(−1)	7.63(−2)
90			7.1(−2)	5.82(−2)
110			7.1(−2)	6.18(−2)
130			8.7(−2)	7.73(−2)
150			1.53(−1)	9.50(−1)
180			–	1.08(−1)

We also display in Figure 2 the results obtained by using the first Born approximation and those corresponding to an exact (partial-wave) treatment of the static potential $V^{(1)}$. We verify that the static potential strongly dominates the scattering outside the small-angle region. It is also worth noting the very large difference between the first Born and optical model curves, even at this rather high energy. The agreement between our calculations and the experimental data remains good down to an incident electron energy of 200 eV, where higher-order terms are required in the construction of the optical potential.

Acknowledgments

It is a pleasure to thank Drs. J. P. Bromberg, F. J. de Heer, R. H. J. Jansen, and J. A. Rees for sending us their experimental data prior to publication. Research for this paper was supported in part by the NATO Scientific Affairs Division under Grant No. 586.

References

Bromberg, J. P. (1974). *J. Chem. Phys.*, **61**, 963.
Byron, F. W. Jr., and Joachain, C. J. (1973a). *Phys. Rev.*, **A8**, 1267.
Byron, F. W. Jr., and Joachain, C. J. (1973b). *Phys. Rev.*, **A8**, 3266.
Byron, F. W. Jr., and Joachain, C. J. (1973c). *Physica*, **66**, 33.
Byron, F. W. Jr., and Joachain, C. J. (1974). *Phys. Rev.*, **A9**, 2559.
Byron, F. W. Jr., Joachain, C. J., and Mund, E. H. (1973). *Phys. Rev.*, **D8**, 2622.
Clementi, E. (1965). *Tables of Atomic Functions*. San Jose Research Laboratory, I.B.M., San Jose, California.
Crooks, G. B. Ph.D. Thesis, University of Nebraska (unpublished).
Crooks, G. B., and Rudd, M. E. (1972). *Bull. Am. Phys. Soc.*, II, **17**, 131.
Fetter, A. L., and Watson, K. M. (1965). In *Advances in Theoretical Physics*. Ed. K. A. Brueckner, Vol. 1, p. 115. New York: Academic Press.
Furness, J. B., and McCarthy, I. E. (1973). *J. Phys. B.*, **6**, 2280.
Goldberger, M. L., and Watson, K. M. (1964). *Collision Theory*, Ch. 11. New York: Wiley.
Gupta, S. C., and Rees, J. A. (1974). *J. Phys. B.*, **8**, 417.
Jansen, R. H. J., de Heer, F. J., Luyken, H. J., van Wingerden, B., and Blaauw, H. J. (1974). FOM Report 35693. Amsterdam: FOM Instituut voor Atoom en Molecuulfysica.
Joachain, C. J. (1974). *Quantum Collision Theory*, Ch. 20. Amsterdam: North-Holland.
Joachain, C. J., and Mittleman, M. H. (1971a). *Phys. Lett.*, **36A**, 209.
Joachain, C. J., and Mittleman, M. H. (1971b). *Phys. Rev.*, **A4**, 1492.
Joachain, C. J., and Quigg, C. (1974). *Rev. Mod. Phys.*, **46**, 279.
Mittleman, M. H. (1961). *Ann. Phys.* (N.Y.), **14**, 94.
Mittleman, M. H., and Watson, K. M. (1959). *Phys. Rev.*, **113**, 198.
Mittleman, M. H., and Watson, K. M. (1960). *Ann. Phys.* (N.Y.), **10**, 268.

27

The Scattering of Electrons from Hydrogen Atoms

J. F. WILLIAMS

Absolute values have been determined for electron–atomic hydrogen cross sections. The method of calibration is based upon the phase shift analysis of relative angular distributions of elastically scattered electrons. Data are reported for the following scattering processes: (1) elastic scattering from 1.2 to 8.7 eV and at 100 eV; (2) resonant elastic scattering below the $n = 2$ states; (3) total 2p excitation near threshold; (4) electron energy loss spectra near the $n = 3$ threshold; (5) differential cross sections for electrons losing 10.2 eV in exciting the $n = 2$ states from 50 to 680 eV. The elastic scattering differential cross section values at low energies agree within the experimental accuracy of 6% with values predicted by the variational, close coupling with pseudostates, and best polarized orbital methods, that is, with those methods that correctly allow for exchange and polarization effects. The ^1S and ^3P resonances are shown to have energies of 9.557 ± 0.010 eV and 9.735 ± 0.010 eV and widths of 0.045 ± 0.005 eV and 0.0060 ± 0.0005 eV, respectively; however, these experimental accuracies are not good enough to distinguish among the many theoretical predictions of the resonance energies and widths. The s, p, and d wave phase shifts for singlet and triplet scattering have been determined from the experimental data. The total 2p excitation cross section values just above threshold are in agreement within the experimental accuracy with the best close-coupling predictions. The resonances below the $n = 3$ level have been detected in the 10.2 eV energy loss electron spectra and angular distributions are reported. From 50 to 680 eV, the differential cross section values for those electrons that have lost 10.2 eV in exciting the $n = 2$ states are found to be not in agreement with any theoretical value. At 680 eV, the angular dependence at large angles is described by a $csc^4(\theta/2)$ relationship.

J. F. WILLIAMS • Physics Department, Queen's University, Belfast, Northern Ireland BT7 1NN.

Introduction

This paper discusses some recent progress in determining accurate, absolute, total, and differential cross sections for the scattering of electrons from atoms, particularly hydrogen atoms. Successive experimental developments in this field have been reviewed by Kieffer and Dunn (1966), Moiseiwitsch and Smith (1968), Smith (1969), Bederson and Kieffer (1971), Schultz (1973), and Andrick (1973). Various theoretical developments have been summarized by Mott and Massey (1965), Burke (1968), Rudge (1968), Gerjuoy (1970), Geltman (1969), and Drachman and Temkin (1970). This paper will concentrate on those collision processes which have been studied at Belfast, especially elastic scattering from the ground state and excitation to the $n = 2$ states of atomic hydrogen. The choice of topics is neither unique nor exhaustive, and time limitations preclude as detailed a discussion as the chosen topics deserve. Our recent work on (e, 2e) and (e, hv) experiments will not be discussed.

1. Elastic Scattering

1.1. Theoretical Background

The theoretical description of electron-scattering processes is generally best understood for elastic scattering at energies below the first inelastic threshold. This understanding arises naturally from the simple nature of the scattering process with only one open channel and from the theoretical development of variational techniques for the calculation of scattering amplitudes and from the methods for including the effect of polarization of the target atom into the scattering approximation. With the Kohn variational method, O'Malley et al. (1961) established upper bounds to the scattering length and, hence, bounds on the phase shifts for the limit of low-energy electrons. This work has been able to provide an estimate of the accuracy of the calculated phase shifts which has not generally been available from the scattering approximations. The variationally calculated s-wave phase shifts of Schwartz (1961), the p-wave phase shifts of Armstead (1968), and the d-wave phase shifts of Gailitis (1965) are considered accurate to better than 0.001 radian. Also a variational foundation has been established for truncated eigenfunction (close-coupling) approximations such that it is possible to determine the direction of the modification required in a given set of phase shifts in order to represent the scattering process more accurately (Percival and Seaton, 1957; Burke and Smith, 1962). In such methods the total wave function for the incident electron and target atom may be expanded as an antisymmetrized (A) sum of products of atomic eigenstates

ψ_i and scattering functions F_i

$$\Psi_{\text{total}} = A \sum_i \psi_i F_i \qquad (1)$$

where the functions F_i are obtained by substituting this expression in the nonrelativistic Schrödinger equation. This procedure leads to a set of coupled integrodifferential equations from which the phase shifts may be calculated. The convergence of the resultant phase shifts becomes slower as more eigenstates are included in Eq. (1); for example, a practical limit is to include up to about six eigenstates of atomic hydrogen (Burke et al., 1963). The inclusion of excited states in the Ψ_{total} expansion is equivalent to allowing for some of the distortion of the target atom in the field of the incident electron. The S states account for spherical distortion while the remaining excited bound states account for some of the nonspherical distortion, i.e., polarization, of the target. The remainder of the polarization arises from the continuum states. Thus, in atomic H, 81.4 % of the polarizability comes from the discrete spectrum of bound states (of which there are an infinite number). In contrast, for He and Ne, the discrete spectrum contributes only about 48 and 20 % to the total polarizability respectively, while for the alkali metals, 98 % of the polarizability is given by just including the first excited state in the expansion (Burke, 1968).

Various methods have been proposed to allow for the effects of higher excited and continuum states; for example, Rotenburg (1962) used a Sturmian basis set of functions, Burke and Taylor (1966) added electron–electron correlation terms, Damburg and Karule (1967) and Damburg and Geltman (1968) used pseudostates, while Burke et al. (1969) used an expansion of Eq. (1) with three eigenstates (1s, 2s, 2p) of H and several pseudostates which were formulated to give the polarizability of H exactly. Somewhat analogous to the pseudostate method of truncating the eigenfunction expansion are the polarized orbital methods (Temkin, 1959, and review by Drachman and Temkin, 1970), in which similar attempts are made to incorporate the basic physics of the scattering process in the form of the wave function. The method has been applied successfully to atomic hydrogen by Temkin and Lamkin (1961).

A general test of these theoretical approximations at low energies is to compare their phase shifts with the variationally calculated phase shifts mentioned above, which are regarded as practically "exact." Tabulated values of the phase shifts which have been calculated in the above approximations by many workers at many energies can be found in the papers by Burke et al. (1969), Drachman and Temkin (1970), and Mott and Massey (1965). Briefly, the pseudostate close-coupling (Burke et al., 1969) and polarized-orbital values (Temkin and Lamkin, 1961) are in best agreement (generally within several tens of milliradians) with the variational values below the first inelastic threshold.

All these theoretical studies produce phase shift values δ_{kl} from which the differential cross sections $I(k, \theta)$ may be calculated. For atomic hydrogen the total spin is $\frac{1}{2}$ then,

$$I(k, \theta) = \tfrac{3}{4}|f^T(k, \theta)|^2 + \tfrac{1}{4}|f^S(k, \theta)|^2 \qquad (2)$$

where

$$f^{T,S}(k, \theta) = \frac{1}{2ik} \sum_{l=0}^{\infty} (2l + 1)[\exp(2i\delta_{kl}^{T,S}) - 1]P_l(\cos \theta) \qquad (3)$$

$P_l(\cos \theta)$ are the normal Legendre functions, k is the incident electron wave number (equal to $E^{1/2}$ a.u., and E, the electron energy, is in rydbergs), θ is the scattering angle, and f^T and f^S are the triplet and singlet scattering amplitudes, respectively. For elastic scattering below the first inelastic threshold, a given theoretical approximation usually estimates only the first few phase shifts. O'Malley et al. (1961) have shown that higher-order phase shifts may be obtained from the effective range expansion

$$\tan \delta_{kl} = \frac{\pi\alpha k^2}{8(l + \tfrac{3}{2})(l + \tfrac{1}{2})(l - \tfrac{1}{2})} + \text{higher-order terms in } k^2 \qquad (4)$$

where α is the polarizability. For $l = 3$, this relationship, at, say, $k = 0.8$ a.u., provides values of δ_{kl} which differ by only milliradians from the close-coupling (with pseudostates) values. The accuracy improves as k decreases.

While the various low-energy elastic-scattering theoretical cross section (or phase shift) values have been compared with the "exact" variationally calculated values, it is only with the recent measurements of Williams (1974) that the first absolute experimental cross section values have become available to test the theoretical predictions. The only previous experimental values (Gilbody et al., 1961) were relative values which were normalized to the Ramsauer and Kollath (1932) molecular hydrogen elastic-scattering cross sections. Their accuracy of $\pm 20\%$ was good for the "state-of-the art" techniques of 1961 but insufficient to distinguish among the various theoretical approximations. A similar appraisal applies to the recent data of Lloyd et al. (1974), who used basically the same method and techniques as Gilbody et al. (1961).

The next section outlines those considerations which have led to the calibration of absolute cross section values.

1.2. Experimental Considerations

While a normal end product of theoretical work may be considered to be the phase shifts, this has not been the case in experimental work until recently (e.g., Andrick, 1973). Usually the experimentalist determines a differential cross section $I(k, \theta)$ from the relationship

$$I(k, \theta) = \frac{N(k, \theta)}{d\Omega} \left[\int_\tau \int_\Omega J_e(x, y, z, k) n_a(x, y, z) \varepsilon(x, y, z, k, \theta, \varphi) \, d\Omega \, d\tau \right]^{-1} \quad (5)$$

where $N(k, \theta)$ is the number of electrons scattered into elements of solid angle $d\Omega$, $J_e(x, y, z, k)$ is the incident electron beam flux density in the interaction volume $d\tau$, $n_a(x, y, z)$ is the target atom beam number density in the interaction volume $d\tau$ for an atom beam velocity distribution function $f(V)$, and $\varepsilon(x, y, z, k, \theta, \varphi)$ is an experimental apparatus function incorporating many factors such as analyzer transmissions, detector efficiency, and dependence of analyzer solid angle of acceptance of the detector upon the scattering angle θ, interaction energy, and interaction volume. The determination of relative differential cross sections requires at least a knowledge of the relative values of all these factors at all times during the experiment.

1.2.1. Cross Section Calibration. The traditional method (Bederson and Kieffer, 1971) of absolute cross section determination then requires the accurate and absolute measurement of at least one value of each of the quantities in Eq. (5). This is a formidable task, which has been thoroughly evaluated and attempted, for example, by Lineberger *et al.* (1966) in a crossed electron–lithium ion beam experiment and by Chamberlain *et al.* (1970), Trajmar (1973), and McConkey and Preston (1973) in static electron–helium gas experiments. Experiments by the present author (Williams and Willis, 1975) present some evidence, but not conclusive proof, that the use of a closed volume (static gas target), with slits to define the interaction volume and localize the target gas, can give rise to spurious scattered electrons and unwanted potentials which distort the angular distributions. Experiments with static gas targets are also made difficult by the variation of interaction volume with scattering angle and by the exponential loss from the scattered beam over the distance from the interaction region to the detector. However, such problems are either avoided or minimized in the crossed modulated beam technique used for atomic hydrogen scattering by Lloyd *et al.* (1974) and Williams and Willis (1974). In this latter work, absolute values and relative variations of all the parameters, except J_a and $f(V)$ for the hydrogen atoms, have been determined. Since a method of determining an absolute value for the density of hydrogen atoms in a beam is not yet available, it is necessary to seek other methods of cross section calibration.

1.2.2. Cross Section Ratios. On the assumption that at least one cross section is known absolutely, either from experimental or theoretical determination, it is possible to measure the ratio of the unknown cross section to that absolute value and so normalize the desired cross section. The method has been widely used (Bederson and Kieffer, 1971; Trajmar, 1973) for stable atomic and molecular gases. For a labile atomic gas, such as atomic hydrogen, this method entails the measurement of a cross section value in atomic hydrogen relative to that in molecular hydrogen for which an absolute value can be independently determined by the method of Section 1.2.1 (usually by another worker in another apparatus, e.g., Gilbody et al., 1961 and Lloyd et al., 1974). However, that method is superseded by the methods of Sections 1.2.3 and 1.2.4.

1.2.3. Phase Shift Analysis of Direct Elastic Scattering. Perhaps the most instructive way of evaluating differential elastic-scattering cross sections (or relative angular distributions of elastically scattered electrons) is to extract from them the phase shifts of all angular momenta states at each energy. If these can be uniquely determined, we have in principle extracted all possible information about the scattering process. In nuclear physics, the reduction of experimental data to phase shifts has received considerable attention (Preston, 1963; MacGregor, 1959; Yang, 1948). Comparable studies in low-energy electron scattering from atoms have only recently been accomplished (Andrick, 1973; McConkey and Preston, 1973; Williams and Willis, 1974), although the way was shown by Holtzmark (1933), Westin (1947), and Hoeper et al. (1968). The method of analysis (McDowell, 1970) is to search for sets of phase shifts which minimize the least-squares sum given by

$$\chi^2 = \frac{1}{n-m} \sum_{i=1}^{n} \left(\frac{I_{\text{expt}}(k,\theta) - I_{\text{calc}}(k,\theta)}{\Delta I(k,\theta)} \right)^2 \qquad (6)$$

in which the n observed quantities $I_{\text{expt}}(k,\theta)$ have an error $\Delta I(k,\theta)$ and the quantities $I_{\text{calc}}(k,\theta)$ are the corresponding values calculated from the parametric fit of m adjustable parameters to the data. In a simple case, n is the number of angles at which $I_{\text{expt}}(k,\theta)$ is measured and m is the number of phase shifts to be fitted. The greater the systematic or statistical errors of the experimental data, the more probable will be the occurrence of additional minima in χ^2. Andrick (1973) and Williams and Willis (1974) have shown for elastic scattering of low-energy electrons from simple atoms (helium and argon) that a set of phase shifts can be obtained by a trial-and-error fit of phase shifts to relative angular distributions. It appears that, with the present experimental accuracy of $\pm 5\%$, four unambiguous phase shifts may be extracted from measured angular distributions. While the phase shift search program may only vary the first four phase shifts, higher-order phase shifts may be assumed from the effective range expansion of Eq. (4). Generally, higher-order ($l \geqslant 3$) phase shifts become significant for angles less than $30°$ and energies greater than 10 eV. The resultant set of phase shifts may then be substituted into equations such as (2) and (3) to obtain an absolute value of the differential cross section, that is, an absolute scale has been determined for the relative angular distributions. This method is more accurate when the differential cross section exhibits structure rather than a flat angular distribution. It was used to calibrate earlier data in atomic hydrogen but has now been surpassed by the more accurate method of the next section.

1.2.4. Phase Shift Analysis of Resonant Elastic Scattering. A significant advance was made with the work of Gibson and Dolder (1969) on resonant and direct scattering of electrons from helium. Their analysis was made possible by the detailed study of resonance profiles which originated with Fano (1961). Fano's work laid the foundations for the simple procedures outlined by Shore (1967) and Comer and Read (1972), which permit an experimental determination of shape parameter, q, amplitude, and width, Γ, of the resonance profile to be performed.

The effect of the resonance on the differential cross section $I(E, \theta)$ as a function of energy is given by the relationship

$$I(E, \theta) = I_a(E, \theta) + \frac{(\varepsilon + q)^2}{(1 + \varepsilon^2)} I_b(E, \theta) \tag{7}$$

where $I_a(E, \theta)$ is the cross section at the resonance minimum, $I_b(E, \theta)$ is the amplitude of the resonance cross section, and $\varepsilon = (E - E_r)/\frac{1}{2}\Gamma$. The total phase shift, δ_{kl}, is equal to the background phase shift, δ_{kb}, plus the resonant phase shift, δ_{kr}, where $\delta_{kr} = \cot^{-1}(-\varepsilon)$. Since it is known that the resonance occurs in only one partial wave and that the resonant portion of that phase shift varies from 0 to π over the width of the resonance, an examination of the resonance profile at various scattering angles enables the phase shifts to be fitted. Once again an absolute value of the elastic differential cross section may be computed from these phase shifts.

This analysis has so far been used mainly for those resonances which occur below the first inelastic threshold such that their only decay channel is the incident elastic channel. Such a resonance is the $^2S(1s2s^2)$ in He. For the heavier rare gases (Weingartshofer et al., 1974), the $^2P_{1/2}$ resonances are located above the first inelastic threshold such that the method of analysis of Franzen and Gupta (1965) and Weingartshofer et al. (1974) has to be slightly modified to determine the full extent of the associated spin–orbit coupling. However, other papers in this volume will describe the analysis of such resonances in more detail.

In atomic hydrogen, preliminary studies have been made by McGowan et al. (1965) at 90°, Schulz (1964) at 0°, and Kleinpoppen and Raible (1965) at 94° of the resonances below the $n = 2$ level. The quality of their data was limited by low signal-to-noise ratios, electron energy resolution, and angular resolution of the incident and scattered electron energy analyzers. McGowan (1967) has analyzed the experimental data and argued that they appear to be consistent, within their respective experimental limitations, with one another and with the theoretical assignments of ^1S and ^3P resonances as predicted by Burke et al. (1963). More detailed studies have been made recently by Sanche and Burrow (1974), who recorded the derivative of the transmitted current, and by Risley et al. (1974), who detected electrons scattered from collisionally excited fast H^- ions. The method of these experiments basically limits the amount of information to be obtained about the resonances to their energies and widths. However, angular distribution studies by Williams (1974) have also provided values of the phase shifts. Figure 1 shows the

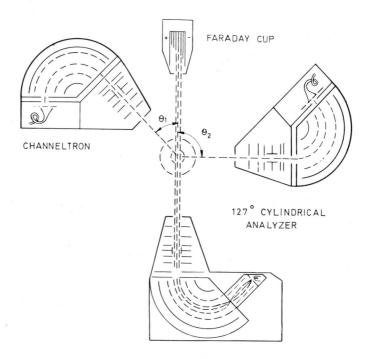

Figure 1. Schematic representation of the apparatus used by Williams (1974). In the plane of the electron scattering for zero azimuthal angle, identical 127° electrostatic analyzers are used to determine the energies of the incident and scattered electrons. Typically, an incident electron beam of 3×10^{-8} A with an energy resolution of 0.020 eV and an angular spread of 1° is used. The modulated atomic beam, which is normal to the plane of the paper, is shielded in the vicinity of the interaction region with a 99.5% transparent tungsten electrostatic shield. A Faraday cup always collects the primary beam, and its physical size restricts the minimum scattering angle which can be seen by the analyzers to about 15°. The figure shows two analyzers which, in the elastic scattering experiments, were used to detect simultaneously the scattered electrons at two scattering angles.

schematic arrangement of incident and scattered electron momentum analyzers used in these latter studies. Figures 2 and 3 clearly show the presence of the ^1S and ^3P resonances. The well-defined shapes of these resonances is attributable to good angular resolution (1°) and good energy resolution (at best 0.012 eV, but typically 0.020 to 0.030 eV). Table 1a shows that all of the experimental values of resonance energies are in agreement within the experimental accuracies (of not better than ±0.010 eV). The most accurate calibration of the electron energy scale is presently about ±0.010 eV, which, in atomic hydrogen, is achieved by determining the threshold energy for 1216 Å photon production from H(2p) state excitation. Unfortunately, such accuracy is not sufficient to differentiate between any of the more recent, refined theoretical estimates (Table 1b). The most accurate calculations appear to be the three-state (1s, 2s, 2p) close coupling, with 20 electron–electron correlation terms, approximation by Burke and Taylor (1966) and the Feshback projection-operator calculations by Bhatia and Temkin (1973) and Chen and Chung (1970). Table 1b lists some theoretical values, which are representative of their method of calculation, for the resonance

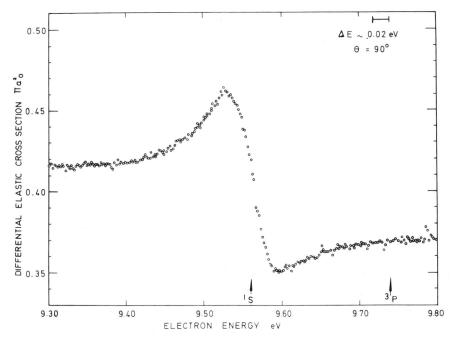

Figure 2. Absolute differential elastic cross section as a function of incident electron energy, showing the ^1S and ^3P resonance structure of atomic hydrogen. The scattering angle was 90°, the angular resolution 1.5°, and the electron energy resolution 0.020 eV.

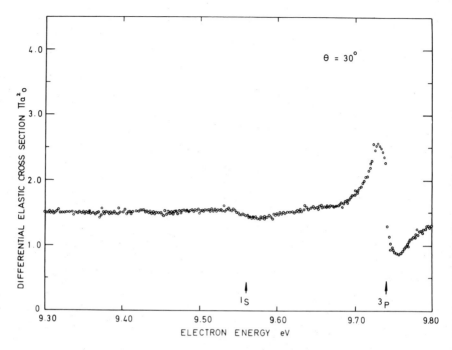

Figure 3. Absolute differential elastic cross section as a function of incident electron energy, showing the ^1S and ^3P resonance structure of atomic hydrogen. The scattering angle was 30°, the angular resolution 1.5°, and the electron energy resolution 0.012 eV.

Table 1a. Experimental Values of Energies (E_r), in eV, and Decay Widths (Γ), in eV, of Resonant States in the Elastic Scattering Channel for e–H Collisions

State	Williams (1974)	McGowan (1967)	Schulz (1964)	Sanche and Burrow (1972)	Kleinpoppen and Raible (1965)	Risley et al. (1974)	Theoretical range
(a) Resonance energies							
$(2s^2)^1S$	9.557(10)	9.56(1)		9.558(10)		9.59(3)	9.545–9.630
$(2s2p)^3P$	9.735(10)	9.71(3)	9.70(15)	9.738(10)	9.73(12)	9.76(3)	9.722–9.767
$(2p^2)^1D$							10.119–10.156
$(2s2p)^1P$		10.130(15)		10.128(10)		10.18(3)	10.170–10.180
$(2p^2)^1S$							10.164–10.198
(b) Resonance widths							
$(2s^2)^1S$	0.045(5)	0.043(6)					0.039–0.109
$(2s2p)^2P$	0.0060(5)	0.009		0.0056(5)			0.0063–0.0091

Table 1b. Theoretical Values of Energies (E_r), in eV, and Decay Widths (Γ), in eV, of Resonant States in the Elastic Scattering Channel of Atomic H below the $n = 2$ Level

State	Energy	Width	Reference	Method
$(2s^2)^1S$	9.5487	0.0406	Bhatia and Temkin (1973)	Feshbach projection–operator
	9.5490	0.0411	Chung and Chen (1972)	Feshbach projection–operator
	9.5518	0.046	Doolan et al. (1974)	Ritz variational method
	9.552	0.050	Matese and Oberoi (1971)	Close coupling method using pseudostates
	9.553	0.051	Burke et al. (1967)	Close coupling with pseudostates
	9.554		O'Malley and Geltman (1965)	Feshbach projection–operator
	9.555	0.048	Burke and Taylor (1966)	Three-state ($1s$, $2s$, $2p$) close coupling with 20 correlation terms
	9.565	0.039	Chen and Chung (1970)	Feshbach projection–operator
	9.582	0.050	Burke et al. (1969)	Close coupling with pseudostates
	9.585	0.054	Burke et al. (1969)	Three-state ($1s$, $2s$, $2p$) close coupling
$(2s2p)^3P$	9.731	0.0091	Drake and Dalgarno (1971)	Perturbation method
	9.773	0.0063	Bhatia and Temkin (1969)	Feshbach projection operator
	9.735	0.0059	Burke (1968)	Three-state ($1s$, $2a$, $2p$) close coupling with 20 correlation terms
	9.735	0.0068	Matese and Oberoi (1971)	Close coupling with pseudostates
	9.754	0.0057	Burke et al. (1969)	Close coupling with pseudostates
	9.767	0.0076	Burke et al. (1969)	Close coupling with pseudostates

energies and widths. A complete list is given by Risley (1974), and an extensive discussion of the methods is given by Chen (1970) and Burke et al. (1967). The experimental accuracy of ± 0.005 eV in determining the decay widths of the resonant states is not good enough to distinguish among the theoretical values. It appears that an order of magnitude improvement in experimental accuracy in establishing the energy and the width of the resonance is required before further guidance or interpretation can be given to the theory.

Since resonant structure occurs in both the singlet and triplet partial waves, which are noncoherent, an angular distribution study of both the ^1S and ^3P resonances permits the determination of both the singlet and triplet phase shifts. Table 2 compares the experimental values with the variationally determined phase shifts of Schwartz (1961) for the s wave, Armstead (1968) for the p wave, and Gailitis (1965) for the d wave. The experimental values (except for the ^3P phase shift) agree, within the experimental accuracy of

Table 2. The Experimentally Determined Phase Shifts (Radians) Are Compared with the Variationally Calculated Values for the Elastic Scattering of Electrons from Atomic Hydrogen at $k = 0.800$ a.u. ($E = 8.7$ eV)

The variational values of the phase shifts are the s-wave values of Schwartz (1961), the p-wave values of Armstead (1968), and the d-wave values of Gailitis (1965).

	Singlet		Triplet	
l	Experiment	Variational theory	Experiment	Variational theory
0	0.904	0.886	1.630	1.643
1	0.000	−0.004	0.461	0.428
2	0.066	0.073	0.062	0.068

±0.021 radian, with the theoretical values. A full description of this experiment is in course of publication (Williams, 1975). It is noted that the determination of the singlet and triplet phase shifts from the angular differential resonance elastic scattering yields information which is obtainable at energies not in the vicinity of a resonance only from an experiment using spin analysis of the electron and atom before and after collision.

These phase shifts allow the angular differential cross section to be calculated and then permit the relative angular distributions to be placed on an absolute scale. All the data presented in this paper have been calibrated against the absolute elastic differential cross section value of $0.727\pi a_0^2$ sr^{-1} at 60° and 8.70 eV. This method of calibration is particularly important in atomic hydrogen and other labile species for which it is not yet possible to measure, by an independent method, an absolute value of the number density of atoms in a modulated beam. However, it is readily seen that a determination of the number density can be made via Eq. (5) now that an absolute cross section value has been deduced from the phase shifts.

It is stressed that this method of cross section calibration only determines the cross section in the vicinity of the resonances. Determination of accurate cross section values at other energies, then, requires that the energy dependence of all parameters in Eq. (5) be known accurately. An independent check on this method of calibration is, of course, provided if angular differential cross sections are obtained over a wide energy range, i.e., the method of Section 2.1.3 can then be used to deduce phase shifts. During the analysis of such data, a further restraint may be placed upon the solution set of phase shifts, for example, that they must fit a simple polynomial based upon an effective range type expansion and hence must be a smooth function of energy except at the resonance. This restraint tends to ensure that the inaccuracy of the phase shifts arising from systematic experimental errors is not much greater than that arising from statistical sources. At least for elastic scattering at low energies, the above considerations constitute some of the advantages of a phase shift analysis of angular differential cross sections. However, further developments must be expected in this relatively new method of analysis in electron scattering.

1.3. Elastic Differential Cross Sections

1.3.1. Low Energies. The recent measurements of the elastic differential cross section at 3.4 and 8.7 eV by Williams (1974) have been extended to an energy of 1.2 eV. The data are shown in Figures 4, 5, and 6. The trends

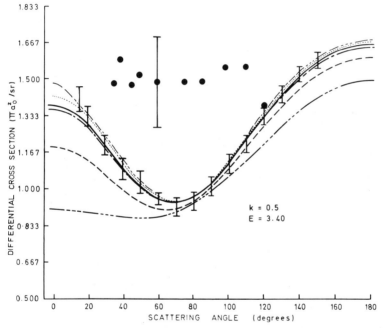

Figure 4. Absolute angular differential cross sections are shown for the elastic scattering of 3.4 eV ($k = 0.5$ a.u.) electrons from atomic hydrogen. The experimental data of Williams (1974) are shown as error bars ($\mathbf{I}$ only), while the measured values are at the center of the error bars. (●), Gilbody et al. (1961); (— — — —), polarized orbital values, Temkin and Lamkin (1961); (— — — — —), variational values (see text); (———) and (— — —), close coupling with pseudostates (Burke, et al. 1969); (— — —), three-state close coupling (Burke and Schey, 1962); (— — —), static exchange (Mott and Massey, 1965).

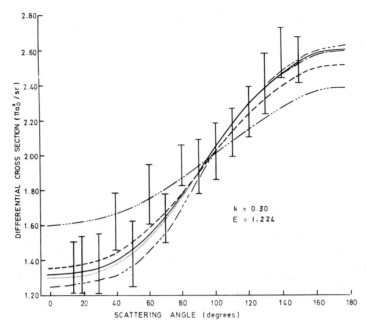

Figure 5. Absolute angular differential cross sections are shown for the elastic scattering of 1.224 ± 0.080 eV ($k = 0.30$ a.u.) electrons from atomic hydrogen. The experimental data of Williams (1974) are shown as error bars (I) with the actual measured value at the center; (------), polarized orbital calculation (Temkin and Lamkin, 1961 and Temkin, 1974); (— — — —), three-state close coupling (Burke and Schey, 1962); (———), close coupling with pseudostates Burke et al., 1969); (- - - - -), variational calculation (see text); (— — —), static exchange (Mott and Massey, 1965).

established at higher energies have been continued, that is, the backward scattering increases and the forward scattering decreases as the electron energy decreases. The minimum in the differential cross section at about 3.4 eV appears to arise from interference among the s, p, and d waves, and the effect is slightly enhanced, since the ^1P phase shift changes sign at about that energy. The theoretical values have been calculated from published phase shifts for only the first three partial waves. The experimental values and their accuracy do not permit a distinction to be made between the variational values, the two-pseudostate close-coupling values (Burke et al., 1969), and the polarized-orbital values (Temkin and Lamkin, 1961; Temkin, 1974). For all forward angles less than about 70°, the three-state ($1s$, $2s$, $2p$) close-coupling and static-exchange values lie outside the experimental limits. At backward scattering angles, all five theoretical approximations give nearly

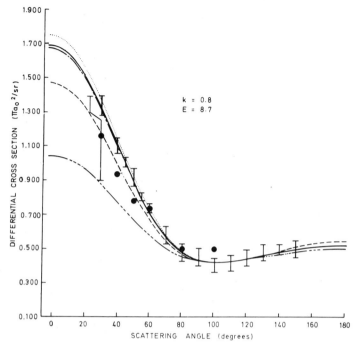

Figure 6. Absolute angular differential cross sections are shown for the elastic scattering of 8.704 ± 0.010 eV ($k = 0.8$ a.u.) electrons from atomic hydrogen. The experimental data of Williams (1974) are shown as error bars (I only), while the measured values are at the center of the error bars. (●), Gilbody et al. (1961); (– – – – –), variational values (see text); (———) and (– – –), close coupling with pseudostates (Burke et al., 1969); (— — —) three-state close-coupling (Burke and Schey, 1962); (– – – –) static exchange (Mott and Massey, 1965).

identical values, which are in good agreement with the experimental values. This agreement confirms that the correct theoretical description of the elastic scattering process below the first inelastic threshold is adequately given by those theories that include the full ground-state polarizability as well as allowing for exchange effects. The present measurements are the first of sufficient accuracy to confirm the generally accepted "exact" variational values and the first to distinguish between the various low-energy scattering approximations. At all energies below the first inelastic threshold, the values predicted by the variational, pseudostate close-coupling, and polarized-orbital methods are in good agreement with the experimental values. The recent relative measurements at 9.4 eV by Lloyd et al. (1974), whose apparatus and method were similar to those of Gilbody et al. (1961), are of insufficient accuracy to distinguish among the various theoretical approximations.

1.3.2. High Energies. From the apparatus function defined in Eq. (5), it is in principle a simple matter, but in practice a very tedious procedure, to extend the absolute calibration of the elastic scattering cross sections at low energies to the relative differential cross sections at high energies. This calibration has been carried out up to 680 eV (Williams, 1975), but this paper reports only the measurements at 100 eV.

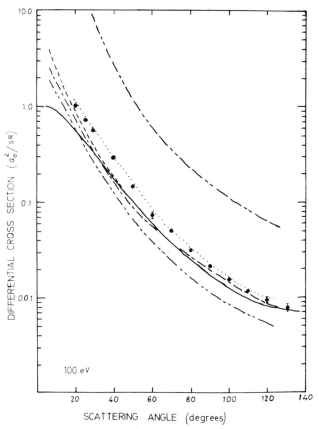

Figure 7. Absolute differential elastic cross section as a function of scattering angle for the elastic scattering of 100 eV electrons from atomic hydrogen. (●), Williams and Willis (1974), (-----), Lloyd *et al.* (1974); (———), Born approximation, Bransden *et al.* (1972); (— — —), Fadeev multiple scattering calculation Sinfailam and Chen (1972); (— — ——), the Glauber approximation with straight line trajectories, Chen *et al.* (1973); (–·–), the Glauber approximation with angle trajectories, Chen *et al.* (1973); (— — —) eikonal–Born approximation, Byron and Joachain (1973).

The only previous measurements of differential elastic cross sections in atomic hydrogen at high energies were the relative values of Lloyd *et al.* (1974), which were placed on an absolute scale by normalization relative to the absolute molecular elastic cross sections of Trajmar *et al.* (1970) at 60° and 50 eV energy and at 100° and 200 eV to the theoretical values of Khare and Moiseiwitsch (1965). These data are compared with the present absolute values and the various theoretical values for an incident electron energy of 100 eV in Figure 7. The angular distributions of Lloyd *et al.* agree well in shape with the present data; however, their normalization leads to their slightly higher cross section values, which implies that the calculation of Khare and Moiseiwitsch is too high at 60° by about a factor of 10%. The Born approximation values agree well in shape with the present data above about 40°, but at smaller angles the Born values become too small. The straight-line Glauber approximation values (Chen *et al.*, 1973) are about a factor of two lower than the present data, while the Glauber approximation values with angle trajectories (Chen *et al.*, 1973) are in best agreement with the present data at around 90° but are also too small in the forward direction. The best fit in the forward direction is obtained by the eikonal Born series approximation used by Byron and Joachain (1973). The multiple scattering expansion based on the Fadeev equations has been used by Sinfailam and Chen (1972); however, their values are about an order of magnitude too high at all angles. Recent theoretical efforts have been directed toward obtaining a better allowance for the continuum in the region from 10 to 100 eV, and, along these lines, a second-order potential calculation by Winters *et al.* (1973) gives an angular distribution which is similar to that calculated by the eikonal Born series expansion.

In summary, it is seen that the eikonal Born approximation almost correctly describes the scattering at around 90°, and all other calculations, except the multiple-scattering approximation, predict cross sections which are too low at all angles.

2. Excitation of the $n = 2$ States of Atomic Hydrogen

2.1. Absolute Total Cross Section Values near Threshold

Experimental and theoretical studies of electron impact excitation of the $n = 2$ states of atomic hydrogen have been discussed by Moiseiwitsch and Smith (1968), Smith (1969), and Geltman and Burke (1970). The only new experimental results on H(2s) state excitation since the review by Massey and Burhop (1969) are the searches by Oed (1971) and Koschmieder *et al.* (1973) for resonance structure below the $n = 3$ level. Their electron energy resolutions of 0.150 and 0.110 eV respectively were not sufficient to resolve the separate resonant state contributions from a slight dip in their excitation function. The separate excitation of the 2s state will not be discussed further in this paper.

The $n = 2$ excitation experiments have studied either the Lyman alpha photons (1216 Å) which arise either from the prompt (10^{-8} sec) decay of the H(2p) atoms or from the delayed quenching of the H(2s) atoms. These photon detection experiments generally determine excitation functions, which, for electron energies greater than the excitation energy of the $n = 3$ states, cannot be compared directly with theoretical predictions of the $n = 2$ excitation cross section until they have been corrected for cascade contributions from higher states which must also be excited by electron impact. Corrections must also be made for the polarization of the detected radiation, unless the observations have been made with a detector of small acceptance angle at scattering angles of either 54.5° or 125.5°, where the observed excitation function is directly proportional to the total cross section (Percival and Seaton, 1957). The most recent values of the 2p excitation cross section, $Q(2p)$, measured by Long et al. (1969) and McGowan et al. (1969) appear to agree well in shape (energy dependence) with the Born approximation values above about 200 eV. Similarly, values of the 2s excitation cross section, $Q(2s)$, measured by Kauppila et al. (1970) appear to agree well in shape with the Born approximation values above about 45 eV. These measurements have then relied upon the Born-approximation-calculated values at some high energy [for example, $0.486\pi a_0^2$ at 200 eV for $Q(2p)$] for normalization and for corrections for cascade contributions [about 2% of $Q(2p)$ at 200 eV (Moiseiwitsch and Smith, 1968)]. These experimental values are then found to be about 20% lower than the best theoretical values [from the close-coupling approximation of Geltman and Burke (1970)] at low (11 eV) energies.

A series of close-coupling calculations (Taylor and Burke, 1967; Burke et al., 1967; Geltman and Burke, 1970) has successively incorporated most of the low-energy physically significant aspects of the $n = 2$ excitation process. Thus, Burke et al. (1967) stated that no important effects were left out of the theoretical description and they suggested that the experimental 2p (and 2s) excitation cross sections could be normalized to the theoretical values just above threshold. Support for the theoretical values was obtained by Ott et al. (1970) and Kauppila et al. (1970), whose values of the ratio of $Q(2s)/Q(2p)$ agreed, within the experimental accuracy of $\pm 10\%$, with the theoretical values over the range 10.4 to 11.4 eV. Direct confirmation of the theoretical values has been provided by the recent absolute cross section values of Williams and Willis (1974). By keeping the incident electron energy below 12 eV cascade processes are avoided. The major advances of the phase shift and resonance analysis methods permit an absolute cross section value to be measured for the elastic scattering cross section, as explained in Section

1.2.4. Then, from Eq. (5), the further measurement of (a) the absolute counting efficiency of the photon detector, (b) the relative angular acceptance of the photon and electron detectors, and (c) the relative elastic cross section values at 11.0 eV (the energy chosen for $Q(2p)$ calibration) and at 8.7 eV (the energy of the elastic cross section calibration) permit an absolute value of $Q(2p)$ to be determined. The measurement of $Q(1s)$ and $Q(2p)$ simultaneously, at the calibration energy of 11.0 eV, ensured that other parameters in the "beam overlap integral" of Eq. (5) were identical for the measurement of $Q(1s)$ and $Q(2p)$. Figure 8 shows that at an energy of 11.02 eV, the calibrated value of $Q(2p)$ of $0.276^{+0.026}_{-0.016} \pi a_0^2$ is about 20% above the previous experimental values of Long et al. (1968) and McGowan et al. (1969) but in excellent agreement with the three-state-plus-20-correlation-terms close-coupling values and the close-coupling with pseudostates values (Geltman and Burke, 1970). All of the quoted experimental error arose from the calibration of the

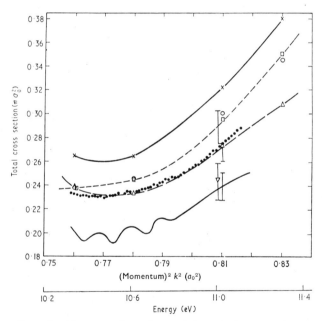

Figure 8. The comparison of several theoretical $Q(2p)$ excitation cross sections is made with experimental data. (●), Williams and Willis (1974); (▽), Long et al. (1969); (————), McGowan et al. (1969). The theoretical curves are close-coupling calculations by Burke et al. (1967); (× — ×), six-state close coupling; (△—△), three-state plus 20 correlation terms close coupling; (○), pseudostate close coupling; (□ - - □) best close-coupling partial cross sections.

photon detector. A further 6% error arose from the elastic cross section determination. A more accurate determination of $Q(2p)$ will necessitate a more accurate method than that of Samson and Cairns (1965) in calibrating the Lyman alpha detector. The implications from the present experimental work are similar to those stated by Geltman and Burke (1970). Further study should be made of the cascade contributions to the Lyman alpha signal over the entire energy range and of the accuracy of the first Born approximation at high energies.

2.2. Resonance Studies

The finite value of the $2p$ excitation cross section at threshold was predicted by Damburg and Gailitis (1963) and confirmed experimentally by Chamberlain et al. (1964), Williams and McGowan (1968), and McGowan et al. (1969). The latter group also confirmed the presence of a large ^1P resonance, of width about 0.015 eV, just above threshold at 10.2 eV, as predicted in a close-coupling calculation by Taylor and Burke (1967). The existence of at least one, possibly two, further resonances at about 10.46 and 10.6 eV was suggested by the measurements of McGowan et al. (1969), and a calculation by Marriot and Rotenburg (1968) indicated that interference between background potential scattering and resonance scattering could produce structure in the ^1P and ^1D partial cross sections. However, as shown in Figure 8, the recent data of Williams and Willis (1974) do not show any structure within the relative accuracy ($\pm 1\%$) of the data points, which is about a factor of five better than that obtained by McGowan et al. (1969).

Further resonant states lying below the $n = 3$ and $n = 4$ levels have been observed in both the photon and electron decay channels, as shown in Figures 9 and 10 respectively and listed in Table 3. Figure 9 shows those resonances observed in the decay of H($2p$) state atoms by photon detection measurements by McGowan et al. (1969) and Williams (1974) and compares their shape with the values calculated in a six-state close-coupling approximation by Burke et al. (1967). These resonances have also been seen in the spectra (Figure 10) of the scattered electrons (Williams, 1975), which must have excited both the $2s$ and $2p$ states. Because of the low energies of the scattered electrons, there is considerable uncertainty (0.080 eV) in the energy scale and poor counting statistics (even after 300 hours counting time). These facts, together with the overlapping nature of the structures, have not yet permitted a detailed analysis of the resonances. However, angular distribution studies (Figure 10) have established that the resonances at 11.75 eV and 11.88 eV may be d- and p-wave resonances as indicated by the theory. There does not appear to be any marked structure at the onset of $n = 3$ excitation processes.

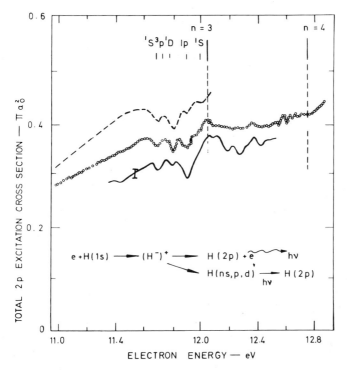

Figure 9. The total $2p$ excitation cross section of atomic hydrogen as a function of electron energy (○), Williams (1974); (———), McGowan et al. (1969); (– – –), a six-state close-coupling calculation by Burke et al. (1967). The electron energy resolution in both experiments was 0.060 eV.

Table 3. Experimental Values for the Energies (E_r), in eV, and Widths (Γ), in eV, of the Resonant States below the $n = 3$ Level as Observed in the 10.2 eV Photon and 10.2 Energy Loss Decay Channels

State	Williams (1975) photon channel	Williams (1975) electron channel	McGowan et al. (1969) photon channel	Risley et al. (1974) electron channel	Theory (Burke et al. 1967)
$(3s^2)^1S$	11.68(3)	11.60(8)	11.65(3)		11.727
$(3s3p)^3P$	–	11.70(8)			11.759
$(3p^2)^1D$	11.78(3)	11.75(8)	11.77(2)	11.86(4)	11.813
$(3s3p)^1P$	11.89(3)	11.88(8)	11.89(2)	(State uncertain)	11.910
$(3s4p)^1S$	–	11.97(8)	–		12.031

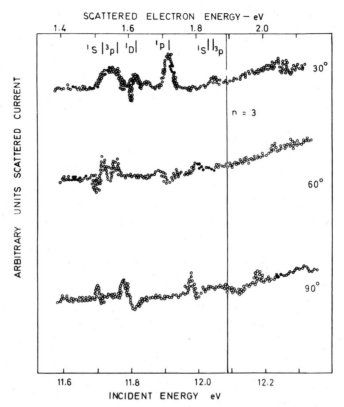

Figure 10. Resonance structure as observed in the scattered electron spectrum of those electrons which have lost 10.2 eV in exciting the $n = 2$ states of atomic hydrogen. The relative scattered current is shown as a function of both the incident and scattered electron energy at scattering angles of 30°, 60°, and 90°. The energy resolution of both the incident and scattered electron energy analyzers was 0.022 eV.

It is clear that an order of magnitude improvement in both the electron energy spread and signal intensities is required in order to clearly resolve the resonance structures.

The above measurements have formed the basis for further studies near threshold of coincident detection of the 1216 Å photon from the H(2p) state decay and the electron which has lost 10.2 eV in exciting the 2p state. Such measurements will be similar to those of Eminyan et al. (1974) on the excitation of the 2^1P state of helium.

2.3. Differential Cross Sections at High Energies

As a prelude to similar coincidence measurements at high energies, the following studies (Williams and Willis, 1974) have been made from 50 to 680 eV of the scattered electrons which have lost 10.2 eV in exciting the $n = 2$ states. A scattered electron energy resolution of 0.095 eV ensured that there were no contributions to the measured signal from the $n = 3$ states. The method of Section 1.2.4 has been used to obtain absolute values of the angular differential cross section $I_{2s+2p}(k, \theta)$, which is the sum of the partial differential cross sections, $I_{2s}(k, \theta)$ and $I_{2p}(k, \theta)$, for exciting the 2s and 2p states, respectively. Figures 11, 12, and 13 show the measured values at

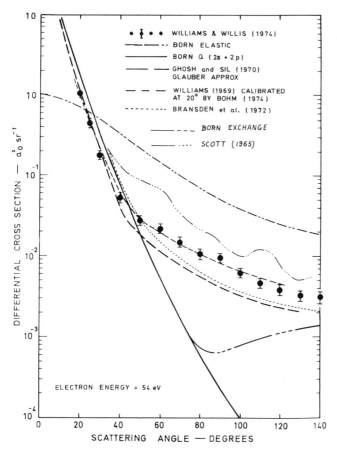

Figure 11. Differential cross section for the excitation of the $n = 2$ states of atomic hydrogen for 54 eV incident electrons.

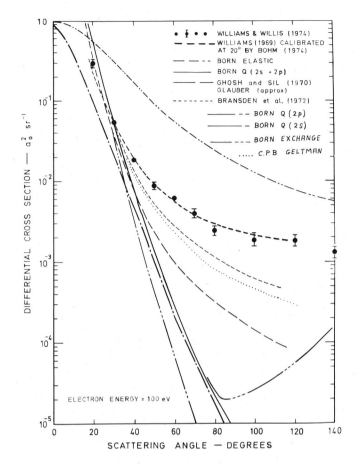

Figure 12. Differential cross section for the excitation of the $n = 2$ states of atomic hydrogen for 100 eV incident electrons.

incident electron energies of 54, 100, and 680 eV. The only other measurements of $I_{2s+2p}(k, \theta)$ are the relative angular distributions of K. G. Williams (1969) at 54 and 100 eV. His data have been recently placed on an absolute scale by Bohm (1974), who made his calibration on the same apparatus as K. G. Williams but with only a single measurement at 20° at each energy. Their data are in agreement with the present values within their respective error limits. It is noted that in the publication of K. G. Williams (1969), his data were normalized to the Born–Oppenheimer approximation values at 21°; however, incorrect cross section values were plotted in Figure 2 of his publication for 100 eV incident electron energy.

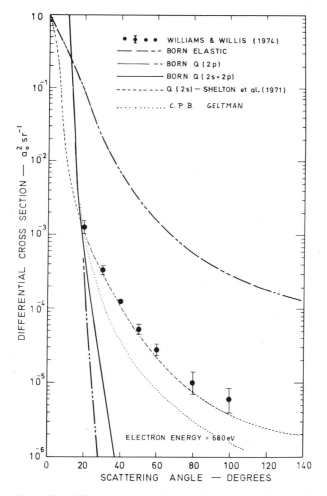

Figure 13. Differential cross section for the excitation of the $n = 2$ states of atomic hydrogen for 680 eV incident electrons.

To obtain an understanding of the significance of these experimental results, it is instructive to study the various theoretical descriptions of the scattering processes. The first Born approximation values (Kingston, 1974, private communication) of $I_{2s}(k, \theta)$, $I_{2p}(k, \theta)$, and the elastic differential cross section $I_{1s}(k, \theta)$, as well as the sum $I_{2s+2p}(k, \theta)$, have been plotted for several electron energies. It is well known (Mott and Massey, 1965) that I_{2s} and I_{2p} decrease as K^{-12} and K^{-14} respectively (where K is the momentum transfer from the incident electron), and that the ratio of I_{1s} to I_{2s+2p} increases rapidly as both the scattering angle and the electron energy increase. These facts are shown by the Born curves in Figure 12. The separate contributions of $I_{2s}(k, \theta)$ and $I_{2p}(k, \theta)$ to the sum $I_{2s+2p}(k, \theta)$ in the Born–Oppenheimer approximation is shown in Figure 12 for an incident electron energy of 100 eV. At angles less than about 30°, I_{2p} becomes dominant, while for angles greater than about 50°, I_{2s} dominates the sum. This relative behavior of I_{2p} and I_{2s}, together with their energy dependence, is characteristic of the Born approximation. The Born approximation is both a high-energy and a small-momentum-transfer approximation. Hence, the calculated angular distributions of the scattered electrons are expected to be accurate over a decreasing angular range about the zero scattering angle as the energy is increased. From Figures 11, 12, and 13 it is seen that the present experimental cross section values are not in agreement with the first Born approximation values at any of the energies; however, there is agreement in the shapes of the angular distributions at 54 eV for $\theta \leqslant 30°$. At all other energies, if there is to be any agreement it must occur at scattering angles less than 20°, the minimum observable angle in the present experiment. Outside these angular ranges, the deviations between experimental and Born values increase rapidly with increasing k and θ. A necessary condition of an experiment which would test the accuracy of the Born approximation in describing differential cross sections at small angles is readily seen to be an angular resolution of not more than 0.5° at 54 eV and better at higher energies.

The scattering event is complicated by three effects: (a) exchange of the incident and target electrons, (b) distortion of the target atom charge distribution by the field of the incident electron, and (c) coupling between open and closed scattering channels. The effect of exchange, which has been calculated in a plane-wave Born–Oppenheimer approximation (Massey and Burhop, 1969), is seen from Figures 11 and 12 to be to increase the scattering in the backward direction by orders of magnitude; however, the cross sections are still much smaller than experimental values. A detailed discussion of the various plane-wave theories of the exchange effects has been made by Truhlar et al. (1968). All of these theories produce cross section values similar to those Born exchange values shown in Figures 11 and 12.

Distortion effects are not included in Born type approximations because of the plane-wave representation of the unbound electron. At high energies, large-angle elastic scattering occurs primarily from the interaction of the incident electron with the proton. Similarly, large-angle inelastic scattering is expected to occur primarily from the same interaction. In the first Born approximation, this inelastic interaction vanishes because of the orthogonality of the initial- and final-state wave functions, and the large-angle scattering arises only from the electron–electron interaction with the consequent rapid decrease of cross section with angle. However, the nuclear interaction has been allowed for in the Glauber approximation calculations of Tai et al. (1970) and Ghosh and Sil (1970). Their calculated values are shown in Figures 12 and 13 by a combined curve which was plotted from the absolute values of Ghosh and Sil in the range $0° < \theta < 40°$ and extended to larger angles from the relative cross section values reported by Tai et al. At small angles their values are in agreement with the Born approximation values. At large angles their values have a different angular dependence and are significantly smaller than the experimental values. Further refinements to their theoretical approach are clearly required.

Geltman and Hidalgo (1971) have included the incident electron–nuclear interaction into a "Coulomb projected" Born (CPB) approximation. Figures 12 and 13 show that the inclusion of this interaction drastically changes the large-angle behavior of the theoretical cross section, which is then approaching the experimental values. The inclusion of exchange was shown to have a negligible ($\sim 5\%$) increase on the CPB-calculated cross section. At small angles the CPB values tend asymptotically toward the Born values. Geltman (this volume) has shown that at high energies the large-angle behavior of I_{2s} and I_{2p}, and hence $I_{2p+2p}(k, \theta)$, is a Rutherford $\csc^4(\theta/2)$ behavior. It is readily seen that such behavior is well satisfied by the experimental data at 680 eV in Figure 13.

However, a further promising theory is a distorted-wave approximation by Shelton et al. (1971). Their approximation differs from the plane-wave approximations in that the potential used to produce the distortion includes static and dynamic (polarization) atomic distortion terms. The effect of polarization is shown to be small at the energies of the present measurements. They have published values of only $I_{2s}(k, \theta)$, which have been plotted in Figure 13. A detailed comparison with the present values of $I_{2s+2p}(k, \theta)$ must await their calculation of $I_{2p}(k, \theta)$; however, the indications are that the angular distributions may be in good agreement with experiment.

The third complicating effect on the scattering process is the coupling between open and closed channels. This effect, which was excluded in the previous approximations, is explicitly included in close-coupling type approximations. A three-state ($1s, 2s, 2p$) close-coupling calculation of the R matrix elements was made at 54 eV by Burke et al. (1963). Scott (1965) determined values of I_{2s} and I_{2p} from those matrix elements, but included only up to a total of seven partial waves. His values of $I_{2p+2p}(k, \theta)$ are shown in Figure 12 to be higher than the experimental values and to contain small oscillations. Brandt and Truhlar (1974) have shown that, at lower energies, the oscillations arise from Scott's calculations and are not contained in Burke's R matrix elements. The error bars of the data of K. G. Williams (1969) were of a size similar to the amplitude of these oscillations, but the present measurements indicate that if any oscillations are present they must be considerably smaller than those predicted by Scott. Further measurements in the region of $0.75 < k < 2.0$ a.u. are in progress to specifically test the close-coupling predictions.

The effect of the large number of open channels upon the $(2s + 2p)$-states excitation process in the energy region below 200 eV has been studied by Bransden *et al.* (1972). They have accounted for several strongly coupled states by a truncated eigenfunction expansion, while the remaining (infinite number) states are allowed for by a second-order potential which is determined by the correct long-range effective potential of the approaching interacting particles. Their data (with units as corrected by Winters *et al.*, 1973) are shown in Figure 12. At small angles their values tend asymptotically toward the Born approximation values, while at large angles their values underestimate the measured cross section but are in closer agreement with experiment than the Glauber, and other, approximation values.

In conclusion, the results of this paper can be arranged into two groups. For both elastic and inelastic ($2p$ excitation) scattering at low energies where there are no competing decay channels, the experimental data are in agreement, within experimental accuracy, with the best theoretical predictions, which are the variational, "best" close-coupling, and polarized-orbital calculations. At high energies, no theory appears to describe all features of the experimental data adequately. These results perhaps reflect the views of the theories which, in the low-energy region, are thought to contain all the important physics of the scattering processes, but which, in the high- (and medium-) energy ranges, are known to not include all of the important physical effects.

Future experimental effort will concentrate upon an elucidation of the electron–atomic hydrogen scattering processes by the use of coincidence techniques in the lowest-energy regions, where multichannel coupling is thought to be important.

References

Andrick, D. (1973). *Adv. Atom. Molec. Phys.*, **9**, 207.
Armstead, R. L. (1968). *Phys. Rev.*, **171**, 91.
Bederson, B., and Kieffer, L. J. (1971). *Rev. Mod. Phys.*, **43**, 601.
Bhatia, A. K., and Temkin, S. (1969). *Phys. Rev.*, **182**, 15.
Bhatia, A. K., and Temkin, A. (1973). *Phys. Rev.*, A8, 2184.
Bohm, B. (1974). Private communication (Ph.D. thesis, University of Adelaide, South Australia).
Brandt, M. A., and Truhlar, G. D. (1974). *Phys. Rev.*, A9, 1188.
Bransden, B. H., Coleman, J. P., and Sullivan, J. (1972). *J. Phys. B.*, **5**, 546.
Burke, P. G. (1965). *Adv. Phys.*, **14**, 521.
Burke, P. G. (1968). *Proc. First Int. Conf. Atomic Physics*, p. 265. New York: Plenum Press.
Burke, P. G., and Schey, H. M. (1962). *Phys. Rev.*, **126**, 147.
Burke, P. G., and Smith, K. (1962). *Rev. Mod. Phys.*, **34**, 458.

Burke, P. G., and Taylor, A. J. (1966). *Proc. Phys. Soc. (Lond.)*, **88**, 549.
Burke, P. G., Schey, H. M., and Smith, K. (1963). *Phys. Rev.*, **129**, 1258.
Burke, P. G., Ormonde, S., and Whitaker, W. (1967). *Proc. Phys. Soc.*, **92**, 319, 336, 345.
Burke, P. G., Gallaher, D. F., and Geltman, S. (1969). *J. Phys. B.*, **2**, 1142.
Byron, F. W., and Joachain, C. J. (1973). *J. Phys. B.*, **6**, L169.
Chamberlain, G., Smith, S. J., and Heddle, D. W. D. (1964). *Phys. Rev. Letters*, **12**, 647.
Chamberlain, G., Mielczarek, S. R., and Kuyatt, C. E. (1970). *Phys. Rev.*, A2, 1905.
Chen, J. C. Y. (1970). *Nucl. Inst. Methods*, **90**, 237.
Chen, J. C. Y., and Chung, K. T. (1970). *Phys. Rev.*, A2, 1892.
Chen, J. C. Y., Hambro, L., Sinfailam, A. L., and Chung, K. T. (1973). *Phys. Rev.*, A7, 2003.
Chung, K. T., and Chen, J. C. Y. (1972). *Phys. Rev.*, A6, 686.
Comer, J., and Read, F. H. (1972). *J. Phys. E.*, **5**, 211.
Damburg, R. J., and Gailitis, M. K. (1963). *Proc. Phys. Soc.*, **82**, 192.
Damburg, R. J., and Geltman, S. (1968). *Phys. Rev. Lett.*, **20**, 485.
Damburg, R. J., and Karule, E. (1967). *Proc. Phys. Soc.*, **90**, 637.
Doolan, G. D., Nuttall, J., and Stagat, R. W. (1974). *Phys. Rev.*, to be published.
Drachman, R. J., and Temkin, A. (1970). *Case Studies in Atomic Collisions*, Vol. 2, p. 401. Amsterdam: North-Holland.
Drake, G. W. F., and Dalgarno, A. (1971). *Proc. Roy. Soc. (Lond.)*, A320, 549.
Eminyan, M., MacAdam, K. B., Slevin, J., and Kleinpoppen, H. (1974). *J. Phys. B.*, **7**, 1519.
Fano, U. (1961). *Phys. Rev.*, **124**, 1866.
Franzen, W., and Gupta, R. (1965). *Phys. Rev. Lett.*, **15**, 819.
Gailitis, M. K. (1965). *Soviet Physics: JETP*, **20**, 107.
Geltman, S. (1969). *Topics in Atomic Collision Theory*. New York: Academic Press.
Geltman, S., and Burke, P. G. (1970). *J. Phys. B.*, **3**, 1062.
Geltman, S., and Hidalgo, M. B. (1971). *J. Phys. B.*, **4**, 1299.
Gerjuoy, E. (1970). *Proc. Second Int. Conf. Atomic Physics*, p. 271, New York: Plenum Press.
Ghosh, A. S., and Sil, N. C. (1970). *Indian J. Phys.*, **44**, 153.
Gibson, J. R., and Dolder, K. T. (1969). *J. Phys. B.*, **2**, 741, 1180.
Gilbody, H. B., Stebbings, R. F., and Fite, W. L. (1961). *Phys. Rev.*, **121**, 794.
Hoeper, P. S., Franzen, W., and Gupta, R. (1968). *Phys. Rev.*, **168**, 50.
Holtzmark, J. (1933). *Det. Kgl. Norske Videnskabers Selskabs. Schrifter*, **6**, No. 53, 1.
Kauppila, W., Ott, W. R., and Fite, W. L. (1970). *Phys. Rev.*, A1, 1099.
Khare, S. P., and Moiseiwitsch, B. L. (1965). *Proc. Phys. Soc.*, **85**, 821.
Kieffer, L. J., and Dunn, G. H. (1966). *Rev. Mod. Phys.*, **38**, 2.
Kleinpoppen, H., and Raible, V. (1965). *Phys. Lett.*, **18**, 24.
Koschmieder, H., Raible, V., and Kleinpoppen, H. (1973). *Phys. Rev.*, A8, 1365.
Lineberger, W. C., Hooper, J. W., and McDaniel, E. W. (1966). *Phys. Rev.*, **141**, 151.
Lloyd, C. R., Teubner, P. J. O., Weigold, E., and Lewis, B. R. (1974). *Phys. Rev.*, A10, 175.
Long, R. B., Cox, D. M., and Smith, S. J. (1969). *J. Res. Nat. Bur. Standards*, **72A**, 521.
MacGregor, M. H. (1959). *Phys. Rev.*, **113**, 1559.
Marriott, R., and Rotenburg, M. (1968). *Phys. Rev. Lett.*, **21**, 722.
Massey, H. S. W., and Burhop, E. H. S. (1969). *Electronic and Ionic Impact Phenomena*, p. 456. London: Oxford University Press.
Matese, J. J., and Oberoi, R. S. (1971). *Phys. Rev.*, A4, 569.
McConkey, J. W., and Preston, A. J. (1973). Eighth ICPEAC, 1, 273 (A. J. Preston, Ph.D. Thesis, Queen's University, Belfast, 1972).
McDowell, M. R. C. (1970). *Proc. Second Int. Conf. Atomic Physics*, p. 289, London: Plenum Press.
McGowan, J. W. (1967). *Phys. Rev.*, **156**, 165.

McGowan, J. W., Clarke, E. M., and Curley, E. C. (1965). *Phys. Rev. Lett.*, **15**, 917; **17**, 66(E).
McGowan, J. W., Williams, J. F., and Curley, E. K. (1969). *Phys. Rev.*, **180**, 132.
Moiseiwitsch, B. L., and Smith, S. J. (1968). *Rev. Mod. Phys.*, **40**, 238.
Mott, N. F., and Massey, H. S. W. (1965). *The Theory of Atomic Collisions*. Oxford: Clarendon Press.
Oed, A. (1971). *Phys. Lett.*, **34A**, 435.
O'Malley, T. F., and Geltman, S. (1965). *Phys. Rec.*, **137**, A1344.
O'Malley, T. F., Spruch, L., and Rosenberg, L. (1961). *J. Math. Phys.*, **2**, 491.
Ott, W. R., Kauppila, W., and Fite, W. L. (1970). *Phys. Rec.*, **A1**, 1089.
Percival, I. C., and Seaton, M. J. (1957). *Proc. Camb. Phil. Soc.*, **53**, 654.
Preston, M. A. (1963). *Physics of the Nucleus*, p. 100. New York: Addison Wesley.
Ramsauer, C., and Kollath, R. (1932). *Ann. Phys.*, **12**, 529.
Risley, J. S., Edwards, A. K., and Geballe, R. (1974). *Phys. Rev.*, **A9**, 1115.
Rotenberg, M. (1962). *Ann. Phys.* (N.Y.), **19**, 262.
Rudge, M. H. R. (1968). *Rev. Mod. Phys.*, **40**, 564.
Samson, J. A. R., and Cairns, R. B. (1965). *Rev. Sci. Inst.*, **36**, 19.
Sanche, L., and Burrow, P. D. (1972). *Phys. Rev. Lett.*, **29**, 1639.
Scott, B. L. (1965). *Phys. Rev.*, **140A**, 699.
Schultz, G. J. (1964). *Phys. Rev. Lett.*, **13**, 583.
Schultz, G. J. (1973). *Rev. Mod. Phys.*, **45**, 378.
Schwartz, C. (1961). *Phys. Rev.*, **124**, 1468.
Shelton, W. H., Leherissey, E. S., and Madison, D. H. (1971). *Phys. Rev.*, **A3**, 242.
Shore, B. W. (1967). *J. Opt. Soc. Am.*, **57**, 881.
Sinfailam, A. L., and Chen, J. C. Y. (1972). *Phys. Rev.*, **A5**, 1218.
Smith, S. J. (1969). *Physics of One and Two Electron Atoms*, p. 574. Amsterdam: North-Holland.
Tai, H., Bassel, R. H., Gerjuoy, E., and Franco, V. (1970). *Phys. Rev.*, **A1**, 1819.
Taylor, A. J., and Burke, P. G. (1967). *Proc. Phys. Soc.*, **92**, 336, 345.
Temkin, A. (1959). *Phys. Rev.*, **116**, 358.
Temkin, A. (1974). Private communication.
Temkin, A., and Lamkin, J. C. (1961). *Phys. Rev.*, **121**, 788.
Teubner, P. J. O., Lloyd, C. R. and Weigold, E. (1973). *J. Phys. B.*, **6**, L134.
Trajmar, S. (1973). *Phys. Rev.*, **A8**, 191.
Trajmar, S., Truhlar, D. G., and Rice, J. K. (1970). *J. Chem. Phys.*, **52**, 9, 4502.
Truhlar, D. G., Cartwright, D. G., and Kupperman, A. (1968). *Phys. Rev.*, **175**, 113.
Weingartshofer, A., Willman, K., and Clarke, E. M. (1974). *J. Phys. B.*, **7**, 79.
Westin, S. (1946). Det. Kgl. Norske Videnshabers Selskabs. Shrifter, **2**, 1.
Williams, J. F. (1974). *J. Phys. B.*, **7**, L56.
Williams, J. F. (1975). *J. Phys. B.*, **8**, 683 and in course of publication.
Williams, J. F., and McGowan, J. W. (1968). *Phys. Rev. Lett.*, **21**, 719.
Williams, J. F., and Willis, B. A. (1974). *J. Phys. B.*, **7**, L51.
Williams, J. F., and Willis, B. A. (1975). *J. Phys. B.*, **8**, 1641.
Williams, K. G. (1969). Sixth ICPEAC, Boston, p. 731. Cambridge, Mass.: M.I.T. Press.
Winters, K. H., Clark, C. D., Bransden, B. H. and Coleman, J. P. (1973). *J. Phys. B.*, **6**, L247.
Yang, C. N. (1948). *Phys. Rev.*, **74**, 764.

28
On the Anisotropy of the Quenching Radiation from Metastable Hydrogen and Deuterium Atoms

G. W. F. DRAKE, A. VAN WIJNGAARDEN, AND
P. S. FARAGO

> When the metastable 2s state of hydrogenic systems is quenched by a static electric field, the emission of the radiation is anisotropic. $R = [I(0) - I(\frac{1}{2}\pi)]/[I(0) + I(\frac{1}{2}\pi)]$, the asymmetry of the radiation emitted parallel and at right angles to the quenching field, was measured for hydrogen and deuterium to a fractional error of 1 part in 1000. The results are in good agreement with theoretical predictions. From the asymmetry R the value of the Lamb shift can be deduced to the same degree of accuracy. The implications of this for the determination of Lamb shift in heavy hydrogenic ions are briefly discussed.

Introduction

Hydrogenic atoms in the metastable 2s state can be induced to radiate by the application of a static electric field. Fite and co-workers (1968) and Casalese and Gerjuoy (1969) first pointed out that this "quenching radiation" is polarized, but it was not appreciated until recently that, in addition, the total radiation summed over the two orthogonal polarizations is anisotropic.

Drake and Grimley (1973) described the quenching of the metastable 2s state of hydrogenic systems by a static electric field as the zero-frequency limit of a two-photon scattering process with a resonance near zero frequency.

G. W. F. DRAKE and A. VAN WIJNGAARDEN • Department of Physics, University of Windsor, Windsor, Ontario, Canada.   P. S. FARAGO • Department of Physics, University of Edinburgh, Edinburgh, Scotland.   Dr. Drake is an Alfred P. Sloan Research Fellow.

The static electric field was considered as a beam of very-low-frequency radiation polarized in the external field direction. It was shown that the transition rate for the emission of quenching radiation of linear polarization **e** in the presence of a static electric field of direction **E** is

$$|Q|^2 \propto |\mathbf{E} \cdot \mathbf{e}|^2 |A|^2 + |\mathbf{E} \times \mathbf{e}|^2 |A'|^2 \tag{1}$$

where A and A' are quantities determined by the fine and hyperfine structure of the atomic system under consideration and are functions of the strength of the quenching field. It should be stressed that $A \neq A'$ only if the Lamb shift is different from zero. Concerning observable phenomena, Eq. (1) has two interesting implications.

(1) If the quenching radiation is observed in a plane perpendicular to the electric field, the intensity of the radiation polarized linearly at an angle ϕ relative to the direction of the electric field will be

$$I(\phi) \propto |A|^2 \cos^2 \phi + |A'|^2 \sin^2 \phi$$

Hence, a comparison of the intensities polarized parallel and at right angles to the direction of the quenching field gives a "polarization"

$$P \equiv \frac{I_\| - I_\perp}{I_\| + I_\perp} = \frac{|A|^2 - |A'|^2}{|A|^2 + |A'|^2} \tag{2}$$

where the subscripts $\|$ and $\perp$ correspond to $\phi = 0$ and $\phi = \tfrac{1}{2}\pi$ respectively.

(2) The total quenching radiation emitted at an angle θ relative to the direction of the quenching field summed over the two orthogonal states of polarization will have an intensity

$$I(\theta) \propto (|A|^2 + |A'|^2) \sin^2 \theta + 2|A'|^2 \cos^2 \theta \tag{3}$$

In other words, the radiation is anisotropic and a comparison of the intensities emitted parallel and at right angles to the quenching field gives an "asymmetry"

$$R \equiv \frac{I(0) - I(\tfrac{1}{2}\pi)}{I(0) + I(\tfrac{1}{2}\pi)} = \frac{|A'|^2 - |A|^2}{3|A'|^2 + |A|^2} \tag{4}$$

Both effects are, in principle, equally suitable for the determination of the ratio $|A'/A|$ and yield identical information about the fine and hyperfine structure of the $n = 2$ levels of hydrogenic systems, since $P = 2R/(R - 1)$. Yet anisotropy measurements are preferable, since this quantity can be determined more accurately than the polarization at the vacuum ultraviolet wavelength of the quenching radiation.

The polarization P was measured by Ott et al. (1970) for hydrogen. In the present paper measurements of the asymmetry R are reported for hydrogen and deuterium. The results in both cases agree with theoretical predictions within the margin of experimental error amounting to 1 part in 1000.

Experimental

In order to improve the accuracy of the results reported previously (van Wijngaarden et al., 1974), the experimental apparatus was modified as illustrated in Figure 1. As before, a monoenergetic (typically 10 keV) beam of protons or deuterons traverses a cesium vapor cell; the emerging beam contains metastable hydrogen or deuterium atoms in the 2s state produced in a near-resonant charge exchange reaction. The remaining ions are deflected out of the beam by a weak (~ 10 V/cm) electric field between the prequenching plates. A collimated beam of neutrals then enters the observation region, where the metastable atoms are quenched in an essentially uniform electrostatic field. In the present arrangement the field is maintained by four cylindrical rods, as in a quadrupole lens, arranged symmetrically with respect to the beam axis. In this case, however, adjacent rods form a pair kept at the same potential. To enhance the symmetry of the system, it is surrounded by a cylindrical mantle (not shown in the sketch) centered on the symmetry axis and held at the same potential as the end plates. The chosen geometry of the system permits the calculation of the field distribution to any prescribed accuracy, including the effects of the finite diameter and length of the rods.

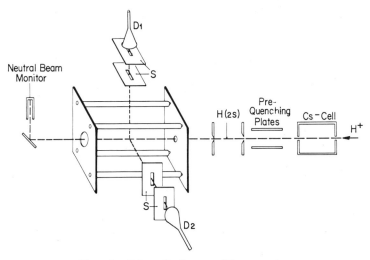

Figure 1. Schematic diagram of the apparatus.

A short section of the beam in the central region of the quenching field is viewed by two channeltrons detecting Ly α radiation emitted in mutually perpendicular directions; for a given choice of electrode potentials, one is parallel and the other is perpendicular to the direction of the quenching field. Equation (3) shows that the total intensity $I(\theta)$ is very insensitive to small errors in angular settings in the neighborhood of $\theta = 0$ and $\frac{1}{2}\pi$, making it easy to correct for the effect of a finite solid angle of observation.

There are two main factors which can introduce an instrumental asymmetry: (i) the two detectors have somewhat different acceptance angles and detection efficiencies, and (ii) the presence of stray magnetic fields, **B**, give rise to a motional electric field $\mathbf{E} = \mathbf{v} \times \mathbf{B}$.

Both these effects can be eliminated to a high degree of accuracy by rotating the quenching field in steps of $\frac{1}{2}\pi$ relative to the directions defined by the line-of-sight of the fixed detectors. This is achieved by a cyclic change of the polarity of the quadrupole rods and by recording the counting rates measured simultaneously by the two detectors at each of the four consecutive field orientations: $N_1(\phi)$ and $N_2(\phi + \frac{1}{2}\pi)$ with $\phi = 0, \frac{1}{2}\pi, \pi, \frac{3}{2}\pi$.

The effect of the motional field could be detected by reversing the direction of the quenching field, yielding a small discrepancy

$$N_1(0)/N_2(\tfrac{1}{2}\pi) \neq N_1(\pi)/N_2(\tfrac{3}{2}\pi).$$

In order to eliminate this effect, the measured quenching radiation intensity is defined as the mean of two values obtained at field directions reversed. Thus

$$N'_1 = \tfrac{1}{2}(N_1(0) + N_1(\pi)) = a_1 N(0)$$
$$N'_2 = \tfrac{1}{2}(N_2(\tfrac{1}{2}\pi) + N_2(\tfrac{3}{2}\pi)) = a_2 N(\tfrac{1}{2}\pi)$$

and, similarly,

$$N''_1 = \tfrac{1}{2}(N(\tfrac{1}{2}\pi) + N_1(\tfrac{3}{2}\pi)) = a_1 N(\tfrac{1}{2}\pi)$$
$$N''_2 = \tfrac{1}{2}(N_2(\pi) + N_2(0)) = a_2 N(0)$$

where $N(0)$ and $N_2(\tfrac{1}{2}\pi)$ are the apparent rates of emission parallel and at right angles to the quenching field and a_1 and a_2 are constants determined by the angle of acceptance and efficiency of the two detectors respectively. From the above equations the asymmetry analogous to that defined by Eq. (4) is obtained in the form

$$R' = (r - 1)/(r + 1)$$

where

$$r^2 = N'_1 N''_2 / N'_2 N''_1$$

The effect of intensity fluctuations on the results which involve pairs of counting rates measured at different times is minimized by monitoring the neutral beam current with the aid of a current-to-frequency converter to define the counting period for each measurement.

The measured counting rates $N_j(\phi)$ contain a contribution from background noise. In order to take this into account, a high (1500 V/cm) prequenching field was applied and the *isotropic* component of the still observable radiation was determined. The noise thus defined amounted to 1–2% of the signal obtained in the absence of the prequenching field and gave a small correction to the directly measured asymmetry R'.

Results and Conclusions

The anisotropy R (corrected for finite solid angle and noise) was determined at a sequence of different quenching field strengths E, and a curve

$$R = \sum_{k=0}^{n} a_k E^{2k}$$

was fitted to the experimental results. In performing the least-squares fitting to single runs, the approximation to $n = 2$ was always found significant. While curves obtained for individual runs were slightly different in details, they always led to extrapolated values $R(E = 0)$ which agreed with one another within their margins of error.

Figure 2 shows the results obtained in a typical run with metastable hydrogen. The statistical error for each measured point is the same and is indicated by one error bar. The dot-dash line is the quadratic least-squares fit and the error bar at $E \to 0$ represents the computed uncertainty in the extrapolated value $R(E = 0)$. The two solid lines represent the results of calculations based on theoretical considerations which will be discussed in detail elsewhere (Drake *et al.*, 1975). It contains a summation of the infinite time-dependent perturbation series for finite field strengths and includes the progressive nuclear spin decoupling with increasing field strength and interaction time. Both curves were obtained on the same assumptions concerning the structural properties of the atoms, namely, the hyperfine-structure interaction was taken into account and the accepted value of the Lamb shift (1057.9 MHz) was used. The lower curve, however, neglects the finite transit time of the atoms between entry to the quenching field and the region of observation, while the upper curve is calculated for a transit time of 25 nsec, appropriate to the kinetic energy of the H atoms in this run. In the extreme case of disregarding hyperfine-structure effects, the calculations give the dashed curve.

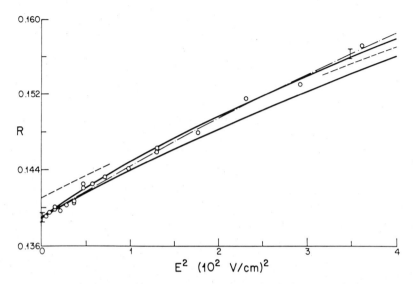

Figure 2. Asymmetry as a function of field strength for hydrogen. *Dash–dot curve:* quadratic least-squares fit to experimental points; *lower solid curve:* theory neglecting finite transit time; *upper solid curve:* theory at 25 nsec transit time; *dashed curve:* theory without hyperfine-structure effects.

Figure 3 demonstrates directly the effect of finite transit time. The two experimental points shown were obtained at the same quenching field (151.6 V/cm) with hydrogen beams of different kinetic energies. The curve represents the theoretical prediction.

The extrapolated zero-field asymmetry $R(E = 0) = 0.13901 \pm 0.00012$ is in good agreement with the theoretical prediction 0.13908 including

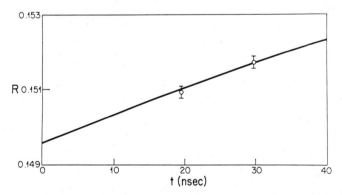

Figure 3. Asymmetry as a function of transit time for hydrogen at constant quenching field strength. *Solid curve:* theory.

hyperfine-structure effects, in contrast to the value 0.14115 obtained by neglecting hyperfine structure. The zero-field polarization of the quenching radiation derived from the asymmetry is $P = 2R/(R - 1) = -0.3229 \pm 0.0003$, and it seems likely that the slightly smaller value -0.30 ± 0.02 measured directly by Ott et al. (1970) reflects some instrumental effect rather than a genuine disagreement with theory.

Figure 4 shows results obtained with metastable deuterium. Circles represent measured values; all have the same error, shown by a single error bar. The dash-dot curve represents the quadratic least-squares fit and the solid curve displays the results of theoretical calculations taking hyperfine coupling and finite transit time into account. The zero-field asymmetry is $R(E = 0) = 0.14121 \pm 0.00014$, to be compared with the theoretical value of 0.14130.

As already pointed out, the anisotropy of the quenching radiation hinges on the existence of a finite Lamb shift; moreover, the asymmetry R and the Lamb shift are approximately proportional to one another. Thus the measurement of the asymmetry can be used for the determination of the Lamb shift to approximately the same fractional accuracy as that of the asymmetry measurement.

Table 1 gives Lamb shift values derived from the results of the present experiments. For hydrogen, three values are listed as derived from the extrapolated zero-field value $R(E = 0)$ and from the two values obtained

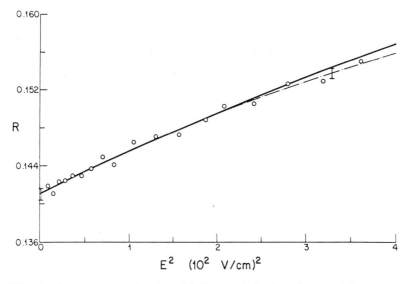

Figure 4. Asymmetry as a function of field strength for deuterium. *Dash-dot curve:* quadratic least-squares fit to experimental points; *solid curve:* theory.

Table 1

Atom	Measurement	Lamb shift (MHz)
H	$R(E = 0)$	1057.4 ± 1.0
	$R(E, t_1)$	1057.1 ± 1.6
	$R(E, t_2)$	1057.4 ± 1.6
H	(Average)	1057.3 ± 0.9
	Microwave res.†	1057.86 ± 0.061
	Microwave res.†	1057.90 ± 0.06
	(Theory)†	1057.911 ± 0.012
D	$R(E = 0)$	1058.7 ± 1.1
	Microwave res.†	1059.28 ± 0.06
	(Theory)†	1059.271 ± 0.025

†Lautrop et al. (1972).

at the same quenching field strength but different transit times, $R(E, t)$. For comparison, high-precision results obtained by microwave resonance techniques and currently accepted theoretical values are also listed.

The accuracy of the present results falls short of those obtained by microwave techniques. There are, however, several reasons which make the asymmetry measurement worthy of serious consideration as an approach to the determination of Lamb shift in heavy ($Z > 3$) hydrogenic ions. For this purpose, at least at present, the application of microwave resonance techniques is not feasible and the comparison with the much less accurate method based on the measurement of the quenching rate of the metastable ions in an electrostatic field appears very favorable. Since the asymmetry is independent of the field strength in the limit of weak quenching fields, a short extrapolation to zero field eliminates all systematic errors which are proportional to powers of the field strength. Thus, no absolute field strength calibration or accurate spatial resolution is necessary. Difficulties arising from the ion beam deflection in the quenching field and background noise are minimized because measurements need be done only at a single position near the region which yields optimum photon counting statistics. Experiments along these lines are currently in preparation.

Acknowledgments

Most of the work reported above was carried out while one of the authors (P.S.F.) was the holder of a Canadian Commonwealth Research Fellowship, and he wishes to express his gratitude for the kind hospitality he enjoyed in the Physics Department of the University of Windsor. The research was supported by the National Research Council of Canada.

References

Casalese, J. S., and Gerjuoy, E. (1969). *Phys. Rev.*, **180**, 327.
Drake, G. W. F., and Grimley, R. B. (1973). *Phys. Rev.*, **A8**, 157.
Drake, G. W. F., Farago, P. S., and van Wijngaarden, A. (1975). *Phys. Rev.*, **A11**, 1621.
Fite, W. L., Kauppila, W. E., and Ott, W. R. (1968). *Phys. Rev. Lett.*, **20**, 409.
Lautrop, B. E., Peterman, A. L., and Rafael, E. de (1972). *Phys. Rev. (Phys. Lett. C)*, **3**, 196.
Ott, W. R., Kauppila, W. E., and Fite, W. L. (1970). *Phys. Rev.* **A1**, 1089.
van Wijngaarden, A., Drake, G. W. F., and Farago, P. S. (1974). *Phys. Rev. Lett.*, **33**, 4.

29
New Measurements of Differential and Integral Cross Sections for Electron Impact Excitation of the n = 2 States of Helium

G. JOYEZ, A. HUETZ, F. PICHOU, AND J. MAZEAU

> *Some basic techniques for obtaining absolute integral and differential inelastic cross sections by electron impact spectroscopy are reviewed. A procedure used in this group is described and was first applied to the $n = 2$ levels of He. In the intermediate impact energy range (30 eV to 50 eV) the measurements made by different groups agree well with each other. New results close to the threshold (0.2 eV to several eV above threshold) are also reported. Within 0.2 eV of the threshold, the many inherent experimental problems which exist in this region are in the process of being solved.*

1. Introduction

The study of electron collision processes through monoenergetic electron beams and high-resolution analysis of the scattered electrons has become an important technique in atomic and molecular physics and is currently termed "electron impact spectroscopy." This technique has seen its resolution much improved in recent years (Joyez et al., 1973), but it still is not possible to separate and study cross sections corresponding to the fine details observed through optical techniques. However, the latter techniques have their application restricted to the excitation of levels which lead

---

G. JOYEZ, A. HUETZ, F. PICHOU, AND J. MAZEAU • Laboratoire de Physique et Optique Corpusculaires, Université de Paris VI, 4 Place Jussieu, 75005 Paris, France.

to optically allowed transitions in a technically accessible range of wavelengths and are also troubled by cascading and radiation trapping. Electron impact spectroscopy is free from many of these problems and allows the comparison, in many cases, of optically allowed and forbidden transitions with equal efficiency. Consequently, the measurement of excitation cross sections for transitions to a great number of levels is only possible by this technique.

Absolute differential (DCS) and integral (ICS) cross sections for excitation processes can be obtained in various ways, and the most important recently used methods are summarized here. An original procedure, used in this group, which is applicable to a wide range of scattering energies and to many targets, is also presented.

The excitation of helium from the ground state to the four $n = 2$ levels is particularly interesting. These transitions are of four different types, namely,

$1^1S \to 2^1P$ optically allowed (21.21 eV)

$1^1S \to 2^3P$ spin forbidden (20.96 eV)

$1^1S \to 2^1S$ symmetry forbidden (20.61 eV)

$1^1S \to 2^3S$ spin and symmetry forbidden (19.81 eV)

and thus represent a particularly severe testing ground for theoretical techniques. A theoretical model may give adequate results for one of these in a given energy range and be inadequate for the others at all energies.

Above an incident energy of 100 eV, the excitation essentially concerns optically allowed transitions and the cross sections are quite well described by the Born and derived approximations. Here we are concerned with energies below 100 eV, which will be divided into three ranges.

First, above 30 eV incident energy, the various experiments yield results which agree closely. Some of these are reported in this paper and are compared with theoretical results.

Second, from 0.2 eV to several volts above threshold, it was possible to apply our procedure and determine absolute cross sections. These are presented and compared with theory.

Finally, from threshold to 0.2 eV above, it was not possible to obtain quantitative results due to the great experimental difficulties which are discussed below. However, qualitative results have been obtained and are presented.

2. Experimental Procedures for Determining Absolute Cross Sections

In order to obtain an absolute DCS utilizing typical electron impact spectrometers, three spatial distributions have to be accurately known, namely, the angular intensity distribution of the incident electron beam, the spatial density of the target, and the detector efficiency. These three parameters have to be convoluted over the scattering volume. It is a very difficult task to control all these parameters, and very often a normalization procedure to accurate experimental or theoretical results is performed.

By using a gas cell, as against a gas beam, the density distribution is uniform and the density is measurable. But, with this geometry, the convolution over the remaining parameters depends strongly on the angle of observation and on the energy being considered. For high energies (above 50 eV), one can suppose that the incident beam intensity and the analysis efficiency are constant over narrow aperture angles defined by a set of slits placed in field-free space; then a trivial integration occurs over the scattering volume. This was done by Crooks et al. (1972), who obtained cross sections for the $n = 2$ levels of helium between 10° and 150° at 50, 100, 200, and 400 eV. Only these types of measurements are truly absolute, as they require no normalization whatsoever.

An overall normalization function may be determined at some energy and extended to other energies by considering that at different energies this function is not altered if all the potential differences encountered by the electron are scaled in a proportional manner. The experimental resolution changes, but this can be taken into account. This technique is only valid for energies which are high enough to neglect stray magnetic and electric field effects. By using this procedure, Opal and Beaty (1972) have obtained cross sections for the helium $n = 2$ manifold at 82 eV and 200 eV in the angular range from 30° to 150°.

Trajmar and co-workers (Trajmar, 1973; Truhlar et al., 1973) have made sure that their experimental device gives an accurate relative DCS for the $n = 2$ manifold between 3° and 140°. By extrapolating their results to 0 and 180°, they determine the integral cross sections (ICS), which are then normalized to the optically measured 2^1P cross section obtained by Donaldson et al. (1972). The absolute scale for the DCSs is then established. This method is limited by the accuracy of the optical measurements as well as the uncertainty due to extrapolation. Also, systematic errors in measuring relative DCSs are difficult to eliminate.

In our group in Paris, the absolute helium elastic DCSs of Andrick and Bitsch (Andrick, 1973) are used for normalization. These workers have measured relative DCSs for elastic scattering with great precision and placed them on an absolute scale by a phase shift fitting technique. This technique is only valid for obtaining elastic cross sections of small atoms below the first excitation threshold.

In our measurements (Hall et al., 1973) a narrow gas beam is used and relative elastic DCSs are measured and then compared to the results of Andrick and Bitsch. The relative angular correction is thus obtained and is usually less than 10% between 20° and 130°. This comparison also yields the overall efficiency of the detector at that analysis energy. This efficiency factor can then be used for any inelastic process which leaves the scattered electrons with this same residual analysis energy, and hence absolute inelastic DCSs can be obtained. This procedure has the advantage of accounting for all spurious effects occurring after the scattering, and thus can be applied for low residual energies (near threshold). The weakness of this technique is that the results depend on the accuracy of the elastic DCS, and, furthermore, the incident beam distribution must not vary in the scattering volume with incident energy. The latter can be verified experimentally.

3. Measurements of the Helium $n = 2$ Excitation Cross Sections

3.1. Intermediate Range 30 eV–100 eV

The accuracy with which the cross sections are obtained depends on the procedure used by the various authors (Crooks et al., 1972; Trajmar 1973; Truhlar et al., 1973; Hall et al., 1973), but, in general, is of the order of 20%.

It is very satisfying to observe the excellent agreement between the various DCS and ICS determinations, considering the completely different methods. The slight discrepancies can always be possibly explained by slight differences in incident energy or in angular resolution.

First-order calculations derived from the Born approximation including or not including polarization or spin exchange, have been intensively studied by Truhlar et al. (1970). But these methods fail to describe the DCS, particularly at large angles. Sometimes a good agreement is obtained for the ICS, but that must be considered fortuitous.

The distorted-wave method of Madison and Shelton (1973) for evaluating the 2^1P excitation is in good agreement for energies above 50 eV, but gets poorer as the incident energy decreases.

The "generalized random-phase analysis" (GRPA) calculations of all four transitions performed by Thomas et al. (1974) are in very good agreement with the observed DCSs, except for the $1^1S \to 2^3P$ transition, but the authors argue that the second-order approximation, which they have not used yet, will give a better angular behavior with the correct magnitude.

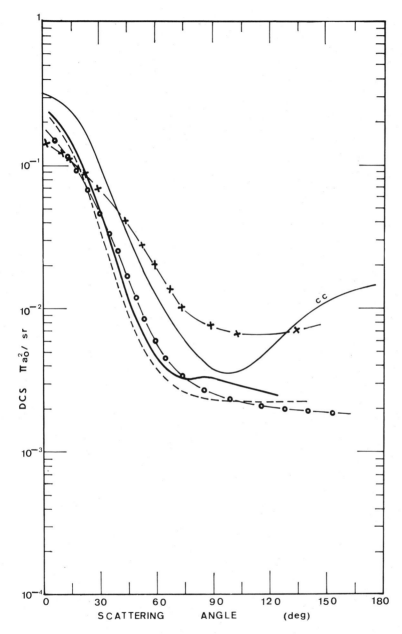

Figure 1. Differential cross section for excitation of the 2^1P level: *Experiment*: (———) Hall et al. (1973) at 29.2 eV; (— — —) Truhlar et al. (1973) at 29.6 eV. *Theory*: (×—×) Madison and Shelton (1973) at 29.2 eV; (—^{cc}—) close coupling at 29 eV Truhlar et al. (1973); (○ — ○) random-phase analysis, Thomas et al. (1974) at 29 eV.

Some striking features have to be underlined for each transition. The excitation of the 2^1P (Figure 1), which is the only optically allowed transition, occurs through many partial waves, as the interaction is through a long-range dipole. This leads, as is observed, to strong forward scattering.

The optically spin-forbidden excitation of the 2^3P (Figure 2) must occur under electron impact through electron exchange and hence through an interaction of very short range, and only a few partial waves are allowed to contribute to its cross sections. Consequently a quite flat DCS is observed. The good angular resolution of Trajmar (1973) has permitted finer details to be observed in the DCS, not actually accounted for in the first-order GRPA calculations. These may not be present in our observations because of poorer angular resolution and wide 10° steps.

The excitation of the 2^1S state (Figure 3) from the ground state is optically forbidden by symmetry arguments. The DCS always exhibits a minimum which is very deep near 50° and particularly pronounced for an incident energy of 40 eV. A fairly good agreement is noticeable with the GRPA method.

The 2^3S DCS (Figure 4) is observed to vary very rapidly with respect to the incident energy. Crooks et al. (1972) have suggested that a very wide resonance may occur at about 50 eV incident energy. GRPA calculations indicate that this may well be a purely dynamic effect.

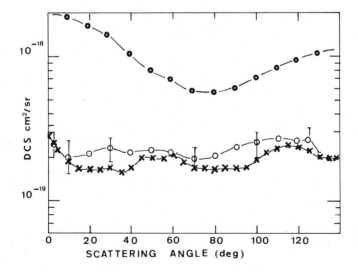

Figure 2. Differential cross section for excitation of the 2^3P level: *Experiment*: ($\tilde{\bigcirc}$ — $\tilde{\bigcirc}$) Hall et al. (1973) at 29.2 eV; (× — ×) Trajmar (1973) at 29.6 eV. *Theory*: (○ — ○) Thomas et al. (1974) at 29 eV.

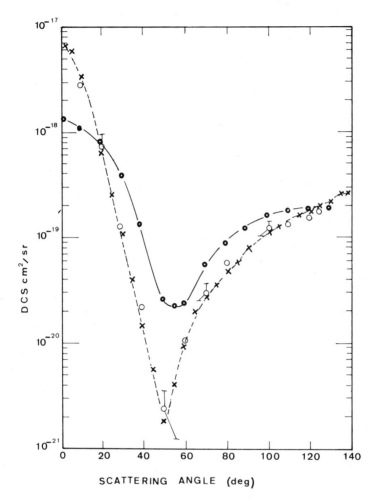

Figure 3. Differential cross section for excitation of the 2^1S level: Experiment: (ϕ — ϕ) Hall et al. (1973) at 39.2 eV; ($\times$ — $\times$) Trajmar (1973) at 40.1 eV. Theory: ($\bigcirc$ — $\bigcirc$) Thomas et al. (1974) at 40 eV.

3.2. 0.2 to 3.0 eV above Threshold

In this range, the close-coupling method is valid, as long as there are only a few channels to be taken into account, and was first fully applied by Burke et al. (1969). The matrix formalism used by Oberoi and Nesbet (1973) gives similar results, but over a larger range of energy.

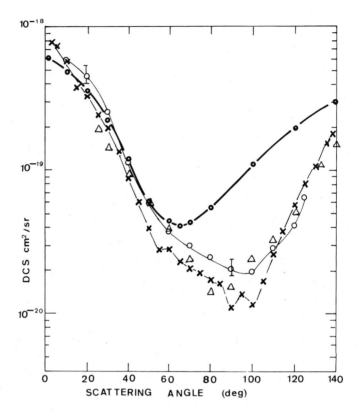

Figure 4. Differential cross section for excitation of the 2^3S level; same as Figure 3 except experiment; △ Crooks (1972) at 40 eV.

Experimentally, this range was previously studied by Ehrhardt et al. (1968), who performed only relative differential observations. Relative ICSs were obtained in our group (Hall et al., 1972) using the RPD and trapped electron method. Here absolute DCSs and ICSs are presented.

The experimental procedure is as follows: excitation functions for each level were studied at several scattering angles. Such an excitation function may be deformed by the transmission efficiency of the electron lenses preceding the analyzer. It is however possible to control this from about 0.2 eV up to several eV above threshold by comparing the shape of our excitation functions for 2^1P of helium and $C^3\Pi_u(v' = 0)$ of N_2 to that given by optical techniques (Jobe and St. John, 1967; Kisker, 1972). It then becomes apparent that a poor focusing efficiency of the electron lenses corresponds to quite constant transmission with respect to the residual energy. By altering these focusing conditions the observed excitation functions can be altered; they never differ one from another by more than 20% in the range considered.

The absolute scale of the excitation functions is then determined. Since large resonances occur in inelastic processes near threshold, cross section determinations at energies where the variation is too large must be avoided, because a slight uncertainty in the energy can strongly affect the results. In this case an energy of 1.6 eV above threshold was chosen, although a slightly lower value has to be used for the 2^3P level in order to avoid a resonance range. An overall analyzer transmission factor is determined at 1.6 eV from the elastic DCS measured by Andrick and Bitsch (Andrick, 1973). It is then possible to determine for each state the excitation cross section 1.6 eV above its respective threshold, provided that the incident beam distribution has not greatly changed in the scattering volume when going from the elastic to the inelastic measurement. This was verified by using various focusing conditions and also by skimming the beam down to a fine pencil. The results were never affected significantly compared to other uncertainties (about 30%).

The general shape of the observed excitation functions and DCSs compares well with the theoretical results of Burke et al. (1969) and Oberoi and Nesbet (1973). The absolute values obtained recently by Oberoi and Nesbet are expected to be quite good, but for the moment only ICSs and the 2^3S DCS are available. For the 2^3S DCS there is a large discrepancy in magnitude, the measured one being some three times smaller than the calculated one. This is rather surprising, as the ICSs agree well (within 30%) with each other and also with the experimental results of Brongersma et al. (1972).

More detailed and precise results will be published elsewhere in the near future.

3.3. Near-Threshold Cross Sections (0–200 meV above)

At threshold, or a few tens of meV above, the stray fields which are very difficult to eliminate, or even to estimate, play an important role. Surface potentials due to impurities, residual magnetic field, and penetrating fields issuing from the electron lenses may alter the detected signal. To this is added the background coming from those incident beam electrons which are not collected and which become thermalized on metal surfaces. However, this latter contribution may be estimated by recording the signal, without introducing the gas beam, but keeping the same background pressure in the vacuum chamber. That was done for all measurements and it revealed itself to be a very important correction at threshold.

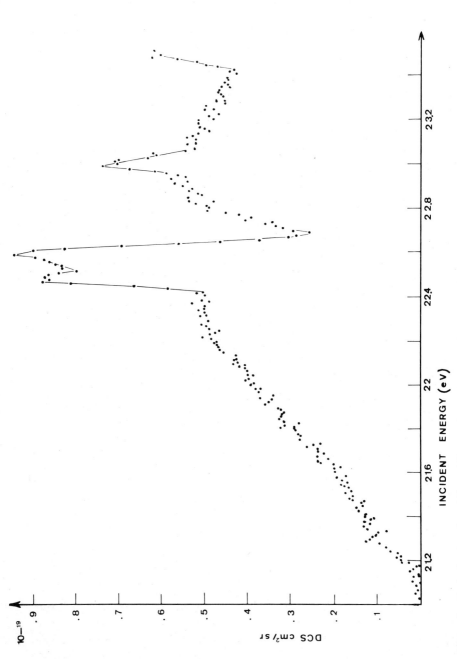

Figure 5. Excitation function of 2^1P at 90° scattering angle. Absolute determination of cross section was obtained at 1.6 eV above threshold.

Nevertheless, it is still impossible to define either the transmission efficiency of the analyzer or the angular resolution which may also change strongly near threshold, as was demonstrated by threshold experiments performed by Cvejanovič and Read (1974a).

However, when the electron lenses are adjusted for high energies, i.e., have a poor focusing efficiency near threshold, peaks do not appear at the thresholds of the P states and the ^3P exhibits only a pronounced rounded step. These were the conditions for the recordings shown in Figures 5–8. However, in perfectly clean operating conditions (i.e., those obtained by baking the metal surfaces surrounding the scattering volume), sharp peaks always appear at the threshold of both S states.

All four threshold onsets have enabled us to locate the elastic ^2S resonance at 19.355 ± 0.010 eV. The peak at threshold of the 2^3S state was not observed at this energy by Ehrhardt et al. (1968), and its interpretation as a perturbation due to the elastic resonance (tail of this resonance) is not really convincing because of its narrowness. If this were the case, it would

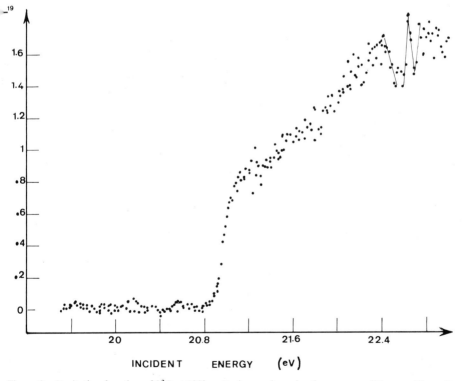

Figure 6. Excitation function of 2^3P at 120° scattering angle under the same conditions as Figure 5.

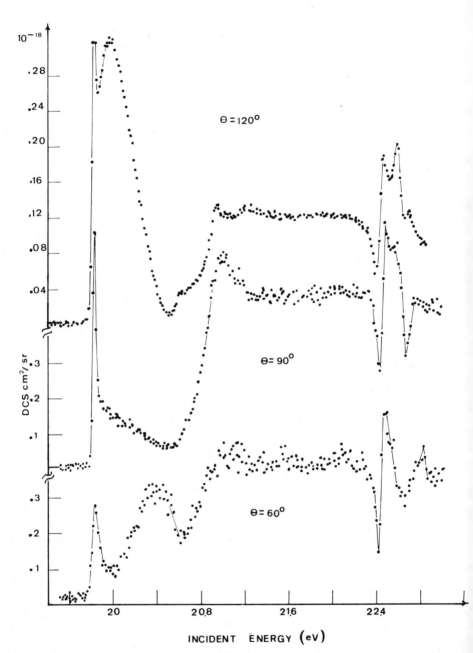

Figure 7. Excitation functions of the 2^3S at 60°, 90°, 120° scattering angles under same conditions as Figure 5.

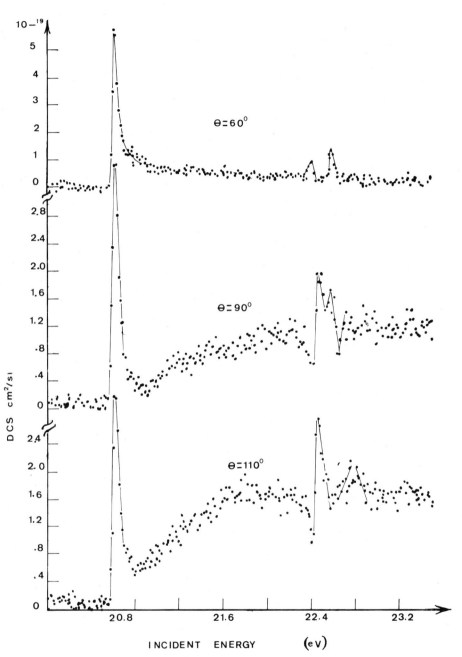

Figure 8. Excitation functions of the 2^1S at 60°, 90°, 110° scattering angles under same conditions as Figure 5.

be expected to be much broader than is observed here. The effect of destructive interference with the p wave advanced by Eliezer and Pan (1970) fails also because at 90°, where the p wave cancels, this narrow peak is still clearly seen, and it is difficult to believe that it is the d wave which provides this interference, as suggested by Herzenberg and Ton That (1975), because the d-wave resonance lies quite far from this energy (1.2 eV).

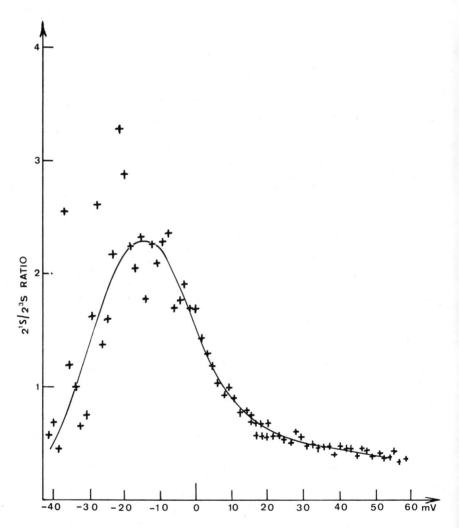

Figure 9. Evolution of $2^1S/2^3S$ ratio, with respect on residual energy, at 90° scattering angle. The zero here indicates the top of 2^1S peak. The uncertainty is large on the left of the diagram because of the weakness of the signals.

The sharp peak at the 2^1S threshold was also observed by Ehrhardt *et al.* (1968), and here the apparent width of this peak is mostly due to our experimental resolution.

The heights of these peaks may be strongly enhanced if the efficiency of collecting very slow electrons rapidly increases as their residual energy decreases. Then the convolution of this transmission efficiency with the excitation functions will depend on the slope of the onset at threshold of these cross sections, as these slopes are expected to be infinity for the s wave and zero for the others.

Some devices were tried in order to give reliable results at threshold, for instance, by shielding the scattering center in various ways. No obvious solutions were found, and it is not possible at the present time to establish either the angular behavior or the magnitude of these processes at threshold.

In the near future, it may be possible to use the observed behavior of ejected and scattered electrons at the ionization threshold to calibrate the transmission efficiency of the instrument down to threshold. In effect, it is predicted (Wannier, 1953) that, from the ionization threshold to at least some tenths of an electron volt above, the electron distribution is isotropic and that the excess energy is randomly distributed between the two outgoing electrons. This prediction has been verified recently by Cvejanovič and Read (1974b).

Nevertheless, it is possible to collect and compare electrons having the same energy after collision under the same conditions, whatever the state being excited, by recording constant residual energy spectra. Under these conditions, the change in peak height ratios for the four states was obtained for residual energies decreasing in 10 meV steps, at a given scattering angle.

The 2^3S to 2^1S ratio varies strongly, as shown in Figure 9. It is hardly conceivable that this is due to a variation of angular collection efficiency and differing DCSs, but, rather, suggests different slopes of the excitation function onsets. Consequently, it is very difficult to speak of a single ratio for exciting these states at threshold, or to give the angular behavior of this ratio, because they depend strongly on the residual energy.

4. Conclusion

There is general agreement among different techniques used to obtain absolute DCS or ICS in the intermediate-energy range, but the theoretical calculations still need to be improved. Nearer threshold, it seems that theory is more successful, and our recent results tend to confirm the validity of these calculations.

At threshold, there are some questions which must now be asked of both experimentalists and theoreticians concerning the verification of threshold laws and their applicability to states of differing symmetries, their range of validity, angular behavior, and how these factors can help us to understand the peaks which very often appear at excitation thresholds.

Acknowledgments

We are deeply grateful to Dr R. I. Hall for his advice and help in preparing this paper and checking the manuscript. This work was supported by the Centre National de la Recherche Scientifique (ERA No. 156).

References

Andrick, D. (1973). *Adv. Atom. Molec. Phys.*, **9**, 207.
Brongersma, B. H., Knoop, F. W. E., and Backx, C. (1972). *Chem. Phys. Lett.*, **13**, 16.
Burke, P. G., Copper, J. W., and Ormonde, S. (1969). *Phys. Rev.*, **183**, 245.
Crooks, G. B. (1972). Ph.D. Thesis, University of Nebraska.
Crooks, G. B., Dubois, R. D., Golden, D. E., and Rudd, M. E. (1972). *Phys. Rev. Lett.*, **29**, 327.
Cvejanovič, S., and Read, F. H. (1974a). *J. Phys. B.*, **7**, 1180.
Cvejanovič, S., and Read, F. H. (1974b). *J. Phys. B.*, **7**, 1841.
Donaldson, F. G., Hender, M. A., and McConkey, J. W. (1972). *J. Phys. B.*, **5**, 1192.
Ehrhardt, H., Langhans, L., and Linder, F. (1968). *Z. Phys.*, **214**, 179.
Eliezer, I., and Pan, Y. K. (1970). *Theor. Chim. Acta*, **16**, 63.
Hall, R. I., Joyez, G., Mazeau, J., Reinhardt, J., and Schermann, C. (1973). *J. de Phys.*, **34**, 827.
Hall, R. I., Reinhardt, J., Joyez, G., and Mazeau, J. (1972). *J. Phys. B.*, **5**, 66.
Herzenberg, A., and Ton That, D. (1975). *J. Phys. B.*, **8**, 426.
Jobe, J. D., and St. John, R. M. (1967). *Phys. Rev.*, **164**, 117.
Joyez, G., Comer, J., and Read, F. H. (1973). *J. Phys. B.*, **6**, 2427.
Kisker, E. (1972). *Z. Phys.*, **257**, 51.
Madison, D. H., and Shelton, W. N. (1973). *Phys. Rev.*, **A7**, 499.
Oberoi, R. F., and Nesbet, R. K. (1973). *Phys. Rev.*, **A8**, 29.
Opal, C. B., and Beaty, E. C. (1972). *J. Phys. B.*, **5**, 627.
Thomas, L. D., Csanak, G., Taylor, H. S., and Yarlagadda, B. S. (1974). *J. Phys. B.*, **7**, 1719.
Trajmar, S. (1973). *Phys. Rev.*, **A8**, 191.
Truhlar, D. G., Rice, J. K., Kupperman, A., Trajmar, S., and Cartwright, D. C. (1970). *Phys. Rev.*, **A1**, 778.
Truhlar, D. G., Trajmar, S., Williams, W., Ormonde, S., and Torres, B. (1973). *Phys. Rev.*, **A8**, 2475.
Wannier, G. H. (1953). *Phys. Rev.*, **90**, 817.

30
Coherent Electron Impact Excitation of Different L States in the n = 3 Shell of Atomic Hydrogen

STEPHEN J. SMITH

> Mahan and his colleagues have recently shown that the Balmer α flux emitted in a direction perpendicular to an electron beam incident on atomic hydrogen is dependent on the value of an applied axial electric field and that this dependence is asymmetric in the direction of the applied field. This result is interpreted in terms of excitation of coherent superpositions of different orbital angular momentum states of the n = 3 shell. The observed asymmetry is specifically related to superpositions of states of opposite parity which represent electronic charge distributions displaced with respect to the nuclear charge by the collision process. This work is related to results of beam foil spectroscopy in which observed beats are ascribed to coherent excitation at the foil. The interpretation of such observations of the effects of coherent excitation processes in terms of the dynamics of the collision processes greatly enhances our physical insight into such processes. The work of Mahan and his colleagues provides the first case of the extension of these ideas to binary collisions.

Observations of effects in the emission of radiation which can be attributed to coherent superpositions of states of the radiating atom are particularly intriguing because of the insight they may provide into the dynamics of excitation and radiation processes. The superposition of two nearly degenerate states, for example, of two states differing by a fine or hyperfine interaction energy, or of two different orbital angular momentum states in the same shell of atomic hydrogen, will in general represent a nonspherical charge

---

STEPHEN J. SMITH • Joint Institute for Laboratory Astrophysics, National Bureau of Standards and University of Colorado, Boulder, Colorado 80302, U.S.A.

distribution which oscillates with a frequency $\Delta E/h$. Thus the spontaneous radiation is anisotropic, and if the period of oscillation is comparable with or shorter than the radiative life of the atom, the observed emission (fluorescence) is modulated. The Hanle effect (Hanle, 1924; Hanle and Pepperl, 1971) is a familiar case in point. The precession of an aligned radiating atom is represented by a superposition of magnetic substrates separated by the Zeeman energy.

In 1965 Bashkin and his colleagues (Bashkin et al., 1965; Bickel and Bashkin, 1967) published an account of the first observations, carried out at the University of Arizona, of modulation effects in the radiation from hydrogen-like ions excited by passage through a thin carbon foil. Ions moving at velocities $\sim 4 \times 10^8$ cm/sec passed through the foil with interaction times of about 10^{-14} sec, very short compared to any periods associated with observable modulations in the radiation process. In the Arizona work, carried out with different ambient electric and magnetic field conditions, radiation in several spectral lines showed oscillations in intensity as a function of distance downstream from the foil, hence as a function of time t of travel from the foil passage at $t = 0$, for an essentially monoenergetic beam.

Sellin and his colleagues from the University of Tennessee (Sellin et al., 1969a,b), working at Oak Ridge, carried out measurements a few years later which greatly clarified the role of external fields. Further clarification was provided in theoretical discussions by Macek (1969), in which he emphasized that oscillatory terms arise naturally in the *absence* of external fields. His calculation of light emission uses superpositions of states u_{JM_J} excited at $t = 0$, i.e.,

$$\Psi(t) = \sum_{JM_J} a_{JM_J} u_{JM_J} \exp\left[-\left(\frac{\gamma_J}{2} + i\frac{E_J}{\hbar}\right)t\right]$$

Nonrandom initial phases may be implicit in the excitation amplitudes a_{JM_J}. The intensity of radiation to a lower state $(J = J_0)$ is

$$\sum_{M_0} |(\Psi(t)|V_q|J_0 M_0)|^2$$

where V_q is a component of the dipole operator. This includes cross terms

$$a_{J'M_J} a^*_{JM_J} (J_0 M_0|V_q|J'M_J)^* (J_0 M_0|V_q|JM_J) \exp\left[i\left(\frac{E_J - E_{J'}}{\hbar}\right)t - \frac{(\gamma_J + \gamma_{J'})}{2}t\right]$$

This entire array of terms, averaged over the beam population, will be referred to here as the *intensity matrix*. The significant point is that each of the cross terms will average to zero unless $M_J = M_{J'}$ assuming axial symmetry in the excitation process. Nonzero cross terms may occur in the intensity matrix, but only for superpositions of states that share the same axial component of angular moment, M_J, and only if the initial phases of those excited states are nonrandom. Such cross terms represent radiation from "coherent" superpositions of states.

Throughout this paper we assume that Russell–Saunders coupling holds. Then a further condition on nonzero cross terms in the intensity matrix is the radiation selection rule $\Delta L = \pm 1$, if we limit discussion to dipole-allowed transitions. Therefore, in a zero-field configuration, the only nonzero cross terms will be those involving two states of the *same parity* (orbital angular moment $L - L' = 0, \pm 2$) decaying to the same lower state. Such a superposition of states of the same parity is *symmetric* about a plane perpendicular to the beam direction at the nucleus. A nonspherical atom with this kind of symmetry is aligned along the beam axis. This represents one aspect of distortion of the atom by the beam foil interaction, but we see that collision dynamics having to do with relative displacement of electron cloud and nuclear charge centers is not represented in the information obtainable in a field-free experiment.

In order to emphasize a further distinction between types of coherent excitation it is useful to note that in view of the 10^{-14} sec time scale of the foil interaction, an uncoupled representation (L, S, M_L, M_S) is appropriate for describing the excitation process. The amplitudes a_{JM_J} are obtained by first calculating corresponding amplitudes $a_{LSM_LM_S}$ and transforming with appropriate Clebsch–Gordan coefficients. A matrix of the $a^*_{LSM_LM_S}, a_{L'S'M'_LM_S}$ contains cross terms analogous to those of the intensity matrix, which also vanish with integration over the beam population unless $M_L = M_{L'}$ and $M_S = M_{S'}$. It can be seen at once that if $L = L'$ (and taking $S = S' = \frac{1}{2}$), the conditions $M_L = M_{L'}$ and $M_S = M_{S'}$ define a pure initial state and the result is trivial. Coherent excitation of a superposition of states does not occur. However, nonzero cross terms arise in the J, M_J representation representing the evolution, from initial $M_L M_S$ states, of spin–orbit coupled states for which $J \ne J'$ but for which $M_J = M_{J'}$. With averaging over the entire beam population, such cross terms are nonzero only if the excitation cross sections $\sigma_{M_L = \pm 1} \ne \sigma_{M_L = 0}$. The radiation must be anisotropic for the modulation to be observable. Thus, the zero-field beats observed by Andrä (1970) corresponded to fine-structure separation of states of the same orbital angular momentum and provided a demonstration of the alignment of excited states by the foil interaction. On the other hand, the first convincing demonstration of alignment involving coherent excitation at $t = 0$ of a superposition of distinct states in the LSM_LM_S representation was the observation of beats corresponding to energy intervals between pairs of J, M_J states, $S_{1/2,1/2}-D_{3/2,1/2}$ and $S_{1/2,1/2}-D_{5/2,1/2}$, by Burns and Hancock (1971).

It remained to demonstrate observational effects of coherent excitation of superpositions representing charge polarization resulting from the collision process. The work I will describe here was carried out by H. Mahan as a part of his doctoral thesis project (Mahan, 1974; also Mahan et al., 1973), in collaboration with A. C. Gallagher and myself. R. V. Krotkov also made important contributions to the interpretation of the results.

This work was carried out in an apparatus (Figure 1) set up for the study of Balmer α excitation by electron impact on a thermal beam of atomic hydrogen. The immediate objective of this work was the separate determination, using time resolution, of the S, P, and D excitation cross sections.

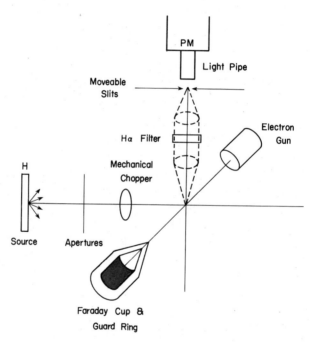

Figure 1. The experimental configuration represented schematically. A thermally dissociated hydrogen beam intersects an electron beam in the field of view of an $f1.2$ optical system with photomultiplier detection (RCA C31034 cooled to dry ice temperature). Atom densities are about 6×10^8 atoms/cm^3. Electron gun currents range from 10 to 50 μA in a beam 1–2 mm in diameter interacting with the atom beam over a path length of 4 mm. Signal count ranges from 40 to 400 counts/sec against a background ~ 100 counts/sec due to light from the hydrogen atom source, which operates at 2500°K to dissociate molecular hydrogen.

Coherent Electron Impact Excitation

The ratios of these cross sections proved to be quite sensitive to electric field mixing, and a careful study of mixing effects was necessary as a basis for obtaining cross sections valid at zero field. Electrodes were installed to permit application of static electric fields in the interaction region along the electron beam axis and in two directions perpendicular to the beam axis. Balmer α intensity was measured along an axis perpendicular to the electron beam by a photomultiplier looking through a Balmer α interference filter.

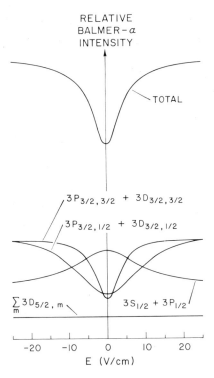

Figure 2. The effects on observed Balmer α emission due to mixing by an axial electric field are represented on the basis of two-state mixing equations and Born excitation cross sections, assuming noncoherent excitation. Hyperfine structure is neglected. In addition to the total, the combined contributions of pairs which are strongly mixed are shown separately. Radiation from the $3D_{5/2}$ state is shown constant in this approximation.

Figure 2 shows the result of a simple calculation of the Balmer α intensity as a function of electric field along the electron beam, based on two-state mixing equations and Born approximation excitation cross sections, and neglecting the possibility of coherent excitations. Because only 1/9 of the 3P population radiates to the $n = 2$ level as indicated in Figure 3, the remainder going to the ground state, Stark-induced mixing of P level populations into D levels will increase the Balmer α radiation. Since the cross section for P excitation is larger than that for S and D levels, and since P–D mixing is relatively efficient, as indicated below, this effect determines the character of the predicted response.

The result of a measurement of the Balmer α intensity, as a function of applied electric field, is shown in Figure 4. Here the electron beam energy is 500 eV. (This is the average collision energy and has been corrected for the gain or loss of energy in the applied axial field.) Atomic deuterium was used to minimize hyperfine structure effects.

The interesting characteristic of this result is the pronounced asymmetry in the Balmer α response with respect to the *direction* of the applied electric field. This requires of the intensity matrix, rewritten with appropriate regard for Stark mixing, that it include terms *linear* in the electric field. These will necessarily be off-diagonal terms (the diagonal terms must be quadratic) and must involve pairs of states excited in nonrandom relative phase.

A glance at the level structure within the $n = 3$ shell, shown in Figure 5, gives a clear indication of what levels will dominate in the mixing process, and why the two-state mixing formulation is reasonable. The mixing

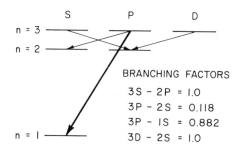

Figure 3. The effects of mixing between the S, P, and D levels of the $n = 3$ shell, as observed in the total Balmer α intensity, are due to net depopulation, by Stark mixing, of the 3P levels, which decay predominantly in Lyman β lines (Bethe and Salpeter, 1957).

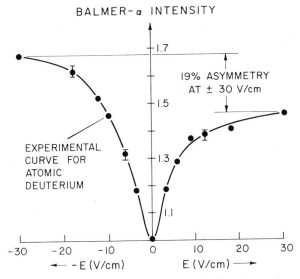

Figure 4. The intensity of Balmer α radiation, observed as indicated schematically in Figure 1, is plotted against applied electric field, normalized to the zero field intensity. The energy of the electron beam is compensated for the effect of the field applied in the interaction region. The results plotted are for deuterium. Similar results obtained using atomic hydrogen show a smaller asymmetry due to the more complex hyperfine structure.

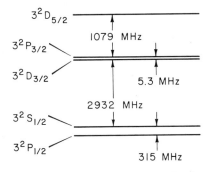

Figure 5. The energy intervals (Garcia and Mack, 1965) within the $n = 3$ shell are represented, with the $P_{3/2}$–$D_{3/2}$ spacing necessarily exaggerated. Only mixing of the $P_{3/2,3/2}$–$D_{3/2,3/2}$, $P_{3/2,1/2}$–$D_{3/2,1/2}$, and $S_{1/2,1/2}$–$P_{1/2,1/2}$ pairs were considered in calculations of mixing effects.

coefficients have an inverse dependence on the separation of any given pair of states, so the very nearly degenerate pairs, notably the $P_{3/2}$–$D_{3/2}$ pair, should dominate. Furthermore, only states with the same axial projection of orbital angular momentum (M_L) and with opposite parity will have non-zero mixing coefficients.

It follows that the asymmetry observed in the Balmer α emission excited by electron impact is the result of coherent excitation of states of *opposite parity*, predominantly the $P_{3/2,3/2}$–$D_{3/2,3/2}$, $P_{3/2,1/2}$–$D_{3/2,1/2}$, and $S_{1/2,1/2}$–$P_{1/2,1/2}$ pairs. As pointed out by Eck (1973), a coherent superposition of states of opposite parity implies a nonzero dipole moment. A calculation of the charge density

$$|\psi_{\text{even}} + \psi_{\text{odd}}|^2$$

yields cross terms that change sign across the plane transverse to the electron beam at the nucleus. The classical picture is of a collision-induced displacement of the electronic charge cloud along the axis of the electron beam. The cross terms oscillate with a frequency corresponding to the energy interval for a pair of states. In this classical context the role of the static external field is to superpose an additional displacement of the charge cloud, which increases or decreases the charge polarization of the upper state during the radiation period. This picture is particularly appropriate for nearly degenerate states such as the $P_{3/2}$–$D_{3/2}$ pair, for which the period corresponding to the energy difference (5 MHz) is long compared to the radiative lifetime.

In order to provide some verification for this interpretation of our observed asymmetry in Balmer α emission, Mahan (1974; Mahan et al., 1973) has carried out calculations of the emission as a function of applied electric field, following from the formulation of Percival and Seaton (1958), using Born excitation cross sections, by inserting terms to represent Stark mixing between excitation and spontaneous emission.

In this formulation the total intensity may be represented by

$$I_{\text{total}} = \int_0^\infty I(t)\,dt \sim \frac{1}{k_i^2} \int \sum_{\alpha\delta\gamma} \left[\sum_{\beta\beta'} \int_0^\infty F_{\beta\beta'}(\alpha, \hat{K}) A_{\beta\delta,\beta'\delta}(\mathbf{E}, t) G_{\delta\gamma}(\hat{e})\,dt \right] K\,dK\,d\phi$$

where $f_\beta(\alpha, \hat{K})$ is the scattering amplitude for excitation of an atom from an initial state α to an excited state β by electron impact with a momentum change vector in the direction $\hat{K}$; $g_{\beta\gamma}(\hat{e})$ is the radiation matrix element giving the transition probability in sec^{-1} for emission of an $\hat{e}$ photon by an atom in an excited state β radiating into the final state γ; $F_{\beta\beta'}(\alpha, \hat{K}) \sim f_\beta(\alpha, \hat{K}) f_{\beta'}^*(\alpha, \hat{K})$, where these JM_J scattering amplitudes have been calculated from the appropriate vector-coupled Born LM_L excitation amplitudes where relative phases of excitation are carefully preserved; $A_{\beta\delta,\beta'\delta}(\mathbf{E}, t) \sim$

$a_{\beta\delta}(\mathbf{E}, t)a_{\beta'\delta}^*(\mathbf{E}, t)$, where the appropriate $a_{\beta\gamma}(\mathbf{E}, t) \sim a_i(t)$ are the amplitudes of states coupled by an external electric field; and $G_\gamma(\hat{e}) \sim g_{\delta\gamma}(\hat{e})g_{\delta\gamma'}^*(\hat{e})$ are the radiation matrix elements connecting these Stark-mixed δ levels and the final $n = 2$ levels resulting in $H\alpha$ radiation.

We considered only the three pairs of states with smallest ΔE mentioned above. This simple calculation, in which the quantum-mechanical phases are carefully followed through the mixing process, predicts an asymmetry and provides some insight into the details of the process. It fails at quantitative prediction of the asymmetry observed, and this may be attributed to the inadequacy of the Born approximation, which does not take into account the substantial charge polarization effect that evidently occurs. Krotkov (1975) has carried out a similar calculation with equivalent results.

Our observations relate to the dynamics of excitation to the $n = 3$ shell of hydrogen by electron impact. Some additional beam foil measurements have been carried out following the appearance of Eck's paper. These measurements demonstrate the longitudinal asymmetry of the atom excited by interaction with a foil. Sellin et al. (1973) applied an electric field parallel and antiparallel to a beam of foil-excited hydrogen atoms and observed the effects of the oscillating charge distribution in the Lyman α signal difference as a function of displacement from the foil. Gaupp et al. (1974) carried out similar measurements, and from the analysis of their data determined the density matrix for foil-excited hydrogen for several energies.

Finally, in a very recent paper, Berry et al. (1974) have shown from analysis of 5016 Å radiation that the surface direction of a foil has a large effect in beam foil excitation of helium atoms. In the plane determined by the surface normal $\hat{n}$ and the beam direction $\hat{z}$, the $3p$-excited electron has a preferential direction of rotation so that the orbital angular momentum is preferentially oriented in the direction $\hat{n} \times \hat{z}$.

From experiments such as these a body of information directly interpretable in terms of coherent excitation and the dynamics of the collision process is being developed which will greatly enhance our physical insight into such processes. The present experiment of Mahan et al. (1973) demonstrates that coherent excitation processes and corresponding dynamical concepts are important for the case of binary collisional excitation by electrons, as well as for the collective effects that occur in excitation by a beam foil.

Acknowledgment

This work was supported by the National Science Foundation through Grant No. GP-39308X.

References

Andrä, H. J. (1970). *Phys. Rev. Lett.*, **25**, 325.
Bashkin, S., Bickel, W. S., Fink, D., and Wangsness, R. K. (1965). *Phys. Rev. Lett.* **15**, 284.
Berry, H. G., Curtis, L. J., Eliis, D. G., and Schechtman, R. M. (1974). *Phys. Rev. Lett.*, **32**, 751.
Bethe, H. A., and Salpeter, E. E. (1957). *Quantum Mechanics of One- and Two-Electron Atoms*, p. 266. New York: Academic Press.
Bickel, W. S., and Bashkin, S. (1967). *Phys. Rev.*, **162**, 12.
Burns, D. J., and Hancock, W. H. (1971). *Phys. Rev. Lett.*, **27**, 370.
Eck, T. G. (1973). *Phys. Rev. Lett.*, **31**, 270.
Garcia, J. D., and Mack, J. E. (1965). *J. Opt. Soc. Am.*, **55**, 654.
Gaupp, A., Andrä, H. J., and Macek, J. (1974). *Phys. Rev. Lett.*, **32**, 268.
Hanle, W. (1924). *Z. Physik*, **30**, 93.
Hanle, W., and Pepperl, R. (1971). *Phys. Bl.*, **27**, 19.
Krotkov, R. V. (1975). *Phys. Rev. A* (in press).
Macek, J. (1969). *Phys. Rev. Lett.*, **23**, 1.
Mahan, A. H. (1974). Ph.D. Thesis, University of Colorado.
Mahan, A. H., Krotkov, R. V., Gallagher, A. C., and Smith, S. J. (1973). *Bull. Am. Phys. Soc.* **18**, 1506 (abstract).
Percival, I., and Seaton, M. (1958). *Phil. Trans. Roy. Soc. (Lond.)*, Ser. A**251**, 113.
Sellin, I. A., Moak, C. D., Griffin, P. M., and Biggerstaff, J. A. (1969a). *Phys. Rev.*, **184**, 56.
Sellin, I. A., Moak, C. D., Griffin, P. M., and Biggerstaff, J. A. (1969b). *Phys. Rev.*, **188**, 217.
Sellin, I. A., Mowat, J. R., Peterson, R. S., Griffin, P. M., Laubert, R., and Haselton, H. H. (1973). *Phys. Rev. Lett.*, **31**, 1335.

31

Details of Collision Dynamics from the Electron Scattering by Laser-Excited Sodium Atoms

I. V. HERTEL

> By using the optical pumping process involved in the laser excitation of atoms, details of the scattering dynamics for the electron scattering by excited atoms, which were hitherto hidden in the averaging procedure, can now be measured when measuring differential inelastic scattering. Details of the experiment are described. Three examples for the new possibilities are discussed and compared with theory and/or experiment. (1) The dependence of superelastic and inelastic collision cross sections on the polarization of the incident light allows one, at least in principle, to distinguish any influence of nuclear spin or spin–orbit interaction on the collision dynamics without detailed calculations. (2) Ratios of scattering amplitudes and their phases can be measured and compared to theoretical calculations. (3) Regions of validity of Born or Glauber approximations can be tested by the asymmetry of the electron cross sections, the atoms being excited by left or right circular polarized light, respectively.

Introduction

Inelastic scattering by ground-state atoms has been investigated for many years. Detection of the inelastically scattered electron with determination of its energy is one standard technique, observation of the fluorescent light of the atom after excitation is another. The most recent development in this field is undoubtedly the coincident measurement of the scattered electron and the photon emitted subsequently from the excited atom as reported by

---

I. V. HERTEL • Fachbereich Physik der Universität, Trier–Kaiserslautern, Kaiserslautern, Germany.

Eminyan et al. (1973), and which is also subject of this volume. Observation of the so-called inverse process has only been made possible very recently (Hertel and Stoll, 1974b). Clearly, to prepare enough target atoms for a successful scattering experiment by an excited state having a lifetime of 10^{-8} sec requires a powerful tunable monochromatic light source. This, however, is available now as the argon laser pumped CW dye laser.

The coincidence experiment and the electron scattering from laser-excited atoms are compared in Figure 1. On the first view they just look like inverse processes. The correlation measured in the coincidence experiment corresponds in the laser experiment to the fact that a deexcitation occurs only when photons are present in the laser beam and reach the scattering region. The angle between outgoing electron and photon beam may be varied in a similar manner as in the inverse experiment. The principle of detailed balance states that the matrix elements describing the two processes are equal, and we have to find out how to deduce the physics from the experimental scattering intensities.

Since this volume is dedicated to Professor Fano, I am delighted to come now to a point where I may mention once again his stimulating influence on the work I am about to report.

In the much discussed recent review article by Fano and Macek (1973), the photon emission after excitation processes has been described in terms of a multipole development. One of the major advantages of this treatment is the disentangling of the dynamic and geometric effects in these processes.

Thus, it was an obvious idea to try to apply this theory also to our inverse process. There was a very fruitful exchange of ideas with Fano on how to do this. Finally, making use of the multipole expansion, Macek outlined the principles of a theory for the laser experiment, which will be published in a forthcoming paper by Macek and Hertel (1974). I am not going to report on this here, since Macek does so elsewhere in this volume. I may, however, point out that one of the major differences between the

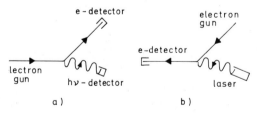

Figure 1. Comparison of electron–photon coincidence experiment (left) and the inverse deexcitation of a laser excited atom (right).

coincidence experiment and the electron scattering by laser-excited atoms seems to be the optical pumping process involving photons in the latter case. Thus, a preparation of the atom in well-defined magnetic substates of the excited state is possible and allows, in principle, for the determination of many more (higher) multipole moments than the inverse experiment, which contains information only up to the alignment tensor of rank 2.

Pumping Process

It may be helpful to say a few words about this pumping process, the details of which have been described by Hertel and Stoll (1974a). The sodium atom is our favorite example since it offers the best compromise for exposure to both an experimental *and* a theoretical treatment. Since we perform a crossed beam experiment where the atom beam displays no Doppler width, we have to excite the atoms by a single-mode laser having a line width of the order of the natural line width. This requires us to consider the hyperfine structure of sodium. The nuclear spin of sodium is 3/2; thus, in the most interesting $3^2P_{3/2}$ state, the total angular momentum may be $F = 0, 1, 2, 3$, while in the $3^2S_{1/2}$ ground state we have $F' = 1, 2$. The selection rule for optical excitation and deexcitation is $\Delta F = 0, \pm 1$. The only transition which leaves us any atoms in the excited state for the stationary case is the transition $3^2S_{1/2}\ F' = 2 \rightarrow 3^2P_{3/2}\ F = 3$. Only in that case, which is the subject of our further discussion, does no pumping to the other ground state occur.

Concerning the population of the magnetic substates, what can we achieve with the laser pumping?

(a) If we pump with right circular polarized (σ^+) light, obeying the selection rule $\Delta M_F = 1$, the positive magnetic quantum number sublevels are populated preferentially, leaving under stationary conditions only the $M_F = 3$ excited state and the $M_F = 2$ ground state populated. Thus, we have produced what one calls *an oriented atom*.

(b) Linear polarized light (π) gives the selection rule $\Delta M_F = 0$, and this obviously leads to no population of the $M_F = \pm 3$ states and, what may not be seen so easily, overpopulates at stationary conditions the states with low magnetic quantum numbers as shown in Figure 2.

Thus, we have produced what one calls an aligned atom. Moreover, multipole moments up to $2F = 6$ are nonzero in this case. The states with different magnetic quantum numbers are here only connected by spontaneous emission and are thus incoherent. The density matrix of the atom thus

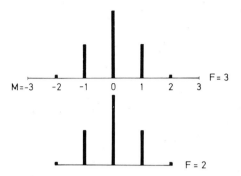

Figure 2. Upper and lower magnetic sublevel populations at stationary conditions.

populated is therefore diagonal in a reference frame parallel to the electric vector of the incident light, henceforth called the *photon frame*. Figure 3 shows the scattering geometry.

The incoming and the scattered electron form the collision plane (paper plane). In this special case the incident light direction lies in the collision plane. We may define a collision frame z_{col}, y_{col}, x_{col} in which to calculate scattering amplitudes. To compare with calculations for the inverse process, we choose z_{col} antiparallel to the outgoing electron. The orientation of the collision frame with respect to the photon frame can be varied in many ways: the *polarization vector* may be rotated through an angle ψ with respect to the collision plane, the incident light direction $\theta_{\hat{n}}$ may be changed, or even off-plane measurements may be performed, changing the light-incidence azimuthal angle $\phi_{\hat{n}}$.

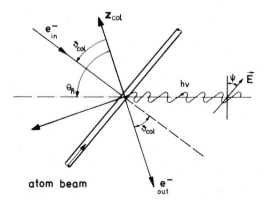

Figure 3. Scattering geometry.

This leaves enough parameters to obtain detailed information on the dynamics. The atom can be turned around with respect to the collision frame, and the laser acts as a projection operator, picking a certain part of information out of the total scattering wave function. In principle, all multipole moments which are induced by the electron excitation process may be obtained in this way.

Experimental Setup

Let us now turn to some experimental details. Figure 4 gives a schematic diagram of the apparatus. We use a standard crossed beam experiment, the atom beam entering perpendicular to the paper plane; 180° electrostatic analyzers are used in both electron gun and detector. Both are independently rotatable around the collision center and with respect to the laser beam. The electric vector of the light is rotatable by a commercial polarization rotator. The fluorescent light is monitored continuously and serves as a reference signal for the laser stabilization. Data accumulation and operation of the experiment is performed by an on-line minicomputer.

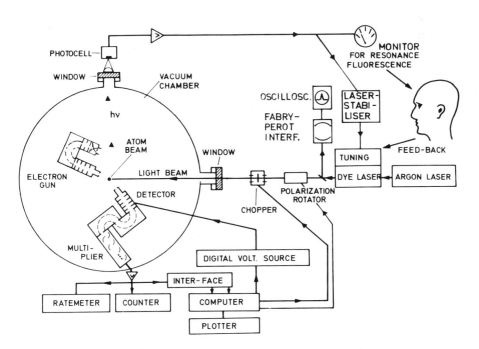

Figure 4. Schematic diagram of the experiment.

The fluorescence of the laser–atom beam intersection in the scattering center can also be observed visually. Radiation is trapped to a very small amount in the atom beam, enough to observe the good parallelism of better than 1:50 of the beam. In Figure 5, energy loss spectra for an incident kinetic energy of a few eV and zero scattering angle are shown. Only a fraction of the atoms is excited and ground state scattering always occurs.

For comparison, the top left of Figure 5 shows an energy loss spectrum with no light. The most prominent loss peak is, of course, the excitation of the resonance 3^2P state. After switching the light on, we observe a number of processes starting from the 3^2P state in the lower left and upper right parts of Figure 5. The 3^2D and 4^2S excitation are most prominent and have to be compared to the superlastic deexcitation process $3^2P \rightarrow 3^2S$ which we will discuss later in detail. The difference of scattering intensities between light on and off (lower right part of Figure 5) displays clearly the processes arising from exciting the atom. The "negative" peak at the position of the $3^2S \rightarrow 3^2P$ inelastic process corresponds to a depopulation of the ground state. Its height reflects approximately the magnitude of the cross section for the excitation of the resonance line from the ground state in the given scale. Even the $3^2P \rightarrow 4^2P$ and $\rightarrow 5^2S$ processes are clearly resolved in this spectrum.

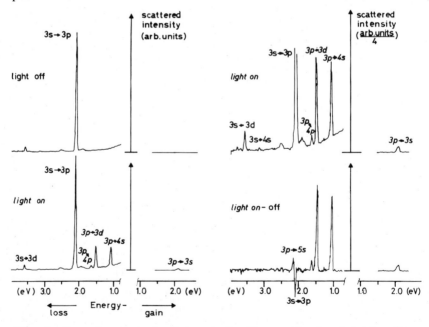

Figure 5. Energy loss spectra for electrons scattered by ground and excited state sodium atoms.

Examples of Possibilities

What can be learned from these experiments, apart from the previously unknown cross sections displayed in Figure 5? I will give three examples from among the many possibilities that have been opened up. Two of these are already comparable with experiments and one will be subject to future study.

(a) The so-called Percival–Seaton hypothesis is, in general, used in calculating electron-scattering phenomena, i.e., total uncoupling of nuclear momentum and electron spin from the orbital angular momentum of the atomic electron during collision is assumed.

Let us have another look at our preferred scattering geometry in Figure 3. The most easily variable parameter is the angle of the polarization vector ψ with respect to the scattering plane.

The theory of Macek and Hertel (1974) predicts that the scattering intensity is given by

$$I(\psi) = c_1 + c_2 \cos 2\psi \qquad (1)$$

if the nuclear spin decouples completely. If not, Eq. (1) has to be modified by adding terms up to $\cos 6\psi$.

Let us now see the experimental results in Figure 6. They are obtained for fixed scattering angle, fixed energy, and fixed angle of the incident light direction. We just measure the scattered intensity of the superelastic peak, varying the direction of the incident electric light vector.

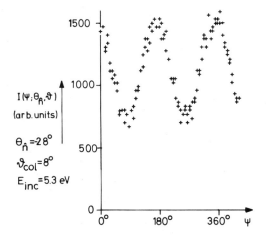

Figure 6. Periodicity of superelastic scattering intensities with the angle of the polarization vector.

The periodicity obviously corresponds to Eq. (1). We have here a case in which a marked difference in the differential scattering cross section exists depending on whether the electric vector is parallel or perpendicular to the scattering plane. Each 90° segment of the curve may be used as an independent measurement of Eq. (1). Let us plot the results as a function of $\cos 2\psi$, which should lead to a straight line if Eq. (1) is a correct description. We normalize to the intensity for $\psi = 90°$. The results are given in Figure 7.

This figure shows two extreme measurements and a best fit curve to a set of eight independent measurements. Within the experimental limits of error, no indication of higher $\cos n\psi$ terms can be discovered. We have made nearly 100 different sets of measurements with the same result: the nuclear spin is decoupled, at least for the energies of 3 eV and more which we have investigated.

In a similar but more complicated fashion, the uncoupling of the electron spin could also be tested. No surprises are to be expected for electron scattering, but in heavy particle collisions the answers may be quite different.

(b) Let us now turn to the next most easily variable parameter, the angle $\theta_{\hat{n}}$ between the incident light direction and the outgoing electron. We may plot as a function of $\theta_{\hat{n}}$ the scattering intensity for $\psi = 0$ normalized to $\psi = 90°$ since a glance at Figure 3 tells us that $I(\psi = 90°)$ is independent of $\theta_{\hat{n}}$ for symmetry reasons.

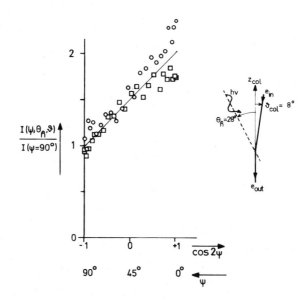

Figure 7. Scattering intensity as a function of $\cos 2\psi$. The circles and squares represent different sets of measurements.

This should lead to information about the scattering multipole moments. They are connected to the more commonly known amplitudes by the following formulas:

$$\langle \tau_0^{[0]} \rangle = (C/\sqrt{3}) \sum_M q_{MM}$$

$$\langle \tau_{1-}^{[1]} \rangle = -C \cdot 2 \, \text{Im}\,(q_{01})$$

$$\langle \tau_{0+}^{[2]} \rangle = -C \cdot \sqrt{2/3} \cdot (q_{00} - q_{11})$$

$$\langle \tau_{1+}^{[2]} \rangle = -C \cdot 2 \, \text{Re}\,(q_{01})$$

$$\langle \tau_{2+}^{[2]} \rangle = -C \cdot \sqrt{2} \cdot q_{11}$$

where

$$q_{MM'} = f_M f_{M'}^* - \frac{f_M g_{M'}^* + f_{M'}^* g_M}{2} + g_M g_{M'}^*$$

$$C = 1 \bigg/ \sum_M q_{MM}$$

$$\langle \tau_{qp}^{[k]} \rangle = v^k(L) \langle T_{qp}^{[k]}(L, \text{col}) \rangle$$

There is one trivial case that involves only geometry—forward scattering, where only the amplitudes for zero angular momentum transfer are non-vanishing. Figure 8 shows the theoretical dependence of the scattering intensity on $\theta_{\hat{n}}$ for $\vartheta_{\text{col}} = 0°$ (dashed line). It is of course symmetric about $\theta_{\hat{n}} = 0°$ and $\pm 90°$.

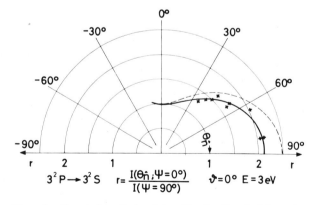

Figure 8. Forward scattering intensities for the electric vector in plane and varying angle of light incidence $\theta_{\hat{n}}$. Theoretical curves for stationary pumping - - - -, for incomplete pumping ———, and experimental points +.

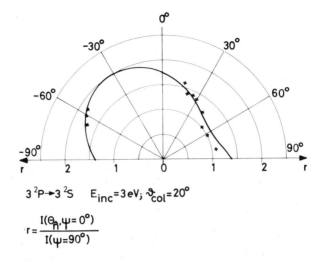

Figure 9. Scattering intensities for 20° scattering angle. Otherwise as Figure 8.

The experimental points do not fit the dashed line; however, if we allow for not quite complete pumping, the experimental points are met by the theory very well. Using the pumping populations deduced in this way, we may now compare calculated amplitudes for nonzero scattering angles. Moores and Norcross (1972) have calculated these and I am very much indebted to them for letting me have their actual complex values. Figure 9 shows a very nice agreement between experiment and theory, although the experimental material is not yet conclusive. A more detailed analysis shows that this agreement implies correct phases and amplitude ratios for f_0, g_0 and f_1, g_1 combinations.

(c) Let me finally, without experimental results, offer a check on the range of validity of the Born or Glauber approximations without any detailed calculations.

Since both theories have an axis of symmetry, no orientation can be produced by them. Thus, for right and left circular polarized light excitation, the same cross sections are predicted by these theories, while, for instance, the close-coupling amplitudes predict a marked difference for off-plane incident σ^+ and σ^- light as shown in Figure 10.

Small scattering angles and high energies correspond to low differences. Systematic measurements of the intermediate energy range would shed new light on this interesting energy range, which still lacks an adequate theoretical treatment.

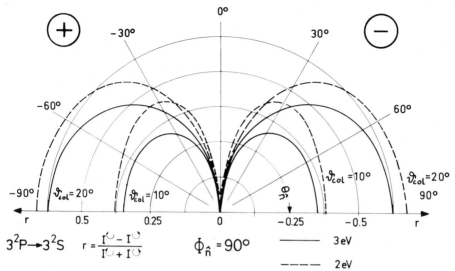

Figure 10. Calculated difference of the scattering intensities for σ^+ and σ^- excitation as a function of the incident light direction $\theta_{\hat{n}}$.

Conclusions

I have tried to illustrate some of the possibilities of the scattering technique by laser-excited atoms. I hope that it will be applied to many more problems, including problems other than the electron collision case. It opens up a number of ways of obtaining very detailed information on the scattering dynamics and, together with similar information from the coincidence experiment, may help to further improve collision theories.

Acknowledgment

A. Stamatović and W. Stoll contributed essentially to the experimental work and interpretation presented here. Financial support of the Deutsche Forschungsgemeinschaft is gratefully acknowledged.

References

Eminyan, M., MacAdam, K., Slevin, J., and Kleinpoppen, H. (1973). *Abstr. VIII ICPEAC* (Beograd Institute of Physics), 318. See also *J. Phys. B.*, **7**, 1519 (1974).
Fano, U., and Macek, J. H. (1973). *Rev. Mod. Phys.*, **45**, 553.

Hertel, I. V., and Stoll, W., 1974a, *J. Phys. B.*, **7**, 570.
Hertel, I. V., and Stoll, W. (1974b). *J. Phys. B.*, **7**, 583.
Macek, J., and Hertel, I. V. (1974). *J. Phys. B.*, **7**, 2173.
Moores, D., and Norcross, D. (1972). *J. Phys. B.*, **5**, 1482.

32

Excitation and Ionization in the Coulomb-Projected Born Approximation

SYDNEY GELTMAN

> A semiclassical description is given of the Coulomb-projected Born (CPB) approximation. The applications of the CPB method to excitation and ionization of hydrogen and helium are reviewed. The improvement in describing large-angle inelastic scattering is discussed semiclassically. Reliability estimates of theoretical results are obtained by comparing Born and CPB predictions.

In this progress report we will describe the physical basis of the Coulomb-projected Born method (CPB), discuss the similarities and differences of its predictions with those of the usual Born approximation, and compare with available experimental data.

For illustration let us consider the semiclassical treatment of an electron–hydrogen atom collision. By "semiclassical" we mean that the motion of the incident electron is treated classically and the target atom is treated quantally. We may write the Hamiltonian in a manner which indicates this separation,

$$H = T_1 - \frac{\hbar^2}{2m}\nabla_2^2 - \frac{Ze^2}{r_2} - \frac{Ze^2}{r_1} + \frac{e^2}{r_{12}} \qquad (1)$$

where T_i or V_i represents a classically treated term and the explicit quantum operators denote the part treated quantally. Thus in (1) only the free-particle

---

SYDNEY GELTMAN • Joint Institute for Laboratory Astrophysics, National Bureau of Standards and University of Colorado, Boulder, Colorado 80302, U.S.A.

motion of the incident electron is treated classically by straight-line trajectories specifying $\mathbf{r}_1(t)$, and the resulting time-dependent Schrödinger equation must be solved to give the description of the excitation of the target as a result of a single impact of specified velocity and impact parameter. The most common method of solving the quantum-mechanical part of the problem is to expand in target eigenstates

$$\Phi_p(\mathbf{r}_2, t) = \sum_q a_{pq}(t) \psi_q(\mathbf{r}_2) e^{-(i/\hbar)E_q t} \tag{2}$$

and obtain the transition amplitudes (a_{pq}) by perturbation theory or the explicit solution of a coupled set of equations resulting from the truncation of the above expansion.

It is clear that our choice of how to split the total Hamiltonian in (1) is quite arbitrary. Consider the alternative of also regarding the interaction between the incident electron and nucleus as part of the classical problem,

$$H = T_1 + V_1 - \frac{\hbar^2}{2m}\nabla_2^2 - \frac{Ze^2}{r_2} + \frac{e^2}{r_{12}} \tag{3}$$

Here our classical trajectory is the Rutherford trajectory of the incident electron in the nuclear field, and the quantum part of the problem is solved in exactly the same way as previously. The same type of approximation used in solving the quantum problem arising from (1) or (3) will in general lead to different transition amplitudes because of the difference in perturbation arising from the different classical trajectories for $\mathbf{r}_1(t)$. The extra term $-Ze^2/r_1$ in the quantal part of (1) *cannot* compensate for its effect in providing a classical Coulomb trajectory in (3). In fact, it can be shown that its only effect in (1) is to change the phase of a_{pq} (for $q \neq p$), and thus it has no effect on transition probabilities.

Proceeding from this semiclassical picture to a fully quantum formulation, one may split the total Hamiltonian in an arbitrary way between the unperturbed part (H_0) and the perturbation (H'). The most common choice for this division is to take the unperturbed part to consist of a noninteracting incident electron and target atom.

$$H_0 = -\frac{\hbar^2}{2m}\nabla_1^2 - \frac{\hbar^2}{2m}\nabla_2^2 - \frac{Ze^2}{r^2}$$

$$H' = -\frac{Ze^2}{r_1} + \frac{e^2}{r_{12}} \tag{4}$$

When this is put into the integral form of the Schrödinger equation containing the scattering boundary condition,

$$\Psi_p^+(\mathbf{r}_1\mathbf{r}_2) = \Psi_0(\mathbf{r}_1\mathbf{r}_2) + (E - H_0 + i\varepsilon)^{-1} H' \Psi_p^+(\mathbf{r}_1\mathbf{r}_2) \tag{5}$$

the excitation amplitude takes the exact integral form

$$f_{pq}(\hat{k}_q) = -\frac{1}{2\pi}\left\langle e^{i\mathbf{k}_q\cdot\mathbf{r}_1}\psi_q(\mathbf{r}_2), \left(-\frac{Ze^2}{r_1} + \frac{e^2}{r_{12}}\right)\Psi_p^+(\mathbf{r}_1\mathbf{r}_2)\right\rangle \quad (6)$$

In complete analogy to the semiclassical development in terms of (3), we may also choose to include the $-Ze^2/r_1$ term in the unperturbed part of the full quantum Hamiltonian, that is, let

$$H_0 = -\frac{\hbar^2}{2m}\nabla_1^2 - \frac{Ze^2}{r_1} - \frac{\hbar^2}{2m}\nabla_2^2 - \frac{Ze^2}{r_2}$$

$$H' = \frac{e^2}{r_{12}} \quad (7)$$

The corresponding exact integral form for the excitation amplitude, obtained through use of the Coulomb Green's function, is

$$f_{pq}(\hat{k}_q) = -\frac{1}{2\pi}\left\langle \psi_{\mathbf{k}_q}^-(\mathbf{r}_1)\psi_q(\mathbf{r}_2), \frac{e^2}{r_{12}}\Psi_p^+(\mathbf{r}_1\mathbf{r}_2)\right\rangle \quad (8)$$

where the final-state plane wave in (6) is replaced by the Coulomb wave $\psi_{\mathbf{k}_q}^-$. For this reason we refer to (8) as the *Coulomb-projected* form of the scattering amplitude. For elastic scattering ($q = p$), this form must be supplemented by the Coulomb amplitude which is contained in the unperturbed solution $\Psi_0(\mathbf{r}_1\mathbf{r}_2)$ in (5).

We then obtain the usual Born approximation and the Coulomb-projected Born approximation by replacing the exact scattering solution Ψ_p^+ by $e^{i\mathbf{k}_p\cdot\mathbf{r}_1}\psi_p(\mathbf{r}_2)$ in (6) and (8), respectively. The roles of the incident electron–nucleus interaction term (Ze^2/r_1) in the semiclassical and Born quantum formulations are very similar. In the free-particle formulations [(1) and (6)] that term makes a vanishing contribution to the direct excitation amplitude (because of the orthogonality of atomic wave functions). In the Coulomb trajectory (3) or Coulomb-projected Born (8) formulations, the Ze^2/r_1 term plays an important, nontrivial role. It thus appears that more of the essential physics of the problem is retained in the Coulomb formulations than in the free-particle formulations.

Applications of the CPB method have been made to the electron-impact excitation and ionization of hydrogen and helium in a series of papers by Geltman and Hidalgo (II, 1971; III, 1972; IV, 1974a; V, 1974b; paper I in this series concerned charge transfer, which is outside the area of this report). Paper II covers the direct excitation of the $2s$, $2p$, and $3s$ states of H, Paper III is concerned with the direct excitation of the 2^1S and 2^1P states of He, and papers IV and V are concerned with the triple differential ionization of H and He, respectively, including exchange effects. We are now extending the excitation calculations to lower energies with the inclusion of exchange. The exchange amplitudes g_{pq} are defined as having the form of the direct amplitudes in (6) and (8), except that $\mathbf{r}_1$ and $\mathbf{r}_2$ are permuted in Ψ_p^+. Thus the Born exchange result is the Oppenheimer approximation, for which analytical expressions have been derived (Corinaldesi and Trainor, 1952; Bell, 1965), and we evaluate the CPB exchange amplitudes by numerical methods.

The most striking departures from the usual Born results are in the differential cross sections, particularly at large scattering angles. This is illustrated in Figure 1 for the $1s$–$2s$ excitation of H. The CPB result is seen to follow the Rutherford form, $\csc^4(\theta/2)$, over a very wide range of the larger angles, while the Born result falls much more rapidly with angle.

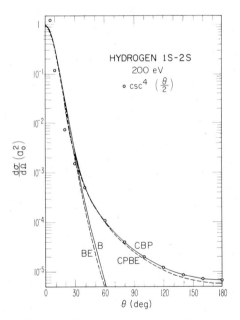

Figure 1. Differential cross section (in units of a_0^2/sr) for the $1s$–$2s$ excitation of hydrogen by 200 eV electrons in the various approximations: B(E), Born (exchange); CPB(E), Coulomb-projected Born (exchange). The points are proportional to the Rutherford angular dependence, $\csc^4(\theta/2)$, and have been normalized to the large-angle CPB results.

Thus it is seen that the quantum-mechanical CPB result is fully consistent with the semiclassical arguments given above. Exchange effects are small at all angles for the CPB result, while they have a major effect at backward scattering angles in the usual Born approximation.

Considerable experimental data are available for the large-angle differential excitation of helium, and in Figure 2 we show a comparison with data of Opal and Beaty (1972). We have not yet evaluated the exchange contributions in helium, but there is no reason to expect those to be appreciably larger than shown for the hydrogen case of Figure 1. The significant improvement in the CPB curve over the Born curve is obvious. Similarly improved large-angle results were also obtained by Madison and Shelton (1973) using distorted wave methods. We use simple uncorrelated exponential wave functions for the helium ground and excited states, and this could give rise to an intrinsic, but we believe small, error in the results of our calculations.

Closely related to the angular variation at a given energy is the variation of $d\sigma/d\Omega$ as a function of energy at fixed angle. It may be recalled that for any nonvanishing scattering angle the Born approximation (without exchange) predicts $d\sigma/d\Omega$ to vary asymptotically as E^{-6} for s–s transitions and as E^{-7} for s–p transitions. We find from the numerical results of all

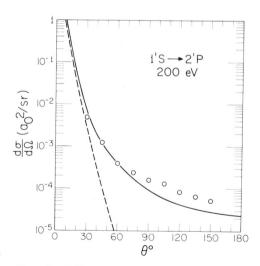

Figure 2. Differential cross section for the 1^1S–2^1P excitation of helium by 200 eV electrons in Born (dashed line) and Coulomb-projected Born (solid line) approximations (exchange not included). Points are the absolute data of Opal and Beaty (1972).

our CPB calculations that for any transition and nonforward scattering angle, the differential cross section asymptotically falls off as E^{-3}. This behavior is radically different from the Born prediction and can easily be understood semiclassically, where the differential cross section for the excitation $p \to q$ may be written as

$$\frac{d\sigma(p \to q)}{d\Omega} = \frac{d\sigma(\text{Rutherford})}{d\Omega} P(p \to q) \qquad (9)$$

Here $P(p \to q)$ represents the probability of a transition due to the traversal of an incident electron on a Rutherford trajectory [as formulated in (3)]. A typical value for this transition probability is the one applicable to the CPB results in Figure 1, or $P(1s \to 2s) = 0.0058$, which would be essentially independent of scattering angle (or impact parameter) from $\theta \cong 40°$ to $180°$.

Examination of the coupled equations for the transition amplitudes shows that in the high energy limit, independent of transition and of classical trajectory (see, for example, Geltman, 1969),

$$P(p \to q) = |a_{pq}(\infty)|^2 \xrightarrow[E \to \infty]{} E^{-1} \qquad (10)$$

Combining this with the well-known E^{-2} energy dependence of the Rutherford cross section gives us the overall E^{-3} dependence, which we have found independently from the fully quantal CPB calculations. It is of interest to note here that an attempt at a similar treatment for the Born approximation would be meaningless for all angles other than forward scattering because the semiclassical trajectories are undistorted straight lines. Only after high-order eikonal procedures or the evaluation of higher-order terms in the Born series could one expect to obtain the correct large-angle behavior.

An illustration of the results for total excitation cross section is given for the 1s–2s case of hydrogen in Figure 3. The detailed numerical results show that both direct approximations (B and CPB) and both exchange approximations (BE and CPBE) become equal to within 1% at about 100 eV, while the total exchange contribution reaches the 1% level at about 600 eV. Thus the large differences between the CPB and Born differential cross sections at large angles (Figure 1) do not show up at all in the integrated cross section at 200 eV. Rather, the main contribution to the total cross sections comes from $\theta = 0°$ to $30°$, where it can be seen that the two direct and the two exchange approximations become indistinguishable. We interpret the near equality of CPB and Born amplitudes as an indication of the correctness of both, because use of the correct Ψ_p^+ in (6) and (8) would yield the same (exact) amplitude. On the other hand, when they are not equal, such as in the large-angle case, we regard the CPB result as the more reliable, although we lack an independent theoretical criterion to predict the degree of reliability.

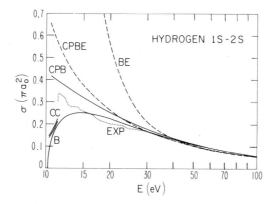

Figure 3. Total cross section for the $1s$–$2s$ excitation of hydrogen. B(E), Born (exchange); CPB(E), Coulomb-projected Born (exchange); CC, close-coupling (lower, Taylor and Burke, 1967; upper, Geltman and Burke, 1970). The curve labeled EXP is an approximate composite of the low-energy data of the four measurements referred to in the text normalized to CPBE at 45 eV.

The experimental curve (as a relative measurement) shown in Figure 3 has been obtained by Lichten and Schultz (1959), Stebbings *et al.* (1961), Hils *et al.* (1966), and Kauppila *et al.* (1970). The three latter measurements also revealed a large (apparent) cascade contribution over a large range of energies starting at about 50 eV. In Figure 3 we have normalized the measurements to the CPBE calculation at about 45 eV. We believe this value to be correct to the order of 3% because of its confluence with the BE value. If we accept this normalization, the extreme low-energy behavior of the cross section is consistent with the close-coupling results just above threshold. It is interesting to note that Lichten and Schultz originally normalized their data to the Born value (without exchange) at this low energy. Although at that time there was little reason to believe the Born approximation to be valid at 45 eV, it is our present conclusion that, except for a 5% exchange contribution, it is essentially correct for this transition. Of course, it does not follow that the Born approximation is valid to such low energies for all transitions.

Our final application of the Coulomb-projected Born method is to the triple differential ionization of H and He. This cross section, $d^3\sigma/d\Omega_a\, d\Omega_b\, dE_a$, is proportional to the probability of ionizing the atom, with the two final electrons going into specified elements of solid angle and energy. Only one of the elements of energy need be specified, since the other will be fixed by

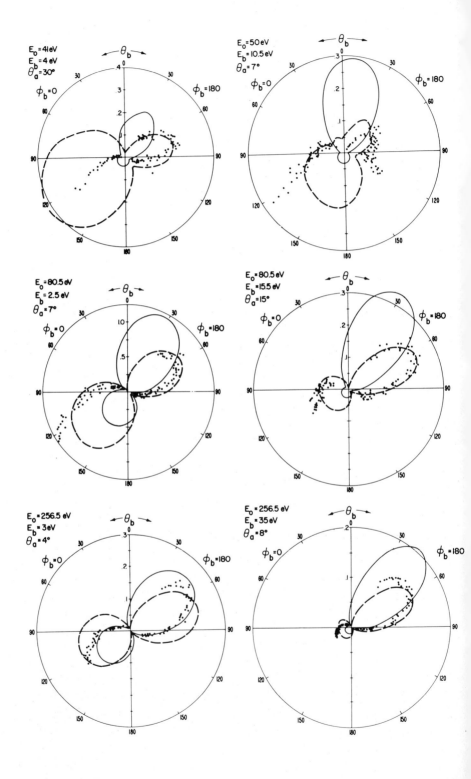

energy conservation. Ehrhardt *et al.* (1972a,b) have made extensive measurements of this quantity for the case in which the incident and the two final electrons are all in the same plane. A sampling of their results and our calculations is given in Figure 4. It can be easily proved that the Born result must be symmetric about the electron momentum transfer vector, completely independent of the sophistication of the atomic wave functions used. In most cases the data show a forward peak which is displaced to larger angles, and also a ratio of forward-to-backward peak magnitude which is in disagreement with the Born values. We note a substantial improvement in representing both of these properties in the CPBE approximation. The effect of exchange in most of the cases is generally small, so that the differences in the theoretical curves in Figure 4 are due to the differences in the basic scattering approximation. The Born approximation is best for the case of high incident energy and small scattering angle ($E_a = 256.5$ eV, $\theta_a = 4°$), as might be expected. Again, the simplest type atomic wave functions were used here to represent the bound and ionized states of helium.

In summary, we are encouraged by the consistently improved agreement with experiment that is attainable with the use of the Coulomb-projected Born approximation. This improvement appears to persist down to energies as low as 40 eV in some cases. We are continuing to investigate the role of exchange and the coherent excitation properties in this approximation.

References

Bell, K. L. (1965). *Proc. Phys. Soc.*, **86**, 246.
Corinaldesi, E., and Trainor, L. (1952). *Nuovo Cimento*, **9**, 940.
Ehrhardt, H., Hesselbacher, K. H., Jung, K., and Willman, K. (1972a). *J. Phys. B.*, **5**, 1559.
Ehrhardt, H., Hesselbacher, K. H., Jung, K., Schulz, M., and Willman, K. (1972b). *J. Phys. B.*, **5**, 2107.
Geltman, S. (1969). *Topics in Atomic Collision Theory*, p. 195. New York: Academic Press.
Geltman, S. (1974). *J. Phys. B.*, **7**, 1994.
Geltman, S., and Burke, P. G. (1970). *J. Phys. B.*, **3**, 1062.
Geltman, S., and Hidalgo, M. B. (1971). *J. Phys. B.*, **4**, 1299.
Geltman, S., and Hidalgo, M. B. (1974). *J. Phys. B.*, **7**, 831.
Hidalgo, M. B., and Geltman, S. (1972). *J. Phys. B.*, **5**, 617.
Hils, D., Kleinpoppen, H., and Koschmieder, H. (1966). *Proc. Phys. Soc.*, **89**, 35.
Kauppila, W. E., Ott, W. R., and Fite, W. L. (1970). *Phys. Rev.* **A1**, 1099.

Figure 4. Triple differential cross sections (in units of $a_0^2/\text{sr}^2/2$ ry) for the coplanar ionization of helium at various incident energies (E_0), fixed angles (θ_a) for one of the emerging electrons, and energies (E_b) of the other emerging electron. Solid line, Born; dashed line, Coulomb-projected Born exchange. The experimental points are the relative data of Ehrhardt *et al.* (1972a,b) normalized to the CPBE calculation in the vicinity of the forward peak.

Lichten, W., and Schultz, S. (1959). *Phys. Rev.*, **116**, 1132.
Madison, D. H., and Shelton, W. H. (1973). *Phys. Rev.*, A7, 499.
Opal, C. B., and Beaty, E. C. (1972). *J. Phys. B.*, **5**, 627.
Stebbings, R. F., Fite, W. L., Hummer, D. G., and Brackmann, R. T. (1961). *Phys. Rev.*, **124**, 2051.
Taylor, A. J., and Burke, P. G. (1967). *Proc. Phys. Soc.*, **92**, 336.

33
Calculations of Triple Differential Cross Sections

D. H. Phillips and M. R. C. McDowell

Triple differential cross sections for ionization of helium by 256.5 eV electrons are computed in the first Born approximation using accurate target and slow electron wave functions. The results are compared with the experiments of Ehrhardt et al. (1972). The first Born approximation fails to describe the recoil peak at this energy. For some angular ranges nondipole contributions are found to be significant.

1. Introduction

The ionization of helium by electron impact has been the subject of a variety of experimental and theoretical investigations, the majority of which have been devoted to the evaluation of accurate total cross sections.

A review of the experimental data for the single and multiple ionization of atoms and ions in their ground states has been presented by Kieffer and Dunn (1966). In several cases there exist significant discrepancies between the different sets of experimental measurements. For the particular case of helium, although the variation between the results of Smith (1930) and Rapp and Englander-Golden (1965) is slight, when these are compared with the results of Schram et al. (1965, 1966) and Gaudin and Hagemann (1967) a considerable divergence is apparent. The effect of this is to produce a 20–30% uncertainty in the cross section curve obtained from the averaged data.

As is the case with the experimental investigations, theoretical treatments of the problem span a considerable period of time. The most commonly

---

D. H. Phillips and M. R. C. McDowell • Mathematics Department, Royal Holloway College, London, England.

used approach to the problem is through the Born approximation, and some of the earliest calculations were made by Massay and Mohr (1933), who used a Coulomb wave representation of the less energetic outgoing electron. Among the most recent theoretical treatments are included the exchange adiabatic calculations of Economides and McDowell (1969), and also calculations, including exchange, based on a Born–Coulomb model (Peach, 1971). In spite of the wide variety of calculations made in Born approximation, which are discussed in a review of ionization theory presented by Rudge (1968), they have all suffered from a common failing. Even after taking due account of the uncertainty in the experimental data, all the calculations have consistently overestimated the cross section at intermediate energies.

Following the publication of the detailed experimental measurements of the triple differential cross section (TDC) made by Ehrhardt *et al.* (1972), there has been renewed interest in the theoretical treatment of the problem. Because of the several angular integrations required for the evaluation of the total cross section, it was not previously possible to form a clear picture of how the different angular regions contribute to the cross section. The TDC data now available provide a critical test of the theoretical models, clearly indicating the points on which they fail. From the characteristic behavior of the TDC it would seem likely that the general agreement between calculations of the total cross section based on different variants of the Born approximation is at least in part due to cancellation of errors as a result of the angular integrations.

By reason of the indistinguishability of the electrons involved in the ionizing collision, there are three distinct processes that contribute to each final state of ionization. In addition to direct scattering, two rearrangement processes, exchange and capture, are possible. Preliminary calculations (Phillips and McDowell, 1973), which examined the relative importance of the three processes, indicate that the capture process makes no significant contribution to the TDC in the energy region considered. This conclusion is confirmed by the results of parallel work undertaken by Schulz (1973), who used Coulomb waves to represent the outgoing electrons. Accordingly, the present calculations of the TDC only take account of the direct and exchange scattering.

In this paper we attempt to evaluate the TDC for electron–helium ionizing collisions within the first Born approximation to high accuracy, and hence to identify more clearly the angular ranges in which the predictions of the first Born approximation are reliable. Unfortunately the experimental measurements are restricted to impact energies below 260 eV, and it is already clear from Bethe–Born plots of the total cross section (Inokuti, 1971; Economides and McDowell, 1969) that the first Born approximation gives an unsatisfactory result for the total cross section at such low energies.

Any attempt to assess the reliability of first Born predictions of the TDC for electron impact on complex atoms requires the use of accurate wave functions for the target and for the slower of the outgoing electrons. Otherwise, little significance can be attached to the degree of agreement with the experimental data.

We shall describe results obtained using a six-configuration CI target wave function (Green et al., 1954) and an adiabatic exchange description of the slower outgoing electron (Temkin and Drachman, 1972; Economides and McDowell, 1969), but we will not attempt to account for correlation between the outgoing electrons (Salin, 1973). A comparable study using a many-parameter Hylleraas ground state and a 1s–2s–2p close-coupling description of the slow outgoing electron has been made by Jacobs, and preliminary results (which agree well with ours) have been presented recently (Jacobs, 1974). A more detailed account of the present work, including results at other energies, will be given elsewhere (Phillips, 1975).

In this paper we have considered only the ionization process for which both the target atom and residual ion are in their ground states. Our calculations, based on the first Born approximation, are confined to cases for which there is an appreciable difference in the energies of the two outgoing electrons.

We briefly outline our theoretical model in Section 2, make a detailed comparison of our results with the experimental measurements at 256.5 eV in Section 3, and state our conclusions in Section 4.

2. Theory

We shall refer to the more energetic outgoing electron as the "fast" electron and the other as the "slow" electron, denoting the respective momenta by $\mathbf{k}_f$ and $\mathbf{k}_s$. Following the notation of Rudge (1968), we introduce the direct and exchange amplitudes, $f(\mathbf{k}_f, \mathbf{k}_s)$ and $g(\mathbf{k}_f, \mathbf{k}_s)$, respectively, defined by

$$f(\mathbf{k}_f, \mathbf{k}_s) = C \langle \overline{\Psi}_{\mathbf{k}_f, \mathbf{k}_s}(1;2,3) | V(1;2,3) | \overline{\Psi}^+_{\mathbf{k}_0}(1,2,3) \rangle$$
$$g(\mathbf{k}_f, \mathbf{k}_s) = C \langle \overline{\Psi}_{\mathbf{k}_f, \mathbf{k}_s}(2;3,1) | V(1;2,3) | \overline{\Psi}^+_{\mathbf{k}_0}(1,2,3) \rangle \quad (1)$$

with

$$C = -(2\pi)^2$$

where $\overline{\Psi}^+_{\mathbf{k}_0}(1,2,3)$ is the solution of the full Schrödinger equation that has developed from the wave function representing the initial state of the system, $\overline{\Psi}_{\mathbf{k}_f, \mathbf{k}_s}(i;j,k)$ corresponds to a specific unperturbed final state of ionization of the system, and $V(1;2,3)$ denotes the interaction potential. The indexes 1, 2, 3 refer to the incident and two atomic electrons respectively.

Neglecting capture, these amplitudes may be combined to give an expression for the TDC (in which full account of indistinguishability is taken),

$$\frac{d^3\sigma}{d\hat{\mathbf{k}}_f\, d\hat{\mathbf{k}}_s\, d\varepsilon} = \frac{2\mathbf{k}_f \mathbf{k}_s}{\mathbf{k}_0} q(\mathbf{k}_f, \mathbf{k}_s)$$

where

$$q(\mathbf{k}_f, \mathbf{k}_s) = |f(\mathbf{k}_f, \mathbf{k}_s)|^2 + |g(\mathbf{k}_f, \mathbf{k}_s)|^2 - \text{Re}\{f(\mathbf{k}_f, \mathbf{k}_s)g^*(\mathbf{k}_f, \mathbf{k}_s)\} \quad (2)$$

here, $\varepsilon = \tfrac{1}{2}k_s^2$, is the energy of the "slow" outgoing electron and $\mathbf{k}_0$ is the momentum of the incident electron.

2.1. The Wave Functions

In accordance with the first Born approximation, we replace $\overline{\Psi}_{\mathbf{k}_0}^+(1, 2, 3)$ appearing in Eq. (1) by $\overline{\Psi}_{\mathbf{k}_0}(1;2,3)$, the wave function describing the state of the system prior to interaction. We express this wave function as a product of a free-electron wave function and a wave function $\Psi_0(\mathbf{r}_2, \mathbf{r}_3)$ describing the target ground state (electrons 2 and 3 being bound):

$$\overline{\Psi}_{\mathbf{k}_0}(1;2,3) = \frac{1}{(2\pi)^{3/2}} e^{i\mathbf{k}_0 \cdot \mathbf{r}_1} \Psi_0(\mathbf{r}_2, \mathbf{r}_3) \quad (3)$$

The simple-one-parameter wave function used in the preliminary calculations (Phillips and McDowell, 1973; Phillips, 1975) is now replaced by a configuration-interaction representation based on an expansion in hydrogenic orbitals which includes the $(1s)^2$, $(1s2s)$, $(2s)^2$, $(2p)^2$, $(2p3p)$, and $(3p)^2$ $^1S_0$ configurations (Green, 1954). Thus,

$$\Psi_0(\mathbf{r}_2, \mathbf{r}_3) = \sum_{i=1}^{6} C_i \chi_i(\mathbf{r}_2, \mathbf{r}_3) \quad (4)$$

where the $\chi_i(\mathbf{r}_2, \mathbf{r}_3)$ are the two-electron configuration wave functions and the C_i are weights.

The final state of ionization is described by the product of a free-electron wave function and a function $\overline{\Psi}_{\mathbf{k}_s}(\mathbf{r}_j, \mathbf{r}_k)$ representing the excited state of helium formed from the "slow" electron and residual ion, antisymmetrization being allowed for by the form of the TDC appearing in (2),

$$\overline{\Psi}_{\mathbf{k}_f, \mathbf{k}_s}(\mathbf{r}_i; \mathbf{r}_j, \mathbf{r}_k) = \frac{1}{(2\pi)^{3/2}} e^{i\mathbf{k}_f \cdot \mathbf{r}_i} \overline{\Psi}_{\mathbf{k}_s}(\mathbf{r}_j, \mathbf{r}_k) \quad (5)$$

where

$$\overline{\Psi}_{\mathbf{k}_s}(\mathbf{r}_j, \mathbf{r}_k) = \psi_{1s}(z = 2, r_j)\psi_{\mathbf{k}_s(\mathbf{r}_k)}$$

The function ψ_{1s} represents the bound electron occuping the 1s state of the residual ion while $\psi_{\mathbf{k}_s}$, which describes the "slow" outgoing electron, is approximated with the exchange adiabatic model of Economides and McDowell (1969). This latter wave function is expressed as a partial wave expansion

$$\psi_{\mathbf{k}_s}(\mathbf{r}) = \frac{1}{(2\pi)^{3/2}k_s^{1/2}} \sum_{l=0}^{\infty} i^l(2l+1) e^{i(\sigma_l + \eta_l)} \frac{u_l(k_s, r)P_l(\hat{\mathbf{k}}_s \cdot \hat{\mathbf{r}})}{r} \quad (6)$$

where the radial functions $u_l(k_s, r)$ satisfy an integrodifferential equation [McDowell *et al.*, 1973, Eq. (11)].

The quantity σ_l is the usual Coulomb phase shift and η_l is the additional phase shift arising from the effects of non-Coulomb potentials, including polarization of the residual ion.

Since $\overline{\Psi}_0$ and $\overline{\Psi}_{\mathbf{k}_s}$ are both taken to represent eigenstates of helium, they are subject to the orthogonality condition

$$\langle \overline{\Psi}_0(j, k) | \overline{\Psi}_{\mathbf{k}_s}(j, k) \rangle = 0 \quad (7)$$

which may be achieved by replacing $\overline{\Psi}_{\mathbf{k}_s}(j, k)$ by

$$\overline{\Psi}^0_{\mathbf{k}_s}(j, k) = \overline{\Psi}_{\mathbf{k}_s}(j, k) - \langle \overline{\Psi}_0 | \overline{\Psi}_{\mathbf{k}_s} \rangle \overline{\Psi}_0(j, k)$$

In view of the complexity of the calculations for the exchange amplitude, which is expected to give only a small contribution to the TDC, the quantity $g(\mathbf{k}_f, \mathbf{k}_s)$ was replaced by means of the relation (Peterkop, 1961)

$$g(\mathbf{k}_f, \mathbf{k}_s) = e^{i\tau} f(\mathbf{k}_s, \mathbf{k}_f) \quad (8)$$

valid if the wave functions are exact.

Since the phase τ is not uniquely specified by the theory we have chosen to set it equal to zero. Other approximations for the exchange amplitude have been examined elsewhere (cf. Rudge, 1968).

The consequent detailed analysis is given elsewhere (Phillips, 1975), together with documented computer programs. The partial wave calculations reported here were checked by specializing to a one-parameter target, together with a plane wave description of the slow outgoing electron, in which case the analysis can be carried out in closed form.

We found that contributions from $l > 6$ were unimportant at the energy considered, the dominant contribution arising from $s \to p$ transitions. However, in some angular regions there was an appreciable quadrupole transition probability ($s \to d$).

3. Results

In Figures 1–6 we compare our calculations of the TDC, for a number of representative cases, with the experimental measurements of Ehrhardt *et al.* at the incident energy $E_0 = 256.5$ eV. We shall use E_f and θ_f to denote the energy and angle of scattering, respectively, of the "fast" outgoing electron. Table 1 shows the sets of collision parameters for which calculations have been made.

Since the experimental measurements are not absolute, it is necessary to normalize them against the calculated TDC curves in some way. In this paper the normalization is achieved by requiring that the maximum of the

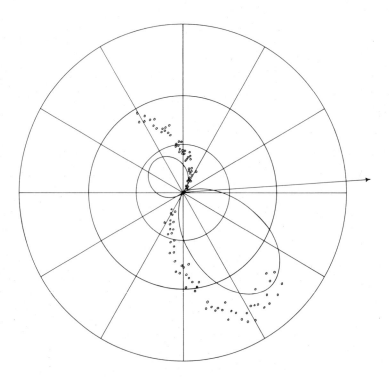

Figure 1. The solid line shows the theoretical results of this work for the TDC (scale: 1 radial division = 1.43 a.u.), to which we have normalized the experimental results of Ehrhardt *et al.* (1972) at the forward peak. Incident energy $E_0 = 256.5$ eV, "fast" electron energy $E_f = 226.0$ eV, "fast" electron scattering angle $\theta_f = 4°$. (The results shown are for the correlated target wave function, exchange and orthogonality contributions included. Angles are measured counterclockwise.)

Triple Differential Cross Sections

Table 1

Figure	E_0 (eV)	E_f (eV)	θ_f
1	256.5	226.0	4°
2	256.5	226.0	6°
3	256.5	226.0	8°
4	256.5	230.5	4°
5	256.5	229.0	6°
6	256.5	182.0	8°

forward peak of the TDC, as measured experimentally, should be in agreement with the calculated value.

For the cases shown in Figures 1–3 a constant normalization factor has been used. The results obtained indicate that the normalization is preserved by the form of the Born approximation used for the calculations. Another point of agreement apparent from these and the remaining cases is the general success of the calculations in reproducing the shape of the forward peak, albeit this peak may not be correctly positioned.

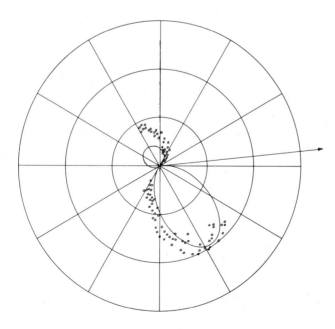

Figure 2. As for Fig. 1, but $E_0 = 256.5$ eV, $E_f = 226.0$ eV, $\theta_f = 6°$.

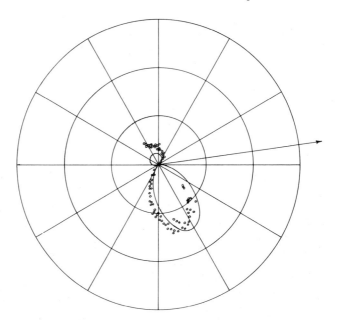

Figure 3. As for Fig. 1, but $E_0 = 256.5$ eV, $E_f = 226.0$ eV, $\theta_f = 8°$.

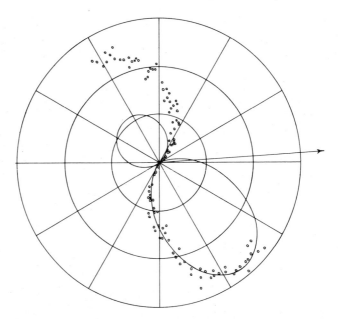

Figure 4. As for Fig. 1, but $E_0 = 256.5$ eV, $E_f = 230.5$ eV, $\theta_f = 4°$. Radial scale: 1 division = 2.09 a.u.

Triple Differential Cross Sections

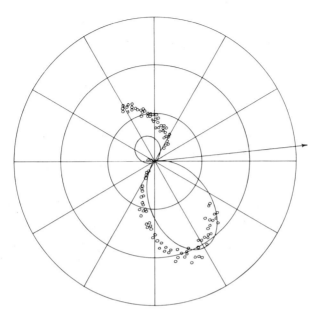

Figure 5. As for Fig. 1, but $E_0 = 256.5$ eV, $E_f = 209.0$ eV, $\theta_f = 6°$. Radial scale: 1 division $= 1.67$ a.u.

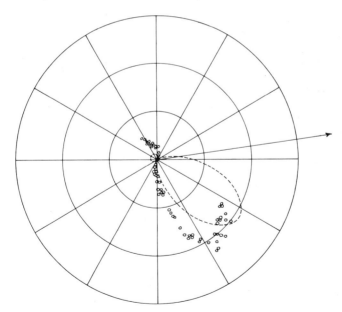

Figure 6. As for Fig. 1, but $E_0 = 256.5$ eV, $E_f = 182.0$ eV, $\theta_f = 8°$. Radial scale: 1 division $= 0.08$ a.u.

The TDC is characterized by a forward and backward peak separated by pronounced minima. These two peaks are referred to as the "binary" and "recoil" peaks respectively. Two points of disagreement arise from the comparison of results: the calculations consistently underestimate the magnitude of the "recoil" peak as measured by Ehrhardt, and the theoretical model tends to produce a "binary" and "recoil" peak with a common axis of symmetry, whereas the experimental results show a definite shift of one axis away from the other. With regard to the second point, Fano (1973), attributing the TDC exclusively to dipole-type ionization for large incident energies, predicts that the "slow" electron distribution will exhibit axial symmetry about the momentum exchange vector $\mathbf{k}_0 - \mathbf{k}_f$, which accurately describes the Born calculations presented here. At lower incident energies the repulsion between the outgoing electrons is expected to displace the peaks, with the consequent loss of symmetry. The neglect of the electron–electron interaction in the final state seems a likely reason for the discrepancy between calculation and experiment as regards the angular positions of the peaks.

Schulz (1973) has made extensive calculation using Coulomb waves to represent the outgoing electrons and varying the effective charges. Although this has resulted in an improvement for the angular position of the "binary" peak, the "recoil" peak is still not satisfactorily described. Also, Salin (1973) has used the impact parameter method to describe the ionization process. These calculations take particular account of the Coulomb interaction between the two outgoing electrons. The results obtained are qualitatively similar to those of Schulz.

The origin of the "recoil" peak has been discussed in the recent publication by Schulz. Vriens (1969) attributed it to ionizing collisions in which an initially bound electron, leaving the ion in a forward direction, is reflected by the potential of the ion and consequently suffers a reversal of direction. It has been suggested by Schulz that another type of collision contributing to the "recoil" peak is that involving the scattering of the incident electron by the nucleus or the electron that remains bound. In this case the mutual interaction of all the atomic particles could result in a nearly uniform distribution for the ejected electron. The contribution from the first type of collision decreases with the energy E_s, while that of the second increases with momentum transfer.

Since calculations based on the first Born approximation exclude the interaction between the incident electron and the nucleus, the explanations suggested by Vriens and Schulz would at least seem compatible with the failure of these calculations to reproduce a "recoil" peak of sufficient magnitude. Confirmation of this conclusion seems to be provided by the TDC calculation of Geltman and Hidalgo (1974). They have used their

Coulomb-projected Born approximation, which takes account of the electron–nucleus interaction, to calculate the TDC of helium for a wide range of collision parameters. The results obtained show a considerable improvement over calculations based on variants of the first Born approximation.

Let us now examine the double differential cross section obtained from the TDC by integrating over the angular coordinates of the "slow" outgoing electron:

$$\frac{d^2\sigma}{d\hat{\mathbf{k}}_f\, d\varepsilon} = \int \frac{d^3\sigma}{d\hat{\mathbf{k}}_f\, d\hat{\mathbf{k}}_s\, d\varepsilon}\, d\hat{\mathbf{k}}_s \qquad (9)$$

where

$$d\hat{\mathbf{k}}_s = \sin\theta_s\, d\theta_s\, d\phi_s$$

Because all of the calculations have been confined to in-plane scattering ($\phi_s = 0°$), we cannot determine exactly the effect of the integration over the angular coordinates. However, we can make a reasonable speculation. In the region about $\phi_s = 0°$, the θ_s integration has the form

$$\int \frac{d^3\sigma}{d\hat{\mathbf{k}}_f\, d\hat{\mathbf{k}}_s\, d\varepsilon} \sin\theta_s\, d\theta_s \qquad (10)$$

Since the "binary" and "recoil" peaks of the TDC are almost diametrically opposite, the presence of the $\sin\theta_s$ factor in the integrand of (10) will result in some cancellation between the peaks. Hence, it seems likely that any model underestimating the magnitude of the recoil peak will tend to overestimate the total cross section, as is the situation regarding Born calculations of the total cross section.

Finally, we have made calculations to investigate how the various partial waves appearing in the exchange adiabatic representation of the "slow" outgoing electron contribute to the TDC. For the purposes of these calculations we have retained only the s-state configurations of the CI wave function describing the target ground state.

We found that for the regions in which the TDC is most significant, the largest contribution arises from the dipole transition ($\Delta l = 1$). However, in other angular regions, where the TDC is less significant, the quadrupole, and some higher-order, contributions assume considerable importance. One such region is $0 < \theta_s < 90°$. Figures 7 and 8 show the dependence of the TDC upon the partial wave contributions in this angular region. Table 2 shows the value of the collision parameters in each case.

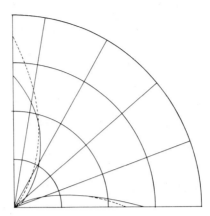

Figure 7. The curves show the calculated TDC, using a correlated target wave function (with p orbitals omitted), and including exchange in the final state for two cases: (———) all contributions $l = 0, \ldots, 6$; (— — —) $l = 0$ and 1 contributions only; and $E_0 = 256.5$ eV, $E_f = 230.5$ eV, $\theta_f = 4°$; $0° \leqslant \theta_s \leqslant 90°$.

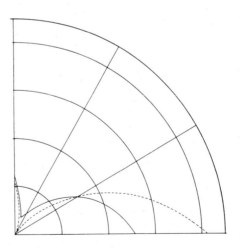

Figure 8. As for Fig. 7, but $\theta_f = 7°$.

Table 2

Figure	E_0 (eV)	E_f (eV)	θ_f
7	256.5	230.5	4°
8	256.5	230.5	7°

4. Conclusions

We have calculated triple differential cross sections for the electron impact ionization of helium in first Born approximation at an incident energy of 256.5 eV. Particular consideration has been given to the use of accurate wave functions to describe the target ground state and the "slow" outgoing electron. The results show the first Born approximation to fail on two points, namely, the prediction of the correct angular positions of the "binary" and "recoil" peaks and the magnitude of the "recoil" peak. Various explanations for the behavior of the TDC are presented, from which it is clear that the discrepancies between the calculations and experimental results arise through faults inherent in the first Born approximation. In this connection we have indicated why Born calculations overestimate the total cross sections in the energy region considered. Further, we have have shown that there are angular regions where the ionization process may be significantly different from a dipole transition.

References

Economides, D. G., and McDowell, M. R. C. (1969). *J. Phys. B.*, **2**, 1323–1331.
Ehrhardt, H., Hesselbacher, K. H., Jung, K., Schulz, M., and Willmann, K. (1972). *J. Phys. B.*, **5**, 2107–2115.
Fano, U. (1973). Review paper, *8th ICPEAC, Belgrade*.
Gaudin, A., and Hagemann, R. (1967). *J. Chim. Phys.*, **64**, 1209–1222.
Geltman, S., and Hidalgo, M. B. (1974). *J. Phys. B.*, **7**, 831. Also contribution to this volume.
Green, L. C., Lewis, M. N., Mulder, M. M., Wyeth, C. W., and Woll, J. W. Jr. (1954). *Phys. Rev.*, **93**, 273–279.
Inokuti, M. (1971). *Rev. Mod. Phys.*, **43**, 247.
Jacobs, V. L. (1974). *Phys. Rev.*, **A10**, 499. Also contribution to this volume.
Kieffer, L. J., and Dunn, G. H. (1966). *Rev. Mod. Phys.*, **38**, 1–33.
Massey, H. S. W., and Mohr, C. B. O. (1933). *Proc. Roy. Soc.*, **A140**, 613–630.
McDowell, M. R. C., Morgan, L. A., and Myerscough, V. P. (1973). *J. Phys. B.*, **6**, 1435–1451.
Peach, G. (1971). *J. Phys. B.*, **4**, 1644–1677.
Peterkop, R. (1961). *Proc. Phys. Soc.*, **77**, 1220–1222.
Phillips, D. H. (1975). Ph.D. Thesis, University of London.

Phillips, D. H., and McDowell, M. R. C. (1973). *J. Phys. B.*, **6**, L165–168.
Rapp, D., and Englander-Golden, P. (1965). *J. Chem. Phys.*, **43**, 1464–1479.
Rudge, M. R. H. (1968). *Rev. Mod. Phys.*, **40**, 564–589.
Saiin, A. (1973). *J. Phys. B.*, **6**, L34–36.
Schram, B. L., de Heer, F. J., van der Wiel, M. J., and Kistemaker, J. (1965). *Physica*, **31**, 94.
Schram, B. L., Boerboom, A. J. H., and Kistemaker, J. (1966). *Physica*, **32**, 185–196.
Schulz, M. (1973). *J. Phys. B.*, **6**, 2580–2599.
Smith, P. T. (1930). *Phys. Rev.*, **36**, 1293–1302.
Temkin, A., and Drachman, R. J. (1972). In *Case Studies in Atomic Collision Physics*, **2**, Eds. M. R. C. McDowell and E. W. McDaniel. Amsterdam: North-Holland.
Vriens, L. (1969). *Physica*, **45**, 400–406.

34
Differential Cross Sections for Electron Impact Ionization of Helium

V. L. Jacobs

> *Differential cross sections for electron impact ionization of helium are calculated in the Born approximation, using a Hylleraas ground state wave function and 1s–2s–2p close-coupling final-state continuum wave functions. The resonant features associated with the lowest ^1S, ^1P, and ^1D autoionizing states are analyzed using the Fano line-shape formula. The angular correlation between the two outgoing electrons is calculated and compared with experimental results.*

Electron impact ionization of atoms is an important process in laboratory and astrophysical plasmas. The most stringent test of existing theories occurs in the analysis of the spectra obtained by simultaneous measurements of the energies and angles of both outgoing electrons. Such measurements have been recently reported by Ehrhardt *et al.* (1972a,b) for electron impact ionization of helium.

In the electron impact ionization process

$$\text{He}(1^1S) + e^-(\mathbf{k}_0) \to \text{He}^+(1s) + e^-(\mathbf{k}_1) + e^-(\mathbf{k}) \tag{1}$$

it is customary to define the ejected electron $e^-(\mathbf{k}_1)$ as the slower of the two outgoing electrons and the scattered electron $e^-(\mathbf{k})$ as the faster one. The cross section which is differential in all independent energy and angular variables is given in the Born approximation by

$$\sigma(\mathbf{k}_1, \mathbf{K}) = \frac{k}{k_0} \frac{4}{K^4} |\langle 1s, \mathbf{k}_1 | e^{i\mathbf{K}\cdot\mathbf{r}_1} + e^{i\mathbf{K}\cdot\mathbf{r}_2} | 1^1S \rangle|^2 \tag{2}$$

where $\mathbf{K} = \mathbf{k}_0 - \mathbf{k}$ is the momentum transfer vector.

V. L. Jacobs • NASA/Goddard Space Flight Center, Greenbelt, Maryland 20771, U.S.A.

The theory developed for differential photoionization cross sections (Jacobs, 1972) has been extended to electron impact ionization. In the Born approximation the angular correlation between the scattered and ejected electrons is given (in terms of the angle ω between the vectors $\mathbf{k}_1$ and $\mathbf{K}$) by the Legendre polynomial expansion

$$\sigma(\mathbf{k}_1, \mathbf{K}) = \sigma(k_1^2, K)\left\{1 + \sum_L \beta_L(k_1^2, K)P_L(\cos \omega)\right\} \qquad (3)$$

and is therefore cylindrically symmetric about the momentum transfer direction. The electron energy loss spectrum and its dependence on the scattering angle is described by the differential cross section $\sigma(k_1^2, K)$.

The explicit expressions which are obtained for the total generalized oscillator strength $df_{nl_1}(K)/dE$ and for the asymmetry parameters $\beta_{nl_1,L}(E, K)$ may be written in the forms

$$\frac{df_{nl_1}(K)}{dE} = \frac{E}{K^2}\frac{1}{2L_0 + 1}\sum_{\lambda l_2 L_T}(2\lambda + 1)|M(\lambda, l_2, L_T)|^2 \qquad (4)$$

and

$$\frac{df_{nl_1}(K)}{dE}\beta_{nl_1,L}(E, K) = \frac{E}{K^2}\frac{2L + 1}{2L_0 + 1}(-1)^{L_0 + l_1}\sum_{\lambda l_2 L_T}\sum_{\lambda' l'_2 L'_T}$$
$$\times (2\lambda + 1)(2\lambda' + 1)\{(2l_2 + 1)(2l'_2 + 1)(2L_T + 1)(2L'_T + 1)\}^{1/2}$$
$$\times \begin{pmatrix} l_2 & l'_2 & L \\ 0 & 0 & 0 \end{pmatrix}\begin{pmatrix} \lambda & \lambda' & L \\ 0 & 0 & 0 \end{pmatrix}\begin{Bmatrix} l_2 & l'_2 & L \\ L'_T & L_T & l_1 \end{Bmatrix}\begin{Bmatrix} \lambda & \lambda' & L \\ L'_T & L_T & L_0 \end{Bmatrix}$$
$$\times M(\lambda, l_2, L_T)M(\lambda', l'_2, L'_T)^* \qquad (5)$$

where

$$M(\lambda, l_2, L_T) = i^\lambda e^{i(\sigma_{l_2} - l_2\pi/2)}(nl_1 l_2, L_T S_0 \| j_\lambda(Kr_1)C^{(\lambda)}(\hat{r}_1)$$
$$+ j_\lambda(Kr_2)C^{(\lambda)}(\hat{r}_2)\| n_0 L_0 S_0) \qquad (6)$$

$E = k_0^2 - k^2$ is the electron energy loss, σ_{l_2} denotes the Coulomb phase shift for the He$^+$ ion, $j_\lambda(Kr)$ is the spherical Bessel function, and $C^{(\lambda)}(\hat{r})$ is the Racah tensor operator.

The generalized oscillator strength for ionization of helium has been calculated using a ground state wave function of the Hylleraas type and 1s–2s–2p close-coupling final-state continuum wave functions (Burke and

Electron Impact Ionization of Helium

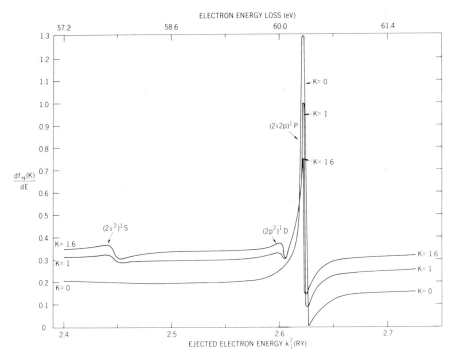

Figure 1. The total generalized oscillator strength (per unit rydberg energy interval) for resonant electron impact ionization of helium.

McVicar, 1962). The effects of the lowest ^1S, ^1P, and ^1D resonances in the electron energy loss spectrum (Figure 1) are found to be accurately represented by the asymmetric line-shape formula introduced by Fano (1961). The line profile parameters q show strong variations with momentum transfer.

In Figure 2 the calculated angular correlation is compared with the experimental results (Ehrhardt *et al.*, 1972b) for an incident electron energy of 256.5 eV. The agreement is satisfactory except near the backward maximum. In addition the observed symmetry axis makes a 4.6° angle with the momentum transfer direction, indicating a further inadequacy of the Born approximation.

A more general analysis of the angular correlation is being carried out in this laboratory in terms of symmetric Euler angle functions of the outgoing electron directions (Bhatia and Temkin, 1964), with particular emphasis on threshold problems in impact ionization and double photoionization.

A fuller account of this work has been submitted to the *Physical Review*.

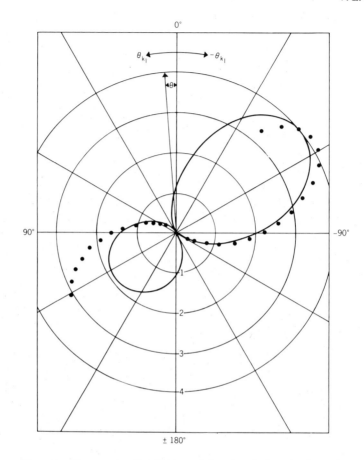

Figure 2. The angular distribution of the ejected electrons from the ionization of helium by 256.5 eV incident electrons. The ejected electron energy is 3.0 eV and the scattering angle θ is 4°. Solid curves, present calculation; dots, Ehrhardt et al. (1972b).

References

Bhatia, A. K., and Temkin, A. (1964). *Rev. Mod. Phys.*, **36**, 1050.
Burke, P. G., and McVicar, D. D. (1962). *Proc. Phys. Soc. Lond.*, **80**, 616.
Ehrhardt, H., Hesselbacher, K. H., Jung, K., and Willman, K. (1972a). *J. Phys. B.*, **5**, 1559.
Ehrhardt, H., Hesselbacher, K. H., Jung, K., Schulz, M., and Willman, K. (1972b). *J. Phys. B.* **5**, 2107.
Fano, U. (1961). *Phys. Rev.*, **124**, 1866.
Jacobs, V. L. (1972). *J. Phys. B.*, **5**, 2257.

35

The Polarized Frozen Target and Polarized Frozen Core Methods in Low-Energy Electron Scattering and in Atomic Structure Calculations

M. LE DOURNEUF, H. VAN REGEMORTER, AND VO KY LAN

> We present consistent calculations of low-energy electron–atom scattering and bound states, based on the polarized frozen target (or core) method. Having described how to build the polarized states of the target (or of the core) for any complex atom, we discuss two variants of the method: the variational version, usually known as the modified close-coupling method using polarization pseudostates, and its adiabatic approximation corresponding to the polarized orbital close-coupling method. The efficiency of these techniques, which include at the outset the predominant long-range correlation effects is assessed by several tests, among which are the binding energies of H^- and O^- and the low-energy $^3P^e$ resonance in $N + e^-$ scattering. Further applications are discussed, as well as the prospect of using the method for excited-states eigenenergy calculations and for bound free processes.

1. Introduction

The polarized orbital methods have been reviewed recently by Drachman and Temkin (1972) and Callaway (1973), and the use of pseudostates for improving the close-coupling method by Damburg (1972) and Geltman (1972). Our aim is not to repeat what has been already discussed, in particular the basic approximations involved in the different polarized-orbital methods. It is rather to underline the efficiency of properly modified close-coupling

M. LE DOURNEUF, H. VAN REGEMORTER, and VO KY LAN • Département d'Astrophysique fondamentale, Observatoire de Paris, 92190 Meudon, France.

methods in many electron–atom scattering problems or atomic-structure calculations.

We are mainly concerned with the low-energy region, where polarization effects are dominant: elastic scattering or near-threshold transitions to a few low-lying atomic states, loosely bound negative ions, and highly excited states of atoms. The basic idea of the methods discussed here is to incorporate from the beginning the essential long-range correlation effects by using frozen-polarization pseudostates constructed *a priori* and independently of the scattering problem to describe the target. The same procedure will be applied to $(N + 1)$ electron bound state calculations using polarized states of the core.

When these predominant long-range correlation effects are taken into account at the outset, the remaining short-range correlation effects are more easily and quickly accounted for by the variational close-coupling method, as well as by iterative techniques discussed below. Thus, the important resonance structure occurring below an inelastic threshold can be obtained by inclusion of a few well-chosen closed channels, and the influence of true or virtual $(N + 1)$ bound states of nearly zero energy on the low-energy scattering can be allowed for by short-range correlation pseudostates.

Therefore, the use of polarized states or pseudostates which incorporate the predominant long-range correlation effects can be considered as a powerful way to improve, and at the same time to simplify, the close-coupling expansion, as has been already underlined by Burke *et al.* (1969), Feautrier *et al.* (1969, 1971), and Calvert *et al.* (1971).

In Section 2, a brief survey is given of applications of the close-coupling approach and of some of the modifications which have been proposed to improve the convergence of the method. The principles of building polarized states or polarization pseudostates to take account of long-range correlation effects are given in Section 3. When these states are constructed, it is possible to examine in Section 4 how to use either the polarized states in an adiabatic polarized frozen target (or core) approximation or the polarization pseudostates in the variationally motivated polarized frozen target (or core) approximation using additional pseudostates. The accuracy of the method is established on the test calculation of the binding energy of H^- and the $^1S^e$ phase shift in $H + e$ scattering. Applications to complex systems are given in Section 6 for two typical examples: the binding energy of O^- and the $^3P^e$ phase shift in $N + e$ scattering.

2. The Close-Coupling Approach to Bound States and Collision Problems

When the Schrödinger equation is solved, using the expansion of the $(N + 1)$-electron system in terms of the target eigenstates,

$$\Psi = \mathscr{A} \sum_i \Phi_i(N) F_i(r_{N+1}) \qquad (1)$$

this gives rise to the well-known infinite set of coupled integrodifferential (ID) equations for the radial part of the scattering functions F_i. In the close-coupling approximation only a few low-lying states are included in expansion (1).

We shall be concerned here with low-energy scattering, mainly elastic scattering, where only one channel is open, and with the bound region, where all channels are closed and the solution of the ID equations of the close-coupling (CC) approximation provides a method for generating bound wave functions for the $(N + 1)$-electron system, starting from exact wave functions for the N-electron system.

From the work of Castillejo et al. (1960), it is clear that it is not possible to completely account for the long-range polarization potential by using a truncated basis of target eigenstates in the total wave function expansion.

The fundamental properties of the target can be represented by a few low-lying states in a very limited number of cases for which the truncated basis of the CC expansion gives a good description of the problem. For example, alkalies can be represented by two-level atoms since the polarizability of the ground state is dominated by the resonance transition. For elastic scattering by alkalies the close-coupling approach must be accurate when good target wave functions are used, as has been proved by many calculations, in particular those of Norcross (1971) and Moores and Norcross (1972). The same conclusion can be drawn from the CC calculations by Dubau and Wells (1973) of the free wave function for the photoionization of two-s-electron systems like Be and Mg, the ejected photoelectron being in the field of an alkali-like core.

Seaton (1973) has recently reviewed the few applications of the CC to bound state calculations in the so-called "frozen core" approximation. This approach, introduced first by Cohen and Kelly (1965), applies when the atomic core can be represented by one or a few low levels. Good results have been obtained for bound states in C^{2+} and for the ground configuration terms in C by Seaton (1972) and Seaton and Wilson (1972), which can be compared favorably with elaborate configuration interaction HF calculations by Moser and Nesbet (1971).

The question of the slow convergence of the CC expansion for hydrogen scattering was first discussed by Burke and Schey (1962). In the case of positron scattering, Perkins (1968) has implemented a modified CC expansion suggested by Damburg and Karule (1967) to take account properly of long-range correlation effects. Burke et al. (1969), in the case of $H + e^-$ scattering, have modified the CC expansion by using *a priori* pseudostates as defined by Damburg and Karule (1967) and Damburg and Geltman (1968), which reproduce the polarizabilities of different H levels.

At the same time, Feautrier *et al.* (1969, 1971) have introduced their multichannel polarized-orbital method, which uses *a priori* polarized orbitals and, at low energies, takes account of the same correlation effects as the Burke *et al.* (1969) method in a more economical way.

In Figure 1 we show the ^1S phase shift obtained by Burke *et al.* (1969) for $H + e^-$ scattering, and in Figure 2 the ^1S phase shift obtained by Feautrier *et al.* (1971) for $H + e^+$ scattering. Both results show an important improvement over ordinary CC results and also demonstrate the importance of the remaining short-range correlations effects for *s*-wave scattering.

The excellent results obtained by Vo Ky Lan (1971) for a highly polarizable system like Li, compared to the elastic scattering $2s$, $2p$ close-coupling results of Burke and Taylor (1969), show that the polarized-orbital method is accurate when properly applied, i.e., without overestimating the dipolar dipolar distortion of the orbitals (thanks to the use of β as discussed below). In fact, in this case the two methods must give the same results at low energies. Results are given in Figure 3.

Matese and Oberoi (1971) have thoroughly investigated the various possible choices of pseudostates for improving the CC formalism. In the case of electron scattering by H, the most elaborate results have been obtained by Burke and Taylor (1966), including 16 short-range correlation terms in addition to the $1s,2s,2p$ expansion. The improvements of different "extended"

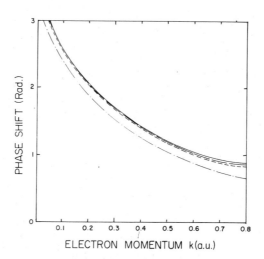

Figure 1. Electron–hydrogen *s*-wave phase shift. (— · —) $1s$; (— — — —) $1s2s2p2\bar{p}$; (— — —) $1svs2\bar{p}vp$-$3\bar{d}vd$; (———) variational calculation of Schwartz (1961).

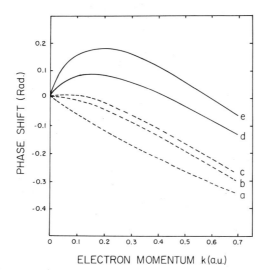

Figure 2. Positron–hydrogen s-wave phase shift. Curve a, $1s$; curve b, $1s2s2p$; curve c, $1s2s2p$-$3s3p3d$; curve d, $1s2s2p$ + polarization; curve e, variational calculation of Schwartz (1961).

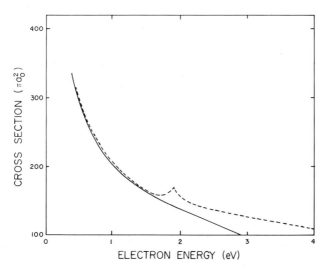

Figure 3. Electron–lithium elastic scattering cross section. (———) Polarized orbital results of Vo Ky Lan (1971); (– – – –) $2s2p$ close-coupling results of Burke and Taylor (1969).

polarized methods applied to elastic scattering have been reviewed by Drachman and Temkin (1972) and Callaway (1973) and will not be discussed any further here.

3. Long-Range Correlation Effects

Two modified close-coupling approaches have been proposed in order to take full account of the polarization of the target. Both methods, however, proceed first to construct polarized wave functions which are eigenstates of the atomic Hamiltonian perturbed by the dipolar part of the static field of the $(N + 1)$th electron.

In order to handle a complex atom one needs to define the procedure used to build up a multiconfigurational pseudostate

$$\bar{\Phi}(L'M_{L'}) = \sum_i a_i \chi_i(\mathbf{r}_1, \ldots, \mathbf{r}_N; L'M_{L'}) \qquad (2)$$

where the χ_i are configurations constructed from HF orbitals and additional polarized pseudoorbitals. If $\Phi_0(LM)$ is a ground configuration of type $ns^2 np^q$ atoms like C, N, O, the full polarization of the target can be represented by a linear combination of all the $nsnp^{q+1}$ and $nsnp^q n\bar{p}$ configurations representing the virtual excitations of the ns core electrons and all the $ns^2 np^{q-1} n\bar{s}$ and $ns^2 np^{q-1} n\bar{d}$ configurations representing the virtual excitations of the np valence electrons.

It is necessary here to have clearly in mind that these polarized states are built to represent the target in the close-coupling expansion of the scattering problem. When one uses these modified states in the close-coupling expansion, if P is the projector onto all the p states included in the CC expansion [see below, Eqs. (9) and (10)], a part of the electronic interaction PVP is taken into account exactly (to all orders) in the scattering problem. Defining $Q = 1 - P$, only QVQ is treated approximately, using the dipolar approximation and first-order perturbation theory to build the polarized states.

The procedure for obtaining the pseudoorbitals nl' has been given by Vo Ky Lan (1972), Feautrier et al. (1971), and Drachman and Temkin (1972). The dipole distortion of an HF orbital φ_{nl} can be represented by

$$\varphi^{pol}_{nl \to l'}(\mathbf{r}_N, \mathbf{r}_{N+1}) = r_{N+1}^{-2} \theta_{nl \to l'}(\mathbf{r}_N, \hat{r}_{N+1}) \qquad \text{if } r_{N+1} \gg r_N \qquad (3a)$$

where $\theta_{nl \to l'}$ is obtained from first-order perturbation theory.

$$[H_{HF} - \varepsilon_{nl}] \theta_{nl \to l'} = -Q v \varphi_{nl}(\mathbf{r}_N) \qquad (4)$$

with $v = 2r_N \cos(\omega_{N,N+1})$.

Using a partial wave expansion of Eq. (4), one obtains the radial part of $\theta_{nl \to l'}$, which has the characteristics of an l' wave, $U_{nl \to l'}(r_N)$, and one defines a *pseudoorbital*, denoted by $\overline{nl'}$,

$$\overline{\varphi}_{nl \to l'}(\mathbf{r}_N) = Y_{l'm'}(\hat{r}_N) r_N^{-1} U_{nl \to l'}(r_N) \tag{5}$$

or a *polarized orbital* by using (3), where $\varphi^{pol}_{nl \to l'}(\mathbf{r}_N, \mathbf{r}_{N+1})$ is a parametric function of r_{N+1}. Pseudoconfigurations are obtained by substituting one pseudoorbital in the Slater determinant, *polarized configurations* by replacing successively each unperturbed orbital by the corresponding polarized orbital.

Once the configurations χ_i have been built, each multiconfiguration pseudostate is a solution of the general first-order perturbation equation

$$[H_A - E_n]\overline{\Phi}(L'M_{L'}) = -V_d \Phi(LM_L) \quad \text{with } V_d = \sum_i r_i \cos(\omega_{i,N+1}) \tag{6}$$

Variation of the coefficients a_i in (2) leads to a system of linear equations which can be solved by standard techniques. This procedure to build *a priori* pseudostates has not yet been applied to the calculation of multiconfiguration pseudostates of complex systems.

An alternative method has been proposed by Burke and Mitchell (1974). Instead of starting by constructing *a priori* pseudoorbitals, they calculate multiconfigurational pseudostates where each pseudoconfiguration is constructed with additional pseudoorbitals of the Slater type, the range coefficients of these being determined by a variational technique in order to obtain the best value of the polarizability α, given for a ground configuration by

$$\alpha = 2 \sum_{L'} \langle \Phi_0(LM_L) | V_d | \overline{\Phi}(L'M_{L'}) \rangle \tag{7}$$

In their original work, Feautrier *et al.* (1971) use the so-called polarized orbital CC method, which can be considered as a generalization of the polarized orbital method of Temkin (1957) to the multichannel case. They have introduced in Eq. (3) an important modulating factor β, defining

$$\varphi^{pol}(\mathbf{r}_N, \mathbf{r}_{N+1}) = \beta(r_{N+1}) \theta_{nl \to l'}(\mathbf{r}_N, \hat{r}_{N+1}) \tag{3b}$$

where $\beta(r_{N+1}) \to r_{N+1}^{-2}$ when $r_{N+1} \to \infty$ is obtained by minimizing the energy of the target, perturbed by the exact electrostatic potential V due to the additional electron in the adiabatic approximation, a method already used by Stone (1966) (cf. Vo Ky Lan, 1972). This procedure can be generalized to polarized multiconfigurational states using

$$\Phi^{pol}(\mathbf{r}_N, \mathbf{r}_{N+1}) = \beta(r_{N+1}) \overline{\Phi}(L'M_{L'}) \tag{8}$$

and is a way to avoid the use of a step function or an overestimation of the dipolar distortion resulting from the first-order perturbation Eq. (6).

4. The Polarized Frozen Target (or Core) Approximation

Once the first problem of determining the polarized wave functions or the polarization pseudostates is solved, they can be used to simplify the scattering problem. We have seen that for complex atoms this first problem is not trivial, but that it gives in itself an interesting representation of a polarized atom which can be tested with the calculation of atomic polarizabilities. As soon as this problem is solved, the reward will be a considerable simplification of the scattering problem. In fact, we shall be able to generalize the efficiency of the method to any complex system without a great deal of labor.

In the first modified close-coupling approach, using polarized states, proposed by Feautrier et al. (1969, 1971), the total wave function is expanded in the form

$$\Psi(\mathbf{r}_N, \mathbf{r}_{N+1}) = \mathscr{A} \sum_{i=1}^{p} [\Phi_i(\mathbf{r}_N) + \Phi_i^{\text{pol}}(\mathbf{r}_N, \mathbf{r}_{N+1})] F_i(r_{N+1}) \qquad (9)$$

where $\mathbf{r}_N$ stands for all atomic electron coordinates, Φ^{pol} is defined by (8), and p open channels are retained in the expansion.

In the second modified close-coupling approach, using polarization pseudostates proposed by Burke et al. (1969), q additional pseudostates are added to represent the omitted channels of the close-coupling expansion:

$$\Psi(\mathbf{r}_N, \mathbf{r}_{N+1}) = \mathscr{A} \left\{ \sum_{i=1}^{p} \Phi_i(\mathbf{r}_N) F_i(r_{N+1}) + \sum_{j=1}^{q} \bar{\Phi}_j(\mathbf{r}_N) F_j(r_{N+1}) \right\} \qquad (10)$$

which leads to a larger system of coupled integrodifferential equations.

The first approach is the more economical way to take proper account of long-range correlation effects. It is appropriate to the study of complex atoms when powerful methods of integration of large systems of coupled equations are not available. In fact, Eqs. (9) and (10) must give the same results at low energy, i.e., when Eq. (10) gives an adequate account of most of the electronic correlation. This is due to the introduction of the β factor in Eq. (8), which gives Eq. (9) most of the flexibility of Eq. (10), β being variationally determined in the adiabatic limit.

When using Eq. (9) to solve the Schrödinger equation, two systems of equations are obtained for the F_i and the Φ_i^{pol}. Because of the r_{N+1} dependence of the latter, the two systems are coupled by dynamic terms, but this coupling is indeed small in the energy region where both Eqs. (9) and (10) are valid and concerns only the weak r_{N+1} dependence of Φ_i^{pol}, due only to the neglected channels in the CC expansion. This partial adiabatic approximation gives rise to Eq. (6) for the polarized wave function and to the following scattering equation:

$$\left[\frac{d^2}{dr^2} - \frac{l_i(l_i-1)}{r_{N+1}^2} + \frac{2z}{r_{N+1}} - k_i^2\right]F_i = 2\sum_{i'}[V_{ii'} + V_{ii'}^{pol} - W_{ii'} - W_{ii'}^{pol}]F_{i'} \quad (11)$$

where in addition to the usual direct and exchange potentials appear the direct and exchange polarization potentials, for which explicit expressions have been given in the general case of complex atoms by Feautrier et al. (1971) and Vo Ky Lan et al. (1972).

On the other hand, the use of expansion (10) has the great advantage of preserving the usual form of the system of ID equations and giving lower bounds on phase shifts while satisfying the variational principle exactly.

For an open shell target, bound functions of the type $\sum_k c_k \psi_k$ for the $(N + 1)$-electron system must be added to expansions (9) or (10) to describe the virtual capture of the incident electron. Additional $(N + 1)$-electron functions or pseudostates can be added to describe short-range correlation effects. The functions F_i and the coefficients c_k are given in this case by two coupled systems of equations. On the other hand, below an inelastic threshold, it is necessary to include explicitly a few closed channels to reproduce resonance effects.

Therefore, using polarized states, the general expression of the wave function is taken of the form

$$\Psi(N+1) = \mathscr{A}\sum_i[\Phi_i(N) + \overline{\Phi}_i(N)\beta_i(r_{N+1})]F_i + \sum_k c_k \Phi^{SR}(N+1) \quad (12)$$

Using polarization pseudostates, it can be written as

$$\Psi(N+1) = \mathscr{A}\left[\sum_i \Phi_i(N)F_i + \sum_j \overline{\Phi}_j(N)F_j\right] + \sum_k c_k \Phi^{SR}(N+1) \quad (13)$$

The first of these expressions has been applied successfully by our group in Li + e scattering (see Figure 3) and in some preliminary calculation for complex systems (see Vo Ky Lan et al., 1972).

Most of the results reported in recent work and presented in this article have been obtained using Eq. (13). More work remains to be done to state exactly the domain of validity of the two expressions, but there is already some evidence that Eq. (12) must be equivalent to Eq. (13) for low-energy electron scattering or for some eigenenergy calculations of loosely bound electrons (as for high Rydberg states).

To integrate the integrodifferential equation to determine the radial functions F_i and the coefficients c_k both the R matrix technique of Burke et al. (1971) and the Seaton and Wilson (1972) code have been used. The latter has been modified for negative ion calculations. The vector-coupling coefficients have been calculated using the Racah technique.

We would like to suggest calling these approximations the polarized frozen target (or core) approximations, because it clears up the difference and emphasizes the simplicity of the method compared to more global approaches in many-body studies. From this point of view, the real meaning of the so-called close-coupling approximation and of its various modified forms is unclear.

The polarized frozen target (or core) approximation refers to the use of Eq. (13), i.e., to the variational version of the method with polarization pseudostates. When one uses Eq. (12), together with the adiabatic approximation described above, one can refer to the "adiabatic frozen target (or core) approximation."

5. Test of the Methods on Two-Electron Systems (H^-, He)

The calculations of the binding energy of H^-, the ^1S phase shift in low-energy $H + e^-$ scattering, and the bound states of the Rydberg series $1s\,ns$ of He can be considered as fundamental tests of the methods described above. The two electrons having opposite spins are allowed to interact at very short distances: short-range correlation (SR) effects are determinant.

Since the eigenstates of the one-electron target (H, He^+) are exact and Eq. (4), giving their dipole distortion, can be solved analytically, the variational polarized frozen core approximation [expansion (13)] will lead to variational bounds of the phase shift and the binding energies, while in the case of complex N-electron targets, the states of which are only known approximately, only the $(N + 1)$-electron energy satisfies the minimum principle. Application of the adiabatic version of the method [expansion (12)] has not yet been applied to this bound state calculation.

The method has been applied to H^- and He, first by Calvert and Davison (1971) and recently by Le Dourneuf and Vo Ky Lan (1974). The accuracy of some results has been improved since, and they are presented in Table 1.

As is well known, the restricted HF approximation is unable to represent the bound state of $H^-(^1S)$. The frozen core approximation (b) already gives 36% of the correlation energy E_c in H^-.

The efficiency of the method is easy to understand, since in the development of the two-electron wave function in the frozen core approximation

$$\Psi^b[1, 2] = \mathscr{A}[\Phi_{1s}F^b_{1s} + c^b_1\Phi_{1s^2}] \quad (14)$$

a complete flexibility is given to the wave function F^b_{1s}, which can then adapt to represent the diffuse electronic cloud of the outer, loosely bound electron.

Table 1. Binding Energies of H$^-$ (^{1}S) and He ($1sns\,^1$S) in Rydbergs

Ref.	Type	H$^-$	He, $n=1$	He, $n=3$	He, $n=5$	He, $n=7$
a	Restricted HF	+0.02414	−1.72336			
b	FC (1s)	−0.02662	−1.74502	−0.12116	−0.042075	−0.021154
c	$1s2s'$	−0.02872	−1.75686	−0.12154	−0.042160	−0.021180
d	$1svs$	−0.02772	−1.75530	−0.12159	−0.042159	−0.021182
e	s limit	−0.02898	−1.75806			
f	$1s2\bar{p}$	−0.04894	−1.77990	−0.12175	−0.042188	−0.021193
g	$1s2s'2\bar{p}2p'$	−0.05276	−1.79896	−0.12236	−0.042320	−0.021124
h	$1svs2\bar{p}vp$	−0.05185				
i	p limit	−0.05294	−1.80102			
j	$1s2\bar{p}3d$	−0.04994	−1.78190	−0.12178	−0.042193	−0.021194
k	$1s2s'2\bar{p}2p'3d$	−0.05368	−1.80066			
l	$1svs2\bar{p}vp3d\nu d$	−0.05304				
m	d limit	−0.05460	−1.80552			
n	$1s2s2p$ + 16 corr.	−0.05528				
o	Exact	−0.05550	−1.80744	−0.12254	−0.042360	−0.021260
p	$1s + (vs2pvp3d\nu d)$	−0.05206	−1.79550	−0.12228	−0.042302	−0.021233

a: Roothan et al. (1960); b, c, g, k: Calvert and Davison (1971); d, j, h, l, p: this work; e: Schwartz (1962). Winter et al. (1970); f: Seaton and Wilson (1972); i, m: Winter et al. (1970); n: Burke and Taylor (1966); o: Pekeris (1962).

The inclusion of the long-range (LR) dipole distortion $2\bar{p}$ of the ground state $1s$ improves considerably the binding energy (83% of E_c is obtained with approximation f for H^-).

Smaller improvements are obtained by including the LR quadrupole pseudostate $3\bar{d}$ [calculation (j)]. Because higher multipoles of the electronic interaction are not specifically LR, there is no sense in going further in this direction. The problem, then, is to account for short-range effects. The remaining s-type SR correlation leading to the s limit of Weiss (1961) has been accounted for by Calvert and Davison (1971) using an SR pseudo-orbital $2s'$ of Slater type, the range of which is obtained by minimizing the energy of the two-electron system.

We investigated an alternative iterative method consisting in reinjecting the frozen core F^b_{1s} function as a second basis function in order to account for the s distortion of the $1s$ function interacting with the outer one. It does not give very good results for the H^- ground state but can be preferred when exploring a large number of excited states, as in He. The inclusion of one SR pseudostate of p or d character improves the binding energies, which approach the p and d limits of Winter et al. (1970); 97% of the correlation energy is obtained in this way [calculation (l)]. The only satisfactory way to handle the small remaining part of the correlation energy is to use explicit correlation functions as Schwartz (1961) and Burke and Taylor (1966) did to compensate for angular correlation effects.

Though not so accurate, but interesting in view of its generalization to complex negative ions, is calculation p, which gives 96% of the correlation in H^- and 86% in $(1s^2)^1 S$ of He, with only one free channel $\Phi_{1s}F_{1s}$ and a superposition of nine optimized $(N + 1)$ correlation configurations:

$$1s^2, 1svs, vs^2, 2\bar{p}^2, 2\bar{p}vp, vp^2, 3\bar{d}^2, 3\bar{d}vd, vd^2$$

Approximation f, with the simple inclusion of $2\bar{p}$, has already been applied by Seaton and Wilson (1972) to the calculation of many He levels, including the ground state. For highly excited states, approximation f is equivalent to the second-order Perturbation theory applied by Poe and Chang (1973). The corresponding (^1S) phase shifts in $H + e^-$ scattering have been calculated and the results corresponding to approximation (l) are given in Figure 1.

6. *Application to Complex Systems* (N, O)

A considerable amount of work has been devoted to electron scattering by np^q atoms and ions. Though it has long been limited to the inclusion of the ground state terms in the CC expansion (Smith et al. 1967; Saraph et al.

1969), recent calculations reviewed in the accompanying paper by Nesbet and Thomas (1975) have pointed out the importance of polarization and of close-correlation effects leading to resonances in the very-low-energy region or real bound states. The method described above, which incorporates *a priori* pseudostates accounting for the whole polarization and the bound function allowing for the capture of the $(N + 1)$th electron in the outer p open shell, is particularly appropriate to study the low-energy properties of these systems. Two examples will be considered here: the calculation of the binding energy of O^-, which is known experimentally, and the up-to-date problem of the $^3P^e$ bound state or resonance of N^-.

(i) *The Binding Energy of O^-*

In the "frozen core" approach, the bound state $O^-(^2P^0)$ is represented by the simple expansion

$$\Psi(^2P^0) = \varphi(^3P^e)F_1 + \varphi(^1D^e)F_2 + \varphi(^1S^e)F_3 + 1s^22s^22p^5 \quad (15)$$

Starting from HF wave functions for O, one obtains a binding energy of 0.0529 Ry, i.e., half of the experimental value (0.108 Ry).

In a "polarized core" scheme, we account for the full dipole polarization of the target ground term $^3P^e$ by starting from the expansion

$$\Psi(^2P^0) = \varphi(^3P^e)F_1 + \varphi(^1D^e)F_2 + \varphi(^1S^e)F_3 + \bar{\varphi}(^3P^0)F_4$$
$$+ \bar{\varphi}(^3D^0)F_5 + 1s^22s^22p^5 + 2०\Phi(N + 1) \quad (16)$$

$\Phi(N + 1)$ represent bound states in which the extra electron is captured in one of the open shells of the φ or $\bar{\varphi}$ orbital (F_i assumed to be orthogonal to the core orbitals). Apart from $1s^22s^22p^5$, which is the dominant configuration of O^-, they can be considered as SR correlation functions. Using this expansion, one obtains 0.11 Ry (i.e., nearly exactly the experimental result). This calculation using the Seaton and Wilson (1972) code, which takes only 50 sec of computer time and 550 K of storage on an IBM 370/168, compares very favorably with much lengthier calculations, based on variational Bethe–Goldstone theory, by Moser and Nesbet (1971) and semiempirical results of Schaeffer *et al.* (1969). It confirms the predominance of polarization effects in the capture of a nonpenetrating p electron. Details of this calculation will be given by Le Dourneuf and Vo Ky Lan (1976).

Comparison with the experimental polarizability of 5.19 a_0^3 gives a test of the accuracy of the wave function. A polarizability of 5.18 is obtained in this way. Allison *et al.* (1972) have obtained 5.20 using the R matrix method.

The polarization effects are also dominant in $O + e$ scattering, for which the adiabatic version of the polarized frozen core approximation applied by Vo Ky Lan et al. (1972) taking account of nearly all the polarization of the target and a very recent calculation by Thomas and Nesbet (1975) are in good agreement with a previous polarized-orbital calculation of Henry (1967) and the experimental data of Sunshine et al. (1967). This question is discussed by Nesbet and Thomas (1975).

(ii) The $^3P^e$ Resonance in $N + e^-$ Scattering

The $^3P^e$ partial wave is the most important contribution to ($N + e^-$) low-energy scattering and is very sensitive to correlation effects. Very accurate calculations must be done near zero energy to locate the ^3P resonance if, as it seems, $N^-(^3P)$ is unbound.

Calculations have been undertaken in collaboration between the Belfast and the Paris groups using the R matrix code. A systematic investigation of configuration interaction (CI) and polarization effects has been described by Burke et al. (1974). Preliminary qualitative results, however, are given in Figure 4.

Calculation A corresponds to the simple frozen core approximation without any configuration interaction:

$$\Psi(^3P^e) = \varphi(^4S)F_1 + \varphi(^2D^0)F_2 + \varphi(^2P^0)F_3 + (1s^22s^22p^4)^3P^e \quad (17)$$

Calculation A_1 is the same as A, with near-degenerate configuration interaction for the term $^2P_0$

$$\varphi(^2P^0) = c_1(1s^22s^22p^3) + c_2(1s^22p^5) \quad (18)$$

Calculation B includes allowance of the polarization of the target ground term with

$$\Psi(^3P^e) = \varphi(^4S^0)F_1 + \varphi(^2D^0)F_2 + \varphi(^2P^0)F_3 + \bar{\varphi}(^4P^e)F_4 + 17\Phi(N+1) \quad (19)$$

where the pseudostate $\bar{\varphi}$ of Burke and Mitchell (1974) takes account of all the $2s$ and $2p$ virtual excitations. Calculation B_1 includes both the CI effect as in A_1 and the polarization of the target. As one sees, the CI effect pushes the resonance toward the lower energies (or modifies the target ground state energy). But the main effect arises from the full allowance for polarization included in calculation B_1, which gives the peak of the ^3P resonance at 0.129 eV, in agreement with previous Bethe–Goldstone calculations by Moser and Nesbet (1971), including three-electron correlations.

In fact, this result must still be considered as preliminary; by improving the accuracy of the integration and including additional CI effects, the ^3P resonance still moves somewhat to lower energies before converging (Burke et al., 1974). This calculation at such a small energy is really a challenge for the most precise methods of integration.

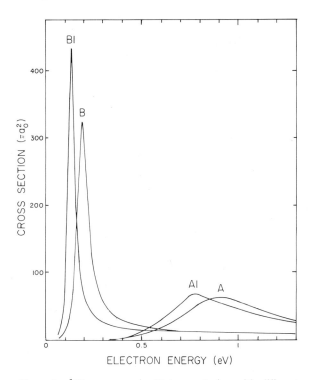

Figure 4. $^3P^e$ resonance in N + e scattering with different representations of the target. (A) $2s^2 2p^3$ (ground configuration); (A1) $2s^2 2p^3 + 2p^5$ (near-degenerate CI in the ground configuration); (B) $2s^2 2p^3$ + polarization pseudostate $^4S^0 \to {}^4\bar{P}$ of the ground term N; (B1) CI and polarization effects.

It is also interesting to study on a larger energy range (0 → 1 Ry) the scattering process N + e⁻, for which there is only one rather qualitative experiment by Neynaber et al. (1961). Four theoretical calculations of the total cross section are compared in Figure 5.

The single configuration CC calculation of Henry et al. (1969) overestimates the cross sections at low energy by several orders of magnitude. The account of near-degenerate correlation effects in the matrix-variational method of Thomas et al. (1974) lowers it partly, but reasonable agreement with experiment can be obtained by full allowance of the ground-configuration polarization, as illustrated by the polarized-orbital calculation by Henry (1968) and by some unpublished results of the Paris group, which account for the whole direct polarization of the atom and which are obtained with the adiabatic version of the polarized frozen target approximation

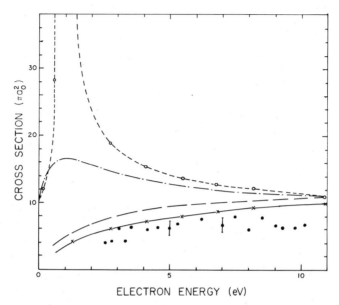

Figure 5. Electron–nitrogen total cross section. (●) Experimental results of Neynaber et al. (1961); (– – –) CC (one configuration) by Henry et al. (1969); (— · —) inclusion of near-degenerate CI effects by Thomas et al. (1974); (— —) polarized orbital by Henry (1968); (———) present work.

[formula (12)]. Results still more accurate are expected by extending to higher energies the variational calculation undertaken to locate the $^3P^e$ low-energy resonance.

The same method should be applied shortly to the study of the true bound state of the carbon negative ion $C^-(^4S^0)$ as well as of the true location of the $C^-(^2D^0)$ and $C^-(^2P^0)$ states, which have been examined by Thomas et al. (1974) using an extrapolation of low-energy $C + e$ phase-shift calculations.

7. Conclusion

The polarized frozen target (or core) method has been introduced to preserve the simplicity of the close-coupling approximation and to avoid the unnecessary efforts and time-consuming calculations of more global approaches.

As is well known, the convergence of the method of superposition of HF configurations is slow, and, on the other hand, the multiconfiguration HF method involves the reoptimization of a great number of orbitals at each step of the calculation. The use of a hierarchy of Bethe–Goldstone equations in matrix-variational form by Nesbet (1973) provides a step-by-step procedure to calculate the correlation energy. Usually limited to the second order of the hierarchy, this approach mainly takes account of the polarization of the target (or core) to all orders.

The same long-range part of the correlation energy is accounted for in the polarized frozen target (or core) method. But the great simplification —and computational advantage—of the latter is due to the use of frozen target (or core) wave functions, which are assumed to be known and whose accuracy can be checked with eigenenergy and polarizability calculations. Moreover, the correlation energy due to the polarization of the target (or core) can be also represented by frozen pseudostates which are constructed at the outset. This combines the advantages of the close-coupling and of the polarized orbital approximation.

Two versions of the method have been described in the present paper. No recent application of the adiabatic version has been given because more work remains to be done to examine the accuracy and advantages of the method. On the other hand, the successful applications of the variational version of the polarized frozen target (or core) method led us to the conclusion that it is a powerful way to handle both electron–atom collision problems and atomic-structure calculations. It gives a method to generate $(N + 1)$-electron bound wave functions and can be used for calculating eigenenergies of excited states, polarizabilities, oscillator strengths, and so forth.

As underlined very often by Seaton (1973), it is important to study in a unified way both the bound and the continuum states. The method described above is particularly suitable for the study of bound free transitions, because bound and free wave functions are calculated on the same footing, using the same program. Applications are in progress to the photoionization of aluminium and to the photodetachment of the negative ion O^-.

References

Allison, D. C. S., Burke, P. G., and Robb, W. D. (1972). *J. Phys. B.*, **5**, 55.
Burke, P. G., and Mitchell, J. B. M. (1974). *J. Phys. B.*, **7**, 665.
Burke, P. G., and Schey, H. M. (1962). *Phys. Rev.*, **126**, 147.
Burke, P. G., and Taylor, A. J. (1966). *Proc. Phys. Soc.*, **88**, 549.
Burke, P. G., and Taylor, A. J. (1969). *J. Phys. B.*, **2**, 869.
Burke, P. G., Gallaher, D. F., and Geltman, S. (1969). *J. Phys. B.*, **2**, 1142.
Burke, P. G., Hibbert, A., and Robb, W. D. (1971). *J. Phys. B.*, **4**, 153. (See also Burke, P. G. (1973). *Comp. Phys. Commun.*, **6**, 288.)
Burke, P. G., Berrington, K. A., LeDourneuf, M., and Vo Ky Lan (1974). *J. Phys. B.*, **7**, L531.
Callaway, J. (1973). *Comp. Phys. Commun.*, **6**, 265.

Calvert, J. Mc. I., and Davison, W. D. (1971). *J. Phys. B.*, **4**, 314.
Castillejo, L., Percival, I. C., and Seaton, M. J. (1960). *Proc. Roy. Soc.*, **A254**, 259.
Cohen, M., and Kelly, P. S. (1965). *Can. J. Phys.*, **43**, 1847.
Damburg, R. J. (1972). In *The Physics of Electronic and Atomic Collisions*, Eds. T. R. Govers and F. J. De Heer, p. 200. Amsterdam: North-Holland.
Damburg, R. J., and Geltman, S. (1968). *Phys. Rev. Lett.*, **20**, 485.
Damburg, R. J., and Karule, E. (1967). *Proc. Phys. Soc.*, **90**, 637.
Drachman, R. J., and Temkin, A. (1972). In *Case Studies in Atomic Collision Physics: II*, pp. 399–481. Eds. E. W. McDaniel and M. R. C. McDowell. Amsterdam: North-Holland.
Dubau, J., and Wells, J. (1973). *J. Phys. B.*, **6**, 1452 and L31.
Feautrier, N., van Regemorter, H., and Vo Ky Lan (1969). *Abstract VI Int. Conf. on the Physics of Electronic and Atomic Collisions.* Cambridge, Mass.: M.I.T. Press.
Feautrier, N, van Regemorter, H., and Vo Ky Lan. (1971). *J. Phys. B.*, **4**, 670.
Geltman, S. (1972). In *The Physics of Electronic and Atomic Collisions.* p. 216. Eds. T. R. Govers and F. J. De Heer, Amsterdam: North-Holland.
Henry, R. J. W. (1967). *Phys. Rev.*, **162**, 56–63.
Henry, R. J. W. (1968). *Phys. Rev.*, **172**, 99.
Henry, R. J. W., Burke, P. G., and Sinfailam, A. L. (1969). *Phys. Rev.*, **178**, 218.
Le Dourneuf, M., and Vo Ky Lan (1974). In *Abstracts IV Int. Conf. on Atomic Physics, Heidelberg* (July 22–26, 1974), pp. 154.
Le Dourneuf, M., and Vo Ky Lan (1976). To be published.
Matese, J., and Oberoi, R. S. (1971). *Phys. Rev.*, **A4**, 569.
Moores, D. L., and Norcross, D. W. (1972). *J. Phys. B.,* **5**, 1482.
Moser, C. M., and Nesbet, R. K. (1971). *Phys. Rev.*, **A4**, 1336.
Moser, C. M., and Nesbet, R. K. (1972). *Phys. Rev.*, **A6**, 1710–1715.
Nesbet, R. K. (1973). *Comp. Phys. Commun.*, **6**, 275.
Nesbet, R. K., and Thomas, L. D. (1975). Contribution to this volume.
Neynaber, R. H., Marino, L. L., Rothe, E. W., and Trujillo, S. M. (1961). *Phys. Rev.*, **123**, 148.
Norcross, C. W. (1971). *J. Phys. B.*, **4**, 1458.
Norcross, C. W. (1974). *Phys. Rev. Lett.*, **32**, 192.
Pekeris, C. L. (1962). *Phys. Rev.*, **126**, 1470.
Perel, J., Englander, P., and Bederson, B. (1962). *Phys. Rev.*, **128**, 1148.
Perkins, J. F. (1968). *Phys. Rev.*, **173**, 164.
Pöe, R. T., and Chang, T. N. (1973). In *Atomic Physics*, 3. Eds. S. J. Smith, S. G. K. Walter, and L. M. Volsky, p. 143. New York: Plenum Press.
Roothan, C. C. J., Sachs, L. M., and Weiss, A. W. (1960). *Rev. Mod. Phys.*, **32**, 186.
Saraph, M. E., Seaton, M. J., and Shemming, J. (1969). *Phil. Trans. Roy. Soc.*, **264**, 77.
Schaeffer, H. F., Klemm, R. A., and Harris, F. E. (1969). *J. Chem. Phys.*, **51**, 4643.
Schwartz, C. (1961). *Phys. Rev.*, **124**, 1468.
Schwartz, C. (1962). *Phys. Rev.*, **126**, 1015.
Seaton, M. J. (1972). *J. Phys. B.*, **5**, L91.
Seaton, M. J. (1973). In *Atomic Physics*, 3. Eds. S. J. Smith, G. K. Walters, and L. H. Volsky, p. 205. New York: Plenum Press; *Comp. Phys. Commun.*, **6**, 247.
Seaton, M. J. and Wilson, P. M. H. (1972). *J. Phys. B.*, **5**, L1 and L175.
Smith, K., Henry, R. J. W., and Burke, P. G. (1967). *Phys. Rev.*, **157**, 51.
Stone, P. M. (1966). *Phys. Rev.*, **141**, 137.
Sunshine, G., Aubrey, B. B., and Bederson, B. (1967). *Phys. Rev.*, **154**, 1.
Thomas, L. D., and Nesbet, R. K. (1975). *Phys. Rev.*, **A11**, 170.
Thomas, L. D., Oberoi, R. S., and Nesbet, R. K. (1974). *Phys. Rev.*, **A10**, 1605.

Temkin, A. (1957). *Phys. Rev.*, **107**, 1004.
Vo Ky Lan (1971). *J. Phys. B.*, **4**, 658.
Vo Ky Lan (1972). *J. Phys. B.*, **5**, 242.
Vo Ky Lan, Feautrier, N., Le Dourneuf, M., and van Regemorter, H. (1972). *J. Phys. B.*, **5**, 1506.
Weiss, A. W. (1961). *Phys. Rev.*, **122**, 1826.
Winter, N. W., Laferriere, A., and McKoy, V. (1970). *Phys. Rev.*, **A2**, 49.

36

Progress Report on the Use of the Many-Body Theory in Inelastic Scattering from Atoms

H. S. Taylor, A. Chutjian, and L. D. Thomas

> The results of the first-order many-body theory (FOMBT) are critically compared with several recent experimental studies of e–He and e–H inelastic scattering. The differential and integral cross sections calculated in the FOMBT are compared to the following measurements: differential and integral cross sections for excitation of the 2^3S, and the $n = 3$ manifold of He at energies near 30, 40, and 50 eV; electron–photon coincidence measurements in the 1^1S–2^1P transition in He; and integral cross sections of the $1^1S \to 2^1S$ and 2^1P transitions in the H atom from threshold to 54 eV. Good agreement is found between the results of these measurements and the FOMBT. Where discrepancies exist, their pattern is noted. Where the first-order theory is lacking, several important second-order physical corrections to the first-order theory are indicated. From the above comparisons, the many-body theory results show that the important physical effects in inelastic e–atom scattering are the proper two-time model, the proper use of transition polarization, and the inclusion of the imaginary part of the second-order transition potential.

1. A Detailed Comparison of the First-Order Many-Body Theory with Recent Experimental and Theoretical Results

In this section the results of the first-order many-body theory (FOMBT) are compared with recent electron-scattering measurements in helium and with recent theoretical calculations on integral electron-scattering cross

---

H. S. Taylor • University of Southern California, University Park, Los Angeles, California 90007, U.S.A.   A. Chutjian • Jet Propulsion Laboratory, California Institute of Technology, Pasadena, California 91103, U.S.A.   L. D. Thomas • IBM Research Laboratory, Monterey and Cottle Roads, San Jose, California 95193, U.S.A.

sections from the hydrogen atom. We assume that the reader is aware of the theoretical basis of the many-body approach (Csanak *et al.*, 1973; Thomas *et al.*, 1974) and with the computational techniques used therein (Thomas *et al.*, 1973). The reader should also be aware of the large body of experimental measurements of the electron-impact excitation of the $n = 2$ states of He (Trajmar, 1973; Hall *et al.*, 1973; Crooks *et al.*, 1972; Truhlar *et al.*, 1973).

Not all of the comparisons between the FOMBT and experiment for the $n = 2$ He differential cross sections (DCS) were published by Thomas *et al.* (1974), and we present in Figure 1 a comparison between the FOMBT and recent measurements of the differential cross section for the He $1^1S \to 2^3S$ transition at incident electron energies (E_0) of 48.2 eV (Hall *et al.*, 1973) and 50 eV (Crooks *et al.*, 1972). The calculations were carried out for an E_0 of 50 eV. From Figure 1, one finds that the FOMBT is in good agreement with the experiments up to a scattering angle $\theta = 40°$. At angles greater than 40°, the experimental DCS drops to a minimum value of 1.6×10^{-21} cm²/sr at 80°, while the theory predicts a shallower minimum, with a minimum

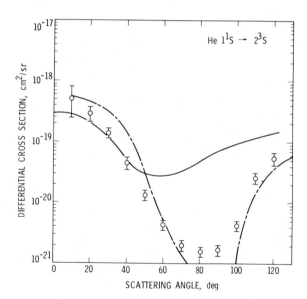

Figure 1. Experimental measurements and theoretical calculations of the differential cross section for the He $1^1S \to 2^3S$ excitation. The experimental points are from Hall *et al.* (1973) ($E_0 = 48.2$ eV); the dot-dash curve from experiments of Crooks *et al.* (1972) ($E_0 = 50$ eV); and the solid line from calculations in the first-order many-body theory ($E_0 = 50$ eV).

cross section of 1.8×10^{-20} cm^2/sr at 65°. The theory and experiment begin to approach one another again at $\theta \gtrsim 120°$.

Another larger body of experimental data from recent results of Chutjian and Thomas (1975) is presented in Figure 2. In these measurements, the electron-impact excitation of the $n = 3$ states of He was observed at $E_0 = 29.2$ and 39.7 eV. It is to be noted that the $n = 3$ states of He are 2–3 eV closer to threshold than the measurements of the $n = 2$ levels at energies near 30 and 40 eV (Trajmar, 1973). Despite this fact, very good agreement is again observed in both magnitude and shape between the measured and calculated $1^1S \to 3^1S$ DCS. As was true in the $n = 2$ case, the FOMBT fails to account for the small modulating oscillations in these DCS, and the probable reason for this will be discussed in Sections 2 and 3. The agreement between the theoretical and experimental *integral* cross sections for these two transitions is excellent. The theoretical integral cross sections average out the fine details, such as modulating oscillations. The experimental–theoretical agreement for the 3^3P DCS again, as in the 2^3P DCS, finds the FOMBT predicting the correct general shape but too large a magnitude of cross section (see Figure 2). This effect may be traced to the neglect, in the first-order theory, of coupling between the nearby 3^3P and 3^3S levels. This effect will also be discussed in Sections 2 and 3.

In outlining the $1^1S \to 3^1P$ results of Figure 2, we note that the 3^1P state has as its neighbors, only 0.014 eV away, the 3^1D and 3^3D states, so that the experimental measurements of Figure 2 include the *sum* of the DCS for the three transitions. Even with this added difficulty, excellent agreement is found between experiment and the FOMBT, which *includes* the $^{1,3}D$ states in the calculated curves. The agreement between the experimental and theoretical integral cross sections is also good, and the comparison implies that the contribution of the ^1D and ^3D states to the integral cross section is less than 6×10^{-19} cm^2. These and other points have been discussed in more detail by Chutjian and Thomas (1975).

Recently, Eminyan *et al.* (1974) have performed difficult electron–photon coincidence measurements in helium. By measuring coincidences between the He $2^1P \to 1^1S$ emitted photons and the electrons having excited the $1^1S \to 2^1P$ transition, they were able to measure the ratio of magnetic sublevel cross sections $\lambda = \sigma_0/(\sigma_0 + 2\sigma_1)$, where σ_m represents the cross section for excitation to the mth sublevel of the $2^1P_{0,\pm1}$ state. The ratio λ was measured at energies between 40 and 200 eV, and at angles between 15° and 30°. The comparison of the FOMBT to these measurements shows good agreement in shape, but the magnitude of the calculated λ appears to be lower than experiment. At the only energy (80 eV) where a comparison was initially possible the experiments show excellent agreement with the distorted wave (DW) theory of Madison and Shelton (1973).

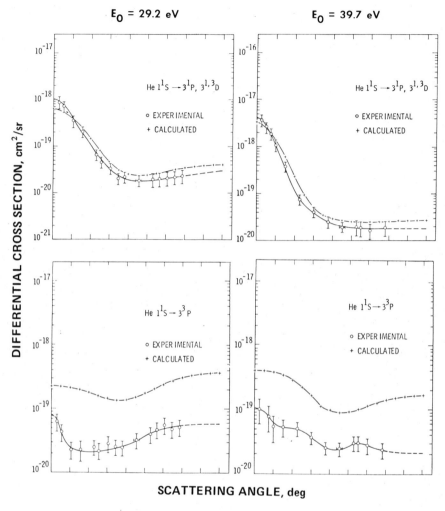

Figure 2. Experimental measurements and theoretical calculations of the differential cross sections for the excitation of the $n = 3$ states of He from Chutjian and Thomas (1975). The incident electron energies E_0 are 29.2 eV and 39.7 eV. The cross sections are calculated in the first-order many-body theory.

Several recent theoretical calculations have been carried out on H atom scattering involving the transitions $1^1S \rightarrow 2^1S$ and 2^1P. Pindzola and Kelly (1975) have calculated the integral electron-impact cross section for these excitations using the key ideas of the FOMBT. In effect, they did a DW calculation in which the outgoing electron is coupled, as in MB theory,

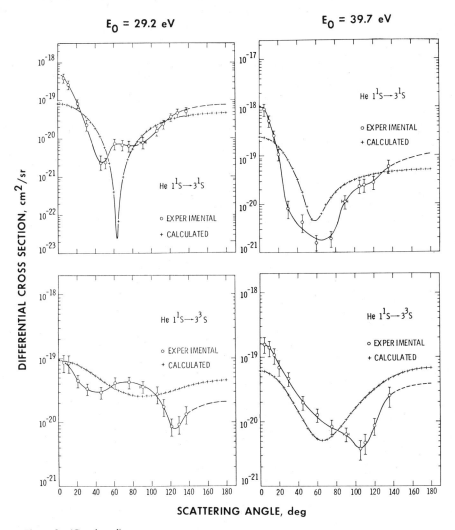

Figure 2 (Continued).

to the 1s ground state, and in another case to the 2s state. In these calculations, one finds that near threshold the first-order MB–DW theory is lower than experiment. This is to be expected, since the MB–DW theory does not take into account resonances, and the threshold region is known to be resonance-dominated. Further above threshold, from 16 to 54 eV, the MB–DW results begin to move above the experiments, and generally lie 20% higher. The agreement in shape between theory and experiment is good at energies greater than 16 eV. The DW results, in which the outgoing electron is coupled to the *final* 2s state, and the 1s–2s close-coupling results are factors of 2 to $2\frac{1}{2}$ above experiment from threshold to ~ 16 eV, and all theoretical results drop to meet experiment at 54 eV.

The recent close-coupling calculations of Callaway and Wooten (1975) on the H atom were carried out at energies between 12 and 20 eV using an eleven-state expansion which included the exact 1s, 2s, 2p, and 3d states, plus seven pseudostates (three of the s-type, three of the p-type, and one d-type). They found excellent agreement, in both magnitude and shape, with experiment.

2. General Features Gathered from the Comparisons of Section 1

He Differential Cross Section Results for the $n = 2$ and $n = 3$ States

One may see (Thomas *et al.*, 1974 and Figure 2) that for both the $n = 2$ and $n = 3$ excitations in helium the FOMBT predicts the correct shape of the DCS and also predicts the correct magnitude of the DCS on the absolute scale. The integral cross section is also generally quite good. When errors arise they show a pattern. For spin-allowed ($\Delta S = 0$) transitions, the scattering at low angles is effected by long-range forces which in turn enhance the small-angle scattering. The theory lies below experiment at these small θ. This can be seen in the comparisons in that paper for the $1^1S \to 2^1P$ and 2^1S excitations at various E_0, and also in Figure 2 for the $1^1S \to 3^1S$ and 3^1P (which includes small $3^{1,3}D$ contributions) excitations.

Moreover, when the final state couples strongly to another nearby final state, and neither state is as strongly coupled to the ground state, the calculated DCS are generally too high for He by a factor of ~ 4. This effect is seen in the transitions $1^1S \to 2^3S$ and 2^3P in the paper by Thomas *et al.* (1974) and in Figure 2 for the $1^1S \to 3^3P$ transition. In the case of the 1s–2s transition in the H atom, the MB(RPA) is a factor 1.2 too high at energies away from threshold (Pindzola and Kelly, 1975).

The FOMBT also fails to account properly for the modulating oscillations in the 2^3S and 2^3P excitations and in the $3^3P(E_0 = 39.7$ eV), 3^1S, and 3^3S excitations (Figure 2). These oscillations appear in those states which are optically forbidden to the ground state, and which are strongly coupled to other nearby excited states.

Electron–Photon Coincidence Measurements in the He $1^1S \to 2^1P$ Transition

The FOMBT, over large ranges of E_0 and θ, is in excellent agreement with experimental DCS measurements of several independent groups. The same good agreement is found between theory and the independent measurements of the integral cross sections. It is also to be noted that the related work of Madison and Shelton (1973) is generally too high in both the DCS and the integral cross section. Since the theoretical 2^1P result is the *sum* of the cross sections of the $2^1P_{0,\pm 1}$ sublevels, and the coincidence experiment gives the *ratio* of these sublevel cross sections, one would expect that if the differential and coincidence experiments are internally consistent, there should also be agreement between the FOMBT and the *ratio* measurements of Eminyan et al. (1974). Such is not the case, and this would imply that the ratio measurements are out of line with both previous experiments and with the MB theory. This comparison is especially important at 40 eV, where Trajmar (1973), Hall et al. (1973), Eminyan et al. (1974), and Thomas et al. (1974) have all compared their results. To add to the confusion, it is difficult to see how the results of Eminyan et al. (1974) agree with Madison and Shelton (1973) at 80 eV when the Madison and Shelton results for the 2^1P integral cross section are $\sim 12\%$ too high. The agreement of the MB theory with DCS over large ranges of E_0 and θ testify to the improbability of a fortuitous cancellation of errors when the sublevel cross sections are added to give the DCS. Further experimental and theoretical work will be needed to untangle this situation.

3. Physical Insights Gained Through the Use of the Many-Body Theory

The first-order many-body formula suggests a two-time model of the following nature for the scattering process (Taylor, 1974). Consider the time necessary for the target to undergo its excitation and the passage time of the incident electron through the target potential. If the former excitation time is less than the latter passage time, then one would expect distorted-wave theories in which the electron sees the *final* state of the atom to be more successful than the first-order MB theories. Since this is manifestly not the case, the FOMBT implies a model in which the passage time is extremely rapid relative to the time needed for an excitation to take place. This result seems plausible, since at incident energies of as low as 1 eV passage times are of the order of 10^{-15} sec, and experience dictates against transition times as short as this. The two-time model, we claim, is essential to any approximate theory that does not attempt to solve large numbers of coupled-channel equations. Clearly, one expects the FOMBT to fail close to threshold where the outgoing electron leaves slowly. This point is difficult to verify experimentally since experimental results will also include effects of resonances which dominate the near-threshold region. The FOMBT does not take resonances into account. It is worth noting that the two-time model is a prime physical idea. Including it, and leaving out polarization as is done in the FOMBT, gives much better results than including polarization and leaving out the two-time model.

The experimental–theoretical comparisons in Section 1 above point to two needed second-order physical corrections. The first is the inclusion of transition polarization to account for the proper long-range forces, (i.e., small-angle scattering) in those cross sections where one expects long-range forces to be dominant (for example, in the spin-allowed transitions $1^1S \to 2^1S$, 2^1P, and 3^1P in He). A comparison of Figures 5 and 10 of Thomas et al. (1974) shows the truth of this statement, and also shows how excellent agreement at small angles can be achieved by the inclusion of transition polarizabilities. This effect is considered in the second-order of MB theory.

Also, since the cross sections for the 2^3S, 2^3P, and 3^3P states of He calculated in the FOMBT, when in error, are too high, we speculate that it is the failure to include the imaginary part of the second-order transition potential that causes this situation. Inclusion of the imaginary part would remove flux from the scattering and thus lower the cross section. We believe that we would only have to include the strongly coupled $n = 2$ levels to properly lower the cross sections for He. Further, we believe that this same imaginary part which couples only the strongly coupled channels will properly account for the modulating oscillations noted in Section 1. As evidence for this statement we note that Thomas and Nesbet (1974), coupling just these $n = 2$ channels and the ground state, were able for the 2^1S and 2^3S states to both lower the cross section and obtain the modulating oscillations. It was not necessary to couple a large number of channels.

In summary, we consider that the MB theory shows that the important physical effects in inelastic scattering are: the proper two-time model; the proper use of transition polarization to account for small-angle scattering; and a second-order calculation of the imaginary part of the second-order transition potential to lower the cross section, to obtain the proper oscillations, and to account for near degeneracies and strong coupling.

A mathematical strong point of this RPA version of the MB theory is the fact that it automatically ensures that the excited target electron moves in the V_{N-1} potential while the other electrons move within V_N potential. Other theories have recognized the importance of this effect (Pindzola and Kelly, 1975; Amusia et al., 1969) but have had to include the mixed basis set in an *ad hoc* way. A mathematical weakness of the MB theory is that resonances do not appear properly in low order. This is physically clear, since the theory is based on the concept of probing the initial states of the target. Since core-excited resonances are usually a property of the excited state, they must be considered to be an *enormous* perturbation of the ground state, and therefore it is not expected that anything but single-particle shape resonances associated with the ground state will be properly described in low order by the MB theory.

The two-time model gives an intriguing and different interpretation of the ionization problem. As is usually assumed (but never explicitly pointed out), the transition time is taken as rapid relative to the passage time. Therefore, one envisions the following sequence for ionization: an incident electron in the field of the initial state, an ionization, and two electrons in Coulomb orbitals leaving simultaneously in the field of the ion. Discrepancies with experiment are blamed on the lack of a mathematical description for the obviously important three charged-particle interactions implicit in this picture. The MB theory assumes that the passage time is rapid compared to the transition time and that the physical sequence is: incident electron in the field of the initial state, *one* outgoing electron again in the field of the initial state (but with the proper outgoing momentum), followed by the emergence of a single electron, in a Coulomb state, and seeing the residual ion. Since one electron leaves after the other, even if they have similar velocities, correlations should not be important. Preliminary calculations based on this physical picture seem to confirm this point (Baluja *et al.*, 1975). This model is again expected to fail near threshold.

Acknowledgments

This work was supported by the National Science Foundation under Grant No. GP-30710X, by the President's Fund of the California Institute of Technology, and by the National Aeronautics and Space Administration under Contract No. NAS7-100 to the Jet Propulsion Laboratory.

References

Amusia, M. Y., Cherepkov, N. A., Chernysheva, L. V., and Sheftel, S. I. (1969). *Zh. Eksp. Teor. Fiz.*, **56**, 1897 [English trans. *JETP*, **29**, 1018 (1969)].
Baluja, K. L., Yarlagadda, B.S., and Taylor, H. S. (1974). Unpublished results.
Callaway, J., and Wooten, J. W. (1975). *Phys. Rev.*, **A11**, 1118.
Chutjian, A., and Thomas, L. D. (1975). *Phys. Rev.*, **A11**, 1583.

Crooks, G. B., Dubois, R. D., Golden, D. E., and Rudd, M. E. (1972). *Phys. Rev. Lett.*, **29**, 327.
Csanak, G., Taylor, H. S., and Tripathy, D. N. (1973). *J. Phys. B.*, **6**, 2040.
Eminyan, M., MacAdam, K. B., Slevin, J., and Kleinpoppen, H. (1974). *J. Phys. B.*, **7**, 1519.
Hall, R. I., Joyez, G., Mazeau, J., Reinhardt, J., and Schermann, C. (1973). *J. de Phys.*, **34**, 827.
Madison, D. H., and Shelton, W. N. (1973). *Phys. Rev.*, **A7**, 499.
Pindzola, M. S., and Kelly, H. P. (1975). *Phys. Rev.*, **A11**, 221.
Taylor, H. S. (1974). To be published.
Thomas, L. D., and Nesbet, R. K. (1974). *Abstracts of Fourth International Conference on Atomic Physics, Heidelberg*, July 22–26, 1974.
Thomas, L. D., Yarlagadda, B. S., Csanak, G., and Taylor, H. S. (1973). *Comp. Phys. Commun.*, **6**, 316.
Thomas, L. D., Csanak, G., Taylor, H. S., and Yarlagadda, B. S. (1974). *J. Phys. B.*, **7**, 1719.
Trajmar, S. (1973). *Phys. Rev.*, **A8**, 191.
Truhlar, D. G., Trajmar, S., Williams, W., Ormonde, S., and Torres, B. (1973). *Phys. Rev.*, **A8**, 2475.

37

Direct Observation of Exchange Scattering by Spin Flip of Polarized Electrons in Excitation of Mercury

G. F. HANNE AND J. KESSLER

> The influence of electron exchange has been directly observed in the excitation of the 6^3P state of mercury with polarized electrons. The electrons scattered in the forward direction show a strong depolarization which is significant between threshold and 8 eV and even leads to a reversal of the polarization vector in a small energy range. The influence of electron exchange decreases rapidly at incident energies above 8 eV. Formulas for the depolarization are given for the case where the $^3P_0$, $^3P_1$, and $^3P_2$ states can be resolved.

1. Introduction

In an electron–atom scattering experiment with unpolarized particles it is impossible to decide whether the incoming electron has been scattered directly or has suffered an exchange collision. Since the electrons are not distinguishable here, the direct and exchange cross sections $|f|^2$ and $|g|^2$ cannot be measured separately. This is different, however, if the colliding electrons are distinguishable, i.e., if their spins are antiparallel and each electron retains its spin direction during the scattering process. In this case, by determining its spin direction, the observed electron can definitely be stated to have suffered a direct *or* an exchange collision.

This is one of the reasons why scattering experiments with polarized particles are so appealing. Such experiments are feasible today, as was

---

G. F. HANNE AND J. KESSLER • Physikalisches Institut, Universität Münster, 44 Münster, Germany.

shown by the pioneer work of Bederson *et al.* (Bederson, 1973) in low-energy electron–atom scattering. The authors used spin-state-selected alkali *atoms* as the target and observed the angular distribution and the change of spin state of the recoil atoms arising from collisions with unpolarized electrons. From these measurements they obtained the exchange cross sections. A group at JILA (Hils *et al.*, 1972), which measured cross sections for direct elastic scattering at a few angles also used spin-state-selected alkali atoms and observed the spin directions of the scattered electrons. Such experiments, which until now have only been made with alkalies can, of course, only be performed with atoms which have unsaturated spins.

Another possibility for observing direct and exchange scattering is to scatter polarized *electrons* on atoms. But—as was stated by Bederson (1969)—"there has not yet been any scattering experiment in low-energy atomic physics employing polarized electrons."

In the first part of this paper an experiment on electron exchange in the excitation of mercury atoms by slow polarized electrons is described. Inelastic exchange scattering leading to the 6^3P state of this "two-electron atom" has directly been observed by measuring the depolarization of the scattered electrons. In the second part, the depolarization is calculated for the case in which the 6^3P_0, 6^3P_1, and 6^3P_2 states can be separated in the experiment.

2. Experimental Study of the 6^3P Excitation

Excitation of a $^3S_1$ State

In the present paper we will neglect all interactions in the scattering process which could cause the spin of the incoming or scattered electron to flip, except for the exchange interaction. Thus we assume that each electron retains its spin direction during the scattering event.

What happens to the spins if, for example, an electron with spin $+\frac{1}{2}$ excites a two-electron atom from the singlet ground state into a triplet excited state?

This process is described in Figure 1. Let the excited state be a $^3S_1$ state, so that angular momentum quantum numbers do not play a role. With the assumptions made, the excitation is only possible due to exchange. In the upper part of the figure, there is an exchange process of the incoming electron with the atomic electron of the same spin direction. This leads to an atomic state with $S = 1$ and $S_z = 0$. The outgoing electron, in this case, has the same spin direction as the incoming electron. This is different in the lower part of the figure. Here the incoming electron exchanges with the electron of opposite spin direction. The resulting atomic state now has

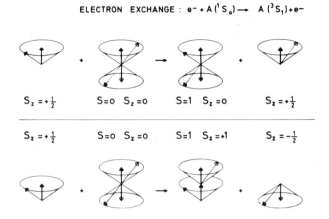

Figure 1. Change of spin state for the excitation of a ^3S state of a "two-electron" atom due to electron exchange.

the quantum numbers $S = 1$, $S_z = +1$, and the spin of the scattered electron has flipped. The probability for this spin-flip process is twice as large as for the non-spin-flip process, as can be shown by a simple calculation, i.e., the ratio of the final to the initial polarization is $P'/P = -1/3$.

It is obvious that for an excitation into a singlet state no spin flip can occur due to exchange scattering.

Spin-Orbit Interaction in the Atom

Now we consider an excitation into a ^3P state. If there is no spin-orbit interaction in the atom we have the three substates with $M_L = \pm 1$ and $M_L = 0$. For each substate—and thus for the whole ^3P state—the ratio of the final to the initial polarization is again $P'/P = -1/3$.

If spin-orbit interaction in the atom is not negligible, there is a fine-structure splitting of the ^3P state into $^3P_0$, $^3P_1$, and $^3P_2$. Then P'/P is different for each fine-structure level and is determined by the scattering amplitudes for the excitation of an atomic state with $M_L = \pm 1$ (g_1 and g_{-1}) and $M_L = 0$ (g_0). This will be shown in Section 3.

For heavy atoms, such as mercury, a further consequence is that Russell–Saunders coupling is no longer valid and a $^1P_1$ contribution is admixed to the $^3P_1$ state:

$$\Psi(^3P_1) = \alpha\Psi^0(^3P_1) + \beta\Psi^0(^1P_1) \tag{1}$$

where $\Psi^0(^3P_1)$ and $\Psi^0(^1P_1)$ denote the pure Russell–Saunders states and the experimental values of the coupling coefficients for mercury are (Lurio, 1965)

$$\alpha = 0.985, \quad \beta = -0.171 \tag{2}$$

The $^3P_0$ and $^3P_2$ states are pure Russell–Saunders states.

Now the cross section consists of two parts coming from the pure triplet states and from the singlet admixture. The electrons resulting from the singlet admixture do not suffer a spin flip. We expected from McConnell and Moiseiwitsch (1968) that at higher incident energies the excitation of the 6^3P_1 state of mercury would be dominated by this non-spin-flip contribution to the $^3P_1$ state so that it could be possible to observe spin flips resulting from the pure exchange transitions only in a certain energy region up to some eV above threshold.

In order to show the influence of the singlet admixture to the scattered electrons it is not necessary to resolve the $^3P_0$, $^3P_1$, and $^3P_2$ states (the peak maximum of the unresolved ^3P state is at about 4.9 eV energy loss).

Experimental Results

Figure 2 is a schematic diagram of the apparatus. Transversely polarized electrons resulting from elastic scattering by a mercury vapor beam (Jost and Kessler, 1966) ($E = 80 \,\text{eV}$, $\theta = 80°$, $P \approx 0.22$, current of polarized electrons $\approx 10^{-10}$ A) pass through a filter lens which removes inelastically scattered electrons. They are then decelerated to 5 to 15 eV and focused on a second mercury vapor beam (target density 10^{-2}–10^{-3} Torr). From the

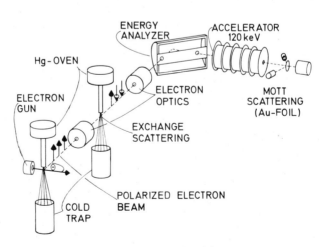

Figure 2. Schematic diagram of the apparatus.

electrons scattered in the forward direction, those which have excited a certain atomic state are selected by a cylindrical mirror analyzer. Their polarization P' is measured by a Mott detector after acceleration to 120 keV. Since only electrons scattered in the forward direction were measured ($\Delta\theta \approx 8°$), the spin-orbit interaction of the scattered electrons could be neglected (Eitel and Kessler, 1971). The main difficulty in such triple scattering experiments is of course the intensity problem. When measuring the curve in Figure 3 we had a counting rate in the Mott detector of about 60–200 pulses per minute if the apparatus was carefully adjusted.

Figure 3 shows the experimental results for the excitation of the 6^3P state. The error bars include statistical errors and reproducibility of the measured values P' and P, as well as the reproducibility of the energy. At incident energies below 8 eV, strong depolarization or even reversal of the direction of polarization has been observed. This agrees with the theoretical prediction that exchange scattering dominates at these energies. Above 8 eV, however, the number of spin-flip processes decreases more rapidly than expected from McConnell and Moiseiwitsch (1968). Above 9 eV incident energy, significant depolarization could no longer be observed.

It should be pointed out that the calculation (McConnell and Moiseiwitsch, 1968) only refers to the *total* excitation cross section. Theoretical predictions of exchange cross sections at small angles are

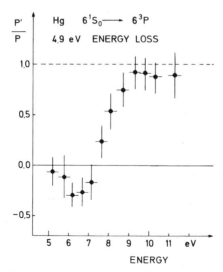

Figure 3. Measured values of the depolarization ratio P'/P for the excitation of the 6^3P state vs. incident energy.

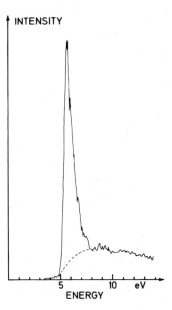

Figure 4. Excitation function for the excitation of the 6^3P state of mercury in forward direction.

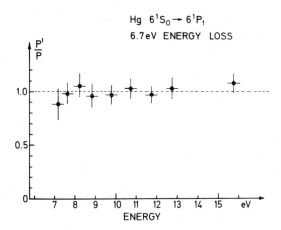

Figure 5. Measured values of the depolarization ratio P'/P for the excitation of the 6^1P state vs. incident energy.

controversial and depend strongly on the approximation made (Steelhammer and Lipsky, 1970; Bonham, 1972). The present measurements show that the influence of exchange scattering is very strong at small angles. This can be seen also from Figure 4, which shows the excitation function of the 6^3P state in the forward direction. There is a high maximum at energies near threshold which is typical for exchange collisions.

The excitation of the ^3P state of mercury is also possible due to a ^4P resonance: $e + Hg \to Hg^- \to e + Hg^*$ (Fano and Cooper, 1965). In this case the total spin is not conserved. This resonance scattering leads also to a depolarization which is different from $P'/P = -1/3$ and may influence the two experimental values near threshold shown in Figure 3.

Measurement of the polarization for the excitation of the 6^1P_1 state did not yield any depolarization as is shown in Figure 5. In the future measurements we hope to resolve the $^3P_0$, $^3P_1$, and $^3P_2$ states. This would yield further details about the scattering amplitudes g_0 and g_1.

3. Theoretical Description of the Excitation of the $^3P_0$, $^3P_1$, and $^3P_2$ States with Polarized Electrons

Calculation of the Density Matrix

We choose the quantization axis (z axis) to be parallel to the momentum of the incident electrons, the scattering plane being the zx plane. If the incident electron is denoted by 1 and the atomic electrons by 2 and 3, we obtain for the wave function of scattering problem asymptotically

$$\Psi \xrightarrow[r_1 \to \infty]{} e^{ikz_1}\chi(1)u_0(r_2, r_3; 2, 3) + \frac{e^{ik_n r_1}}{r_1}$$
$$\times \left[\left(\sum_v a_v(\theta_1, \phi_1)u_v(\mathbf{r}_2, \mathbf{r}_3; 2, 3) \right) \alpha_1 \right.$$
$$\left. + \left(\sum_v b_v(\theta_1, \phi_1)u_v(\mathbf{r}_2, \mathbf{r}_3; 2, 3) \right) \beta_1 \right] \quad (3)$$

The summation must be carried out over all atomic substates which have the same excitation energy. $\chi(1)$ denotes an arbitrary spin state of the incident wave and α, β are the eigenstates of the Pauli matrix σ_z. The u_0 and u_v are the antisymmetrized wave functions of the ground state and of the excited states of the two-electron atom, respectively, which include the spins of the atomic electrons. The $a_v(\theta, \phi)$ and $b_v(\theta, \phi)$ are the scattering amplitudes.

For their calculation one has to use a wave function Ψ which is antisymmetrized in all three electrons. From (3) we obtain for the density matrix

$$\rho = \begin{pmatrix} |\Sigma a_\nu u_\nu|^2 & (\Sigma a_\nu u_\nu)(\Sigma b_\nu^* u_\nu^*) \\ (\Sigma a_\nu^* u_\nu^*)(\Sigma b_\nu u_\nu) & |\Sigma b_\nu u_\nu|^2 \end{pmatrix} \quad (4)$$

As we only observe the scattered electrons, we have to integrate over the space and the spin coordinates of the atomic electrons. Thus the density matrix (4) reduces to

$$\rho = \begin{pmatrix} \Sigma |a_\nu|^2 & \Sigma a_\nu b_\nu^* \\ \Sigma a_\nu^* b_\nu & \Sigma |b_\nu|^2 \end{pmatrix} \quad (5)$$

The quantities of interest can now be calculated from

$$\frac{d\sigma}{d\Omega} = \operatorname{tr} \rho = \Sigma(|a_\nu|^2 + |b_\nu|^2); \qquad \mathbf{P}' = \frac{\operatorname{tr} \rho \boldsymbol{\sigma}}{\operatorname{tr} \rho} \quad (6)$$

where σ_i ($i = x, y, z$) are the Pauli matrices.

We want to point out that the scattering amplitudes $a_\nu(\theta, \phi)$ and $b_\nu(\theta, \phi)$ depend on the spin state of the incident wave $\chi(1)$. In an experiment there are two cases of interest, namely longitudinal polarization ($P = P_z$, $P_x = P_y = 0$) and transverse polarization ($P = P_y$, $P_x = P_z = 0$) of the incident beam.

Calculation of the $^3P_0$, $^3P_1$, and $^3P_2$ States

We have to distinguish between two limiting cases: (i) if the spin–orbit relaxation time of the atom is much larger than the excitation time, the fine-structure splitting is too small to be resolved in the electron scattering experiment, and (ii) if the energy resolution is good enough to separate the fine-structure levels, then according to the uncertainty principle we can no longer assume the excitation time to be short compared with the spin–orbit relaxation time.

These two limiting cases are schematically shown in Figure 6. In case (i) we can break the problem in two parts (Bederson, 1969). First the substates of ^3P with $M_L = \pm 1$ and $M_L = 0$ are excited. Thus the u_ν in (3) depend on the quantum numbers M_L and M_S. The excitation of the states

$$u_\nu = u_{M_S M_L} = |S, L; M_S, M_L\rangle$$

is described by scattering amplitudes $a_{M_S M_L}$ which are proportional to the so-called exchange amplitudes g_1, g_{-1}, or g_0 corresponding to atomic states with $M_L = 1$, $M_L = -1$, and $M_L = 0$, respectively. After the excitation the ^3P states relax into the $^3P_0$, $^3P_1$, and $^3P_2$ states.

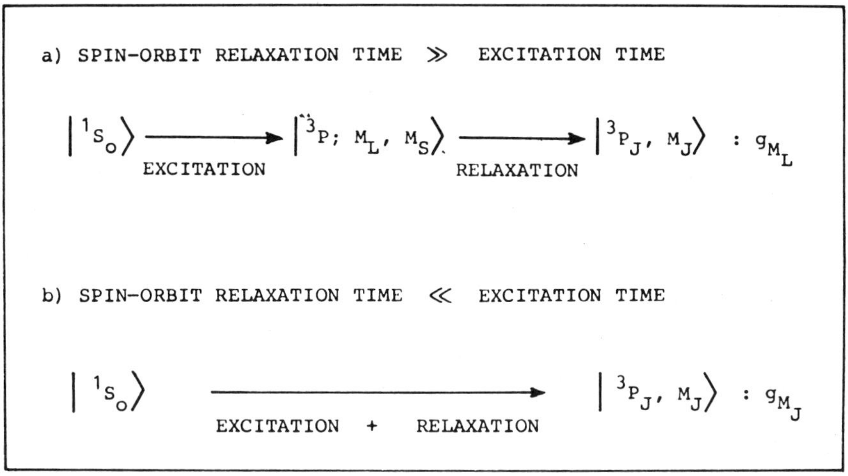

a) SPIN-ORBIT RELAXATION TIME $\gg$ EXCITATION TIME

$$|{}^1S_0\rangle \xrightarrow{\text{EXCITATION}} |{}^3P;\ M_L,\ M_S\rangle \xrightarrow{\text{RELAXATION}} |{}^3P_J,\ M_J\rangle\ :\ g_{M_L}$$

b) SPIN-ORBIT RELAXATION TIME $\ll$ EXCITATION TIME

$$|{}^1S_0\rangle \xrightarrow{\text{EXCITATION + RELAXATION}} |{}^3P_J,\ M_J\rangle\ :\ g_{M_J}$$

Figure 6. Schematic diagram of the time dependence for two limiting cases in the process $e + \text{Hg}({}^1S_0) \to e + \text{Hg}({}^3P_{0,1,2})$.

In this case the density matrix can be calculated with the relations

$$\sum a_\nu u_\nu = \sum_{M_J} \sum_{M_L, M_S} a_{M_S M_L} C^{SLJ}_{M_S M_L M_J} |S, L; M_S, M_L\rangle \tag{7}$$

and a corresponding relation for $\sum b_\nu u_\nu$, where $C^{SLJ}_{M_S M_L M_J}$ are the Clebsch–Gordan coefficients which mix the $|S, L; M_S, M_L\rangle$ states to $|S, L; J, M_J\rangle$. Using relations (6) and (7) we found the depolarization ratios for transversely polarized electrons $[P = P_y$, i.e., $\chi(1) = (1/\sqrt{2})(\alpha_1 + i\beta_1)]$ to be

$$
{}^3P_0: \quad \frac{P'_y}{P} = -\frac{\frac{1}{3}|g_0|^2}{\frac{2}{3}|g_1|^2 + \frac{1}{3}|g_0|^2} \tag{8a}
$$

$$
{}^3P_1: \quad \frac{P'_y}{P} = -\frac{|g_1|^2}{2|g_1|^2 + |g_0|^2} \tag{8b}
$$

$$
{}^3P_2: \quad \frac{P'_y}{P} = -\frac{|g_1|^2 + \frac{2}{3}|g_0|^2}{\frac{10}{3}|g_1|^2 + \frac{5}{3}|g_0|^2} \tag{8c}
$$

If the polarization is parallel to the quantization axis, the rotation of the coordinates transforms $|g_0|^2$ and $|g_1|^2$ into $2|g_1|^2 - |g_0|^2$ and $|g_0|^2$, respectively.

This case (i) is not well established for mercury, where the fine-structure splitting of the 6^3P state is larger than 0.7 eV. Thus one has to consider the above-mentioned case (ii) (see Figure 6) and then obtains, instead of (7),

$$\Sigma a_\nu u_\nu = \sum_{M_J} a_{M_J} \sum_{M_L, M_S} C^{SLJ}_{M_S M_L M_J} |S, L; M_S, M_L\rangle \qquad (9)$$

and a corresponding relation for $\Sigma b_\nu u_\nu$. It can be seen from (9) that the amplitude a_{M_J} can be expressed in terms of g_1, g_{-1}, and g_0. For mercury one has also to include the 6^1P_1 admixture to the 6^3P_1 state. Now there is interference between the amplitudes g_1 and g_0 (from parity arguments it follows that $g_{-1} = -g_1$). We found that the differential cross section for each fine-structure level here depends on the initial transverse polarization (scattering asymmetry) and that there is polarization of the scattered electrons even when the electrons are initially unpolarized. We want to point out that this is due to spin–orbit interaction for the atomic electrons (for electrons 2 and 3). A more sophisticated calculation, of course, has to take into account the spin–orbit interaction for the continuum electron too.

The purpose of this section was to take a very first step toward obtaining theoretical results. We hope that we may have stimulated some theoreticians to undertake a further analysis of the problem, which seems particularly necessary for the case (ii) discussed above.

A more detailed presentation is in press (*J. Phys. B, Atom. Molec. Phys.*). See also G. F. Hanne and J. Kessler, *Phys. Rev. Lett.*, **33**, 341 (1974).

References

Bederson, B. (1969). *Comm. Atom. Molec. Phys.*, **1**, 65.
Bederson, B. (1973). *Proc. III Int. Conf. Atomic Physics*, **3**, 401.
Bonham, R. A. (1972). *J. Chem. Phys.*, **57**, 1604.
Eitel, W., and Kessler, J. (1971). *Z. Physik*, **241**, 355.
Fano, U., and Cooper, J. W. (1965). *Phys. Rev.*, A **138**, 400.
Hils, D., McCusker, M. V., Kleinpoppen, H., and Smith, S. J. (1972). *Phys. Rev. Lett.*, **29**, 398.
Jost, K., and Kessler, J. (1966). *Z. Physik*, **195**, 1.
Lurio, A. (1965). *Phys. Rev.* A**140**, 1505.
McConnell, J. C., and Moiseiwitsch, B. L. (1968). *J. Phys. B*, **1**, 406.
Steelhammer, J. C., and Lipsky, S. (1970). *J. Chem. Phys.*, **53**, 1445.

38
Electron–Photon Coincidence Technique for Electron Impact on Atoms

M. Eminyan, H. Kleinpoppen, J. Slevin,
and M. C. Standage

> A brief introduction is given to the general theory of the electron–photon coincidence technique, and the specific application to ^1P and ^3P excitations in helium is described. The paper emphasizes the relationship between the complex excitation amplitudes which characterize the collision process and the alignment and orientation of the excited atoms. The data from the first electron–photon angular correlation measurements are presented. These data yield values for the ratio of differential cross sections for exciting the degenerate sublevels and the relative phase of the corresponding amplitudes, or, equivalently, the alignment and orientation parameters. The results are obtained in dimensionless form and are free from absolute calibration or normalization difficulties. They are compared with various theoretical approximations. The application of the coincidence technique to a measurement of threshold polarization is described and results for 3^1P excitations are discussed.

1. Introduction

Within the last few years there has been a considerable increase in the level of sophistication of experimental techniques in atomic collision physics. For instance, Rubin et al. (1969) used a polarized atom beam in a recoil experiment to determine the ratio of spin-flip to total differential cross sections for electron–alkali inelastic scattering. Hils et al. (1972)

M. Eminyan • U.E.R. de Physique, Université de Paris VII, 2, Place Jussieu, 75005 Paris, France.   J. Slevin • Department of Physics, University of Stirling, Stirling, FK9 4LA, Scotland.   H. Kleinpoppen and M. C. Standage • Institute of Atomic Physics, University of Stirling, Stirling FK9 4LA, Scotland.   M. Eminyan's present address is: Ecole Polytechnique, Laboratoire de Physique de la Matière Condensée, Plateau de Palaiseau, 91120 Palaiseau, France.

measured the polarization of electrons scattered from polarized alkali atoms, and Hanne and Kessler (1974) have recently reported the first use of polarized electrons in a scattering experiment (electron scattering on mercury atoms). Hertel and Stoll (1974) have made measurements of superelastic collisions between low-energy electrons and laser-excited sodium atoms. This paper describes a new experimental method for studying inelastic electron–atom collisions in which electrons and photons are detected in delayed coincidence. All of these experiments have one thing in common—they explore the elementary collision process in more detail than do the usual measurements of differential and total cross sections. Although all still fall short of the perfect scattering experiment described by Bederson (1969), each method nevertheless reduces the number of fundamental collision parameters which must be averaged over before a comparison with theory can be made. The information obtained by these experiments is then expected to provide very sensitive tests of the various collision theories.

However, before describing the technique of electron–photon coincidences in detail, it is instructive first to consider briefly the traditional kinds of experiment which have been used hitherto in the study of excitation processes. The experimental techniques fall into two main categories.

The first concerns experiments where observations are made on either the scattered electrons or the recoiling atoms. An electron beam of given energy is passed through an atom beam or, in some cases, a gas-filled cell. Using an appropriate energy analyzer, measurements are made on the angular distribution of electrons scattered with an energy loss corresponding to the excitation energy of the state being studied. In a recoil experiment, certain of the kinematic properties of the recoiling atoms are used to distinguish between elastic and inelastic events. Both kinds of experiment yield values of the differential cross section. The determination of an absolute value for this quantity has posed, perhaps, the greatest difficulty for these experiments.

The second technique involves measurements on the electric dipole radiation, which results from the decay of the excited state. These experiments are of two types: (i) those which measure the intensity of the emitted light and (ii) those which measure its polarization. In the former, the intensity of light emitted at a particular angle is measured as a function of incident electron energy. When the results are corrected for cascade population of the excited state and polarization of the emitted radiation, and a suitable normalization procedure is devised, values for the total excitation cross section are obtained as a function of the incident energy. Measurements of the polarization of the emitted light, usually observed at 90° to the electron beam axis, give information on the relative population of the magnetic sublevels excited by electron scattering at all angles.

In summary, all of the above experimental techniques yield cross sections, differential and total. The differential cross section is a measurement of the probability of scattering in different directions, and does not provide explicit information on the dynamics of the collision process. The total cross section is a sum of the scattering into all angles and represents an average of the complete interaction for electrons at given energy. In each case, the experimental results involve averages over fundamental collision parameters. For example, measurements of differential cross sections do not distinguish between excitation to the different degenerate sublevels of the excited atom. The experiments involving measurements of line polarization do, however, separate excitations to the magnetic sublevels, but since the analysis of the radiation takes place without regard to the electrons, these measurements of necessity involve an average over all electron scattering angles. Again, important detail is lost in the averaging process.

Thus, it is clear that if an experiment can be devised which permits an analysis of only that radiation emitted from atoms which are excited to a given state by electrons scattered in a *particular* direction, new information on the "details" of the collision process can be obtained. The detection of photons emitted in a given direction in delayed coincidence with electrons scattered into a small solid angle provides a technique for precisely this kind of measurement. At intermediate electron energies (tens of eV), the collision time is short compared to any characteristic time of the excited state, and the excitation process is *coherent*. An analysis of the polarization of the coincident light or alternatively its angular distribution (angular correlation measurements) allows one to reconstruct the dipole composition of the atomic source and hence determine the complex excitation amplitudes which characterize the excitation process. It will be shown, furthermore, in the next section, how a study of the excitation into different magnetic sublevels is directly related to the way in which angular momentum is transferred to the atom during the collision process, thus providing *explicitly* details of the collision dynamics.

Coincidence techniques have been used in experimental physics for over 20 years, in the earlier years only in nuclear physics. They have been used extensively for the last 10 years in heavy-particle collisions, where, for example, they yield information on the charged state of both scattered and recoiling atoms. More recently, these techniques have permitted the high resolution available in optical spectroscopy to separate excitations to levels spaced too closely to be resolved by the usual energy loss analysis (see, for example, Barat, 1973; Jaecks, 1973). The number of applications of the coincidence technique in atomic physics has now greatly increased and the necessary electronic hardware is readily available commercially.

As discussed above, we are concerned in this paper with showing how the coincidence technique has been extended to include an analysis of the emitted radiation when the scattered electron and photon are detected in delayed coincidence. We will study the nature and theory of these kinds of measurements for excitation of P states of helium. The particular experiment which we have done at Stirling University will also be described. We review the results reported in detail in Eminyan et al. (1974, 1975) for the excitation of 2^1P and 3^1P states of helium in the energy range 40–200 eV. The experimental data are compared with the results of various theoretical calculations. In the case of 2^1P excitations, the 58.4 nm line in the ultraviolet was monitored. For 3^1P excitation, observations were made independently on both the 501.6 nm in the visible and the 53.7 nm line in the ultraviolet. We also describe the experiment of King et al. (1972), in which the coincidence technique was used to measure threshold polarization of impact line radiation.

2. Theory of Coincidence Experiment

General Theory

Consider the following process involving three sets of energy levels of an atomic system. The atom, initially in the ground state, is excited to a set of degenerate, or nearly degenerate, upper states by electron impact. The atomic system in turn decays from the upper levels to a set of closely spaced lower levels. In a collision at intermediate energies, the excited states are populated in a time of the order of 10^{-15} sec, which is much shorter than any characteristic time of these states [e.g., life time or reciprocals of transition frequencies of spin–orbit and hyperfine structure interactions as pointed out by Macek and Jaecks (1971)]. The important feature of this excitation is that the upper states are excited *coherently* even if they are completely resolved and can interfere in the decay. At time $t = 0$, the collision occurs producing a set of excited states whose angular momenta are given by SM_S, LM_L, where S and L are, respectively, the total spin and orbital momenta and M_S, M_L their projection along the incoming electron beam axis z. The wave function of the excited atom is then

$$|\psi(t=0)\rangle = \sum_{M_S M_L} a(SLM_S M_L)|SLM_S M_L\rangle \tag{1}$$

where $|SLM_S M_L\rangle$ is a state vector describing a particular atomic state, and $a(SLM_S M_L)$ is the probability amplitude for exciting this atomic state from the ground state. $a(SLM_S M_L)$ is a function of the incident electron energy and scattering angle (θ_e, ϕ_e). After the collision the upper states evolve with time under the influence of fine and hyperfine interactions (when present) and of the radiation field. The time evoluation operator can be obtained by solving the Schrödinger equation. The calculation of the probability of photon emission for a given transition proceeds from the quantum theory of radiation and has been worked out by Percival and Seaton (1958). Thus they obtain expressions relating the partial cross sections to the polarization of the atomic line radiation. Their results, however, are averaged over all directions of the scattered electrons and integrated over time. More recently Wykes (1972) has extended the Percival and Seaton theory and expressed the probability of observing polarized photons as a function of direction of emission for a given electron scattering angle. This theory is given in a form suitable for coincidence experiments. In this paper we follow the alternative but equivalent way to calculate the probability of light emission taken by Macek and Jaecks (1971). Their approach is similar to that used by Franken (1961) to derive the Breit formula under condition of short light pulse excitation. At time t after the collision, the atom "relaxes" under the influence of the internal and radiation fields into a state described by the following wave function (it is assumed that there are no external fields present):

$$|\psi(t)\rangle = \sum_{JFM_F} a(JFM_F)|JIFM_F\rangle e^{-(\gamma/2 + iE_{JF}/\hbar)t} \qquad (2)$$

where I is the nuclear spin, J the electronic angular momentum, F the total angular momentum, and E_{JF} the energy of a particular atomic state. The factor $\exp(-\tfrac{1}{2}\gamma t)$ expresses the radiation damping of the upper levels of mean life time $1/\gamma$.

Assuming that the atomic levels are adequately described by LS coupling, the amplitudes can be written in both representations by the usual coupling rules in the following way:

$$a(JFM_F) = \sum_{M'_L M'_S M_I M_J} a(LM'_L SM'_S IM_I)$$
$$\times (LM'_L SM'_S | LSJM_J)(JM_J IM_I | JIFM_F) \qquad (3)$$

In the electric dipole approximation, the probability of observing a photon with momentum $\mathbf{k}$ and polarization $\hat{\varepsilon}$ in a time interval Δt after the collision, while the electrons are scattered in a direction (θ_e, ϕ_e), is proportional to the square of the electric dipole matrix elements. Thus,

$$w(E, \theta_e, \phi_e; \mathbf{k}, \hat{\varepsilon}, t) = C \sum_\mu \int_0^{\Delta t} |\langle \psi_\mu | \hat{\varepsilon} \cdot \mathbf{D} | \psi(t) \rangle|^2 \, dt \qquad (4)$$

where $|\psi_\mu\rangle$ is the wave function of the states reached in the decay, $\mathbf{D}$ is the electric dipole moment operator, and C is an overall normalization constant.

Expanding Eq. (4), we obtain

$$w(E, \theta_e, \phi_e; \mathbf{k}, \hat{\varepsilon}, t) = C \sum_{J_0 F_0 M_{F_0}} \left| \sum_{JFM_F} \langle J_0 F_0 M_{F_0} | \hat{\varepsilon} \cdot \mathbf{D} | JFM_F \rangle \right.$$

$$\left. \times a(JFM_F) \right|^2 \int_0^{\Delta t} e^{-(\gamma + i(E_{JF} - E_{J'F'})/\hbar)t} \, dt \quad (5)$$

This expression contains three parts:
  (a) The amplitudes $a(JFM_F)$ describing the scattering process whose determination is the purpose of the experiment.
  (b) The dipole matrix elements describing the decay. They depend on the direction of emission, on the two independent polarization states of the photons, and on the line intensity, Their angular dependence can be made explicit and in a form suitable for the particular measurement to be done: angular correlation, linear or circular polarization.
  (c) A time-dependent factor which contains an exponential damping factor and an oscillatory term describing the interference in the decay of the different excited levels of the fine and hyperfine structure multiplet. There is an extensive discussion of an expression similar to Eq. (5) in the papers of Macek and Jaecks (1971) and Wykes (1972).

Excitation of the n^1P States of Helium

For excitation of the n^1P states of He⁴ from the ground 1^1S state, both fine and hyperfine interactions are absent, and the initial state [Eq. (1)] is represented by

$$|\psi(t=0)\rangle = a(1)|11\rangle + a(0)|10\rangle + a(-1)|1-1\rangle \quad (6)$$

If we restrict our discussion to $^1P \to {}^1S$ transitions, Eq. (5) becomes

$$w(E, \theta_e, \phi_e; \mathbf{k}, \hat{\varepsilon}, t) = C \left| \sum_{M_L} \langle 0 | \hat{\varepsilon} \cdot \mathbf{D} | 1 M_L \rangle a(M_L) \right|^2 \int_0^{\Delta t} e^{-\gamma t} \, dt \quad (7)$$

The polarization unit tensor $\hat{\varepsilon}$ can be expressed in terms of two linearly independent vectors $\hat{\varepsilon}^{(\alpha)}$, $\alpha = 1, 2$, such that

$$\hat{\varepsilon} = \hat{\varepsilon}^{(1)} \cos \beta + \hat{\varepsilon}^{(2)} \sin \beta \quad (8)$$

with $\beta = (\hat{\varepsilon}, \hat{\varepsilon}^{(1)})$. $\hat{\varepsilon}^{(1)}$ is chosen in a plane defined by the $\hat{z}$ axis and $\mathbf{k}$. $\hat{\varepsilon}^{(2)}$ is such that the vectors $\hat{\varepsilon}^{(1)}$, $\hat{\varepsilon}^{(2)}$, and $\mathbf{k}$ form a right-handed frame (Figure 1). The $\hat{\varepsilon}^{(\alpha)}$ may be expressed in terms of the spherical unit vectors,

$$\hat{\varepsilon}^{(\alpha)} = \sum_q (-1)^q \varepsilon^{(\alpha)}_{-q} \hat{e}_q \quad (9)$$

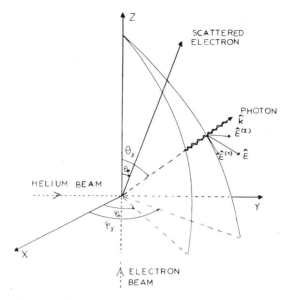

Figure 1. The general coordinate system. $\hat{\varepsilon}$ is the polarization vector of the light emitted at $(\theta_\gamma, \phi_\gamma)$.

with $q = 0, \pm 1$. The spherical components of the polarization vectors are represented in Table 1, where $(\theta_\gamma, \phi_\gamma)$ is the direction of photon emission. Substituting Eq. (9) into (7), we obtain

$$w(E, \theta_e, \phi_e; \mathbf{k}, \hat{\varepsilon}^\alpha, t) = C \sum_{MM'qq'} (-1)^{q+q'} a(M)a^*(M') \varepsilon^{(\alpha)}_{-q} \varepsilon^{(\alpha)*}_{-q'}$$

$$\times \langle 0|D_q|1M\rangle\langle 1M'|D_{q'}|0\rangle \int_0^{\Delta t} e^{-\gamma t}\, dt \quad (10)$$

with D_q being the qth spherical component of the dipole moment operator. The matrix elements can be factored out by use of the Wigner–Eckart theorem

Table 1

q	-1	0	1
$\varepsilon^{(1)}_q$	$\dfrac{1}{\sqrt{2}} \cos\theta_\gamma\, e^{-i\phi_\gamma}$	$-\sin\theta_\gamma$	$-\dfrac{1}{\sqrt{2}} \cos\theta_\gamma\, e^{i\phi_\gamma}$
$\varepsilon^{(2)}_q$	$-\dfrac{i}{\sqrt{2}} e^{-i\phi_\gamma}$	0	$-\dfrac{i}{\sqrt{2}} e^{i\phi_\gamma}$

as follows:

$$w(E, \theta_e, \phi_e; \mathbf{k}, \hat{\varepsilon}^{(\alpha)}, t) = C \frac{|\langle 0\|D\|1\rangle|^2}{3} \sum_{MM'} a(M) a^*(M') \varepsilon_M^{(\alpha)} \varepsilon_{M'}^{(\alpha)*} \int_0^{\Delta t} e^{-\gamma t} \, dt \quad (11)$$

A coincidence experiment to determine the excitation amplitudes can be performed in two almost equivalent ways. Measurements can be made of the angular correlation between the scattered electrons and emitted photons or, alternatively, the polarization of the radiation can be determined. Although only angular correlation measurements are presented in this paper, experiments are in progress in Stirling and elsewhere to measure the polarization of the coincident light. It is instructive, therefore, to develop the formulation for both kinds of experiment. Before applying Eq. (11) to each of the experimental situations, it is useful to consider the amplitudes in some detail.

The excitation amplitudes are functions of incident electron energy and electron scattering angle. The dependence on ϕ_e can be factored out as follows:

$$a(M; E, \theta_e, \phi_e) = a_M(E, \theta_e) e^{-iM\phi_e} \quad (12)$$

The mirror symmetry of the scattering process in the plane of scattering implies that $a_{-1} = -a_1$. The wave function $|\psi\rangle$ of (6) can be normalized so that the amplitudes are related to differential cross sections as follows:

$$|a_1|^2 = \sigma_1$$
$$|a_0|^2 = \sigma_0 \quad (13)$$
$$2|a_1|^2 + |a_0|^2 = \sigma$$

Here σ is the differential cross section for exciting the $n^1 P$ state and σ_M is the differential cross section for exciting the M sublevel. The amplitudes a_M are, in general, complex numbers. However, since $|\psi\rangle$ is defined only up to an overall phase factor, a_0 may be assumed real and positive. The relative phase χ between a_1 and a_0 is defined by

$$a_1 = |a_1| e^{i\chi} \quad (14)$$

Therefore $|\psi\rangle$ is described completely for given $E, \theta_e,$ and ϕ_e by the parameters

$$\sigma = |a_0|^2 + 2|a_1|^2$$
$$\lambda = |a_0|^2/\sigma \quad \text{with } 0 \leq \lambda \leq 1 \quad (15)$$
$$\chi \quad \text{where } -\pi \leq \chi \leq \pi$$

Thus a measurement of σ, λ, and χ constitutes a complete determination of the scattering. σ is the probability of electron scattering in a particular direction, while the dimensionless quantities λ and χ describe the state of the atom after the collision. The parameterization of the scattering in terms of σ, λ, and χ is more convenient than amplitudes for analyzing the experimental data.

Consider now the theory for each kind of measurement, i.e., angular correlations and measurements of polarization.

(a) *Angular Correlations.* Substituting the spherical components of the polarization vectors (Table 1) into Eq. (11), integrating over the resolution time of the coincidence circuit Δt ($\Delta t \gg 1/\gamma$), and summing over the photon polarizations, one obtains

$$\frac{d^2 P_c}{d\Omega_e \, d\Omega_\gamma} = \frac{\sigma}{\Sigma} \frac{dP_c}{d\Omega_\gamma} \tag{16}$$

where $d^2 P_c/d\Omega_e \, d\Omega_\gamma$ is the probability density for scattering of the electron in direction (θ_e, ϕ_e) in any $n^1 P$ excitation, with subsequent emission of the photon in direction $(\theta_\gamma, \phi_\gamma)$. Σ is the total (integrated) cross section for excitation of a $n^1 P$ at energy E, and $dP_c/d\Omega_\gamma$ is the probability density for photon emission after electron scattering in a particular direction upon which λ and χ depend. It is given by

$$\frac{dP_c}{d\Omega_\gamma} = \frac{3}{8\pi} \left\{ \lambda \sin^2 \theta_\gamma + \left(\frac{1-\lambda}{2}\right)(\cos^2 \theta_\gamma + 1) \right.$$

$$- \left(\frac{1-\lambda}{2}\right) \sin^2 \theta_\gamma \cos 2(\phi_\gamma - \phi_e)$$

$$\left. + [\lambda(1-\lambda)]^{1/2} \cos \chi \sin 2\theta_\gamma \cos (\phi_\gamma - \phi_e) \right\} \tag{17}$$

where the normalizing condition

$$\iint_{4\pi} \frac{d^2 P_c}{d\Omega_e \, d\Omega_\gamma} d\Omega_e \, d\Omega_\gamma = 1 \tag{18}$$

has been used.

Since the angular correlation depends on $\cos \chi$, the sign of χ cannot be found from this kind of measurement. χ is the phase between the excitation amplitudes, and its sign will determine whether the emitted light is right or left circularly polarized. Only a measurement of circular polarization can give this information.

In the Stirling experiment, the electron beam is incident in the z direction on the target located at the origin of coordinates. Scattered electrons are collected by an analyzer whose position defines the scattering plane, which is taken to be the xz plane. Therefore, $\phi_e = 0$ for all detected scattering events. Photons are counted without regard to polarizations by a detector placed in the xz plane on the opposite side of the electron beam from the analyzer, i.e., $\phi_\gamma = \pi$. In this case, Eq. (17) takes the form

$$\frac{dP_c}{d\Omega_\gamma} = \frac{3}{8\pi} N \qquad (19)$$

where N is the angular correlation defined by

$$N = \lambda \sin^2 \theta_\gamma + (1-\lambda)\cos^2 \theta_\gamma - 2[\lambda(1-\lambda)]^{1/2} \cos\chi \sin\theta_\gamma \cos\theta_\gamma \qquad (20)$$

Therefore the measurement of the angular distribution of time-correlated photons can be used with (20) to extract λ and $|\chi|$ at a particular scattering angle.

(b) *Linear and Circular Polarization.* Consider a linear polarizer set at angle β in front of the photon detector. The probability density for scattering an electron in a given direction with subsequent emission and observation of the photon is

$$\frac{d^2 P_c}{d\Omega_e\, d\Omega_\gamma}(\beta) = \frac{\sigma}{\Sigma}\left\{\left(\frac{dP_c}{d\Omega_\gamma}\right)_{\hat{\varepsilon}^{(1)}} \cos^2\beta + \left(\frac{dP_c}{d\Omega_\gamma}\right)_{\hat{\varepsilon}^{(2)}} \sin^2\beta\right\} \qquad (21)$$

Here $(dP_c/d\Omega_\gamma)_{\hat{\varepsilon}^{(\alpha)}}$ is the probability density for photon emission polarized along $\hat{\varepsilon}^{(\alpha)}$ which can be obtained from (11). Thus

$$\left(\frac{dP_c}{d\Omega_\gamma}\right)_{\hat{\varepsilon}^{(1)}} = \frac{3}{8\pi}\left\{\lambda \sin^2\theta_\gamma + \left(\frac{1-\lambda}{2}\right)\cos^2\theta_\gamma[1 + \cos 2(\phi_\gamma - \phi_e)]\right.$$
$$\left. + [\lambda(1-\lambda)]^{1/2} \cos\chi \sin 2\theta_\gamma \cos(\phi_\gamma - \phi_e)\right\} \qquad (22)$$

and

$$\left(\frac{dP_c}{d\Omega_\gamma}\right)_{\hat{\varepsilon}^{(2)}} = \frac{3}{8\pi}\left\{\left(\frac{1-\lambda}{2}\right)[1 - \cos 2(\phi_\gamma - \phi_e)]\right\} \qquad (23)$$

Therefore, for the photon detector set at a fixed angle *out of* the plane of scattering [i.e., $(\phi_\gamma - \phi_e) \neq 0, \pi$], the measurement of

$$\frac{d^2 P_c}{d\Omega_e\, d\Omega_\gamma}(\beta)$$

obtained by rotating the polarizer axis yields values for λ and $|\chi|$. Thus, the information obtained from a measurement of the linear polarization is identical to that given by a photon angular distribution.

Let us now consider a measurement of circular polarization. When the elliptically polarized photons are to be observed, the polarization tensor is complex and can be written

$$\hat{\varepsilon}_\pm = \frac{1}{\sqrt{2}}(\hat{\varepsilon}^{(1)} \mp i\hat{\varepsilon}^{(2)}) \qquad (24)$$

where ε_+ and ε_- represent, respectively, right and left circular polarization. We use the convention in optics that right circularly polarized light has negative helicity. The circular polarization fraction of the correlated photons is defined by

$$P^{\text{circ}} = \frac{I^{\sigma^+} - I^{\sigma^-}}{I^{\sigma^+} + I^{\sigma^-}} \tag{25}$$

where I^{σ^+} and I^{σ^-} are the intensities of right and left circularly polarized coincidence photons.

Substituting Eq. (24) into Eq. (11), one obtains

$$P^{\text{circ}} = \frac{3}{4\pi} \sin \theta_\gamma \sin (\phi_\gamma - \phi_e)[\lambda(1-\lambda)]^{1/2} \sin \chi \bigg/ \frac{dP_c}{d\Omega_\gamma} \tag{26}$$

where $dP_c/d\Omega_\gamma$ is defined in Eq. (17). For coincidence photons emitted perpendicular to the plane of scattering (i.e., $\theta_\gamma = \phi_\gamma = \pi/2$ and $\phi_e = 0$),

$$P^{\text{circ}} = 2[\lambda(1-\lambda)]^{1/2} \sin \chi \tag{27}$$

It is then obvious that a combination of linear and circular polarization (which gives the sign of χ) measurements determine the collision parameters λ and χ completely.

An alternative description of the preceding theory in terms of the alignment and orientation parameters of the excited states, commonly used in optical pumping work, is given by Fano and Macek (1973) and Blum and Kleinpoppen (1975) and in this volume. A measurement of the circular polarization yields a value for the orientation, while the angular correlation results determine the direction and magnitude of the axes of the alignment tensor. Thus, for n^1P excitations, the angular correlation function Eq. (17) takes the form in the collision frame used above:

$$\frac{dP_c}{d\Omega_\gamma} = \frac{3}{8\pi} \{ \tfrac{2}{3} + A_{2+}^{\text{col}} \sin^2 \theta_\gamma \cos 2(\phi_\gamma - \phi_e) + A_{1+}^{\text{col}} \sin 2\theta_\gamma \cos (\phi_\gamma - \phi_e)$$
$$+ \tfrac{1}{3} A_0^{\text{col}}(3 \cos^2 \theta_\gamma - 1) \} \tag{28}$$

where A_{q+}^{col} are the components of the alignment tensor of rank 2, given in this case by

$$A_0^{\text{col}} = \tfrac{1}{2}\langle 3L_z^2 - L^2 \rangle = (1 - 3\lambda)/2$$
$$A_{1+}^{\text{col}} = \tfrac{1}{2}\langle L_xL_z + L_zL_x \rangle = [\lambda(1-\lambda)]^{1/2} \cos \chi \tag{29}$$
$$A_{2+}^{\text{col}} = \tfrac{1}{2}\langle L_x^2 - L_y^2 \rangle = \tfrac{1}{2}(\lambda - 1)$$

Equation (28) gives the special angular dependence associated with each component of the alignment tensor.

Similarly, the circular polarization which represents, in units of $\hbar$, the expectation value $\langle L_y \rangle$ of the atomic orbital angular momentum perpendicular to the plane of scattering can be related to the orientation vector $\mathbf{O}^{\text{col}}$ in the following way:

$$\langle L_y \rangle = -P^{\text{circ}} = 2O_{1-}^{\text{col}} \tag{30}$$

where

$$O_{1-}^{\text{col}} = -[\lambda(1-\lambda)]^{1/2} \sin \chi \tag{31}$$

Furthermore, it can be easily verified that $\langle L_x \rangle = \langle L_z \rangle = 0$. It is then apparent that the net orbital angular momentum ($\Delta \mathbf{L}$) transferred to the atom in the collision is restricted to a component perpendicular to the scattering plane, i.e., a "net" selection rule $\langle \Delta \mathbf{L} \cdot \hat{n} \rangle = 0$ applies to the scattering plane ($\hat{n}$ is any unit vector in this plane). This restriction is a direct consequence of the parity invariance of the excitation process.

Due to the fact that the excitation/deexcitation process $^1S-^1P-^1S$ for helium represents a rather straightforward case, the relations for the linear and circular polarization of photons detected in coincidence can be obtained in the following way. Taking the incident direction of the electron beam along the z axis and the scattering plane as the xz plane, we introduce the following set of basis functions:

$$\psi_x = -\frac{1}{\sqrt{2}}(|11\rangle - |1-1\rangle)$$

$$\psi_y = \frac{i}{\sqrt{2}}(|11\rangle + |1-1\rangle) \tag{32}$$

$$\psi_z = |10\rangle$$

Noting that $a_1 = -a_{-1}$, the excited-state wave function [Eq. (6)] can be rewritten in terms of this new basis set as

$$|\psi(t=0)\rangle = a_0 \psi_z - \sqrt{2} a_1 \psi_x \tag{33}$$

The physical significance of Eq. (33) can be seen from the fact that for $^1P-^1S$ deexcitations we can associate with each basis function ψ_x, ψ_y, ψ_z a classical linear oscillator parallel to the x, y, and z axes respectively. Thus the excited atom can be represented by an elliptical dipole consisting of two coherently excited classical linear oscillators which lie in the scattering plane parallel to the x and z axes, with amplitudes proportional to $\sqrt{2} a_1$ and a_0 respectively (see Figure 10 and further discussion in the next section). The radiation emitted perpendicular to the scattering plane by these oscillators has linearly polarized components $I_{x,y,z}^\pi$ which are proportional to the moduli squared of their amplitudes. Thus $I_x^\pi \propto 2|a_1|^2 = 1 - \lambda$, $I_z^\pi \propto |a_0|^2 = \lambda$, and, trivially, $I_y^\pi = 0$.

To obtain the circular polarization perpendicular to the scattering plane, Eq. (33) is expanded in terms of the combination of basis functions $\psi_z \mp i\psi_x$ (corresponding to left (+) and right (−) circular polarization along the y direction):

$$|\psi(t=0)\rangle = \tfrac{1}{2}(a_0 - i\sqrt{2}a_1)(\psi_z - i\psi_x) + \tfrac{1}{2}(a_0 + i\sqrt{2}a_1)(\psi_z + i\psi_x) \quad (34)$$

The circularly polarized components are $I_y^{\sigma\pm} \propto |a_0 \mp i\sqrt{2}a_1|^2$ and the circular polarization perpendicular to the scattering plane is

$$P_{zx}^{\text{circ}} = \frac{I_y^{\sigma+} - I_y^{\sigma-}}{I_y^{\sigma+} + I_y^{\sigma-}} = 2[\lambda(1-\lambda)]^{1/2} \sin\chi \quad (35)$$

Equation (33) is identical to Eq. (27).

Excitation of n^3P States of Helium

The excitation of the n^3P states of helium from the ground state is interesting since it involves a rearrangement collision, i.e., the n^3P states can only be excited by exchange, assuming magnetic interactions are negligible. Thus, whereas the n^1P collisions involve the determination of seven real numbers corresponding to direct and exchange scattering (see Bederson, 1970), for the ^3P case only three real numbers are required for a complete determination of the scattering process: $|g_0|$, $|g_1|$, and their relative phase. g_i is the amplitude for excitation of the i magnetic sublevel. A coincidence experiment will determine two of these: the phase and the ratio

$$|g_0|^2/(|g_0|^2 + 2|g_1|^2).$$

The coincidence rate for n^3P excitation is more complicated than for n^1P because of the presence of fine structure. The angular correlation for ^3P excitations can be obtained from Eq. (5) of this paper or Eq. (15) of Macek and Jaecks (1971). For the case of the photon detector located in the plane of scattering, it is given by

$$\frac{dP_c}{d\Omega_\gamma} = \lambda(\tfrac{41}{54} - \tfrac{5}{18}\cos^2\theta_\gamma) + 0.5(1-\lambda)(\tfrac{67}{54} + \tfrac{5}{18}\cos 2\theta_\gamma)$$
$$- \tfrac{5}{18}[\lambda(1-\lambda)]^{1/2}\cos\chi \sin 2\theta_\gamma \quad (36)$$

In this expression we have assumed that the experimental resolution time is longer than any beat period involved in the decay of the fine-structure levels. The fractions 41/54, 5/18, 67/54 represent the depolarization caused by the fine-structure interaction, which causes the angular distribution to smear out, $\lambda = |g_0|^2/\sigma$ and $\sigma = |g_0|^2 + 2|g_1|^2$.

3. Experimental Methods and Results

The basic experimental method for electron–photon coincidence measurements (for photon polarization and angular correlations) has been described in detail elsewhere (King et al., 1972; Eminyan et al., 1974). Briefly, the method consists in crossing an atom beam with an energy-selected electron beam and observing delayed coincidences between electrons scattered inelastically at a particular angle and the photons which result from the decay of the excited states. Only those electron–photon pairs originating from a single scattering event are correlated in time. The uncorrelated events form a uniform background of chance coincidences.

In the experiment of King et al. (1972), electrons scattered in the forward direction having excited He atoms into the 3^1P state were detected in coincidence with 501.6 nm photons emitted perpendicular to the incident beam direction. The aim of their experiment was to investigate the threshold polarization of impact line excitation which is governed by the selection rule $\Delta m_L = 0$. This selection rule should hold exactly along the beam axis of the incoming and inelastically scattered electrons in the forward direction. Accordingly, the photon radiation from the $3^1P \to 2^1S$ transition detected in coincidence with the forward-scattered electrons should be completely polarized parallel to the incoming beam direction. Figure 2 shows the results of such measurements by King et al. (1972), which approach the expected threshold polarization of 100% at the lowest energy values (Percival and Seaton, 1958). The deviation from this value at higher energies has been explained by the authors on the grounds that both the initial and the scattered beam directions are not confined to the symmetry axis of the incoming beam. As a result of this, the momentum transferred to the atom will have a range of angles about the beam axis which results in a reduction of the polarization of the radiation. Figure 2 includes an estimate of this depolarization (dotted line) based upon the beam divergence and the Born approximation. King et al. (1972) pointed out that this theoretical estimate cannot be expected to be very good at these rather low energies (where the Born approximation is poor), but it qualitatively explains the trend of the depolarization as observed in the experiment. The importance of this experiment lies in the fact that it yielded the expected threshold polarization of 100% for the above transition, compared to a value of about 60% obtained by the traditional type of measurement of Soltysik et al. (1967). As we do know now from recent near-threshold polarization measurements (e.g., Heddle et al., 1974; Ottley and Kleinpoppen, 1975), resonance phenomena near the excitation thresholds strongly affect line polarization data near threshold.

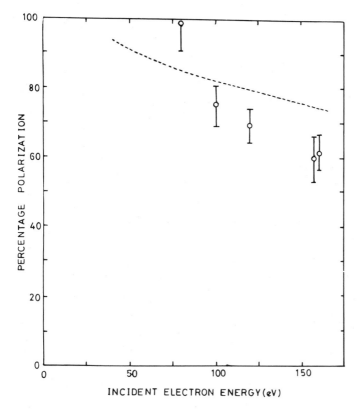

Figure 2. The circles show the linear polarization for the $3^1P \to 2^1S$ transition in helium as measured in the electron–photon coincidence experiment of King et al. (1972). The broken curve shows values of the polarization calculated by means of the Born approximation.

In the experiment of Eminyan et al. (1974), measurements were made of the angular correlation between photons and the inelastically scattered electrons. The basic apparatus along with a block diagram of the coincidence circuit is shown in Figure 3. Figure 4 shows a typical coincidence spectrum. The 6.4 nsec represents the resolution time of the apparatus. The effect of the 0.58 nsec lifetime of the 2^1P state is visible on the trailing edge of the coincidence peak.

Two kinds of photon detector were used. The ultraviolet photons from the decay of both 3^1P and 2^1P to the ground state were detected by a channeltron. A separate study of the 3^1P excitation was made by oservation of the 501.6 nm line in the visible, which results from the decay of 3^1P to 2^1S. This line was detected using a photomultiplier and narrow-band interference

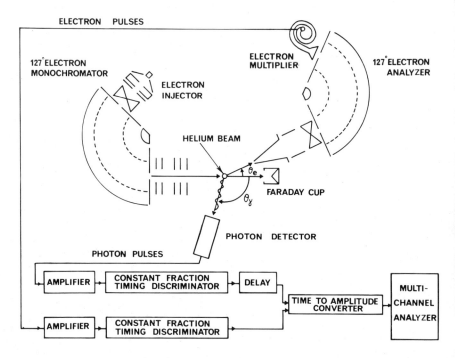

Figure 3. Schematic diagram of the apparatus of Eminyan *et al.* (1974). The helium beam emerges perpendicular to the plane of the diagram, which represents the plane of scattering.

filter. The photon detector was mounted on a table which could be rotated through the angular range $\theta_\gamma = 30°$ to $130°$.

The total number of real coincidences counted in time T is normalized to the total number of electron counts accumulated in the same period. This renders the results free from error due to any variation in electron beam intensity and target atom density. The experiment then consists in measuring the angular correlation function N [see Eq. (20)] as a function of θ_γ. A least-squares fit of N, suitably convoluted over the photon detector aperture, yields values of the parameters λ and $|\chi|$ upon which N depends.

The experimental results obtained from the electron–photon coincidence experiments of Eminyan *et al.* (1974, 1975) for the 2^1P and 3^1P levels of helium are summarized in tables in these earlier publications. Here the data are only presented in graphic form. Figure 5 shows the angular correlation curves for three scattering angles at 60 eV and for 2^1P excitations. 2^1P and 3^1P data for λ are presented in Figure 6 and for $|\chi|$ in Figure 7. Figure 8 shows the data for the orientation $|O_{1-}^{col}|$. The results of theoretical calculations are included where available.

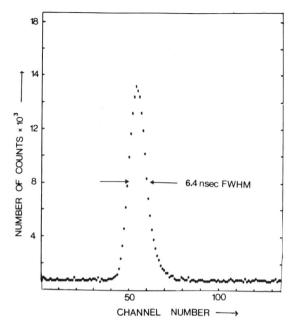

Figure 4. Delayed coincidence spectrum from the helium $2^1P \rightarrow 1^1S$ transition for an electron energy of 80 eV, scattering angle $\theta_e = 16°$, and photon angle $\theta_\gamma = 126°$. Accumulation time ~12 hours (Eminyan et al., 1974).

The results of the experiment are best discussed by considering the angular correlation curves shown in Figure 5. The dashed curves represent the prediction of the first Born approximation (FBA) averaged over the detector aperture. Two features are apparent: (1) the minimum of the radiation pattern, i.e., of the angular correlation curve, becomes less pronounced at larger scattering angles, and (2) the angle of the minimum shifts markedly from that of the FBA at the largest scattering angles studied.

The interpretation of these features is straightforward in a semiclassical model where the radiating atom is described as a superposition of two linearly polarized electric dipoles oscillating coherently and 90° out of phase in the plane of scattering. As the dipoles oscillate their resultant traces out an ellipse in the scattering plane (see Figure 10). In any theory for which $\chi \equiv 0$, e.g., the FBA, the ellipse degenerates into a straight line. In the case of the FBA, this dipole is oriented along the direction of linear momentum transfer, θ_K, and the radiation intensity falls to zero in this direction. Figure 5 shows that even for small θ_e, the minimum of the radiation pattern is nonzero. Interpreted in terms of the semiclassical model, the

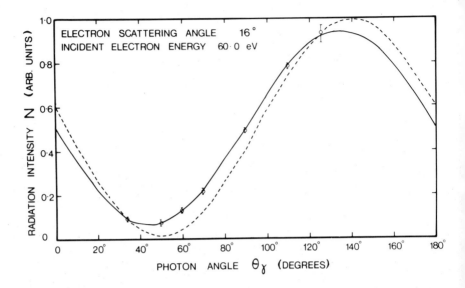

Figure 5. Electron–photon angular correlations from $2^1P \to 1^1S$ transition in helium for different scattering angles (Eminyan *et al.*, 1974). Error bars represent one standard deviation. Solid line, least-squares fit of angular correlation function to the data; dashed line, first Born approximation.

elliptical dipole has a high degree of ellipticity at small scattering angles and tends progressively toward a circular dipole with increasing scattering angle, together with some rotation of the major axis away from the momentum transfer axis. As the scattering angle increases from small values, the greater degree of circular polarization of the atomic dipole reflects the increasing angular momentum imparted to the atom perpendicular to the plane of

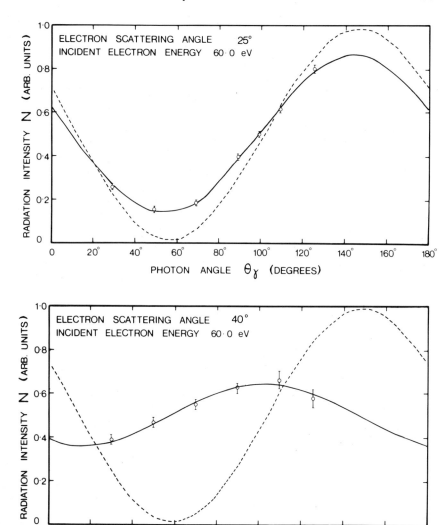

scattering. The orientation provides a direct measure of this effect since $2O^{col}_{-1}$ is the expectation value of the perpendicular component of the excited-state angular momentum [see Eq. (30)].

For larger angles, the orientation data for both 2^1P and 3^1P follow the same trend, rising steadily toward a maximum value of 0.5, which corresponds to a 100% circularly polarized atomic dipole, with the higher-

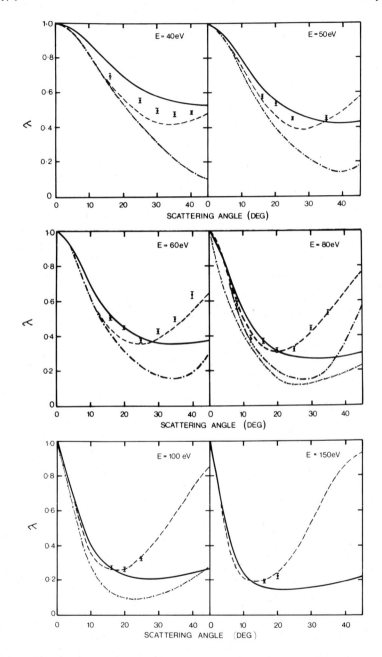

Figure 6. Variation of λ with electron energy and scattering angle for 2^1P and 3^1P excitations, from the experiments of Eminyan *et al.* (1974, 1975).

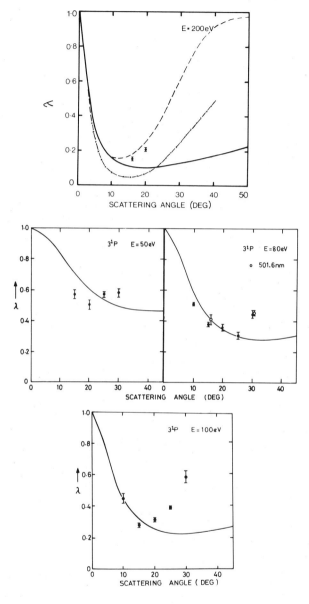

Error bars represent one standard deviation. ——— first Born approximation; ——— Distorted-wave theory (Madison and Shelton, 1973); —·— Many-body theory (Csanak et al., 1973); ×—× eikonal distorted-wave Born approximation (Joachain and Vanderporten, 1974): ● data for $2^1P \to 1^1S$ and $3^1P \to 1^1S$ transitions; ◯ data for $3^1P \to 2^1S$ transition in helium.

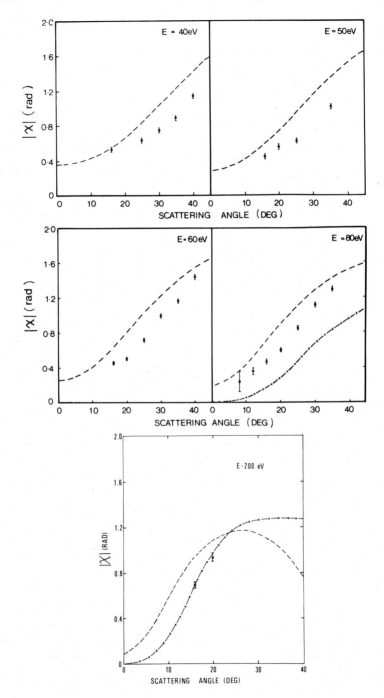

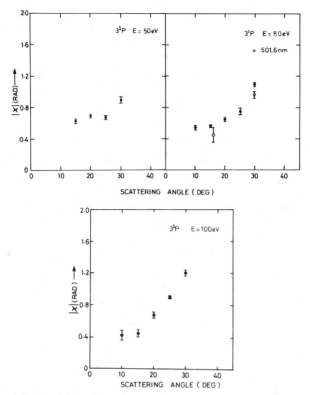

Figure 7. Variation of $|\chi|$ with electron scattering angle for 2^1P (left) and 3^1P (above) excitations. Eminyan *et al.* (1974, 1975). Error bars are one standard deviation. Legend as in Figure 5. ● data for $2^1P \to 1^1S$ and $3^1P \to 1^1S$ transitions; ○ data for 501.6 nm line.

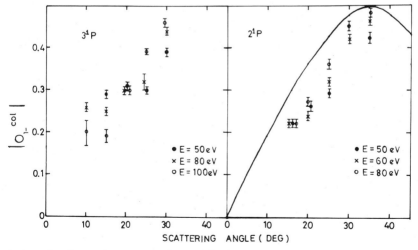

Figure 8. Orientation data for 2^1P and 3^1P. Solid line is distorted wave calculation of Madison and Shelton (1973) at 78 eV.

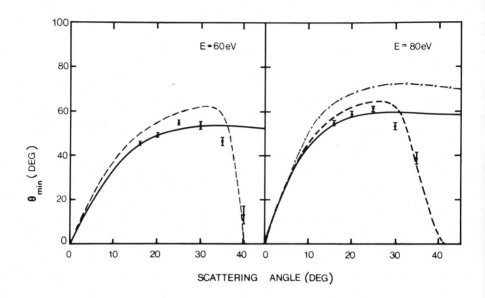
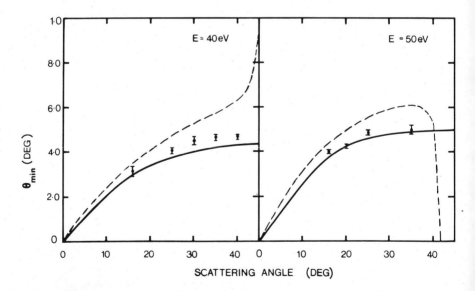

Figure 9. Variation of θ_{min} with electron energy and scattering angle for 2^1P and 3^1P excitation of helium, Eminyan *et al.* (1974, 1975). Legend as in Figure 6.

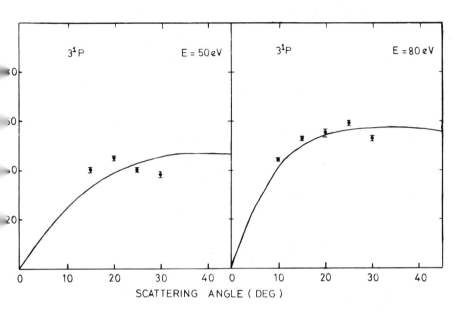
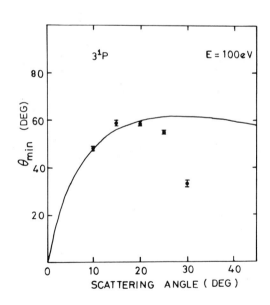

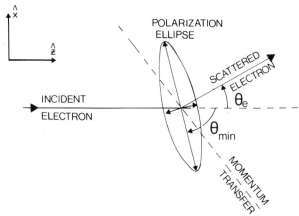

Figure 10. Semiclassical model of the radiating atom. The two dipoles oscillating 90° out of phase are represented as vectors along the major and minor axes of the ellipse.

energy data rising most rapidly. However, for smaller angles, the behavior of the 2^1P data appears to be different from that for the 3^1P data. Whereas for 2^1P, the orientation data converge to a value of 0.22 at $\theta_e = 16°$, over the energy range 40–200 eV, the 3^1P data have a crossover point at about $\theta_e = 20°$, and for fixed scattering angle at $\theta_e < 20°$, the 3^1P data decrease with increasing energy and are nearly constant at small angle and fixed energy. This difference in behavior may be more apparent than real since it may well be that $\theta_e = 16°$ is the crossover point for the 2^1P data and that at smaller angles they would show the same behavior as the 3^1P data.

A parameter of importance in understanding the collision process is the angle θ_{min}, which is related to λ and χ in the following way:

$$\tan 2\theta_{min} = 2[\lambda(1 - \lambda)]^{1/2} \cos \chi/(2\lambda - 1) \qquad (37)$$

In the semiclassical model of the radiating atom (see Figure 10) θ_{min} is the direction of the dominant dipole excited in the collision process. In this same sense it is a direction of symmetry of the excited state, and also, one might expect, of the collision process itself. Figure 9 shows the variation of this angle with electron energy and scattering angle. The striking feature of the 2^1P data at 60 and 80 eV and the 3^1P data at 100 eV is the sudden shift of θ_{min} away from the momentum transfer axis. For the 2^1P data this shift is in good agreement with the distorted-wave results.

Finally, the comparison between the experimental data and the various theories, as shown in Figure 6 and 7, for two of the three parameters which determine the scattering, allows us to make the following observations:

(i) *FBA*. Although at the smaller angles the agreement with λ is good, this theory always gives trivial results for the phase χ, i.e., $\chi \equiv 0$, in considerable disagreement with experiment. It is worth emphasizing that the comparisons with the FBA depend only on the kinematics of the collision and are independent of the particular choice of atomic wave functions.

(ii) *Many-body theory*. The agreement with this theory is poor at larger scattering angles. This approximation does yield values for χ, but these have not been published by the authors.

(iii) *Distorted-wave theory*. Except at the very lowest energies, the agreement with this calculation is very good. Data and theory are in excellent accord for λ, although the experimental phase χ lies about 20% below the theoretical phase.

(iv) *Eikonal DWBA*. This calculation gives excellent agreement with the experimental phase at the largest energy. At lower energies, the agreement is not good, but this calculation is not expected to be valid below about 200 eV.

4. Conclusion

The results of the electron–photon angular correlation experiment are significant for several reasons. The relative magnitude and phase of complex excitation amplitudes have been measured for the first time. The data pertain to identically prepared upper states at a definite incident energy and scattering angle, and permit direct comparison with theory without the need for angular integration or summation over degenerate sublevels and without loss of the information carried by the complex nature of the scattering amplitudes. Furthermore, the data are obtained in dimensionless form and are free from absolute calibration or normalization difficulties.

The difficulties involved in doing coincidence experiments are relatively well known, especially with regard to the long integration times required to collect the data, and it has been a matter of some speculation whether or not they could be performed on a sufficiently systematic basis to be useful in testing the various theoretical models used to describe the scattering. The Stirling experiment has, however, demonstrated that not only is it possible to do an experiment which will yield values for the complex excitation amplitudes, but that the range of data is sufficiently extensive to test the validity of the different theoretical approximations. It is evident, therefore, that these kinds of coincidence experiments will, within the next few years, permit the development of theoretical models which better account for the details of the collision.

Future developments of electron–photon coincidence experiments with oriented atoms and polarized electrons will certainly result in the complete analysis of electron–atom collision processes, giving not only detailed information on the magnetic sublevel excitation but also distinguishing between direct, exchange, and spin–orbit interactions.

Acknowledgments

We acknowledge the support of the Science Research Council. We also acknowledge the important scientific contribution of Dr. K. B. MacAdam to the electron–photon coincidence experiment carried out at Stirling University.

J. Slevin wishes to thank Dr. Barat and the other members of the Laboratoire des Collisions Atomiques, Université de Paris-Sud, Orsay, for their hospitality during his stay there. It was during this stay that a substantial part of the manuscript was prepared. J. Slevin is also grateful to the Royal Society for a Fellowship held during this period. We would also like to thank Drs. K. Blum and I. McGregor for critical reading of the manuscript.

References

Barat, M. (1973). In *The Physics of Electronic and Atomic Collisions*, Ed. B. C. Cobic and M. V. Kurepa, p. 43. Belgrade: Institute of Physics.
Bederson, B. (1969). *Comm. Atom. Molec. Phys.*
Bederson, B. (1972). *Comm. Atom. Molec. Phys.*
Blum, K., and Kleinpoppen, H. (1974). *Phys. Rev.*, **A9**, 1902.
Blum, K., and Kleinpoppen, H. (1975). *J. Phys. B.*, **8**, 922. See also this volume.
Csanak, G., Taylor, H. S., and Tripathy, D. N. (1973). *J. Phys. B.*, **6**, 2040 and private communication.
Eminyan, M., MacAdam, K. B., Slevin, J., and Kleinpoppen, H. (1974). *J. Phys. B.*, **7**, 1519.
Eminyan, M., MacAdam, K. B., Slevin, J., Standage, M., and Kleinpoppen, H. (1975). *J. Phys. B.*, **8**, 2058.
Fano, U., and Macek, J. H. (1973). *Rev. Mod. Phys.*, **45**, 553.
Franken, P. (1961). *Phys. Rev.*, **121**, 508.
Hanne, G. F., and Kessler, J. (1974). *Phys. Rev. Lett.*, **33**, 341. See also this volume.
Heddle, D. W. O., Keesing, R. G. W., and Watkins, R. D. (1974). *Proc. Roy. Soc. Lond.*, A **337**, 443.
Hertel, I. V. (1975). Contribution to this volume.
Hertel, I. V., and Stoll, W. (1974). *J. Phys. B.*, **5**, 583.
Hils, D., McCusker, M. V., Kleinpoppen, H., and Smith, S. J. (1972). *Phys. Rev. Lett.*, **29**, 398.
Jaecks, D. H. (1973). In *The Physics of Electronic and Atomic Collisions*. Eds., B. C. Cobic and M. V. Kurepa, p. 43. Belgrade: Institute of Physics.

Joachain, C. J. and Vanderporten, R. (1974). *J. Phys. B.*, **7**, 817, and private communication.
King, G. C. M., Adams, A., and Read, F. H. (1972). *J. Phys. B.*, **5**, 254.
Macek, J., and Jaecks, D. H. (1971). *Phys. Rev.*, **A4**, 2288.
Madison, D. H., and Shelton, W. N. (1973). *Phys. Rev.*, **A7**, 499.
Ottley, T. W., and Kleinpoppen, H. (1975). *J. Phys. B.*, **8**, 621. See also Ottley, T. W. (1974). Thesis, University of Stirling.
Percival, I. C., and Seaton, M. J. (1958). *Phil. Trans. Roy. Soc.*, A **251**, 113.
Rubin, K., Bederson, B., Goldstein, M., and Collins, R. E. (1969). *Phys. Rev.*, **182**, 201.
Soltysik, E. A., Fournier, A. Y., and Gray, R. L. (1967). *Phys. Rev.*, **153**, 152.
Wykes, J. (1972). *J. Phys. B.*, **5**, 1126.

39
Theory of Measurement of Impact Radiation on Atoms

JOSEPH MACEK

The theory of angular distribution and polarization of light excited by atomic collisions is reviewed. Particular attention is given to the effect of small forces which perturb the initial alignment and orientation.

1. Introduction

Excitation of an atom leaves it generally in an anisotropic state. This anisotropy is manifest in the polarization and angular distribution of the decay radiation. The connection between the geometrical properties of the radiation and the source parameters constitutes one aspect of the theory of measurement. An equally important aspect is the selection of an appropriate set of parameters to characterize atomic sources. Both of these aspects have been considerably influenced over the past twenty years by the contributions of Professor Fano (1957). I believe that Professor Fano will also review the theory of measurements at this conference; therefore, I will give only a brief review of the general concepts, then discuss specific aspects emphasizing the role of "weak" forces in determining the magnitude of measured parameters.

When speaking of a quantum system we immediately think of eigenstates and eigenvalues analogous to the normal modes and frequencies of a classical oscillator. Just as for a classical oscillator, we quickly realize that normal modes or eigenstates are inadequate to specify the state of a system, since in general such systems oscillate in coherent superpositions of normal

JOSEPH MACEK • Behlen Laboratory of Physics, University of Nebraska–Lincoln, Lincoln, Nebraska 68508, U.S.A.

modes. Such states cannot be described by the probability that the system is in a specific eigenstate; rather, one considers the amplitude that the system is in a specific eigenstate. Accordingly, we write the wave function for the system as

$$\Psi = \sum_m a_m \psi_m \tag{1}$$

where ψ_m is an eigenfunction of the system. If all components of a system are described by the same wave function Ψ the system is said to be in a pure state. Often, however, our knowledge of a system is incomplete. This more general mixed state is described by a weighted average of the various pure states which is not conveniently represented in the amplitude language.

Fano, in his celebrated 1957 paper, showed how the density matrix language avoided the semantic difficulties of the eigenstate representation for describing both microscopic and macroscopic quantum systems. He gave the example of an unpolarized light beam where one can say that it is an incoherent mixture of two linear polarizations or an incoherent mixture of two circular polarizations. Clearly, representation in terms of specific polarizations is nonunique and rather cumbersome. This contrasts with the succinct and adequate statement that the light is unpolarized. One seeks a mathematical language similarly succinct.

Physical quantities, that is, quantities measured in an actual experiment, relate to the square modules $|\Psi|^2$ averaged over the components of the system. In terms of the eigenstate expansion (1), the averaged $|\Psi|^2$ becomes

$$\langle |\Psi|^2 \rangle = \sum_{mm'} \langle a_m^* a_{m'} \rangle \psi_m^* \psi_{m'} \tag{2}$$

The quantities $\langle a_m^* a_{m'} \rangle$ define the density matrix ρ:

$$\rho_{m'm} = \langle a_m^* a_{m'} \rangle \tag{3}$$

The density matrix describes both pure and mixed states, so it represents an improvement over the language of amplitudes.

Fano (1951) pointed out that when m and m' denote magnetic quantum numbers certain linear combinations of density matrix elements offered considerable advantages for the description of atomic systems. These parameters, called state multipoles, are defined by

$$\rho_q^{[k]} = \sum_{m'm} \rho_{m,m} (-1)^{k-j-m'} (j'm'j - m|kq) \tag{4}$$

where $(j'm'j - m|kq)$ are Clebsch–Gordan coefficients. The state multipoles are seen to be irreducible components of the density matrix and therefore have more simple transformation properties under coordinate rotations than do the nonreduced components in Eq. (3). The theory of angular correlations is conventionally expressed in terms of the state multipoles of Eq. (4).

The language of state multipoles is entirely adequate for describing atomic systems, but it is rather abstract, and, since it is framed in terms of matrix elements, is closely married to the quantum theory. Accordingly, it does not make direct contact with intuitive physical quantities and possible alternative fundamental atomic theories such as the classical theory. One would hope that a description of measurements would transcend any particular microscopic theory of atoms. A development in this direction, based on the mean values of a standard set of operators, was indicated by Fano (1957).

The mean value of an operator O is given by

$$\langle(\Psi|O|\Psi)\rangle = \sum_{m',m=-j}^{j} \rho_{m'm} O_{mm'} = \text{Tr}\,\rho O \qquad (5)$$

If the mean values of a set of $N = (2j + 1)^2$ different operators $O_1, O_2, \ldots, O_N$ are known from measurements, then Eq. (5) represents a set of N linear equations for the N density matrix elements $\rho_{m,m}$ and can be inverted, provided the N equations are linearly independent. The set of operators is said to be complete* if the corresponding Eqs. (5) are independent. Alternatively, one could use the mean values of a complete set of operators rather than the density matrix to characterize atomic states. The theory of measurements then seeks an appropriate set of operators O_n and relates them to experiment.

For a multiplet of levels characterized by angular momentum quantum numbers, a complete set of operators $T^{[k]}q$ are constructed from the components of the angular momentum operator $\mathbf{J}$. The mean values of $T^{[k]}q$ equal, to within a normalization constant, the state multipoles $\rho^{[k]}q$. This is not to say that the parameters are equivalent. By no means! The mean values of angular momentum components represent a more general language. For example, in those theories (Burgess and Percival, 1970) which represent atomic states as ensembles of classical orbits, the mean values of $T^{[k]}q$ can be calculated by averaging the tensors constructed from the electronic angular momentum components over an ensemble of classical orbits, whereas construction of the density matrix elements requires a representation in terms of eigenstates, i.e., it is a language specific to the quantum theory. Such a classical calculation could prove useful in the limit of large principal quantum numbers near the ionization threshold (Wannier, 1953; Rau, 1971). In any event the mean values $\langle T^{[k]}q \rangle$ tell us something about the angular momentum transferred to the atom during a collision.

---

*Note that the complete set of operators discussed here differs quite substantially from the complete set of commuting observables used to characterize eigenstates.

2. Examples

Relating the mean values $\langle T^{[k]}q \rangle$ to measurements represents the main task of angular correlation theory. This is accomplished by (a) representing the detected quantity as the mean value of an operator, (b) decomposing this "measurement" operator into its irreducible components, (c) applying the Wigner–Eckart theorem to relate mean values of irreducible components of the "measurement" operator to mean values of $T^{[k]}q$, and (d) transforming to a preferred reference frame where the mean values take their simplest form. I briefly sketch two examples of this procedure.

Collision-Induced Fluorescence

The first example concerns the measurement of collision-induced fluorescence. Here the detected intensity of dipole radiation for the transition $i \to f$ is given by (Fano and Macek, 1973)

$$I = C \sum_{m_f} \langle (i'|\hat{\varepsilon} \cdot \mathbf{r}'|f)(f|\hat{\varepsilon}^* \cdot \mathbf{r}|i) \rangle \qquad (6)$$

where $\hat{\varepsilon}$ is the polarization vector selected by the detector and C is a constant. In a reference frame (ξ, η, ζ) defined by the incident light direction and polarization we have $\hat{\varepsilon} = (\cos \beta, i \sin \beta, 0)$, where $\beta = 0$ for linear polarization and $\pi/4$ for circular polarization. The measured intensity is written in this frame as

$$I = \tfrac{1}{3}\{\langle (i'|\mathbf{r}' \cdot \mathbf{r} P_f(\mathbf{r}', \mathbf{r})|i) \rangle - \tfrac{1}{2}\langle (i'|(3\xi'\xi - \mathbf{r}' \cdot \mathbf{r})P_f(\mathbf{r}', \mathbf{r})|i) \rangle$$
$$+ \tfrac{3}{2}\langle (i'|(\xi'\xi - \eta'\eta)P_f(\mathbf{r}', \mathbf{r})|i) \rangle \cos 2\beta$$
$$+ \tfrac{3}{2}\langle (i'|i^{-1}(\mathbf{r}' \times \mathbf{r}) \cdot \hat{\zeta} P_f(\mathbf{r}', \mathbf{r})|i) \rangle \sin 2\beta \qquad (7)$$

where the scalar operator P_f is defined by

$$P_f(\mathbf{r}', \mathbf{r}) = \sum_{m_f} |f)(f|$$

Equation (7) represents the measured intensity as a sum of four terms, the first representing the matrix elements of a scalar operator having the same value for all states i. The second term represents the anisotropy of emission averaged over polarizations while the third and fourth terms represent the anisotropy with linear and circular polarization respectively.

Each term of Eq. (7) is proportional to the mean value of a real irreducible tensor operator $S^{[k]}qp$, where $p = \pm$ indicates the parity of the operator under reflections in the $\xi\zeta$ plane. These operators are more convenient than the more conventional complex quantities since two symmetry properties, hermiticity, manifest by the real nature of the operators, and reflection symmetry, manifest by the index p, are explicitly exhibited. Note here that the operators $S^{[k]}qp$ have nonstandard normalization.

The Wigner–Eckart theorem is now used to relate the mean values of $S^{[k]}qp$ to the mean values of operators $T^{[k]}qp$ constructed from angular momentum components and therefore independent of the dynamics of the transition. We have

$$\langle(i'|S^{[k]}_{qp}|i)\rangle = \langle(i'|T^{[k]}_{qp}|i)\rangle(i\|S^{[k]}\|i)/(i\|T^{[k]}\|i) \tag{8}$$

When substituted into Eq. (7), Eq. (8) represents the light intensity in terms of the fundamental dimensionless parameters $\langle T^{[k]}qp \rangle$ expressed in a frame defined by the detector axis and polarization selector. For convenience we define the alignment and orientation parameters A_0^{det}, A_{2+}^{det}, and O_0^{det} in this reference frame as (Fano and Macek, 1973)

$$A_0^{\text{det}} = \langle T^{[2]} \rangle/\langle T_0^{[0]} \rangle = \langle(i'|3J_\zeta^2 - J^2|i)\rangle/j_i(j_i + 1)$$
$$A_{2+}^{\text{det}} = \langle(i'|J_\zeta^2 - J_\eta^2|i)\rangle/j_i(j_i + 1) \tag{9}$$
$$O_0^{\text{det}} = \langle(i'|J_\zeta|i)\rangle/j_i(j_i + 1)$$

Our final step transforms these quantities to a frame (x, y, z) defined by the collision geometry. The z axis is along the incoming beam direction, and the x axis is in the plane defined by the incident and scattered beam directions. Then reflection symmetry in this plane implies that $\langle T^{[k]}qp \rangle$ vanishes unless $p(-1)^k$ is even. We have

$$O_0^{\text{det}} = O_{1-}^{\text{col}} \sin\theta \sin\phi$$
$$A_0^{\text{det}} = A_0^{\text{col}}\tfrac{1}{2}(3\cos^2\theta - 1) + A_{1+}^{\text{col}}\tfrac{3}{2}\sin 2\theta \cos\phi + A_{2+}^{\text{col}}\tfrac{3}{2}\sin^2\theta \cos 2\phi$$
$$A_{2+}^{\text{det}} = A_0^{\text{col}}\tfrac{1}{2}\sin^2\theta \cos 2\psi + A_{1+}^{\text{col}}\{\sin\theta \sin\phi \sin 2\psi \tag{10}$$
$$\quad + \sin\theta \cos\theta \cos\phi \cos 2\psi\}$$
$$\quad + A_{2+}^{\text{col}}\{\tfrac{1}{2}(1 + \cos^2\theta)\cos 2\phi \cos 2\psi - \cos\theta \sin 2\phi \sin 2\psi\}$$

where θ, ϕ are the polar coordinates of this light detector and ψ is the third Euler angle required to specify the orientation of a linear polarization analyzer.

Equation (10) is appropriate to analyze collision experiments in which the collision geometry contains only a plane of symmetry, rather than an axis of symmetry. Interest in such experiments in atomic collision physics is a fairly recent development. To date there exist the coincidence measurements of McKnight and Jaecks (1971) on Lyman α excitation by charge transfer, the measurements of Eminyan et al. (1973) on He 2^1P and 3^1P excitation by electrons, and a recent measurement of orientation and alignment by heavy-ion collisions with thin carbon foils tilted with respect to the beam direction by Berry *et al.* (1974). Conventional experiments, such as polarization measurements without detection of the scattered particle, identify only an axis of symmetry, the beam axis. Because of rotational symmetry about this axis, only one tensor component A_0^{col} in Eq. (10) is nonzero.

These symmetry arguments are illustrated by the measurements of Imhof and Read (1971), who made coincidence measurements at forward scattering angles $\theta = 0°$. Here an axis of symmetry exists and only A_0^{col} is nonzero. The discovery by Berry and co-workers (1974) that beam–foil collisions with tilted foils can be used to orient atoms provides another illustration of our symmetry arguments. The geometry is sketched in Figure 1. The foil is tilted so that its normal **n** makes an angle α with respect to the beam direction. The coordinate system—which differs from the one used in reference (4)—and the direction of the detector axis are shown. The light emitted in this direction determines the polarization ellipse in Figure 1b. The light properties are characterized by four Stokes parameters: I, the total intensity, M and C, two linear polarizations, and S, the circular polarization. Six measurements at each tilt and observation angle, three with and three without a quarter-wave plate, with the linear polarizer at 0°, 45°, and 90° to the beam-detector plane, provide the input data to extract the Stokes parameters in Table 1. Also listed is the polarization fraction

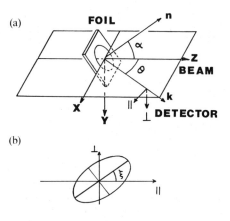

Figure 1. (a) The beam viewing geometry. The foil normal **n** is tilted at an angle α to the beam axis $\hat{z}$. The light vector **k** is in the $\hat{x}\hat{z}$ plane, perpendicular to the $\hat{n}\hat{y}\hat{z}$ plane, at an angle θ to the z-axis. (b) The polarizarion ellipse is tilted at an angle ξ to the $\parallel$ direction.

Table 1. Stokes Parameters for the 5016 Å, ^{4}HeI, $2s^1S–3p^1P$ Transition at 130 keV Beam Energy (see text for definition of symbols)

View angle θ	Foil angle (deg)	M/I	C/I	S/I	fp	$\xi = -\tfrac{1}{2}\tan^{-1}\theta$
90	0	0.158(12)	−0.016(40)	0.007(58)	0.158(12)	0.0
	20	0.132(22)	−0.082(13)	0.042(22)	0.160(23)	16 ± 4
	30	0.123(29)	−0.042(25)	0.114(68)	0.171(78)	10 ± 6
	45	0.084(28)	−0.140(23)	0.105(10)	0.194(38)	30 ± 5
53	0	0.127(15)	–	–	0.127(15)	0.0
	20	0.106(10)	−0.045(18)	0.033(20)	0.120(29)	12 ± 5
	30	0.087(15)	−0.069(31)	0.093(29)	0.145(45)	22 ± 12
	45	0.059(18)	−0.077(07)	0.107(37)	0.144(42)	40 ± 15

$p = (M^2 + C^2 + S^2)^{1/2}/I$ and the angle ξ of the polarization ellipse with respect to the beam-detector plane. The parameters C, S, and ξ should equal zero at $\alpha = 0°$ on the basis of cylindrical symmetry. The alignment and orientation parameters (source parameters) extracted from the Stokes parameters (light parameters) are given in Table 2. The error bars on the data are rather large, but even so there is definite proof of nonzero orientation at nonzero tilt angles. At $\alpha = 0°$ all parameters except A_0^{col} should be zero on the basis of cylindrical symmetry. That A_{2+}^{col} is nonzero possibly indicates experimental error larger than quoted, or some departure of carbon foils from strict cylindrical symmetry.

Electron Scattering from Optically Pumped Atoms

The second example of the relation of measured intensities to alignment and orientation parameters is provided by Hertel and Stoll's (1974a, b) measurement of electron scattering from optically pumped sodium atoms. Here optical pumping by circularly polarized light prepares the atom in a particular eigenstate of total angular momentum FM_F with $M_F = F$,

Table 2. Alignment and Orientation Parameters

Foil tilt angle	A_0^{col}	A_{2+}^{col}	A_{1+}^{col}	O_{1-}^{col}
0°	−0.090(36)	0.016(9)	—	—
20°	−0.081	0.012	−0.024(7)	−0.013(11)
30°	−0.072	0.008	−0.021	−0.038
45°	−0.054	−0.0002	−0.040	−0.040

and the Z axis is along the direction of the circular polarization vector. Collision excitation of the same atomic level by the time-inverse scattering process prepares a mixed state characteristic of the collision process. The scattered electron intensity is proportional to the degree of overlap of the two states, that is,

$$I = C\langle (i'|FM_F)(FM_F|i)\rangle \tag{11}$$

Here the measurement operator is simply the projection operator $\tau = |FM_F)(FM_F|$. It is expressed in terms of irreducible components $\tau^{[k]}q$ defined by

$$\tau^{[k]}q = \sum_{M'_F M''_F} |FM'_F)(FM''_F|(F - M''_F FM'_F|kq)(-1)^{k-F-M''_F} \tag{12}$$

as

$$\tau = \sum_{kq} \tau^{[k]}q(kq|F - M_F FM_F)(-1)^{k-F-M_F} \tag{13}$$

Substituting Eq. (13) into Eq. (11) using the Wigner–Eckart theorem then gives the scattered electron intensity in terms of the mean values $\langle T_0^{[k]}(ph)\rangle$ in a frame defined by the incident light direction:

$$I = C \sum_{kq} (kq|F - F, FF)(-1)^k \frac{(F\|\tau^{[k]}\|F)}{(F\|T^{[k]}\|F)} \langle i'|T_0^{[k]}(ph)|i\rangle \tag{14}$$

The final step transforms the parameters $T_0^{[k]}(ph)$ to the collision frame

$$\langle T_0^{[k]}(ph)\rangle = \sqrt{4\pi}\,(2k+1)^{-1/2} \sum_{pq=0}^{k} \langle T_{qp}^{[k]}(\text{col})\rangle Y_{qp}^{[k]}(\theta, \phi) \tag{15}$$

where θ, ϕ are the spherical coordinates of the laser light axis and $Y_{qp}^{[k]}$ are real spherical harmonics. Reflection symmetry in the scattering plane implies that

$$\begin{aligned} \langle T_{q-}^{[k]}(\text{col})\rangle &= 0 \quad \text{for } k = \text{even} \\ \langle T_{q+}^{[k]}(\text{col})\rangle &= 0 \quad \text{for } k = \text{odd} \end{aligned} \tag{16}$$

Note here that all multipole moments in Eq. (13) and Eq. (15) can be determined by suitable measurements with a variety of incident light directions. This contrasts with the coincidence experiments where only $k = 0, 1,$ and 2 enter. Applications of this theory will be discussed in a future publication (Macek and Hertel, 1974) and by Professor Hertel in this volume.

3. Effect of Weak Forces

In the second part of this paper I will discuss some of the physical interactions that affect the magnitude of the various alignment and orientation parameters. In a complete theory, such as the close-coupling theory (Burke and Smith, 1962), one can in principle incorporate every interaction, including the spin–orbit interaction, hyperfine interactions, and possibly external fields, into the calculation of scattering amplitudes. With these scattering amplitudes one can calculate all of the orientation and alignment parameters. Such a procedure is unnecessarily complicated and does not emphasize the rather different roles played by "small" magnetic and electric forces as compared to the stronger electrostatic forces responsible for collision excitation. That these "small" forces play a rather different role than the Coulomb force is illustrated by a simple example. Placing an atom in a magnetic field does not alter the cross section for exciting atom levels by electrons of moderate (a few eV) energy. On the other hand, because of the Hanle effect, the polarization of the collision-induced fluorescence may be substantially altered. Clearly, the "small" magnetic force affects the alignment parameters but has little effect on the excitation cross section. We expect in general that weak forces which can be safely ignored when calculating total excitation cross sections cannot be ignored when calculating alignment and orientation. One of the tasks of atomic collision theory will be to incorporate the "small" forces such as spin–orbit forces, the hyperfine interaction, external fields, and possibly the effect of outgoing charged particles on a collision-excited atom. Fortunately, the effect of the truly small forces due to external fields and the hyperfine interaction can be treated rather completely.

The "small" forces induce splittings of atomic levels. Transitions between split levels are characterized by a transition frequency v or a transition period $T = 1/v$. If the period T is large compared to the collision time T_{col} the small forces play no essential role during the collision; rather, they disturb the initial alignment and orientation after the collision. This disturbance is often slow enough to be observed by modern time-resolved spectroscopy. The beam-foil experiments by Andrä and co-workers (Wittman et al., 1973), Figure 2, on He 3^3P excitation provides a nice example of such measurements. The light intensity polarized parallel to the beam is seen to oscillate in time owing to the small fine structure splitting of the 3^3P level. Experiments by Burns and Hancock (1971), Figure 3, on this same transition show that it is the alignment rather than the total intensity which oscillates.

If $H(t)$ is the Hamiltonian operator corresponding to the small interaction, then the mean values of $S_{qp}^{[k]}$ (or $\tau_q^{[k]}$) vary with time according to

$$\langle S_q^{[k]}(t) \rangle = \left\langle (i' \left| \exp\left(i \int_0^t \mathscr{H}(t')\,dt'/\hbar\right) S_q^{[k]} \times \exp\left(-i \int_0^t \mathscr{H}(t')\,dt'/\hbar\right) \right| i) \right\rangle \quad (17)$$

Experiments performed without high time resolution measure the mean values $S_{qp}^{[k]}(t)$ averaged over the decay of the excited state. We are

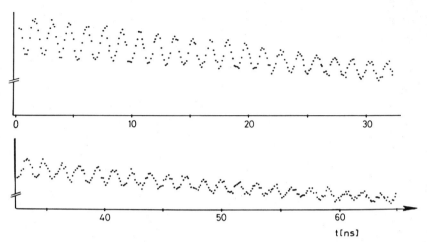

Figure 2. He $3^3P_1-3^3P_2$ zero-field quantum beats at 659 MHz. The lower part is a continuation of the upper part. From Wittman et al. (1973).

interested, however, in the mean values at $t = 0$ right after the collision. It is desirable, if possible, to factor the time modulation explicitly from the mean values. This can be done conveniently in two important situations, namely, when H represents a purely internal interaction and when H represents the effect of an external field with an axis of symmetry.

Internal Fields

Consider first the case when H represents an internal field. For definiteness, suppose H represents the hyperfine interaction. Then H is a scalar under transformations of atomic coordinates and the operator $S_{qp}^{[k]}(t)$ is still an irreducible tensor component. Normally, the nuclear spins are unpolarized in atomic collisions, and we are interested in the mean value of $S_{qp}^{[k]}(t)$ averaged over nuclear coordinates. This operator $\bar{S}_{qp}^{[k]} = \text{Tr}_{\text{nuc}}(I|S_q^{[k]}(t)|I)$ is still an irreducible tensor acting on electronic coordinates only. We then apply the Wigner–Eckart theorem to this new tensor operator to relate its mean values to the mean values of the operator $T_{qp}^{[k]}$, constructed from components of the electronic angular momentum $\mathbf{J}$ rather than the total angular momentum $\mathbf{F}$:

$$\langle (i'|S_{qp}^{[k]}(t)|i) \rangle = \frac{(J \| \bar{S}^{[k]}(t) \| J)}{(J \| T^{[k]} \| J)} \langle (i'|T_q^{[k]}|i) \rangle \qquad (18)$$

We see from Eq. (18) that for unpolarized nuclei, the hyperfine interaction modulates the magnitude of the parameters $T_{qp}^{[k]}$ but does not interlink different irreducible tensor components.

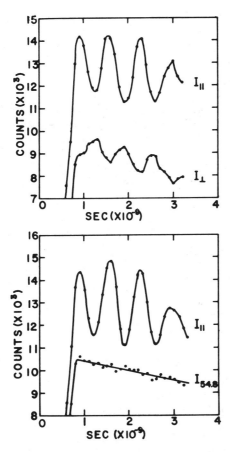

Figure 3. Zero-field modulations in the decay of foil-excited 3^3P states of helium, illustrating the zero phase of the oscillations and the absence of oscillations when the polarization analyzer is oriented at 54° 44′. From Burns and Hancock (1971).

The time-dependent factor in Eq. (18) has been written in essentially closed form by many authors (Fano and Macek, 1973, and references therein). It is

$$\frac{(J\|\bar{S}^{[k]}(t)\|J)}{(J\|T^{[k]}\|J)} = g^{(k)}(t)\frac{(J\|S^{[k]}\|J)}{(J\|T^{[k]}\|J)} \tag{19}$$

where

$$g^{(k)}(t) = \sum_{FF'} \frac{(2F+1)(2F'+1)}{(2I+1)}\begin{Bmatrix} F' & F & k \\ J & J & I \end{Bmatrix}^2 \cos\omega_{F'F}t \tag{20}$$

This factor, with $k = 2$, and F replaced by J and I by S to represent spin–orbit coupling, describes the oscillations shown in Figures 2 and 3. For low time resolution, $g^{(k)}(t)$ is averaged over the exponential decay $e^{-\gamma t}$, an operation which effectively replaces $\cos \omega_{F'F} t$ in Eq. (20) with the constant with $\gamma^2/(\gamma^2 + \omega_{F'F}^2)$. The resulting factor $\langle g^{(k)} \rangle$ was derived by Percival and Seaton (1958) for fluorescence induced by electron collisions. The time-averaged factor is always less than unity; thus the weak spin–orbit or hyperfine interactions only reduce the absolute magnitudes of the collision-induced anisotropy.

External Fields

A complete treatment for external fields analogous to Eqs. (20) or (18) has not been given. Various special cases, however, serve to indicate the effects that occur. We consider here the important case of an external field with an axis of symmetry. Now the operator $S_q^{[k]}(t)$ no longer transforms as an irreducible tensor component under coordinate rotations. We will use the tensor notation, however. For fields with an axis of symmetry it is natural to evaluate the matrix element of $S_q^{[k]}(t)$ in a frame with Z axis along the field axis. Then we have

$$(j_i m_i' | S_q^{[k]}(t) | j_i m_i) = \exp\left(i \int_0^t [E_{m_i'}(t') - E_{m_i}(t')] \, dt'/\hbar\right)(j_i m_i' | S_q^{[k]} | j_i m_i) \qquad (21)$$

where $E_{m_i}(t)$ is the instantaneous eigenvalue of $H(t)$. Owing to the assumed axis of symmetry of $H(t)$, m_i will be a good quantum number. If the difference $E_{m_i'}(t') - E_{m_i}(t')$ is proportional to $m_i' - m_i$, as for the weak-field Zeeman effect, then we see that $S_q^{[k]}(t)$ is merely multiplied by the phase factor $q\chi(r)$, where $\chi(t)$ is a phase angle given by $q^{-1} \int_0^t [E_{m'}(t') - E_m(t')] \, dt'$. The phase factor does not depend upon m' and m individually but only on q; consequently it does not enter into the average defining $S_q^{[k]}(t)$. The only effect of the field is to multiply each mean value by $\exp[iq\chi(t)]$, equivalent to a rotation through $\chi(t)$ about the field axis. Each of tensorial components $S_q^{[k]}$ calculated at $t = 0$ in the collision frame remains unchanged in a rotating frame which coincides with the collision frame at $t = 0$. Note here that the field interlinks different q components of the irreducible tensors, but leaves the rank k unchanged.

The situation is quite different when $E_{m_i'} - E_{m_i}$ depends explicitly on both m_i' and m_i. Then the effect of the phase factor in Eq. (21) cannot be represented as a rotation and the alignment or orientation at $t = 0$ depends upon octapole and higher-rank state multipoles. This can be seen directly by attempting to evaluate the reduced matrix element of Eq. (21),

$$\sum_{m_i'm_i} \begin{pmatrix} j_i & k' & j_i \\ -m_i' & q & m_i \end{pmatrix} (j_i m_i | S_q^{[k]}(t) | j_i m_i)$$

$$= \sum_{m_i'm_i} \begin{pmatrix} j_i & k' & j_i \\ -m_i' & q & m_i \end{pmatrix} \exp[\chi_{m_i'm_i}(t)] \begin{pmatrix} j_i & k & j_i \\ -m_i' & q & m_i \end{pmatrix} (j_i \| S^{[k]} \| j_i) \quad (22)$$

where

$$\chi_{m_i'm_i}(t) = \int_0^t [E_{m_i'}(t') - E_{m_i}(t')] \, dt'$$

Owing to the factor $\exp \chi_{m_i'm_i}(t)$], the sum in Eq. (22) is nonzero even when $k' = k$. If $\chi_{m_i'm_i}(t)$ depended only upon the difference $m_i' - m_i$, then the exponential would factor out and the product of the $3 - j$ symbols would vanish unless $k' = k$.

Lombardi (1969) has made an interesting application of this effect of external fields to produce orientation by placing an electric field crosswise on helium atoms excited and aligned but not oriented in one of the P states. The emitted light was found to be circularly polarized, showing the orientation arising from the coupling of an initial alignment in an external field.

The external electric field also generates nonzero components of the alignment tensor with $q \neq 0$. For P states aligned by collision excitation with an electric field at an angle α to the initial alignment, the induced orientation and alignment $O(\alpha)$ and $A(\alpha)$ are given by

$$O_{1-}^{col}(\alpha) = -A_0^{col}(o)\tfrac{1}{2} \sin \chi \sin 2\alpha$$

$$A_0^{col}(\alpha) = A_0^{col}(o)[1 - \tfrac{3}{2} \sin^2 2\alpha \sin^2 \chi/2]$$

$$A_{1+}^{col}(\alpha) = A_0^{col}(o)\tfrac{1}{2} \sin 4\alpha \sin^2 \chi/2 \quad (23)$$

$$A_{2+}^{col}(\alpha) = A_0^{col}(o)\tfrac{1}{2} \sin^2 2\alpha \sin^2 \chi/2$$

where $\chi = \chi_{01}(t = \infty)$.

The induction of orientation by electric fields is a matter of some importance for collision theory. The Lombardi effect is measurable with fields as low as 100 V/cm. An outgoing ion that has excited an atom by collision must be further than 10^{-5} cm away from the atom before its field falls below 100 eV/cm. If the ion moves slowly enough the integrated effect of the field could alter the alignment and orientation.

This might serve to explain Berry's results (Berry et al., 1974), where orientation is produced by tilted foils. Carbon foils presumably have quite rough surfaces so that it is difficult to see immediately how the surface direction can produce a net effect. We know that foils must be charged to some extent, if only because electrons are knocked out by the incident ions. If we suppose that microfields extend into space normal to the foil on the average, then these fields could induce orientation according to Eq. (23). Note that $A_0^{col}(\alpha)$ must decrease with increasing α while $O_1^{col}(\alpha)$ increases. This trend is in qualitative agreement with the data of Berry et al. (1974), although quantitative agreement with all of the data is lacking. In particular, $A_{1+}^{col}(\alpha)$ must vanish at 45°, whereas Berry's data show a maximum at that angle.

4. The Glauber Theory

A complete scattering theory calculation such as the close-coupling approximation (Burke and Smith, 1962) or the R-matrix approach (Fano and Lee, 1973; Burke et al., 1973) incorporates the long-range forces explicitly by fitting the scattering wave function to an asymptotic expansion incorporating all the long-range interactions. More approximate formulations which leave out the long-range potentials could be quite incorrect for purposes of calculating alignment and orientation. One might allow for this using formulas similar to Eq. (23); however, it is difficult to select a time (or a distance) when the collision is over in the expression for the phase factor in Eq. (21). One really needs a theory incorporating the phase distortion into the calculation of scattering amplitudes. Distorted-wave approximations are one possibility, and it is interesting to note that such calculations by Madison and Shelton (1973) for electron excitation of helium agree well with the data of Kleinpoppen and co-workers (Eminyan et al., 1973).

The Glauber (1959) theory represents another distorted-wave approximation currently enjoying some popularity in atomic physics. One might hope that it would treat the long-range or at least the intermediate-range forces well enough to give some insight into orientation and alignment at energies below the region of validity of the Born theory. Unfortunately, the Glauber transition operator has two planes of symmetry, the scattering plane and a plane perpendicular to the incident beam direction. Because of this additional plane of symmetry, the orientation parameter O_{1-}^{col} vanishes as does A_{1+}^{col}.

This additional plane of symmetry originates from the assumption of purely transverse momentum transfer in the conventional Glauber approximation. Byron (1971) introduced a version without this latter assumption. His calculation required the numerical evaluation of a six-dimensional integral for electron–hydrogen scattering. We have recently expressed Byron's approximate amplitudes in terms of double integrals (Macek and Gau, 1974) and have evaluated the predicted orientation of Lyman α

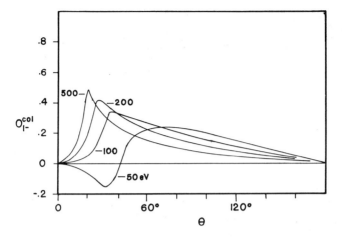

Figure 4. Orientation of hydrogen 2p states excited by electron collision in Byron's version of the Glauber approximation.

radiation excited by electron collisions with atomic hydrogen. Figure 4 shows our results. One can see that nonzero orientation is predicted by Byron's version of the Glauber approximation. Note also that near zero degrees, where either the momentum transfer is purely transverse or where there is approximate cylindrical symmetry about the beam axis as in Imhof and Read's (1971) experiment, the orientation vanishes. This example illustrates the importance of phase distortions when calculating orientation. Whether the predicted orientation is correct must await experiment or more soundly based theoretical calculations.

Acknowledgment

This work was supported by National Science Foundation Grant No. GP-39310.

References

Berry, H. G., Curtis, L. J., Ellis, D. G., and Schectman, R. M. (1974). *Phys. Rev. Lett.*, **32** 751.
Burgess, A., and Percival, I. C. (1970). In *Advances in Atomic and Molecular Physics*, Vol. 4, p. 709. Eds. D. R. Bates and I. Esterman, New York: Academic Press.
Burke, P. G., Hibbert, A., and Robb, W. D., *Proc. Phys. Soc. Lond.*, **4**, 153.
Burke, P. G., and Smith, K. (1962). *Rev. Mod. Phys.*, **34**, 458.

Burns, D. J., and Hancock, W. H. (1971). *Phys. Rev. Lett.*, **2F**, 370.
Byron, F. W. (1971). *Phys. Rev.*, **A4**, 1907.
Eminyan, M., MacAdam, K. B., Slevin, J., and Kleinpoppen, H. (1973). *Phys. Rev. Lett.*, **31**, 576; see also *J. Phys. B.*, **7**, 1519 (1974).
Fano, U. (1951). National Bureau of Standards, Report No. 1214.
Fano, U. (1957). *Rev. Mod. Phys.*, **29**, 74.
Fano, U., and Lee, C. M. (1973). *Phys. Rev. Lett.*, **31**, 1573.
Fano, U., and Macek, J. (1973). *Rev. Mod. Phys.*, **45**, 553.
Glauber, R. J. (1959). In *Lectures in Theoretical Physics*, Eds. W. E. Britten and L. G. Duncan, New York, Vol. 1: Interscience.
Hertel, I. V., and Stoll, W. (1974a). *J. Phys. B.*, **7**, 570.
Hertel, I. V., and Stoll, W. (1974b). *J. Phys. B.*, **7**, 583.
Imhof, R. E. (1969), and Read, F. H. (1971). *J. Phys. B.*, **4**, 450.
Lombardi, M. (1969). *J. Phys. Radium*, **30**, 631.
McKnight, R., and Jaecks, D. H. (1971). *Phys. Rev.*, **A4**, 2281.
Macek, J., and Gau, J. (1974). *Phys. Rev.*, **A10**, 522.
Macek, J., and Hertel, I. V. (1974). *J. Phys. B.*, **7**, 2173.
Madison, D. H., and Shelton, W. N. (1973). *Phys. Rev.*, **A7**, 499.
Percival, I. C., and Seaton, M. J. (1958). *Phil. Trans. Roy. Soc. Lond.*, A **251**, 113.
Rau, A. R. P. (1971). *Phys. Rev.*, **A4**, 207.
Wannier, G. H. (1953). *Phys. Rev.*, **90**, 817.
Wittman, W., Tillmann, K., and Andrä, H. J. (1973). *Nucl. Instr. & Meth.*, **110**, 1.

40
On the Theory of Electron–Photon Coincidence Experiments

K. BLUM AND H. KLEINPOPPEN

The theory of electron–photon angular correlations previously given by other authors has been reformulated in terms of density matrices in the helicity representation. The formalism is developed in some detail because the methods used and most of the formulas obtained are quite general and can be applied to atoms excited by electron impact as well as other mechanisms. The special example for the excitation of singlet P states of helium is discussed in detail and reference is made to beam–foil excitation of the same states.

1. Introduction

The theory of coincidence experiments has been given by various authors (Macek and Jaecks, 1971; Wykes, 1972; Fano and Macek, 1973). We are in particular interested in the formulation given in Section 1 of Fano and Macek's paper. It is our main purpose to give an equivalent but more abstract description in terms of density matrices in the helicity representation. This method gives a compact formulation of the theory and, in particular, is useful for the discussion of more complex cases. Furthermore, the polarization state of the emitted photons is not a quantity characteristic of the emission process itself but of the radiation detectors which distinguish definite polarizations of the photons. In the density matrix formalism (with only the emission direction fixed) it is not necessary to specify the polarization from the beginning. Thus, one obtains a more "natural" description of the radiation process.

K. BLUM and H. KLEINPOPPEN • Institute of Atomic Physics, University of Stirling, Stirling, Scotland.

We present the formalism in some detail because the methods used and most of the formulas obtained are quite general and can be applied to atoms excited by electron impact as well as other mechanisms. Special examples will be discussed in order to illustrate the more abstract concepts.

Before starting with the general theory in Section 3, we give in Section 2 an introduction into the theory of density matrix and state multipoles and into the description of atomic ensembles in terms of these quantities. Some examples are discussed in detail.

In Section 3 we discuss angular distribution and polarization of the radiation resulting from the decay of an ensemble of excited atoms. This information is included in the polarization density matrix of the photons. It is our main purpose to derive an expression for the elements of this matrix in which the quantities characterizing the excited atoms, the quantities describing the dynamics of the decay, and the geometrical elements are separated. In order to show how special information can be extracted from this general equation, we derive expressions for the Stokes parameters and for the intensity of photons emitted in a fixed direction with a fixed polarization vector. As an illustration we discuss some recent experiments.

Throughout this section we write the density matrix in the helicity representation (the helicity is the component of the angular momentum in the direction of motion). This method has been applied frequently, particularly in elementary particle physics (see, for example, Gasiorowicz, 1966). It allows one to take full advantage of angular momentum techniques in an elegant manner. In particular, it is the invariance of the helicity under rotation which simplifies the arguments (Jacob and Wick, 1959).

2. Description of Atomic States in Terms of Density Matrices and Multipole Moments

The Density Matrix

We consider an ensemble of atoms in a state with sharply defined angular momentum j but in an arbitrary superposition of magnetic sublevels M. In the following we are only interested in the angular momentum properties of the system and we suppress all other indices. The formulas we are going to derive can be used only in connection with questions relating to these particular properties under consideration and give no information about the suppressed variables.

If all atoms of the ensemble are described by a single wave function $|\psi\rangle$, the ensemble is said to be in a pure state. Choosing the angular momentum states $|j, M\rangle$ as a basis, we express $|\psi\rangle$ in terms of these states:

$$|\psi\rangle = \sum_M a_M |j, M\rangle \tag{1}$$

In general the ensemble must be described as a statistical mixture of the pure states $|\psi^{(n)}\rangle$ in which the atoms may be found. It is convenient to describe this system in terms of a density operator ρ, which is by definition (Fano, 1957)

$$\rho = \sum_n W_n |\psi^{(n)}\rangle\langle\psi^{(n)}| \tag{2}$$

where W_n is the probability of finding the pure state $|\psi^{(n)}\rangle$ in the mixture.

Expressing each state $|\psi^{(n)}\rangle$ in terms of the basis $|j, M\rangle$, we write

$$|\psi^{(n)}\rangle = \sum_M a_M^{(n)} |j, M\rangle \tag{3}$$

and from Eq. (2) follows

$$\rho = \sum_{nMM'} W_n a_{M'}^{(n)} a_M^{(n)*} |j, M'\rangle\langle j, M| \tag{4}$$

Choosing the states $|j, M\rangle$ as a basis we obtain a special $(2j + 1)$-dimensional matrix representation for ρ. We obtain the element in the m'th row and mth column of this density matrix by multiplying Eq. (4) from the left by $\langle j, m'|$ and from the right by $|j, m\rangle$ and by using the orthogonality relation $\langle j, m|j, M\rangle = \delta_{mM}$:

$$\rho_{m'm} \equiv \langle jm'|\rho|jm\rangle = \sum_n W_n a_{m'}^{(n)} a_m^{(n)*} \tag{5}$$

From the definition follows that ρ is hermitian: $\rho_{M'M} = \rho_{MM'}^*$. Perhaps more explicitly, the matrix representation (5) can be obtained in the following way. We write $|\psi^{(n)}\rangle$ in the form of a standard column vector

$$|\psi^{(n)}\rangle = \begin{pmatrix} a_j^{(n)} \\ a_{j-1}^{(n)} \\ \vdots \\ a_{-j}^{(n)} \end{pmatrix}$$

and the hermitian adjoint state $\langle\psi^{(n)}|$ as the row vector

$$\langle\psi^{(n)}| = (a_j^{(n)*}, a_{j-1}^{(n)*}, \ldots, a_{-j}^{(n)*})$$

where the asterisk denotes complex conjugation. We calculate the direct products,

$$|\psi^{(n)}\rangle\langle\psi^{(n)}| = \begin{pmatrix} a_j^{(n)} \\ \vdots \\ a_{-j}^{(n)} \end{pmatrix} (a_j^{(n)*}, \ldots, a_{-j}^{(n)*})$$

$$= \begin{pmatrix} a_j^{(n)} a_j^{(n)*} & a_j^{(n)} a_{j-1}^{(n)*} & \cdots & a_j^{(n)} a_{-j}^{(n)*} \\ a_{j-1}^{(n)} a_j^{(n)*} & \cdots & & \\ \vdots & & & \\ a_{-j}^{(n)} a_j^{(n)*} & \cdots & & a_{-j}^{(n)*} a_{-j}^{(n)*} \end{pmatrix}$$

and, multiplying by W_n and summing over all n, we obtain the representation (5).

The expectation value $\langle Q \rangle$ for any operator acting on the states $|\psi^{(n)}\rangle$ is defined by

$$\langle Q \rangle = \sum_n W_n \frac{\langle \psi^{(n)} | Q | \psi^{(n)} \rangle}{\langle \psi^{(n)} | \psi^{(n)} \rangle} \tag{6a}$$

This can be expressed in terms of the density matrix (Fano, 1957):

$$\langle Q \rangle = \frac{\text{Tr}\,\rho Q}{\text{Tr}\,\rho} \tag{6}$$

Spherical Tensor Operators and State Multipoles

Instead of decribing the atomic ensemble by its density matrix, the atoms can be characterized by the expectation values of a complete, linearly independent set of operators acting on the states $|j, M\rangle$ (Fano, 1957). If the angular symmetries of the system are of interest it is convenient to use a set of spherical tensor operators T_{KQ} (where K denotes the rank of the tensor and Q the spherical component). We start with the definition and general properties of these quantities and then discuss some examples as an illustration.

For a given ensemble of atoms with definite value j of angular momentum, we define the related tensor operators by making use of the angular momentum addition rules (Fano, 1953):

$$T(j)_{KQ} = (2j+1)^{1/2} \sum_{M'M} (-1)^{j-M} (jM', j, -M|KQ)|jM'\rangle\langle jM| \tag{7}$$

The Clebsch–Gordan coefficient $(jM', j - M|KQ)$ "projects" out of the products $(-1)^{j-M}|jM'\rangle\langle jM|$ the irreducible parts, and thus the tensor operators of rank K transform according to the irreducible representation $D^{(K)}$ of the rotation group. From the angular momentum addition rules follow the restrictions $K \leqslant 2j$, $-K \leqslant Q \leqslant K$. We note that the quantities (7) are not hermitian but satisfy the condition

$$T_{KQ}^{+} = (-1)^{Q} T_{K,-Q} \tag{8}$$

The operators are normalized according to

$$\operatorname{Tr} T(j)_{KQ} T(j)_{K'Q'}^{+} = (2j+1)\delta_{KK'}\delta_{QQ'} \tag{9}$$

Using the states $|jM\rangle$ as a basis, we can derive a special matrix representation for $T(j)_{KQ}$, analogously to the derivation of Eq. (5). The element in the m'th row and mth column is given by the expression

$$\langle jM'|T(j)_{KQ}|jM\rangle = (2j+1)^{1/2}(-1)^{j-M}(jM', j-M|KQ) \tag{10}$$

It is sometimes convenient to express the tensor operators T_{KQ} in terms of angular momentum operators. By making use of the angular momentum addition rules, we write ($K \leqslant 2$)

$$T(j)_{KQ} = N_K \sum_{q_1 q_2} (1q_1, 1q_2|KQ) J_{q_1} J_{q_2} \tag{11}$$

where we used that the angular momentum J transforms as a rank-1 tensor. J_q denotes its qth spherical component, defined by

$$J_0 = J_z, \quad |J_{\pm 1} = \mp(2)^{-1/2}(J_x \pm iJ_y) \tag{12}$$

In Eq. (11) and the following formulas, the operators J_q are understood to be the standard angular momentum operators for fixed j (characterized by their algebraic properties or written as $(2j+1)$-dimensional matrices).

The normalization factor N_k is determined by calculation of the matrix elements of both T_{KQ} and J_q, and comparing them we obtain

$$N_K = \left[(-1)^{K+2j} \begin{Bmatrix} 1 & 1 & K \\ j & j & j \end{Bmatrix} (2j+1)^{1/2}(j+1)j \right]^{-1} \tag{13}$$

For convenience we give the explicit expressions for the lower tensors. By specializing Eqs. (11) and (13) to $K = 1$ and using the commutation relations of the operators J_q, we obtain

$$T(j)_{1Q} = \left[\frac{3}{j(j+1)}\right]^{1/2} J_q \tag{14a}$$

Specializing to $K = 2$ and turning to the Cartesian components by using Eq. (12), we obtain

$$T(j)_{20} = N_2(6)^{-1/2}(3J_z^2 - \mathbf{J}^2)$$
$$T(j)_{2\pm 1} = \mp N_2(2)^{-1/2}[(J_x J_z + J_z J_x) \pm i(J_y J_z + J_z J_y)] \tag{14b}$$
$$T(j)_{2\pm 2} = N_2(2)^{-1/2}[J_x^2 - J_y^2 \pm i(J_x J_y + J_y J_x)]$$

where N_2 is given by

$$N_2 = (30)^{1/2}[(2j + 3)(j + 1)j(2j - 1)]^{1/2}$$

The expectation values $\langle T_{KQ} \rangle$ are referred to as statistical tensors or state multipoles (Fano, 1953). From Eq. (14) follows that $\langle T_{1Q} \rangle$ is proportional to the average angular momentum component $\langle J_q \rangle$ of the atomic ensemble. The quantities $\langle T_{2,Q} \rangle$ are proportional to the averages of second-rank tensors constructed from angular momentum components. Using Eq. (14), the combinations $\langle T_{KQ} \pm T_{K-Q} \rangle$ may be calculated. These quantities are proportional to the components of orientation vector and alignment tensor defined by Fano and Macek (1973).

The various electric and magnetic multipole operators Q_{KQ} of atoms are related to the corresponding operators T_{KQ}. Applying the Wigner–Eckart theorem to the matrix elements of both operators and eliminating the common $3j$ symbol, we obtain

$$\langle jM'|Q_{KQ}|jM \rangle = \frac{\langle j\|Q_K\|j \rangle}{\langle j\|T_K\|j \rangle} \langle jM'|T(j)_{KQ}|jM \rangle \tag{15}$$

Using the conventions of Edmonds (1957) we obtain for the reduced matrix element $\langle j\|T_K\|j \rangle = [(2j + 1)(2K + 1)]^{1/2}$. As an example, the electric quadrupole moment Q is defined as the matrix element:

$$Q = \langle jj|Q_{20}|jj \rangle \tag{16}$$

Applying the Wigner–Eckart theorem to this element, the reduced matrix element $\langle j\|Q_2\|j \rangle$ may be expressed in terms of Q. Using Eqs. (6a) and (15), we obtain a relation for the expectation values of Q_{2Q} and T_{2K} (for atomic states with fixed j):

$$\langle Q_{2Q} \rangle = \frac{Q}{\begin{pmatrix} j & j & 2 \\ j & -j & 0 \end{pmatrix}[5(2j + 1)]^{1/2}} \langle T_{2Q} \rangle \tag{17}$$

Density Matrix Expansion

The set $T(j)_{KQ}$ is complete, that is, any operator acting on atomic states with fixed j can be written as a linear combination of the operators $T(j)_{KQ}$. That is in particular true for the density matrix (5). In order to expand ρ in terms of the tensor operators, we make the ansatz (Fano, 1957)

$$\rho = \sum_{KQ} b_{KQ} T(j)_{KQ} \tag{18}$$

where the sum runs over all $K \leq 2j$ and all Q satisfying $-K \leq Q \leq K$.

From Eqs. (6) and (9) follows

$$\langle T^+_{KQ}\rangle = \frac{\text{Tr}\,\rho T^+_{KQ}}{\text{Tr}\,\rho} = \frac{(2j+1)b_{KQ}}{\text{Tr}\,\rho}$$

and thus

$$\rho = \frac{(\text{Tr}\,\rho)}{(2j+1)}\sum_{KQ}\langle T(j)^+_{KQ}\rangle T(j)_{KQ} \qquad (19)$$

Equation (19) expresses ρ in terms of the $(2j+1)^2$ state multipoles $\langle T(j)_{KQ}\rangle$. (Note that due to the hermiticity condition $\rho_{M'M} = (\rho_{MM'})^*$, just $(2j+1)^2$ real independent parameters are required to characterize ρ.) In particular, Eq. (19) is useful in all cases where the angular symmetries of the system are of interest, and we use it as starting point in the next section.

Up to this point the discussion has been quite general and independent of the excitation process. All the information on how the atoms have actually been excited are carried by the quantities $\langle T(j)_{KQ}\rangle$. Let us now discuss an ensemble of atoms excited by electrons scattered in a fixed direction. The excited atoms can be described as a coherent superposition of magnetic sublevels as in Eq. (1) (Macek and Jaecks, 1971):

$$|\psi\rangle = \sum_M a_M |jM\rangle \qquad (1)$$

where the amplitudes a_M are functions of energy and scattering angle of the electrons and an average over unobserved variables. For simplicity we have assumed a fixed j value. We use a Cartesian coordinate system in which the z axis (quantization axis) coincides with the direction of the ingoing electrons and where the xz plane is the scattering plane (collision system). We normalize to

$$\langle\psi|\psi\rangle = \sum_M |a_M|^2 = \sigma \qquad (20)$$

where σ is the differential electron–atom scattering cross section.

The elements of the density matrix, characterizing the pure state (1), can be constructed from the amplitudes a_M as in Eq. (5):

$$\rho_{M'M} = a_{M'} a^*_M \qquad (21)$$

With the normalization (20) the trace is given by

$$\text{Tr}\,\rho = \sigma \qquad (22)$$

The state multipoles $\langle T_{KQ}\rangle$ are functions of the excitation amplitudes a_M. From Eqs. (6) and (22) follows

$$\sigma\langle T(j)_{KQ}\rangle = \sum_{M'}\langle jM'|\rho T(j)_{KQ}|jM'\rangle = \sum_{M'M}\langle jM'|\rho|jM\rangle\langle jM|T(j)_{KQ}|jM'\rangle$$

and using Eqs. (10) and (21), we obtain

$$\sigma \langle T(j)_{KQ} \rangle = (2j+1)^{1/2} \sum_{M'M} a_{M'} a_M^* (-1)^{j-M'} (jM, j - M'|KQ) \qquad (23)$$

This relation can, of course, be obtain immediately from the definition

$$\langle T_{KQ} \rangle = \frac{\langle \psi | T_{KQ} | \psi \rangle}{\langle \psi | \psi \rangle} \qquad (24)$$

Example

As an example we discuss the experiment of Eminyan et al. (1974). (See also the paper of Eminyan et al. in this volume.) In this experiment a 2^1P state of helium is produced from the ground state by electron impact. The photons emitted in the following decay process and the scattered electrons are measured in coincidence. The state of the excited atoms is characterized by Eqs. (1), (21), or (23) (with $j = 1$). Due to reflection invariance in the scattering plane we have the relation (Macek and Jaecks, 1971)

$$a_1 = -a_{-1} \qquad (25)$$

or

$$\langle T_{KQ} \rangle = (-1)^{K+Q} \langle T_{K-Q} \rangle = (-1)^K \langle T_{KQ} \rangle^* . \qquad (26)$$

Thus $\langle T_{10} \rangle$ is zero, $\langle T_{1, \pm 1} \rangle$ imaginary, and the quantities $\langle T_{2,Q} \rangle$ real. We calculate the imaginary part of $\langle T_{1, \pm 1} \rangle$ and the real part of $\langle T_{2,Q} \rangle$ and turn to Cartesian tensors with the help of Eq. (14). We use Eq. (20) and the same notation as in Eminyan's paper (that is, we choose a_0 real and define the quantities Λ and $\varkappa$ by $|a_0|^2 = \Lambda \sigma$, $a_1 = |a_1|e^{i\varkappa}$). The results are

$$\begin{aligned}
\langle J_y \rangle &= -2[\Lambda(1-\Lambda)]^{1/2} \sin \varkappa &&= 2O_{1-} \\
\langle J_x^2 - J_y^2 \rangle &= \Lambda - 1 &&= 2A_{2+} \\
\langle J_x J_z + J_z J_x \rangle &= [\Lambda(1-\Lambda)]^{1/2} \cos \varkappa &&= 2A_{1+} \\
\langle 3J_z^2 - \mathbf{J}^2 \rangle &= 1 - 3\Lambda &&= 2A_{0+}
\end{aligned} \qquad (27)$$

Equation (27) may be compared with Eqs. (15) to (18) in Eminyan's paper (1974) (note that in Eqs. (16) and (17) of this paper a factor $\frac{1}{2}$ is missing on the right side). On the right side in Eq. (24) we expressed the tensors in terms of the components of orientation vector and alignment tensor (Fano and Macek, 1973).

We note that the quantities (27) can be derived simply as expectation values of the standard angular momentum-1 matrices J_i and the symmetric, traceless second-rank tensors $J_{ik} = \frac{3}{2}(J_i J_k + J_k J_i) - \delta_{ik} \mathbf{J}^2$ ($i, k = x, y, z$).

$j = 1$ states can be completely characterized in terms of this set. The algebraic properties of the J_i and J_{ik} may be found in the literature (see, for example, Blum and Kleinpoppen, 1973).

3. Characterization of the Emitted Radiation

We consider an ensemble of atoms excited by electron impact and the subsequent decay by photon emission. Scattered electrons and emitted photons are detected in coincidence. We are going to derive an expression for the polarization density matrix of the photons.

We choose a fixed right-handed orthogonal xyz coordinate system in such a way that the z axis coincides with the direction of the ingoing electrons and the fixed momentum of the scattered electrons lies in the xz plane (collision system). The emitted photons may be observed in the direction of the unit vector **n** with polar angles θ, φ.

It is essential that by fixing the directions of incident and scattered electrons a plane (scattering plane) is selected. It is with respect to this plane that the polar angles θ, φ can be defined in a physically reasonable way and the dependence of photon intensity and polarization on these angles can be observed. Without observing the scattered particles, only a single axis (beam axis of the incident electrons) is defined. The first consistent theory of this type of experiment was given by Percival and Seaton (1958). Due to the rotational symmetry of the system about the beam axis the intensity is isotropic in the plane perpendicular to this axis. Thus, one obtains only an average over the φ dependence, and a certain amount of information is lost.

The polarization vector of the photons is restricted to the plane perpendicular to **n**. Two orthogonal unit vectors $\mathbf{e}_\alpha$, $\mathbf{e}_\beta$ may be chosen to "span" this plane in such a way that $\mathbf{e}_\alpha$ has polar angles $(\theta + 90°, \varphi)$ and $\mathbf{e}_\beta$ the angles $(0, \varphi + 90°)$. From these vectors we construct the circular unit vectors

$$\mathbf{e}_{\pm 1} = \mp (2)^{-1/2}(\mathbf{e}_\alpha \pm i\mathbf{e}_\beta) \tag{28}$$

These vectors describe left or right circularly polarized photons or, using a more modern terminology, photons with helicity $\lambda = \pm 1$. It is convenient to choose a coordinate system where the basis vectors are **n** and the helicity vectors $\mathbf{e}_\lambda$, defined in Eq. (28), and where **n** is the quantization axis. In the following we use this "helicity system."

The polarization density matrix ρ of the photons is a two-dimensional matrix in the plane perpendicular to **n**. ρ contains all information about angular distribution I and polarization of the radiation. A discussion of the density matrix description of photon beams may be found in the literature (for example, Akhiezer and Berestetski, 1963; McMaster, 1954). For example, consider a completely polarized photon beam with polarization vector $\boldsymbol{\varepsilon}$, which may be expanded in terms of the helicity vectors $\boldsymbol{\varepsilon} = \sum_{\lambda = \pm 1} a_\lambda \mathbf{e}_\lambda$. The density matrix can then be represented in terms of the expansion coefficients a_λ [analogously to Eq. (21)]. The element in the λ'th row and λth column are given by $\rho_{\lambda'\lambda} = a_{\lambda'} a_\lambda^*$.

We discuss transitions between sharply defined atomic levels $j_i \to j_f$, where $j_i(j_f)$ is the angular momentum of the excited (deexcited) atom. The state of the excited atom may be described by a density matrix ρ_{in}, which we expand in terms of the spherical tensor operators (7). This expansion is given by Eq. (19):

$$\rho_{\text{in}} = \frac{\sigma}{2j_i + 1} \sum_{KQ} \langle T(j_i)_{KQ}^+ \rangle T(j_i)_{KQ}$$

normalized to

$$\text{Tr}\,\rho = \sigma$$

After the atomic decay the combined photon–atom system is described by a density matrix ρ_{out}, connected with ρ_{in} by the expression

$$\rho_{\text{out}} = \hat{M} \rho_{\text{in}} \hat{M}^+ \qquad (29)$$

$\hat{M}$ is the operator which transforms the initial state (excited atom) into the final one (atom and photon). In the dipole approximation its elements are given by

$$\langle j_f M_f, \mathbf{n}\lambda | \hat{M} | j_i M_i \rangle = \frac{e\omega^2}{(2\pi c^3)^{1/2}} \langle j_f M_f | \mathbf{e}_\lambda^* \mathbf{r} | j_i M_i \rangle \qquad (30)$$

where $\mathbf{r}$ is the transition dipole operator. $|j_f M_f, \mathbf{n}\lambda \rangle$ describes the combined state of atoms in the angular momentum state $|j_f M_f \rangle$, and photons emitted in the direction $\mathbf{n}$ with helicity λ (we suppressed the dependence on all other variables). ρ_{out} is normalized to $\text{Tr}\,\rho_{\text{out}} = I$. Expanding $\mathbf{r}$ in terms of the basis $\mathbf{e}_\lambda$, we obtain $\mathbf{e}_\lambda^* \mathbf{r} = \mathbf{r}_\lambda^+$.

The elements of the photon polarization density matrix in the helicity representation are given by "sandwiching" ρ_{out} between helicity states and summing over the magnetic quantum number of the unobserved atoms:

$$\rho_{\lambda'\lambda} = \sum_{M_f} \langle j_f M_f, \mathbf{n}\lambda' | \rho_{\text{out}} | j_f M_f, \mathbf{n}\lambda \rangle$$

$$= \sum_{M_f M_i M_i'} \langle j_f M_f, \mathbf{n}\lambda' | \hat{M} | j_i M_i' \rangle \langle j_i M_i' | \rho_{\text{in}} | j_i M_i \rangle \langle j_i M_i | \hat{M}^+ | j_f M_f, \mathbf{n}\lambda \rangle$$

Electron–Photon Coincidence Experiments

From Eqs. (16) and (30) and from the Wigner–Eckart theorem follows

$$\rho_{\lambda'\lambda} = \overline{W} \cdot \sigma \sum_{K'Q'} (2K'+1)^{1/2} \langle T^+_{K'Q'} \rangle \sum_{M_f M_i M'_i} (-1)^{j_i - M_i + Q'} \\ \begin{pmatrix} j_f & 1 & j_i \\ -M_f & -\lambda' & M'_i \end{pmatrix} \begin{pmatrix} j_f & 1 & j_i \\ -M_f & -\lambda & M_i \end{pmatrix} \begin{pmatrix} j_i & j_i & K' \\ M'_i & -M_i & -Q' \end{pmatrix} \tag{31}$$

$$= \overline{W} \sigma \sum_{K'Q'} (2K'+1)^{1/2} \langle T(j_i)_{K'Q'} \rangle (-1)^{j_i + j_f + \lambda'} \\ \times \begin{pmatrix} 1 & 1 & K' \\ -\lambda & \lambda' & Q' \end{pmatrix} \begin{Bmatrix} 1 & 1 & K' \\ j_i & j_i & j_f \end{Bmatrix} \tag{31a}$$

where we used the notation

$$\overline{W} \equiv \frac{e^2 w^4 |\langle j_f \| \mathbf{r} \| j_i \rangle|^2}{2\pi c^3 (2j_i + 1)^{1/2}} \tag{32}$$

The sum over the 3j symbols in Eq. (31) has been related to a 6j symbol by using standard techniques [the relevant expression is given, for example, by Eq. (6.2.8) of Edmonds (1957)]. Equation (31a) gives the elements $\rho_{\lambda'\lambda}$ in the helicity frame (n as quantization axis). We transform to the collision system (z as quantization axis) by means of a rotation, using the transformation properties of the tensor operators. We obtain

$$\rho_{\lambda'\lambda} = \overline{W}\sigma(-1)^{j_i + j_f + \lambda'} \sum_{K'Q'} (2K'+1)^{1/2} \begin{pmatrix} 1 & 1 & K' \\ -\lambda & \lambda' & Q' \end{pmatrix} \begin{Bmatrix} 1 & 1 & K' \\ j_i & j_i & j_f \end{Bmatrix} \\ \times \sum_Q \langle T(j_i)_{KQ} \rangle (D^{(K)}_{Q'Q}(0\theta\varphi))^* \tag{33}$$

Equation (33) is our main result. It contains all information about angular distribution I and polarization in a compact form. The formula consists of several parts different in nature. The expectation values $\langle T_{KQ} \rangle$ carry the information on how the atoms have actually been excited. The dynamics of the decay process is contained in the reduced matrix elements. All other factors are connected with the geometry. The angular dependence is given explicitly by the elements $D^{(K)*}_{Q'Q}$ of the rotation matrix, where we used the conventions of Edmonds (1957). Equation (33) is equivalent to the formulation in Section 1 of Fano and Macek's paper (1973).

It is sometimes convenient to express the information contained in $\rho_{\lambda'\lambda}$ in terms of the Stokes parameter, $I, I\eta_1, I\eta_2, I\eta_3$. I is the unpolarized intensity as before, η_3 gives the degree of linear polarization in the direction $\mathbf{e}_\alpha$, η_1 gives the degree of linear polarization at angles $\pm \pi/4$ to the axis $\mathbf{e}_\alpha$, and η_2 represents the degree of circular polarization.

Expressing the elements $\rho_{\lambda'\lambda}$ in terms of these parameters, we obtain (see, for example, Akhiezer and Berestetski, 1963)

$$\rho_{11} = \frac{I}{2}(1 + \eta_2) \qquad \rho_{1-1} = \frac{I}{2}(-\eta_3 + i\eta_1)$$

$$\rho_{-11} = \frac{I}{2}(-\eta_3 - i\eta_1) \qquad \rho_{-1-1} = \frac{I}{2}(1 - \eta_2) \qquad (34)$$

Thus, $\eta_2 = \rho_{11} - \rho_{-1-1}$, etc., and the Stokes parameters can easily be calculated from Eq. (33).

In order to show how special information can be extracted from (33) and the Stokes parameters, and to relate our equations to some other formulations, we write down an expression for the intensity $I_\varepsilon(\theta, \varphi)$ of photons emitted in the directions θ, φ and with a fixed polarization vector ε. We expand ε in terms of the Cartesian basis vectors $\mathbf{e}_\alpha$, $\mathbf{e}_\beta$:

$$\varepsilon = a_\alpha \mathbf{e}_\alpha + a_\beta \mathbf{e}_\beta$$

The expansion coefficients satisfy the normalization condition $|a_\alpha|^2 + |a_\beta|^2 = 1$. We parametrize in the following way

$$\varepsilon = \cos\beta \mathbf{e}_\alpha + e^{i\vartheta}\sin\beta \mathbf{e}_\beta \qquad (35)$$

where $\cos\beta(\sin\beta)$ is the magnitude of $a_\alpha(a_\beta)$ and ϑ is the phase necessary to describe elliptical polarization. The intensity I is then given by the expression:

$$I_\varepsilon = \frac{I}{2}[1 + \eta_3 \cos 2\beta + \eta_1 \sin 2\beta \cos\vartheta + \eta_2 \sin 2\beta \sin\vartheta] \qquad (36)$$

Inserting Eq. (34) for the Stokes parameters, we obtain the intensity I_ε expressed in terms of the atomic moments $\langle T_{(j)KQ}\rangle$. The expression obtained may be related to Fano and Macek's formulation with the help of the formulas (11) and (13).

As an illustration we discuss the experiment of Eminyan et al. (1974). Specializing our general results to this case, we calculate the Stokes parameters. We express the state multipoles $\langle T_{KQ}\rangle$ in terms of the quantities σ, Λ, $\varkappa$ defined in Section 1 and obtain

$$I = \overline{W}\sigma(3)^{-1/2}\left[\frac{1-\Lambda}{2}(1 + \cos^2\theta - \sin^2\theta \cos 2\varphi) + \Lambda \sin^2\theta \right.$$
$$\left. + [\Lambda(1-\Lambda)]^{1/2} \cos\varkappa \sin 2\theta \cos\varphi\right] \qquad (37a)$$

$$I\eta_3 = \overline{W}\sigma(3)^{-1/2}\left[\frac{1-\Lambda}{2}(-\sin^2\theta + (1+\cos^2\theta)\cos 2\varphi) \right.$$
$$\left. + \Lambda \sin^2\theta + [\Lambda(1-\Lambda)]^{1/2}\cos\varkappa \sin 2\theta \cos\varphi\right] \qquad (37b)$$

$$I\eta_1 = -\overline{W}\sigma(3)^{-1/2}[(1-\Lambda)\cos\theta\sin 2\varphi$$
$$+ 2[\Lambda(1-\Lambda)]^{1/2}\cos\varkappa\sin\theta\sin\varphi] \tag{37c}$$

$$I\eta_2 = -\overline{W}\sigma(3)^{-1/2} \cdot 2[\Lambda(1-\Lambda)]^{1/2}\sin\varkappa\sin\theta\sin\varphi \tag{37d}$$

Note that in the convention used in this paper η_2 has the opposite sign as that in the paper by Eminyan et al. (1974).

From Eq. (37d) follows that the degree of circular polarization is determined by the phase $\varkappa$ (and *vice versa*, measurements of η_2 allow a direct determination of $\varkappa$). If one detects the photons in the y direction of the collision system ($\theta = \varphi = 90°$), it follows that the degree of circular polarization is given by

$$\eta_2 = -2[\Lambda(1-\Lambda)]^{1/2}\sin\varkappa \tag{38}$$

which may be compared with Eq. (27).

Detecting the photon in the scattering plane ($\varphi = 0$) we obtain from Eqs. (37a) and (37b)

$$\eta_3 = +1 \tag{39}$$

Thus, the photon is completely linear polarized in $\mathbf{e}_x$ direction. This result may be understood in terms of a classical model (Macek and Jaecks, 1971).

The polarization of the emitted line radiation has been discussed in some detail in Kleinpoppen (1974). (See also the paper of Eminyan et al. in this volume.)

As a second example we mention the experiment by Berry et al. (1974). In this experiment, measurements of the photon polarization are made of the light emitted in He $2s^1S-3p^1P$ transitions after excitation in a tilted beam–foil experiment. The results are discussed in terms of a Stokes parameter which can be easily calculated from our general formulas.

References

Akhiezer, A., and Berestetski, V. B. (1963). *Quantum Electrodynamics*. New York: Wiley.
Berry, H. G., Curtiss, L. J., Ellis, D. C., and Schectman, R. M. (1974). *Phys. Rev. Lett.*, **32**, 751; see also this volume.
Blum, K., and Kleinpoppen, H. (1973). *Phys. Rev.*, **179**, 1902.
Edmonds, A. R. (1957). *Angular Momentum in Quantum Mechanics*. Princeton, N.J.: Princeton University Press.
Eminyan, M., MacAdam, K. B., Slevin, J., and Kleinpoppen, H. (1974). *J. Phys. B.*, **7**, 1519.
Fano, U. (1953). *Phys. Rev.*, **90**, 577.
Fano, U. (1957). *Rev. Mod. Phys.*, **29**, 74.
Fano, U., and Macek, J. (1973). *Rev. Mod. Phys.*, **45**, 553.
Gasiorowicz, S. (1966). *Elementary Particle Physics*. New York: Wiley.
Jacob, M., and Wick, G. C. (1959). *Ann. Phys.*, **7**, 451.

Kleinpoppen, H. (1974). Invited Paper for the *IV Int. Conf. Atom. Physics*, Heidelberg.
Macek, J., and Jaecks, O. H. (1971). *Phys. Rev.*, **174**, 2288.
McMaster, W. (1944). *Am. J. Phys.*, **22**, 357.
Percival, I. C., and Seaton, N. J. (1958). *Phil. Trans. R. Soc.*, **A251**, 113.
Wykes, J. (1972). *J. Phys. B.*, **5**, 1126.

41
Spatial Asymmetries in Atomic Collisions

H. G. BERRY, L. J. CURTIS, D. G. ELLIS, AND R. M. SCHECTMAN

Anisotropic excitation of atoms is discussed in terms of the subsequently emitted radiation. Examples are given of radiation from beam–foil excited atoms: (a) quantum beats between levels of the same parity are observed from aligned atoms; (b) quantum beats between levels of opposite parity are observed from excited atoms with net electric dipole moments; and (c) circularly polarized light is observed from atoms oriented in a tilted-foil geometry which does not have cylindrical symmetry.

A system of atoms which has been excited isotropically is statistically populated among the different magnetic substates of orbital angular momentum. That is, the atoms are not oriented or aligned and, consequently, the light emitted in the decay of an excited state is isotropic and unpolarized. Departures from isotropy or spherical symmetry can be represented in terms of the alignment and orientation of the atoms. We shall discuss how the atomic alignment and orientation can be measured in terms of the linear and circular polarization of the emitted radiation.

The first measurements of the breakdown of spherical symmetry in excited atoms through observations of the polarization of emitted light involved excitation by beams of electrons or photons or other atoms and ions. In most such cases, cylindrical symmetry was retained. The beam axis can be defined as the traditional z axis for orbital angular momenta of the excited atoms. Then the cross sections for different absolute values of angular

---

H. G. BERRY • Department of Physics, The University of Chicago, Chicago, Illinois 60637, U.S.A.   L. J. CURTIS, D. G. ELLIS, and R. M. SCHECTMAN • Department of Physics and Astronomy, The University of Toledo, Toledo, Ohio 43606, U.S.A.

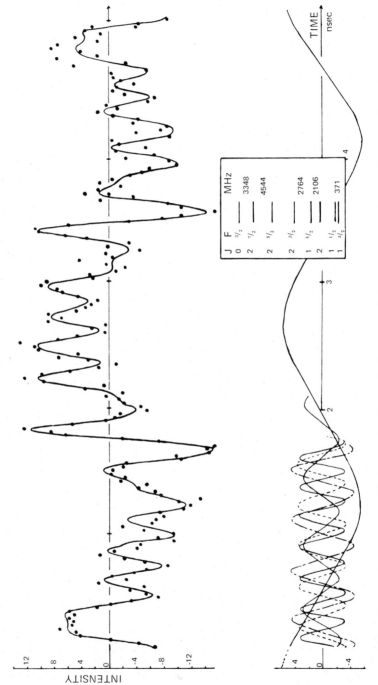

Figure 1. Modulations in the decay $4p(^3P)$, ^{7}Li II. The five contributing frequencies and the energy levels are shown below.

momentum projections with respect to this axis, $\sigma(|m_L|)$, are not necessarily equal. Their departures from equality have been set in terms of the fractional linear polarization of the light emitted during decay of the excited state (Percival and Seaton, 1958) for interactions which are independent of nuclear and electronic spins. The departure from spherical symmetry is observed as a net electric dipole along or perpendicular to the beam axis (depending on whether low or high $|m_L|$ are populated). Thus, the alignment, a single parameter, can be measured from either two total intensity measurements at two angles or the linear polarization fraction at one angle.

In beam–foil spectroscopy, the added factor of good time resolution of 10^{-11} sec or better after the impulsive excitation (10^{-14} sec) allows observation of excited states which are not energy eigenstates of the atomic Hamiltonian. The fractional linear polarization, observed in a particular direction, is then modulated in time with the frequencies corresponding to the energy differences of the coherently superposed energy eigenstates. Figure 1 is an example of such a time-resolved decay (Berry et al., 1974b), and the theory of these modulations under conditions of cylindrical symmetry is well

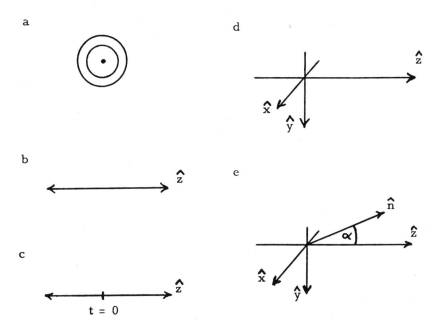

Figure 2. Schematic representation of increasing source asymmetry: (a) spherical symmetry, (b) cylindrical symmetry about z axis, (c) time-resolved cylindrical symmetry, (d) reflection asymmetry in xy plane, and (e) orientation axis $\hat{n} \times \hat{z}$.

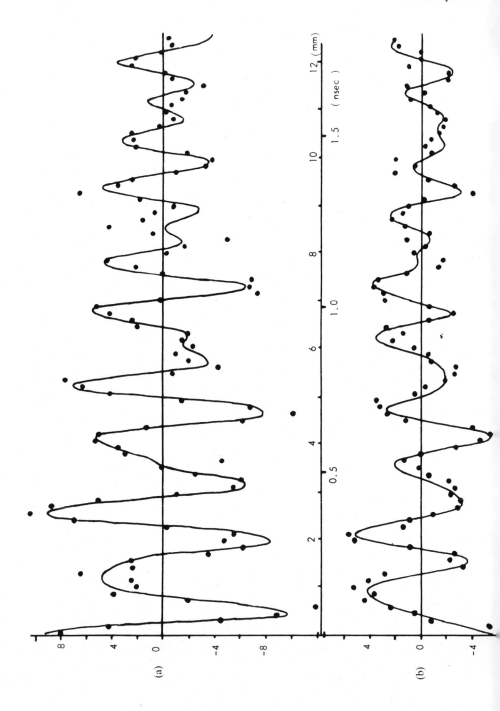

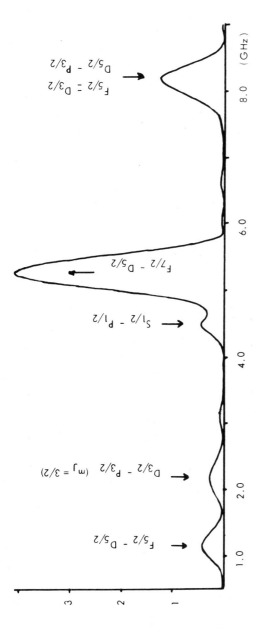

Figure 3. Modulations in the decay of He$^+$($n = 4$) in electric fields of ± 465 V/cm (a, b), and the Fourier transform of the difference curve (c). The modulation frequencies correspond to $m_J = \frac{1}{2}$ Stark-shifted energy separations, except where noted.

developed (Macek and Jaecks, 1971; Berry et al., 1972) in terms of the density operators of the excited system.

In Figure 2, we indicate schematically the gradual breakage of excitation symmetry. The pure cylindrical symmetry of 2b has a definite time $t = 0$ imposed for the beam–foil excitation in 2c. This could equivalently be any time-resolved beam excitation by electrons or photons, for example. We next look for breakage of the cylindrical symmetry through reflection asymmetry in the xy plane perpendicular to the beam z axis, as in 2d. The $+z$ and $-z$ directions are then distinguishable, and a coherent superposition of states of opposite parity with respect to reflection in the xy plane can be formed. Eck (1973) suggested such $2s, 2p$ coherence in beam–foil excited Lyman α, which has since been verified (Sellin et al., 1973; Gaupp et al., 1974).

We have observed mixed-parity coherence between many states of different orbital angular momentum in Balmer α of hydrogen ($n = 3$, SPD) and in 4686 Å of He$^+$ ($n = 4$, SPDF). An external electric field is used to mix the atomic eigenstates of opposite parity which are part of the coherent superposition so that time-dependent intensity modulations can be observed in the photon decay, the frequencies being proportional to the Stark-shifted energy differences. If a coherent superposition occurs, the amplitudes and phases of these modulations will vary with the external field direction. The difference between the field parallel and antiparallel to the beam direction will give twice the amplitude of the modulations due to the excitation coherence of the mixed-parity states, as shown in Figure 3. We find that coherence is exhibited between essentially all possible mixed-parity combinations and is a rather general attribute of beam–foil collisions, which thus produce states both of nondefinite energy and nondefinite parity.

We have shown (Berry et al., 1974a) that the beam–foil interaction is a function of the foil surface direction. Thus, there occurs a further breaking of the cylindrical symmetry as shown in Figure 2e when the foil normal n is tilted relative to the beam axis z. The alignment and orientation of such a collision which retains only reflection symmetry in the yz plane can be described by four parameters: three for the alignment and one for the orientation. Two of many possible such sets of four parameters are suggested: in spherical tensor notation the three second-rank components, $\rho_0^2, \rho_1^2, \rho_2^2$, describe the alignment and the first-rank component, ρ_1^1, describes the orientation (Berry et al., 1974a). These are proportional respectively to the parameters $A_0^c, A_{1+}^c, A_{2+}^c$, and O_{1-}^c of Fano and Macek (1973). Ellis (1973) has discussed the general case for time-resolved impulsive excitation of atoms.

The polarization state of the emitted light is a measure of the source anisotropy. The four Stokes parameters describe the polarization state and, in the viewing geometry shown in Figure 4, are related to the alignment and

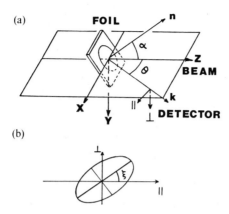

Figure 4. Beam and viewing geometry. The foil normal n is tilted at an angle α to the beam axis z.

orientation parameters by

$$I = I_0[2 + (\tfrac{3}{10})^{1/2}\rho_0^2(2 - 3\sin^2\theta) + (\tfrac{9}{2})^{1/2}\rho_2^2 \sin^2\theta]$$
$$M = I_0[-3(\tfrac{3}{10})^{1/2}\rho_0^2 \sin^2\theta - (\tfrac{9}{2})^{1/2}\rho_2^2(1 + \cos^2\theta)] \quad (1)$$
$$C = I_0 \cdot \tfrac{6}{5}\sqrt{5}i\rho_1^2 \sin\theta, \quad S = I_0 2\sqrt{3}\rho_1^1 \sin\theta$$

where I_0 is a normalization constant. Thus, the polarization state of the emitted light completely determines the alignment and orientation of the excited state. However, we must note that the Stokes parameters are insensitive to tensor components ρ_q^k with $k > 2$, and hence excitation amplitudes are completely determined only for S and P states. In Eq. (1), M/I is the standard fractional linear polarization $(I_\parallel - I_\perp)/(I_\parallel + I_\perp)$, which is proportional to the alignment ρ_0^2 for the cylindrical case only (when $\rho_1^2 = \rho_2^2 = 0$), while S is the fractional circular polarization.

We have observed the ^{4}He I $2s(^1S)-3p(^1P)$ transition at 5016 Å excited in the beam–foil geometry Figure 4 and measured its Stokes parameters to obtain the linear polarization fraction M/I and the circular polarization fraction S/I. We find that $S > 0$, which corresponds classically to the emitting electron preferentially orbiting clockwise, as viewed in Figure 4. The foil-tilt angle dependence was measured at 130-keV beam energy and is quite well fitted to a $\sin\alpha$ dependence as shown in Figure 5a. This suggests that the atom is set spinning by a torque fixed in direction relative to the foil surface. The very different energy dependences of the "alignment," Figure 5b, and the orientation, Figure 5c, suggest that the interaction mechanisms for the

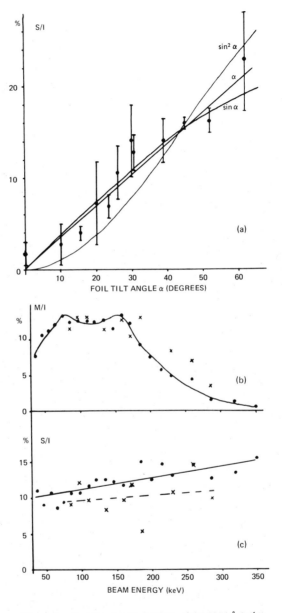

Figure 5. Fractional polarizations of the 5016 Å $2s(^1S)$–$3p(^1P)^4$ He I transition: (a) Circular polarization as a function of foil tilt angle α; (b, c) Linear and circular polarization fractions respectively, as functions of beam energy for a carbon foil (●) and an aluminum foil (×).

two are at least partially independent. However, an initial Born approximation calculation of asymmetric charge capture at the foil surface predicts no orientation (Band, 1974).

The excitation amplitudes are straightforwardly derived from the Stokes parameters, as detailed elsewhere (Berry et al., 1974a). In conclusion, we note that the asymmetric surface interaction has induced a relatively large orientation of the angular momentum of the atom. It would be interesting to compare the radiation produced in the analogous experiments of nonperpendicular crossed beams of atoms and electrons or photons or other atoms.

References

Band, Y. (1974). Private communication.
Berry, H. G., Subtil, J. L., and Carré, M. (1972). *J. Phys.* (Paris), **33**, 947.
Berry, H. G., Curtis, J. L., Ellis, D. G., and Schectman, R. M. (1974). *Phys. Rev. Lett.*, **32**, 751.
Berry, H. G., Pinnington, E. H., and Subtil, J. L. (1974). *Phys. Rev.*, **A10**, 1065.
Eck, T. G. (1973). *Phys. Rev. Lett.*, **31**, 270.
Ellis, D. G. (1973). *J. Opt. Soc. Am.*, **63**, 1232.
Fano, U., and Macek, J. (1973). *Rev. Mod. Phys.*, **45**, 553.
Gaupp, A., Andrä, H. J., and Macek, J. (1974). *Phys. Rev. Lett.*, **32**, 268.
Macek, J., and Jaecks, O. H. (1971). *Phys. Rev.*, **A5**, 2288.
Percival, I. C., and Seaton, M. J. (1958). *Phil. Trans. Roy. Soc.*, **A251**, 113.
Sellin, I. A., Mowatt, J. R., Peterson, R. S., Griffin, P. M., Lambert, R., and Hazelton, H. H. (1973). *Phys. Rev. Lett.*, **31**, 1335.

42
Atoms in Intense Electromagnetic Fields

P. LAMBROPOULOS AND MELISSA LAMBROPOULOS

> Laser have provided a light source whose brightness is great enough and spectral width narrow enough to allow observation of multiphoton processes. These are processes in which a single transition of the atom or molecule involves the absorption, emission, or scattering of more than one photon. When the light intensities are not too high, Nth order perturbation theory allows calculation of many new effects, with N the number of photons involved. These effects, in addition to their intrinsic interest, can be used as probes of atomic structure. In some case they will allow straightforward determination of atomic properties which are experimentally inaccessible with conventional techniques. Thus, long after its novelty wears off, multiphoton physics is likely to take its place beside conventional spectroscopy as a useful tool for understanding nature.

1. Introduction

Before the advent of the laser, interaction of light with matter usually meant absorption, stimulated emission, and scattering. In all of these processes each atomic transition involved the interaction of an electron with one photon. The large light intensities available with lasers, however, make it possible to observe processes in which each electronic transition involves the absorption, emission, or scattering of more than one photon. Thus an electron can, for example, be excited from its ground state to an excited state via the absorption of three photons. By intense fields, we shall mean

---

P. LAMBROPOULOS • Physics Department, Texas A & M University, College Station, Texas 77843, U.S.A. MELISSA LAMBROPOULOS • Joint Institute for Laboratory Astrophysics, University of Colorado and National Bureau of Standards, Boulder, Colorado 80302, U.S.A. P. Lambropoulos is on leave during 1975–76 to the Physics Department of the University of Southern California, Los Angeles, California 90007, U.S.A. M. Lambropoulos's present address is Department of Physics, UCSB, Santa Barbara, California 93106, U.S.A.

fields for which such multiphoton transitions are significant (or observable). Of course it ought to be mentioned that in the microwave region, multiphoton transitions (Winter, 1959) had been studied before the appearance of lasers. For optical frequencies, the two-photon spontaneous decay of metastable (2S) hydrogen and helium is an example of a multiphoton process which is independent of the existence of lasers and which in fact was first discussed in connection with astrophysical processes (Novick, 1969).

As the laser light intensity is increased beyond a certain limit, the perturbation introduced by the light becomes stronger than the binding of the electron to the nucleus. In this regime of intensity, not only do we have multiphoton transitions, but we also encounter the complication that, since the coupling of the light to the electron is stronger than the binding to the nucleus, the conventional perturbation theory of quantum electrodynamics is of dubious validity. As a rough rule of thumb, laser light intensities equivalent to about 10^{15} to 10^{16} W/cm^2 represent this demarcation line. What is often overlooked, however, is that not only the laser power but also the spectral line width is important in determining this demarcation line. And since ultra-high-power lasers tend to also have large line widths, the breakdown of perturbation theory in a given process may not actually occur as soon as one might expect on the basis of intensity alone.

Before embarking upon further discussion of multiphoton transitions, it should be mentioned that static fields can also be intense. The most familiar case is that of the Stark effect in a strong electric field (Bakshi et al., 1973; Cohn et al., 1972). Relatively recently, interest has focused on the behavior of atoms in strong magnetic fields (of the order of 10^6–10^7 G) or superstrong magnetic fields (Henry et al., 1974; Smith et al., 1972; Sokolov et al., 1973) (of the order of 10^{12} G). These problems, as well as problems in synchrotron radiation, are of considerable current interest in connection with astrophysical problems.

2. General Features of Multiphoton Ionization

Let us now turn to an examination of some of the features of multiphoton processes, and for the time being, assume that the laser intensity is such that conventional radiation perturbation theory is valid. This means that the Hamiltonian of the system "atom plus field" can be written as $H = H^A + H^R + V$, where H^A is the Hamiltonian of the atom, H^R the Hamiltonian of the field, and V the coupling between the two. For more compact notation, we write (P. Lambropoulos, 1974b)

$$H = H^A + H^R + V \equiv H^0 + V \tag{1}$$

which defines H^0, the unperturbed Hamiltonian. For optical and ultraviolet laser frequencies, the electric-dipole approximation is, in most cases, quite adequate, and thus we can write

$$V = -e \sum_j \mathbf{r}_j \cdot \mathbf{E} \qquad (2)$$

where $\mathbf{E}$ is the electric field at the position of the nucleus, $\mathbf{r}_j$ the position operator of the jth electron (the summation being over all electrons of the atom), and $-e$ is the electronic charge. The main phenomena in multiphoton processes can be elucidated in terms of one-electron atomic models, in which case V simplifies to $V = -e\mathbf{r} \cdot \mathbf{E}$. The electric field operator $\mathbf{E}$ is usually written as $\varepsilon E + \varepsilon^* E^*$, where ε is, in general, a complex polarization vector and E is the amplitude of the field. Then the coupling V assumes the form $-e[(\mathbf{r} \cdot \varepsilon)E + (\mathbf{r} \cdot \varepsilon^*)E^*]$. For light linearly polarized, ε is real. In semiclassical treatments of the problem, E is a time-dependent amplitude, while in fully quantized formulations, E is expanded in terms of creation and annihilation operators.

To be specific, let us first consider multiphoton ionization which has been dealt with in a number of recent experiments (Fox et al., 1971; Kogan et al., 1971; Bakos et al., 1972a,b,c; Delone et al., 1972; Agostini et al., 1972; Evans the Thoneman, 1972; Held et al., 1972a,b, 1973; LuVan and Mainfray, 1972; LuVan et al., 1972; M. Lambropoulos et al., 1973; Cervenan and Isenor, 1974; Agostini and Bensoussan, 1974). In Figure 1a, two-photon ionization is represented schematically. The electron makes a transition from the initial bound state $|1\rangle$ to the continuum by absorbing two laser photons each of frequency ω; $\mathbf{K}$ is the wave vector of the outgoing photoelectron. If it happens that the photon energy matches the difference between the energies of states $|1\rangle$ and $|3\rangle$, for example, as in Figure 1a, and if the dipole transition $1 \to 3$ is allowed, one can write the rate of two-photon ionization as the probability for excitation from 1 to 3 (given by the relevant cross section σ_{13} times the light intensity I) multiplied by the time the electron stays in state $|3\rangle$ (i.e., the lifetime τ_3 of state $|3\rangle$) times the probability per unit time for ionization from state $|3\rangle$ (given by the product of the relevant cross section $\sigma_{3\mathbf{K}}$ times the light intensity I). Thus we can write the equation

$$W_{1 \to \mathbf{K}} = (\sigma_{13} I) \tau_3 (\sigma_{3\mathbf{K}} I) \equiv \hat{\sigma} I^2 \qquad (3)$$

which defines $\hat{\sigma}$ obtained by lumping all atomic parameters into one. The quantity $\hat{\sigma}$ shall be referred to as the generalized cross section (GCS).

If the photon frequency does not match any energy difference as in Figure 1b, we still can have two-photon ionization, except that higher light intensity will be necessary. Although the foregoing reasoning will no longer apply, the transition rate can be written (P. Lambropoulos, 1974b) as

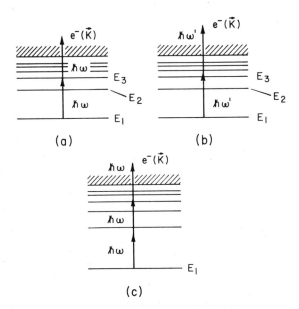

Figure 1. Schemes for multiphoton ionization: (a) resonant two-photon ionization; (b) nonresonant two-photon ionization; (c) three-photon ionization.

$W_{1 \to \mathbf{K}} = \hat{\sigma} I^2$, where $\hat{\sigma}$ is now given by

$$\hat{\sigma} = \frac{\alpha^2 m}{\hbar} \omega^2 K \left| \sum_n \frac{\langle \mathbf{K} | \boldsymbol{\varepsilon} \cdot \mathbf{r} | n \rangle \langle n | \boldsymbol{\varepsilon} \cdot \mathbf{r} | 1 \rangle}{\omega_n - \omega_1 - \omega} \right|^2 \qquad (4)$$

where α is the fine-structure constant, m the electronic mass, and ω the photon frequency, and the summation extends over all atomic states (including the continuum in principle); ω_n and ω_1 are the energies of the respective atomic states divided by $\hbar$. The light intensity I occurring in the previous equations is measured in number of photons per square centimeter per second. It is important to note that the summation is performed before the square of the absolute value is taken. The transition from the initial to the final state can be thought of as taking place via virtual transitions to all possible atomic states (intermediate states), or via a nonstationary intermediate state which is a superposition of all the stationary states of the atom. Owing to the presence of the energy differences in the denominators, the expression for $\hat{\sigma}$ diverges when the photon energy approaches one of these differences. Then we have a resonance with an intermediate state, and Eq. (4) should reduce to Eq. (3). For this, Eq. (4) must be corrected, as we shall discuss subsequently.

Three-photon ionization is shown schematically in Figure 1c. The transition rate for this process can be written as $W_{1\to K} = \hat{\sigma}_3 I^3$, where $\hat{\sigma}_3$ is given by

$$\hat{\sigma}_3 = 2\pi\alpha^3 \frac{m}{\hbar}\omega^3 K \left|\sum_m \sum_n \frac{\langle K|\boldsymbol{\varepsilon}\cdot\mathbf{r}|m\rangle\langle m|\boldsymbol{\varepsilon}\cdot\mathbf{r}|n\rangle\langle n|\boldsymbol{\varepsilon}\cdot\mathbf{r}|1\rangle}{(\omega_m - \omega_1 - 2\omega)(\omega_n - \omega_1 - \omega)}\right|^2 \tag{5}$$

Here we have a double summation over atomic states and a double resonance denominator.

For N-photon ionization we have a transition rate given by $W^{(N)}_{1\to K} = \hat{\sigma}_N I^N$, where

$$\hat{\sigma}_N = \frac{(2\pi\alpha)^N}{4\pi^2}\frac{m}{\hbar} K\omega^N |M^{(N)}_{fi}|^2 \tag{6}$$

and $M^{(N)}_{fi}$ is an $(N - 1)$-fold summation of the type of Eq. (5), involving N atomic matrix elements and a product of $(N - 1)$ resonance denominators of the form $[\omega_{n_{N-1}} - \omega_1 - (N - 1)\omega]\ldots[\omega_{n_2} - \omega_1 - 2\omega][\omega_{n_1} - \omega_1 - \omega]$, where $|n_1\rangle, |n_2\rangle, \ldots, |n_{N-1}\rangle$ are the intermediate atomic states over which the summation is performed.

If the generalized cross section $\hat{\sigma}_N$ is plotted as a function of photon frequency, it will have a dispersive character with peaks and valleys (Bebb and Gold, 1966; Bebb, 1966, 1967). If the logarithm of the transition rate $\ln W^{(N)}_{1\to K}$ is plotted as a function of $\ln I$, one should obtain a straight line with slope N. The plots shown in Figure 2 represent typical experimental results (Agostini et al., 1970) of multiphoton ionization of the noble gases with the second harmonic of the Nd-glass laser. These curves represent from 6- to 15-photon ionization. It appears therefore that for higher intensities, the process does not follow the I^N law but varies more slowly with intensity. If one used either portion of the curve to deduce an experimental value for the exponent, this value would turn out to be smaller than the theoretical exponent. But in some experiments the opposite has also happened, as shown in Figure 3, representing five-photon ionization of metastable $2^3 S$ helium (Bakos et al., 1972a). For a certain range of laser frequencies, N_{exp} has been found to be larger than N_{theor}, i.e., the curve of $\ln W^{(N)}$ as a function of $\ln I$ would bend over backward.

Attempts have also been made to measure $W^{(N)}_{1\to K}$ and compare it with calculated values. Such comparisons have consistently shown the experimental values to be much higher than the calculated ones, and the discrepancy increases with N, the order of the process. For example, experimental values of $W^{(4)}$ for potassium atoms (Held et al., 1972a) have been found to be at least two orders of magnitude higher than theoretical values. Such experiments have been performed with laser intensities sufficiently low for perturbation theory to be valid.

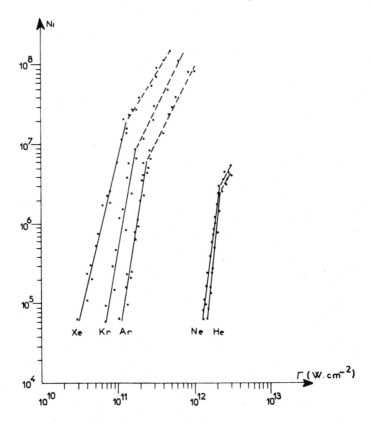

Figure 2. Experimental results by Agostini *et al.* (1970) on the multiphoton ionization of noble gas atoms. The number of ions produced is plotted against the intensity of the incident $\lambda = 0.53\,\mu$ light.

It seems, therefore, that there are serious discrepancies between experimental results and the predictions of perturbation theory, and this has led to suggestions that such discrepancies signify the breakdown of perturbation theory in the above-discussed experiments. As we shall see shortly, however, this is not necessarily the case.

If we take a closer look at the conditions of the experiments which show the departure from the law $W^{(N)} \sim I^{(N)}$, we will find that in all such cases there was at least one atomic intermediate state in resonance with some number of photons less than N. Assume for example that the first $N - 1$ photons bring the atom to some excited state, and thus the Nth photon ionizes that excited state of the atom. Clearly, above a certain light intensity, all atoms that reach that excited state are ionized. Then further increase of

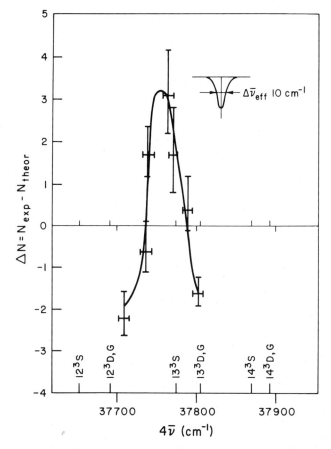

Figure 3. Dependence ΔN on the frequency of the laser radiation $\bar{\nu}$: $\Delta N = N_{exp} - N_{theor}$. Data from work of Bakos *et al.* (1972a) on the five-photon ionization of metastable triplet helium.

the intensity will not lead to an increase of the rate of the last step of the process and as a result the curve will have an I^{N-1} dependence on laser intensity. In other words, we have saturation of the last step of the transition and this is indeed what has happened in some cases as, for example, in four-photon ionization of Cs observed by Mainfray and co-workers (Held *et al.*, 1962b) using a Nd-glass laser. This involves a three-photon resonance with the $6f$ state of Cs, and for intensities larger than about 10^7 W all atoms excited to $6f$ are ionized; consequently, above this intensity the rate is proportional to the third power of the intensity, as it should be.

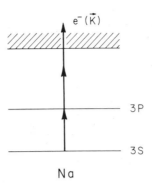

Figure 4. Scheme for three-photon ionization of sodium via a resonance in the first step.

In general the situation is more complicated than that. To consider a somewhat different case, assume that in a three-photon ionization process it is the first step (rather than the last) which is resonant. Specifically, let us consider three-photon ionization of Na via the $3S$–$3P$ resonance transition as in Figure 4. When we have a resonance with an intermediate state, one of the denominators in the expression for the generalized cross section vanishes. In this particular case the summation over n in Eq. (5) reduces to a single term, namely $n = 3P$, and the denominator $\omega_{3P} - \omega_{3S} - \omega$ vanishes. We must then take into account the fact that the intermediate atomic state (in this case $3P$) has a finite lifetime and hence a nonzero width. Moreover, the width of $3P$ is not determined by just spontaneous emission but in the present case there are two additional contributions to the width, namely, stimulated emission back to the ground state due to the resonant laser light, and transitions upward which lead to ionization via the absorption of two more photons. Thus the denominator $\omega_{3P} - \omega_{3S} - \omega$ is replaced by $\omega_{3P} + S_{3P} - \omega_{3S} - S_{3S} - i\Gamma_{3P}$, with $\Gamma_{3P} = \Gamma_{3P}^0 + \Gamma_{3P \to 3S}^I + \Gamma_{3P \to K}^I$, where Γ_{3P}^0 is the spontaneous decay width (natural lifetime) of $3P$, $\Gamma_{3P \to 3S}^I$ is the stimulated emission width, and $\Gamma_{3P \to K}^I$ the width due to ionization. The last two contributions depend on the intensity and spectrum of the incident light, often not in a simple manner. Note also that shifts of the energies of $3P$ as well as $3S$ must be included. In addition to the vacuum shift contributions—which can always be assumed to be included in the unperturbed energies—there will be induced contributions depending on the light intensity. In our present example the dominant parts of the shift and width above a certain light intensity will be those due to the resonance transition $3P \leftrightarrow 3S$. These shifts and widths have a resonance structure themselves and as a result exhibit saturation. Shifts and widths of intermediate resonant atomic states will generally increase with laser intensity, and consequently the dependence of the process on laser power will deviate from I^N (I^3 in this case). Another related effect that can be illustrated with this example is the following. If the photon is not on exact resonance with the 3P level but is, say, two or three laser line widths away, the induced width is zero but not the shift. In fact, the shift can be quite substantial and in general will increase as the laser is tuned closer to resonance, but still not overlapping with the level. An example (P. Lambropoulos, 1974a) of the dependence of the shift on detuning is shown in Figure 5. The shift reaches a maximum while the level 3P is still outside the laser line shape. Therefore, even in a near-resonance situation, where the width is zero, the presence of the intensity-dependent shift in the denominator can have a significant effect on the transition rate. An example of this is shown in Figure 6, where the transition rate with the shift included is compared to the rate without the shift, and the rates are plotted as functions of detuning from the resonance. Clearly, one can make an error from one to two orders of magnitude by neglecting the shift. Moreover the line shape of the transition around the resonance is not simple. This result has been obtained (P. Lambropoulos, 1974a) by considering only one resonant intermediate state. If there are two such states, say two fine-structure states whose distance from each other is a few laser line widths, then there will be interference between the amplitudes and the situation will be more complicated. It must also be stressed that near or on resonance the

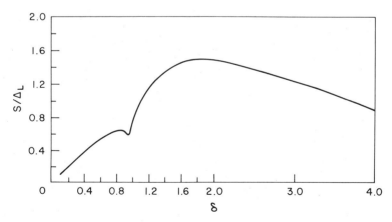

Figure 5. The resonance shift S/Δ_L in two-photon ionization as a function of detuning $\delta = (\omega_L - \omega_0)/\Delta_L$, where ω_L is the laser frequency, ω_0 the resonance frequency, S the resonance shift, and Δ_L the laser halfwidth (assuming a square laser line shape.)

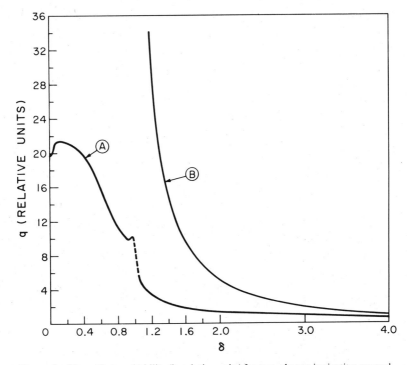

Figure 6. Transition probability (in relative units) for two-photon ionization around a resonance with an intermediate state. Curve A represents the transition probability with all saturation effects included. Curve B has been obtained using the off-resonance formula with all shifts neglected. The quantity δ is defined by $\delta = (\omega_0 - \omega_{21})/\Delta_L$, where ω_0 is the laser center frequency, ω_{21} the frequency of the atomic transition in resonance with the laser, and Δ_L the laser line half-width.

details of the laser spectrum are important. The above results have been obtained with a square line shape (P. Lambropoulos, 1974a; also see Chang and Stehle, 1973).

Near-resonance experiments have been performed in cases where these complexities are negligible, as well as in cases where they play a significant role. An experiment on the three-photon ionization of sodium (M. Lambropoulos et al., 1973) has measured relative total cross sections in the region of single-photon and two-photon resonances. Results from ionization with the light from a single dye laser tuned so that $2\omega_L \simeq \omega(4d) - \omega(3s)$ are shown in Figure 7. In this case the dominant broadening mechanisms are ionization and stimulated two-photon emission, for which there is a very low probability. Because of the low rate of decay of the $4d$ state, one would expect the resonance in the cross section to exhibit a Lorentzian profile, as in fact it does.

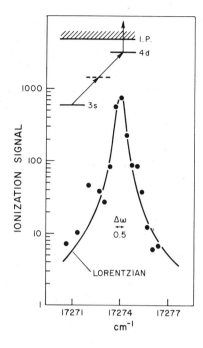

Figure 7. Relative total cross section for three-photon ionization of sodium via a two-photon resonance with the 4d state.

Additional results from this experiment, using two tunable lasers coincident upon the sodium beam, demonstrate the effect of broadening on the resonance profile.

The measurement illustrated in Figure 8 was made using one laser tuned to the vicinity of the $3p^2\ P_{3/2}$ transition and a second laser tuned so that $\omega_{L1} + \omega_{L2} = \omega(5s) - \omega(3s)$. In the figure, the total cross section is plotted against ω_{L1}. It displays a non-Lorentzian shape. The measured linewidth of laser 1 was less than 0.15 Å, including jitter. The second laser had a linewidth (FWHM) of about 5 Å and an intensity about 20 times greater than that of laser 1. The observed broadening effect has as probable cause the saturation of the two resonance transitions.

Now what about the discrepancy between measured and calculated values of multiphoton transition rates? First, we must take into account the fact that an N-photon process is not proportional to I^N but to an Nth-order photon correlation function (P. Lambropoulos et al., 1966; P. Lambropoulos, 1968; Mollow, 1968; Agrawal, 1970). Derivations that lead to the I^N dependence neglect the photon statistics (correlations) aspect. Qualitatively

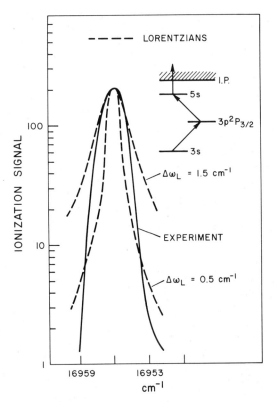

Figure 8. Relative total cross section for three-photon ionization of sodium via two-photon resonance with 5s and near-single-photon resonance with $3p\ ^2P_{3/2}$.

speaking, this comes about because an N-photon process can be viewed as a coincidence measurement. Hence, it is going to make a difference whether the photons arrive bunched or not. In other words, the photon statistics characteristic of the light will influence the rate. Quantitative analysis of the problem reveals that chaotic (Gaussian statistics) light is more efficient than purely coherent (Poisson statistics) light by a factor of $N!$ for an N-photon process. Since calculations have invariably assumed purely coherent light (which leads to I^N dependence) and since actual pulsed lasers are not such ideal sources, a good part of the discrepancy is undoubtedly due to this effect. The influence of photon correlations in multiphoton processes has been investigated experimentally by Shiga and Imamura (1967) and Lecompte et al. (1974). However, we still do not have clean experimental results with lasers of known mode structure which could be compared with reasonable theoretical models of the laser field. Perhaps advances in cw tunable dye lasers will offer the possibility of more quantitative experiments in this area.

3. Special Effects in Multiphoton Processes

Let us now turn to some of the phenomena that are peculiar to multiphoton processes. One of these interesting aspects is that the total rate of a multiphoton transition depends on the polarization of the incident light. It is easy to see why by just looking at the channels available for two different light polarizations (Figure 9). Consider for the moment a hydrogen-like atom (neglecting fine structure), and as an example look at a four-photon process. The selection rules for linearly polarized light are $\Delta l = \pm 1$ and $\Delta m = 0$, while for circularly polarized light they are $\Delta l = \pm 1$ and $\Delta m = \pm 1$ (+ corresponding to right and − to left circular polarization). As a consequence of the Δm selection rule, for linearly polarized light all channels indicated by solid arrows are open, but for circularly polarized light only the channels indicated by the dashed arrows are open. Obviously, the total rates in the two processes will be different, and this was first observed by Robinson and co-workers (Fox *et al.*, 1971; Kogan *et al.*, 1971) in two- and three-photon ionization of Cs. They found that in both cases circularly polarized light gave a higher rate, by a factor of 1.28 in two-photon ionization and by a factor of 2.15 in three-photon ionization. The reason circular polarization gave higher rates, although it has only one channel, is that the photon frequencies were in near resonance with intermediate states favored by circular polarization, and the interference between channels reduced the rate for linear polarization. Thus, in addition to the selection rules, interference will also affect the ratio of the rates. In general, one must calculate the transition probability in detail in order to decide the value of the ratio for circular to linear polarization. Needless to say, this ratio will depend on the

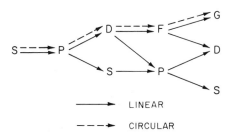

Figure 9. Channels available for four-photon ionization with linearly or circularly polarized light.

photon frequency, and it can vary by several orders of magnitude (P. Lambropoulos, 1972a, 1973c; Gontier and Trahin, 1973a,b). For example, if the photon frequency ω in three-photon ionization of an alkali atom is such that the transition goes via the near-resonance $nS + 2\omega \to n'D$, the ratio of circular to linear will typically be about 2.4; if ω is such that $nS + 2\omega \to n'S$, the ratio will be much smaller than unity, because for circular polarization the transition $nS + 2\omega \to n'S$ is forbidden. The contribution then will be from the $n'D$ states which are farther away.

As the order N of the process increases, the number of channels available to linear polarization increases very rapidly, while for circular polarization there is never more than one channel. Thus, for large N, linear polarization will dominate in most cases, and for this reason, the maximum for the ratio of circular to linear given by Klarsfeld and Maquet (1972) is of only theoretical interest. In this respect, the calculation by Reiss (1972) is more realistic, albeit of somewhat uncertain validity.

What can one learn from this effect? As an example, consider near-resonance two-photon ionization in an alkali atom and let nP be the near-resonant intermediate state. One can show (P. Lambropoulos, 1973c) that then the ratio μ of the rates for circular to linear is a function of the ratio $x \equiv \langle KS|r|nP\rangle/\langle KD|r|nP\rangle$ of the radial matrix elements for ionization into the S and D partial waves. This is to be expected, since the S channel can only be reached with linearly polarized light. In Figure 10, μ is shown as a

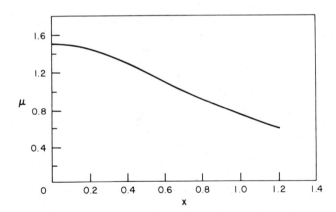

Figure 10. The ratio μ of the rate of near-resonant two-photon ionization of an alkali atom using circularly polarized light to the rate using linearly polarized light (plotted against x, where x is the ratio of the radial matrix elements for ionization into the S and D partial waves).

function of x. As long as x is in the sensitive region of $x \gtrsim 0.1$ (which will usually be the case), $|x|$ can be determined from measurement of μ. If we take, for example, Robinson's (Kogan et al., 1971) result for Cs, and recognize that in his experiment there was near-resonance with the 10P state, from his value 1.28 ± 0.2 we can deduce $0.105 < |x| < 0.564$, which is in the region of what one would have expected. It should be realized that Robinson's experiment was not designed to determine x, hence the large uncertainty. This could now be measured more accurately with a tunable dye laser.

A similar analysis can be used in near-resonance three-photon ionization (P. Lambropoulos, 1973c, 1974c). If the photon frequency ω is such that $nS + 2\omega \to n'D$, then the ratio μ is a function of the ratio of the radial matrix elements for going from $n'D$ to the P or F partial wave, i.e., a function of $x' \equiv \langle KP|r|n'D\rangle/\langle KF|r|n'D\rangle$. In Robinson's (Fox et al., 1971) experiment with Cs, 2ω was in near resonance with 9D. Using his experimental value 2.15 ± 0.4 for the ratio, we obtain $|x'| = 0.396$, which again is within the range of what we would have expected.

One can therefore use these effects to measure ratios of bound–free matrix elements. The attractive feature of this method is that one needs to measure only ratios and not absolute magnitudes of transition rates. An alternative method is to measure angular distributions of the photoelectrons. Preliminary work on sodium and titanium has already been performed to measure angular distributions of electrons from resonant two-photon ionization (Edelstein et al., 1974; M. Lambropoulos and Berry, 1973). Angular distributions in multiphoton ionization are an interesting subject in themselves and are covered in some detail in another paper (by R. Stephen Berry) in this volume.

The previous polarization effects were discussed in the context of a simple hydrogenic model without spin. Let us now include spin–orbit coupling and consider an alkali atom. The ground state is an $S_{1/2}$ state while the P excited states are split into $P_{1/2}$ and $P_{3/2}$ fine-structure states. In Figure 11 we show schematically three-photon ionization of Na via near-resonance with the first excited state 3P. The separation between $3P_{1/2}$ and $3P_{3/2}$ is about 17 cm^{-1}. We assume that we are using a dye laser with line width about 1 to 2 cm^{-1}, which can be tuned around the resonance transition $3S \to 3P$. If the light were circularly polarized one might expect the outgoing photoelectrons to have some net spin polarization. Physically, this would come about because a circularly polarized photon has a definite projection of angular momentum on its direction of propagation, or, more precisely, definite helicity; since there is spin–orbit coupling, the photon angular momentum could cause the photoelectron to have a preferential spin orientation along the direction of photon propagation. To determine exactly what will happen one must go through the angular momentum

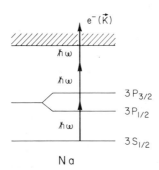

Figure 11. Scheme for three-photon ionization in sodium via near-resonance with the spin–orbit split 3p doublet.

algebra. For light right circularly polarized we have the following channels:

$$S_{1/2}(\tfrac{1}{2}) \to P_{3/2}(\tfrac{3}{2}) \to D_{5/2}(\tfrac{5}{2}) \to F_{7/2}(\tfrac{7}{2})\uparrow$$

$$S_{1/2}(-\tfrac{1}{2}) \begin{matrix} \nearrow P_{3/2}(\tfrac{1}{2}) \to D_{5/2}(\tfrac{3}{2}) \to F_{7/2}(\tfrac{5}{2})\uparrow\downarrow \\ \searrow P_{1/2}(\tfrac{1}{2}) \to D_{3/2}(\tfrac{3}{2}) \to F_{5/2}(\tfrac{5}{2})\uparrow\downarrow \end{matrix}$$

The numbers inside parentheses are the m_j values of the states that participate. The upper channel will lead to photoelectrons spin polarized along the direction of propagation of the photon (indicated by arrows pointing upward) while the other channels will also lead to spin polarization opposite to the photon propagation (arrows pointing downward), that is, to a mixture. Since the final result involves the superposition of all open channels, it will depend on the photon frequency (because the frequency appears in the energy differences in the denominators and that is what decides how much each channel contributes). Obviously, interference will play a substantial role here. After performing the necessary calculations, one obtains the result shown in Figure 12 (P. Lambropoulos, 1973a), which also includes the case of two-photon ionization. These curves have been obtained without the inclusion of shift and width effects discussed earlier, which will occur exactly on resonance with the $P_{1/2}$ and $P_{3/2}$ states. Such effects are bound to influence the photoelectron polarization for sufficiently high intensities, the reason being that they appear in the denominators and hence modify the contributions of the various channels.

Experimentally, measurements of this effect have thus far been confined, for reasons of laser intensity and available signal, to cases involving at least

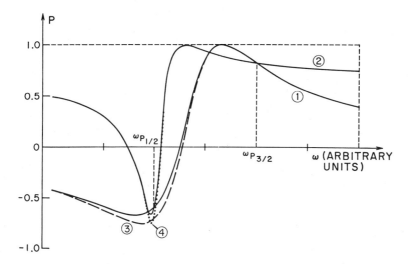

Figure 12. Photoelectron polarization P for two- and three-photon ionization via a P near-resonance as a function of photon frequency, ω, in the vicinity of the P levels. Curve 1, two-photon ionization for $\rho \equiv R(n'P^{\frac{3}{2}}; nS^{\frac{1}{2}})/R(n'P^{\frac{1}{2}}; nS^{\frac{1}{2}}) = 1$. Curve 2, two-photon ionization for $\rho = 5$. Curve 3, three-photon ionization for $\rho = 1$. Curve 4, three-photon ionization for $\rho = 5$.

one resonance. It is therefore difficult to compare them to theory in detail. The previously mentioned experiment on two- and three-photon ionization of sodium atoms (M. Lambropoulos et al., 1973) has also yielded preliminary polarization results.

In Figure 13 the data are plotted with their statistical error bars. A complete analysis of systematic errors has not been performed, and probable systematic errors are not indicated. To facilitate visual comparison of the measured shape to the theoretically predicted dispersion curve, the empty circles plot the data as normalized to the maximum of the theoretical polarization. These data were obtained under the following conditions: (a) laser 1 was tuned to the vicinity of the sodium D lines (and it is the frequency of laser 1 against which polarization is plotted); (b) laser 2 was tuned so that $\omega_{L1} + \omega_{L2} = \omega(4d) - \omega(3s)$; (c) the line width of laser 2 was sufficiently great that $\Delta\omega_{L2} > \omega(3p^2P_{3/2}) - \omega(3p^2P_{1/2})$, so that stimulation emission from $4d$ to both $3p$ levels probably occurred at all measured points; and (d) the intensity of laser 2 was at least two orders of magnitude greater than the intensity of laser 1.

The polarizations were observed to be strongly power dependent, as illustrated in Figure 14. This is as one would expect in a regime where the incident radiation contains appreciable power on an atomic resonance

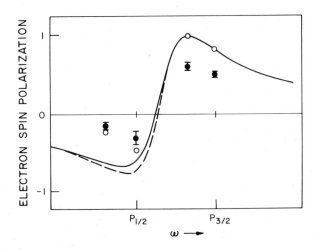

Figure 13. Experimentally determined electron spin polarization plotted against laser frequency. The solid and dashed lines are curves 1 and 3, respectively, from Figure 12. The solid circles are the experimental points. The open circles represent the data as normalized to the maximum of the theoretical curve. The experimental conditions are described in the text.

frequency. An experiment thus has yet to be performed under conditions which provide a clear test of the theory.

These are not the only multiphoton processes that will yield spin-polarized photoelectrons. Other processes have been proposed by Farago and Walker (1973) and by P. Lambropoulos (1974a). Such processes have some similarity with the Fano effect (Fano, 1969), through which spin-polarized electrons have been obtained by usual single-photon ionization. But there are also differences. The effects of spin–orbit coupling could, for the purposes of this discussion, be divided into two classes: splitting of the energy levels and modification of the wave function. The Fano effect is based on the modification of the wave function in the continuum, which results in the dependence of the radial matrix elements for the bound-free transition on the total angular momentum j. The spin polarization in multiphoton ionization is the result of the splitting of the energy levels and does not require the modification of the continuum wave function. Of course, if the latter exists, it will have some effect. But it is not necessary. In Figure 12, two of the curves show the effect of the modification of the nP wave functions due to spin–orbit coupling in Cs. An example of the effect of spin–orbit perturbation of the continuum wavefunction is given in Figure 15 (P. Lambropoulos, 1975).

Atoms in Intense Electromagnetic Fields

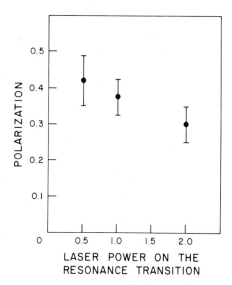

Figure 14. Power dependence of the electron spin polarization in three-photon ionization via the $3p\,^2P_{3/2}$. Experimentally determined polarization is plotted against laser power on the resonance transition.

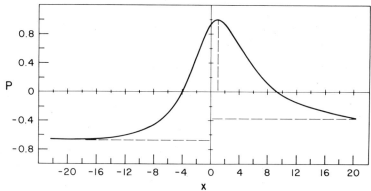

Figure 15. Effect upon electron spin polarization of spin–orbit perturbation in the continuum. Polarization is plotted as a function of x for the case of near-resonant two-photon ionization of an alkali atom, and photon frequency

$$\omega = \omega(nP_{1/2}) + 0.66[\omega(nP_{3/2}) - \omega(nP_{1/2})].$$

$$x \equiv \langle KD(5/2)|r|nP\rangle / \langle KD(3/2)|r|nP\rangle.$$

For $x = 1$, $P = +1.0$, which is the case derived previously without spin–orbit perturbation of the continuum wave function.

The spin polarization, as well as the total transition rate with circularly polarized light, will also be affected by the following optical pumping type of effect. For light perfectly (100%) right circularly polarized, and exactly on resonance with, for example, the $nP_{1/2}$ state, only the induced transition $S(m_j = -\frac{1}{2}) \rightarrow P_{1/2}(m_j = +\frac{1}{2})$ can take place. But the $P_{1/2}(+\frac{1}{2})$ can also decay spontaneously to the $S(+\frac{1}{2})$ by emitting a linearly polarized photon. If this part of the spontaneous lifetime is shorter than the laser pulse duration, atoms are pumped into the $S(+\frac{1}{2})$ state out of which they cannot be taken with right circularly polarized light. The end result is that the total photoelectron production is less than one would predict by ignoring this effect. In other words the "effective" pulse duration of the laser is smaller than the actual duration.

Consider now the experimentally more realistic case in which the light is slightly elliptically polarized, say 98% right circularly polarized and 2% left. The polarization vectors for right ($+$) and left ($-$) circular polarization are $\varepsilon^{\pm} = \mp(1/\sqrt{2})(\hat{x} \pm i\hat{y})$, where $\hat{x}, \hat{y}$ are unit vectors along the x, y axes. For elliptically polarized light, the polarization vector can be written as $\varepsilon = \eta_+ \varepsilon^+ + \eta_- \varepsilon^-$, where $|\eta_+|^2 + |\eta_-|^2 = 1$. If I is the total light intensity, the right and left intensities will be $I^+ = |\eta_+|^2 I$ and $I^- = |\eta_-|^2 I$, respectively. Now, for two-photon ionization, we have a mixing of channels, and, as shown in Figure 16, the ejected photoelectron is not in a D wave, but there is also an S wave present. The relative contribution of the two channels depends on the relative magnitude of the intensities I^+ and I^-, as well as on the relative strength of the $P \rightarrow D$ and $P \rightarrow S$ transitions, i.e., on the ratio

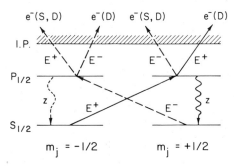

Figure 16. Scheme representing channel contributions in the case of elliptically polarized light. (Specifically, resonant two-photon ionization of an alkali atom via a $P_{1/2}$ state is shown.) The solid arrows represent the dominant pathway when the process is below saturation.

$x = \langle KS|r|nP\rangle/\langle KD|r|nP\rangle$ of the respective bound–free matrix elements. In usual single-photon processes, a small deviation from perfect polarization will have a correspondingly small effect on the process. When we deal with intense fields, however, a small percentage of the opposite polarization can represent a very large intensity which may have significant effects, as we shall see subsequently. It should be *a priori* evident that with elliptically polarized light we will have to account for the mixing of channels as well as the disturbance of the equilibrium populations of the various states due to optical pumping.

Let us first consider the case in which both I^+ and I^- are large enough to saturate the respective $S \to P$ transitions and, in fact, assume that they are well above saturation. Then the spontaneous lifetime will be much longer than the induced lifetime. As a result, the optical pumping effect on the populations could be neglected as a very rough approximation. We have calculated the photoelectron spin polarization for this case, and the results are shown in Table 1. We recall that for pure right circular polarization and a $P_{1/2}$ resonance, the spin polarization is -0.6. Now it is a function of x and can, in principle, vary between -0.18 for $x = 0$ and $+0.99$ for $x \to \infty$. Usually one would expect x to vary between 0.1 and 0.6. Note that polarization -0.6 cannot be achieved now even when $x = 0$, i.e., even when there is no S-wave contribution. This is because, although there is no S wave, the D wave can be reached via different combinations of polarizations, which leads to different spin polarization than would the pure circular polarization.

If we consider the case in which there is significant redistribution of populations as a result of optical pumping, the calculation becomes

Table 1. Photoelectron Spin Polarization in Resonant Two-Photon Ionization via the Channel $S_{1/2} \to P_{1/2} \to \mathbf{K}$ as a Function of $x = \langle KS|r|nP\rangle/\langle KD|r|nP\rangle$. Light elliptically polarized, 98% right and 2% left. Both I^+ and I^- above saturation.

$|x|$	P
0	-0.18
0.1	-0.16
0.25	-0.06
0.316	0.0
0.50	$+0.18$
0.75	$+0.40$
1.00	$+0.58$
∞	$+0.99$

considerably more involved because the populations of the states $S_{1/2}(\pm\frac{1}{2})$ and $P_{1/2}(\pm\frac{1}{2})$ must be calculated as a function of time. This has been done by P. Lambropoulos and Stuart (1975). One again finds significant effects on the spin polarization as well as the transition rate.

4. Other Approaches to the Theory of Multiphoton Processes

All effects discussed so far have been derived using the conventional perturbation theory, in which the coupling between atom and radiation is the expansion parameter. Let us now review briefly two other approximation schemes. There are several ways to discuss them. We shall present them here from a somewhat unified point of view which shows their similarities as well as their differences.

They both are semiclassical methods in which the unperturbed part of the Hamiltonian is the atomic Hamiltonian $H^{\text{atom}} \equiv H^0$, while the interaction in the dipole approximation is $V(t) = e\mathbf{r} \cdot \mathbf{E}(t)$, where $\mathbf{E}(t)$ is the classical electric field and is a function of time. It can be shown quite generally (Power, 1965; Power and Zienau, 1959) that in the dipole approximation, $e\mathbf{r} \cdot \mathbf{E}$ is equivalent to $-(e/mc)\mathbf{p} \cdot \mathbf{A} + (e^2/mc^2)A^2$, where $\mathbf{A}(t)$ is the vector potential. It should be kept in mind that $\mathbf{E} = -(1/c)(\partial \mathbf{A}/\partial t)$. The total Hamiltonian is time dependent in this formalism and is given by

$$H = H^0 + V(t) = H(t) \tag{7}$$

while the formal solution of the Schrödinger equation

$$\frac{\partial}{\partial t}\Psi(t) = -\frac{i}{\hbar}H(t)\Psi(t)$$

is

$$\Psi(t) = \exp\left[-\frac{i}{\hbar}\int_{t_0}^{t} H(t')\,dt'\right]\Psi(t_0) \tag{8}$$

Suppose now that we construct an operator $\Lambda(t)$, defined by

$$\Lambda(t) \equiv \exp\left[\frac{i}{\hbar}\int_{t_0}^{t} V(\tau)\,d\tau\right] \tag{9}$$

which is a unitary operator, i.e. $\Lambda\Lambda^\dagger = 1$. If we define a new wave function $\tilde{\Psi}(t) \equiv \Lambda(t)\Psi(t)$, the Schrödinger equation becomes

$$\frac{\partial}{\partial t}\tilde{\Psi}(t) = -\frac{i}{\hbar}\tilde{H}^0\tilde{\Psi}(t)$$

or, integrated formally,

$$\tilde{\Psi}(t) = \tilde{\Psi}(t_0) - \frac{i}{\hbar} \int_{t_0}^{t} d\tau \, \tilde{H}^0(\tau)\Psi(\tau) \tag{10}$$

where $\tilde{H}^0 \equiv \Lambda(t)H^0\Lambda^+(t)$. In other words, we have absorbed V in $\tilde{H}^0$. Of course, so far we have done nothing more than rewrite the Schrödinger equation.

Considering the explicit form $H^0 = -(\hbar^2/2m)\nabla^2 - e^2/r$ of the one-electron atomic Hamiltonian and taking the $\mathbf{p} \cdot \mathbf{A}$ form for V, one can show that the transformed Hamiltonian $\tilde{H}^0$ becomes

$$\tilde{H}^0 = -\frac{\hbar^2}{2m}\nabla^2 - \frac{e^2}{|\mathbf{r} + \boldsymbol{\alpha}|} \tag{11}$$

where $\boldsymbol{\alpha} \equiv (e/m\omega c)\mathbf{A}_0 \sin \omega t$ and $\mathbf{A}(t) = A_0 \sin \omega t$ is the vector potential, which is taken as a pure monochromatic wave in this method. A_0 is the amplitude of the wave. This method was proposed by Henneberger (1968) and was most recently rediscovered by Faisal (1973). Its roots go back to Kramer in the forties. Its possible usefulness lies in the fact that $e^2/|\mathbf{r} + \boldsymbol{\alpha}|$ can be expanded in a series of the form

$$-e^2/|\mathbf{r} + \boldsymbol{\alpha}| = V_0(\mathbf{r}, \boldsymbol{\alpha}_0) + \sum_{n \neq 0} V_n(\mathbf{r}, \boldsymbol{\alpha}_0) e^{in\omega t} \tag{12}$$

where

$$\boldsymbol{\alpha}_0 \equiv (e/m\omega c)\mathbf{A}_0, \quad V_0 = (-e^2/2\pi) \int_0^{2\pi} d\phi |\mathbf{r} + \boldsymbol{\alpha}_0 \sin \phi|^{-1}$$

and $V_n = (-e^2/2\pi^2) \int e^{i\mathbf{q}\cdot\mathbf{r}} q^{-2} J_n(\boldsymbol{\alpha}_0 \cdot \mathbf{q}) d^3q$, with J_n being a Bessel function. Since V_0 is time independent it can be incorporated in the unperturbed Hamiltonian by taking $\bar{H}^0 \equiv -(\hbar^2/2m)\nabla^2 + V_0$ and treating the remainder of the series as the perturbation. Thus, $\bar{H}^0$ contains the field, and if we use its eigenstates in the perturbation expansion we are in a sense using the eigenstates of the dressed atom. Presumably this approach may be more appropriate for high field intensities. Using this formalism, Henneberger (1968) obtained a shift of the ionization potential due to the intense field. It must be emphasized that one still has to use perturbation theory to obtain transitions using the series as the perturbation. The procedure becomes fairly complicated for high-order processes and for nonhydrogenic models, and the method has not received much attention so far. Note that owing to Eq. (11) this method could be called a space-translation method.

Suppose now that we make the same transformation $\Lambda(t)$ but use $e\mathbf{r} \cdot \mathbf{E}$ for V instead of $\mathbf{p} \cdot \mathbf{A}$. To distinguish this transformation from the previous one, call it $R(t)$; i.e.,

$$R(t) \equiv \exp\left[-(ie/\hbar)\int_{t_0}^{t} d\tau\, \mathbf{E}(\tau)\cdot\mathbf{r}\right] = \exp\left[(ie/\hbar c)\mathbf{A}(t)\cdot\mathbf{r}\right] \qquad (13)$$

Again we have a Schrödinger equation for the transformed wave function $\tilde{\Psi} \equiv R\Psi$ governed by the transformed Hamiltonian $\tilde{H}^0 \equiv RH^0R^+$. In integral form we have

$$\tilde{\Psi}(t) = \tilde{\Psi}(t_0) - (i/\hbar)\int_{t_0}^{t} d\tau\, \tilde{H}^0(\tau)\tilde{\Psi}(\tau) \qquad (14)$$

Since $R(t)$ contains $\mathbf{A}\cdot\mathbf{r}$ in the exponent, the previous procedure that enabled us to obtain Eq. (12) does not apply. Instead, one can express $\tilde{H}^0$ as $\tilde{H}^0 = H^0 + [R, H^0]R^+$. If we regard now the commutator part as the perturbation we can write

$$\tilde{\Psi}(t) = \tilde{\Psi}_i^0(t) - \frac{i}{\hbar}\int_{t_0}^{t} d\tau\, e^{-(i/\hbar)H^0(t-\tau)}[R(\tau), H^0]R^+(\tau)\tilde{\Psi}(\tau) \qquad (15)$$

where

$$\tilde{\Psi}_i^0(t) \equiv \exp[-(i/\hbar)H^0(t-t_0)]\tilde{\Psi}_i$$

and $\tilde{\Psi}_i$ is the initial $\tilde{\Psi}$ state. At this point it appears that we have done nothing more than add complexity. However, if we note that $R^+(\tau)\tilde{\Psi}(\tau) = \Psi(\tau)$ and let $t_0 \to -\infty$ and $t \to +\infty$ we can write the equation

$$\tilde{\Psi}(+\infty) = \tilde{\Psi}_i(+\infty) - \frac{i}{\hbar}\int_{-\infty}^{+\infty} d\tau\, e^{-(i/\hbar)H^0(t-\tau)}[R(\tau), H^0]\Psi(\tau) \qquad (16)$$

Let ϕ_j be the unperturbed atomic states in the absence of the field. If we replace $\Psi(\tau)$ in the integrand by $e^{-(i/\hbar)H^0\tau}\phi_i$, which is the unperturbed initial atomic state at time τ, and also use the fact that $\tilde{\Psi}(+\infty) = \Psi(+\infty)$ (because at time $t = +\infty$ the field has been switched off), we obtain the approximate result

$$\Psi(+\infty) \cong \phi_i - \frac{i}{\hbar}\int_{-\infty}^{+\infty} d\tau\, e^{-(i/\hbar)H^0(t-\tau)}[R(\tau), H^0]e^{-(i/\hbar)H^0\tau}\phi_i \qquad (17)$$

Using this to calculate transitions to a final state ϕ_f via the scattering matrix $S_{fi} = \langle\phi_f|\Psi(+\infty)\rangle$ gives us the momentum translation approximation (Reiss, 1970a,b, 1971). This is equivalent to retaining the first term in the expansion of Eq. (16) in terms of the commutator. After some mathematical manipulations, S_{fi} can be expressed as $\langle\phi_f|e^{ie\mathbf{A}(t)\cdot\mathbf{r}}|\phi_i\rangle$, where ϕ_f and ϕ_i are the final and initial unperturbed atomic states, respectively. If we take $\mathbf{A}(t) = \boldsymbol{\alpha}\sin\omega t$, the exponential above can be expanded in a series of harmonics of ω involving Bessel functions as coefficients. The various harmonics represent the multiphoton transitions. This approximation was originally proposed by Reiss (1970a) and was obtained in a somewhat different way. The necessary conditions for its applicability as given by Reiss (1970a,b) are $\omega/2E \ll 1$ and $eaa_0\omega/E \ll 1$, where a_0 is the Bohr radius, a the amplitude of **A**, and E a "characteristic" atomic energy. The implication is that the order N of the multiphoton transition is much larger than one and that the field intensity a is smaller than a certain number. This upper limit, however, turns out to be above the expected breakdown of perturbation theory. A typical prediction of this theory is that for sufficiently high light intensities higher-order processes become more probable than lower-order processes. This is one of the results that Reiss (1970a) has called intensity effects.

The momentum translation approximation has been used by a number of authors (see, e.g., references quoted in Decoster, 1974) in a variety of calculations of multiphoton processes. Recently it has also been criticized by some authors (Cohen–Tannoudji *et al.*, 1973; Decoster, 1974; Herman, 1973). Basically this method represents the first term of an expansion. This first term contains the field to all orders, and so do the infinitely many terms that are omitted. At this point there is no general proof that the omitted terms are in fact negligible. On the other hand, certain calculations based on the first term alone have agreed fairly well with perturbation theory results. Thus it seems that the approximation contains some truth, but the complete delineation of its applicability remains an open question. It is clear that the method using the first term only is invalid when there are resonances with intermediate atomic states. Then one must include some of the other terms of the expansion (Haque, 1973) to obtain better agreement with perturbation theory. It should be noted that for high-order photoionization ($N \gg 1$) it is almost certain that there will be intermediate resonances since the last few photons will be making transitions in the dense part of the atomic spectrum. Thus the requirement $N \gg 1$ and the absence of resonances would seem to be contradictory, which implies that the method must in most cases include the higher-order terms. Most of the results using this technique have been obtained with hydrogenic atomic models, which leads to considerable analytic simplicity. This advantage is lost, however, as one begins including higher-order terms and considers atoms other than hydrogen. Moreover, as we have seen, shifts and widths play an important role in resonance and near-resonance multiphoton transitions. According to Reiss (1974) this can also be included in the momentum translation method. Obviously, it can be done phenomenologically. Whether it can be done in the systematic fashion offered by perturbation theory remains to be seen. Finally, it has never been established that as the light intensity increases, the dominance of the first term of the expansion increases. Hence, the high-field behavior of the method is not clear.

In summary, both the space-translation (Henneberger, 1968) and the momentum-translation (Reiss, 1970b) methods may offer possible alternatives to perturbation theory in some cases, provided their validity can be proved and their mathematical complexity can be shown to be preferable to that of conventional radiation perturbation theory. Both alternatives, however, are themselves approximation schemes involving an expansion in a series and are therefore perturbative in some sense.

Finally, space does not permit us to elaborate on a number of other treatments intended for the high-intensity limit where perturbation theory is in doubt. We simply mention the work of Keldysh (1965) and Ritus (Nikishov and Ritus, 1967). More recently, Geltman and Teague (1974) and Mittleman (1974) have proposed methods in which the problem of a free electron in a classical EM field is solved first—thus obtaining the Volkoff solutions—and then the atomic potential is used as the perturbation. These methods are applicable in the very intense field limit where the atomic potential is much weaker than the external field. At this point it is too early to attempt to evaluate the predictions and usefulness of these new methods.

Note Added in Proof

The experiments that led to the results of Figure 2 have now been repeated by the Saclay group, using a single-mode, picosecond Nd–YAG laser. The new results were reported very recently at the Second Conference on Interaction of Electrons with Strong Electromagnetic Field, Budapest, Hungary, October 6–10, 1975. They have obtained straight lines whose slopes, within the experimental errors, are as expected by perturbation theory, up to powers of the order of 10^{15} W/cm^2. Owing to the short pulse duration in the new experiments, instrumental saturation seems to have been avoided; hence no bending of the curves has been observed. Since the previous experiments had been performed with a laser of considerably larger bandwidth, it would seem that there might have been some overlap of resonances which would have caused the initial slope to be smaller than the theoretical slope.

References

Agostini, P., Barjot, G., Bonnal, J. F., Mainfray, G., Manus, C., and Morellec, J. (1970). IEEE, *J. Quantum Electronics*, **QE-6**, 782.
Agostini, P., and Bensoussan, P. (1974). *Appl. Phys. Letts.*, **24**, 216.
Agostini, P., Bensoussan, P., and Boulassier, J. C. (1972). *Opt. Commun.*, **5**, 293.
Agrawal, G. S. (1970). *Phys. Rev.*, **A1**, 1445.
Bakos, J., Kiss, A., Szabo, L., and Tendler, M. (1972a). *Phys. Letts.*, **A39**, 283.

Bakos, J., Kiss, A., Szabo, L., and Tendler, M. (1972a). (1972b). *Phys. Letts.*, A**39**, 317.
Bakos, J., Kiss, A., Szabo, L., and Tendler, M. (1972c). *Phys. Lettrs.*, A**41**, 163.
Bakshi, P., Kalman, G., and Cohn, A. (1973). *Phys. Rev. Letts.*, **31**, 1576.
Bebb, H. B. (1966). *Phys. Rev.*, **149**, 25.
Bebb, H. B. (1967). *Phys. Rev.*, **153**, 23.
Bebb, H. B., and Gold, A. (1966). *Phys. Rev.*, **143**, 1.
Cervenan, M. R., and Isenor, N. R. (1974). *Opt. Commun.*, **10**, 280.
Chang, C. S., and Stehle, P. (1973). *Phys. Rev. Letts.*, **30**, 1283.
Cohen-Tannoudji, C., Dupont-Roc, J., Fabre, C., and Grynberg, G. (1973). *Phys. Rev.*, A**8**, 2747.
Cohn, A., Bakshi, P., and Kalman, G. (1972). *Phys. Rev. Letts.*, **29**, 324.
Decoster, A. (1974). *Phys. Rev.*, A**9**, 1446.
Delone, G. A., Delone, N. B., and Piskova, G. K. (1972). *Sov. Phys.-JETP*, **35**, 672.
Edelstein, S., Lambropoulos, M., Duncanson, J., and Berry, R. S. (1974). *Phys. Rev.*, A**9**, 2459.
Evans, R. G., and Thoneman, P. C. (1972). *Phys. Letts*, A**39**, 133.
Faisal, F. H. M. (1973). *J. Phys. B.*, **6**, L89.
Fano, U. (1969). *Phys. Rev.*, **178**, 131.
Farago, P. S., and Walker, D. W. (1973). *J. Phys. B.*, **6**, L280.
Fox, R. A., Kogan, R. M., and Robinson, E. J. (1971). *Phys. Rev. Letts.*, **26**, 1416.
Geltman, S., and Teague, M. R. (1974). *J. Phys. B.*, **7**, L22.
Gontier, Y., and Trahin, M. (1973a). *Phys. Rev.*, A**7**, 1899.
Gontier, Y., and Trahin, M. (1973b). *Phys. Rev.*, A**7**, 2069.
Haque, S. N. (1973). *Phys. Rev.*, A**8**, 3227.
Held, B., Mainfray, G., and Morellec, J. (1972a). *Phys. Letts.*, A**39**, 57.
Held, B., Mainfray, G., Manus, C., and Morellec, J. (1972b). *Phys. Rev. Letts.*, **28**, 130.
Held, B., Mainfray, G., Manus, C., Morellec, J., and Sanchez, F. (1973). *Phys. Rev. Letts.*, **30**, 423.
Henneberger, W. C. (1968). *Phys. Rev. Letts.*, **21**, 838.
Henry, R. J. W., O'Connell, R. F., Smith, E. R., Chanmugam, G., and Rajagopal, A. K. (1974). *Phys. Rev.*, D**9**, 329.
Herman, R. M. (1973). Unpublished.
Keldysh, L. V. (1965). *Sov. Phys.-JETP*, **20**, 1307.
Klarsfeld, S., and Maquet, A. (1972). *Phys. Rev. Letts.*, **29**, 79.
Kogan, R. M., Fox, R. A., Burnham, G. T., and Robinson, E. J. (1971). *Bull. Am. Phys. Soc.*, **16**, 1411.
Lambropoulos, M., and Berry, R. S. (1973). *Phys. Rev.*, A**8**, 855.
Lambropoulos, M., Moody, S. E., Lineberger, W. C., and Smith, S. J. (1973). *Bull. Am. Phys. Soc.*, **18**, 1514.
Lambropoulos, P., Kikuchi, C., and Osborn, R. K. (1966). *Phys. Rev.*, **144**, 1081.
Lambropoulos, P. (1968). *Phys. Rev.*, **168**, 1418.
Lambropoulos, P. (1972a). *Phys. Rev. Letts.*, **28**, 585.
Lambropoulos, P. (1972b). *Phys. Rev. Letts.*, **29**, 453.
Lambropoulos, P. (1972c). *Phys. Letts.*, A**40**, 199.
Lambropoulos, P. (1973a). *Phys. Rev. Letts.*, **30**, 413.
Lambropoulos, P. (1973b). *Bull. Am. Soc.*, **18**, 709.
Lambropoulos, P. (1973c). *J. Phys. B.*, **6**, L319.
Lambropoulos, P. (1974a). *J. Phys. B.*, **7**, L33.

Lambropoulos, P. (1974b). *Phys. Rev.*, **A9**, 1992.
Lambropoulos, P. (1975). To be published.
Lambropoulos, P., and Stuart, K. M. (1975) to be published; Stuart, K. M. Thesis, Texas A & M University.
Lecompte, C., Mainfray, G., Manus, C., and Sanchez, F. (1974). *Phys. Rev. Lett.*, **32**, 265.
LuVan, M., and Mainfray, G. (1972). *Phys. Lett.*, **A39**, 21.
LuVan, M., Mainfray, G., Manus, C., and Tugov, I. (1972). *Phys. Rev. Lett.*, **29**, 1134.
Mittleman, M. H. (1974). *Phys. Lett.*, **A47**, 55.
Mollow, B. R. (1968). *Phys. Rev.*, **175**, 1555.
Nikishov, A. I., and Ritus, V. I. (1967). *Sov. Physics–JETP*, **25**, 145.
Novick, R. (1969). In *Physics of One- and Two-Electron Atoms*. Eds., F. Bopp and H. Kleinpoppen. Amsterdam: North-Holland.
Power, E. A. (1965). *Introductory Quantum Electro-dynamics*. New York: American Elsevier Publishing Co.
Power, E. A., and Zienau, S. (1959). *Phil. Trans. R. Soc. Lond.*, **A251**, 427.
Reiss, H. R. (1970a). *Phys. Rev. Lett.*, **25**, 1149.
Reiss, H. R. (1970b). *Phys. Rev.*, **A1**, 803.
Reiss, H. R. (1971). *Phys. Rev.*, **D4**, 3533.
Reiss, H. R. (1972). *Phys. Rev. Lett.*, **29**, 1129.
Reiss, H. R. (1974). Unpublished; private communication.
Shiga, F., and Imamura, S. (1967). *Phys. Lett.*, **A25**, 706.
Smith, E. R., Henry, R. J. W., Surmelian, G. L., O'Connell, R. F., and Rajagopal, A. K. (1972). *Phys. Rev.*, **D6**, 3700.
Sokolov, A. A., Zukovskii, V. Ch., and Nikitina, R. (1973). *Phys. Lett.*, **A43**, 85.
Winter, J. M. (1959). *Ann. de Phys.*, **4**, 755.

43
The Effect of Multimode Laser Operation on Multiphoton Absorption by Atoms

JOEL GERSTEN AND MARVIN MITTLEMAN

The nonlinear interaction of intense light with atoms is found to be strongly affected by the mode structure of the laser beam. In many situations the rate for nonlinear processes may be strongly amplified by employing the proper mode structure. A statistical analysis is presented which explores this possibility for an arbitrary nonlinear process. Particular attention is paid to the case where many equivalent modes are present and to the case where two inequivalent modes are present. An experiment related to the present theory is discussed.

The probability of multiphoton absorption by an atom (or molecule) from a laser beam is critically dependent upon the mode structure of the laser. This has been observed experimentally (Lecompte *et al.*, 1974) and predicted theoretically (Agarwal, 1970). The theory relies upon a quantum-electrodynamic description of the photon statistics of the beam and thus is rather complicated, the complication increasing rapidly as the mode structure becomes more intricate.

The interaction of lasers with matter is, in most cases, accurately obtained by a classical description of the laser light, so we present such a description of the multiphoton absorption from a multimode laser here. Moreover, laser discharges are frequently not reproducible from one laser to another and even from pulse to pulse in the same laser. It therefore seems reasonable

JOEL GERSTEN and MARVIN MITTLEMAN • Department of Physics, The City College of the City University of New York, New York, N.Y., U.S.A.

to try to treat the mode structure of a laser statistically. That is, we shall neglect the strong-mode coupling effects in many lasers and treat the power in each mode as a statistical variable. The question of phase relations between the modes can also be treated, but we shall only deal with the situation where there are no fixed phase relations between modes. We also treat only spatially homogeneous lasers.

We pose the problem as follows: Suppose that a single-mode laser induces some atomic process with a probability $F(E^2)$, where E is the amplitude of the electric field in that mode. We now ask for the same transition probability when the laser is in multimode operation with the same total power. We assume that F is not a rapidly varying function of frequency in the band of the laser.

The (classical) laser field may then be written as

$$E = \sum_i E_i \cos(\omega_i t + \phi_i) \tag{1}$$

where the sum runs over all modes. This may be rewritten as

$$E = E(t) \cos(\bar{\omega} t + \phi(t)) \tag{2}$$

where $\bar{\omega}$ is an average frequency of the laser and the functions $E(t)$ and $\phi(t)$ are slowly varying, provided that the average ω is much greater than the band width of the laser*

$$\frac{E^2(t)}{\bar{\omega}^2} = \sum_i \frac{E_i^2}{\omega_i^2} + 2 \sum_{i>j} \frac{E_i E_j}{\omega_i \omega_j} \cos[(\omega_i - \omega_j)t + \phi_i - \phi_j] \tag{3}$$

$$\tan \phi(t) = \frac{\sum_i (E_i/\omega_i) \sin(\Delta_i t + \phi_i)}{\sum_i (E_i/\omega_i) \cos(\Delta_i t + \phi_i)} \tag{4}$$

where

$$\Delta_i = \omega_i - \bar{\omega} \tag{5}$$

Equation (2) describes field of frequency $\bar{\omega}$ with a complicated beat structure at the frequencies $\omega_i - \omega_j \ll \bar{\omega}$.

The calculation which led to the transition probability $F(E^2)$ may now be repeated with the "single-mode" field of Eq. (2), provided only that the beat frequencies are small compared to the atomic frequencies relevant to this transition. The result must then be averaged over the time occurring in $E^2(t)$ to get the transition probability in the field of the multimode laser. The result will, of course, depend upon the phase and amplitude parameters

---

*A proper choice of $\bar{\omega}$ can minimize the error associated with this approximation. The choice $\sum E_i^2 \Delta_i = 0$ makes the error of order $(\Delta/\bar{\omega})^2$.

E_i and ϕ_i. However, we now perform averages over these quantities because of the presumed statistical nature of the laser pulses. We constrain the total power of the laser to be fixed from pulse to pulse; that is, the relation

$$\sum_i E_i^2 = E_0^2 \tag{6}$$

is maintained.

The average result for an N-mode laser is

$$\bar{F}_N = C_N \int F(E^2(t)) \, \delta(\sum E_i^2 - E_0^2) E_1 \ldots E_N \exp(-\alpha_1 |E_1^2| \ldots -\alpha_N(|E_N|^2) \tag{7}$$
$$\times dE_1 \ldots dE_N \, d\phi_1 \ldots d\phi_N$$

where C_N is a normalization constant and where the last factors allow for a shaping of the frequency profile of the laser by an appropriate choice of the α_s. Here $(2\alpha_i)^{-1}$ is the most probable field in the ith mode.

It is immediately apparent that the time average occurring in Eq. (7) is redundant once the phase averaging has been done. If we set

$$\xi_i = \phi_i + \Delta_i t \tag{8}$$

and define the complex variables

$$z_i = E_i \, e^{i\xi_i} \tag{9}$$

then if terms of order $\Delta_i/\bar{\omega}$ are neglected, Eq. (7) becomes

$$\bar{F}_N = C_N \int F\left(\left|\sum_i z_i\right|^2\right) (d^2 z_i)^N \delta\left(E_0^2 - \sum_i |z_i|^2\right) \exp(-\sum_i \alpha_i |z_i|^2) \tag{10}$$

If we define the Fourier transform of F

$$F(|z^2|) = \int dk \, e^{-ik|z^2|} \tilde{F}(k) \tag{11}$$

Table 1. Amplification Factor for Various Values of the Integers

		n		
N	2	3	4	5
2	4/3	2	16/5	16/3
3	3/2	27/10	27/5	81/7
4	8/5	16/5	256/35	128/7
5	5/3	25/7	125/14	3125/126

and also transform the δ function, this can be written

$$\bar{F}_N = C_N \int \frac{ds\, dk}{2\pi} \tilde{F}(k)\, e^{isE_0^2} \int (d^2 z_i)^N \exp\left[-ik|\sum_i z_i|^2 - \sum_i (\alpha_i + is)|z_i|^2 \right] \quad (12)$$

The z integrations can each be broken into an integral over the real part of z_i and over the imaginary part of z_i. They are equal, so that

$$\bar{F}_N = C_N \int \frac{ds\, dk}{2\pi} \tilde{F}(k)\, e^{isE_0^2} \left\{ \int (dx)^N \exp\left[-ik\left(\sum_i \alpha_i\right)^2 - i\sum_i s_i x_i^2 \right] \right\}^2 \quad (13)$$

where

$$s_i = s - i\alpha_i \quad (14)$$

The integral appearing in Eq. (13) may be expressed as

$$I = \int dx_i \ldots dx_N \exp\left(-i \sum_{ij} \chi_i M_{ij} \chi_j \right) \quad (15)$$

where $M_{ij} = k + S_i \delta_{ij}$. The matrix is symmetric and is thus diagonalizable by an orthogonal transformation. Thus,

$$I = (\pi/i)^{N/2} (\det M)^{-1/2} \quad (16)$$

The determinant of M is easily evaluated and we find

$$\det M = \left[1 + \sum_{i=1}^N \frac{k}{s_i} \right] \prod_{j=1}^N s_j \quad (17)$$

Consequently, Eq. (13) becomes

$$\bar{F}_N = C_N \int \frac{ds\, dk}{2\pi} \tilde{F}(k)\, e^{isE_0^2} \frac{\pi^N}{\det M} \quad (18)$$

The constant C_N' can be obtained from the condition that $\bar{F}_N = F$ when $F(E^2)$ is independent of E^2. Thus

$$C_N^{-1} = -(\pi)^N \sum_{j=1}^N e^{-\alpha_j E_0^2} \left[\prod_{\substack{l=1 \\ l \neq j}}^N (\alpha_j - \alpha_l)^{-1} \right] \quad (19)$$

We may use the inverse of Eq. (11) to eliminate $\tilde{F}$ and then perform the k integral, obtaining

$$\bar{F}_N = iC_N \pi^N \int \frac{ds}{2\pi} \int_0^\infty dt\, \frac{F(t) \exp[isE_0^2 - it(\sum_i s_i^{-1})^{-1}]}{s_1 \ldots s_N [\sum_i s_i^{-1}]} \quad (20)$$

The remaining integrals become difficult as N gets large. We now turn to some simple examples.

Effect of Multimode Laser Operation

1. If the pulse is square, i.e., if all the α_i are equal: The s integral may be performed and the t integral scaled with the result

$$\bar{F}_N = \frac{N-1}{N} \int_0^N dt\, F(E_0^2 t)\left(1 - \frac{t}{N}\right)^{N-2} \quad (21)$$

In obtaining Eq. (21) it is simpler to use the criterion that led to Eq. (19) than to use Eq. (19) and the limiting process where the α_i all approach α.

For a multiphoton absorption process where the field is not too high

$$F(E_0^2) \sim (E_0^2)^n \quad (22)$$

where n is the number of photons absorbed.

Then the amplification factor due to the multimode operation is

$$A = \frac{\bar{F}_N}{F} = N^{n-1} \frac{n!N!}{(N+n-1)!} \quad (23)$$

This factor is given for low values of the two integers in the table. It is seen to be a rising function of both. This can be explained as follows: F is a rapidly rising function of its argument so that the averaging process will get its maximum contribution from the large values of $E(t)$. Thus the beat phenomenon, although giving the same average E^2, will provide much higher averages since the peak values of E^2 will be much higher.

For $N = 2$ the integral may be reduced to

$$\bar{F}_2 = \frac{\chi}{\sinh \chi} \int_0^1 dt\, F(2E_0^2 t) I_0[2x(t(1-t))^{1/2}] \quad (24)$$

where $\chi = \frac{1}{2}|\alpha_1 - \alpha_2|E_0^2$.

For large X (essentially only one mode), this returns to $F(E_0^2)$. For $F(E^2) \sim E^{2n}$ this is a monotonically falling function of X so that the maximum amplification factor is obtained at $X = 0$. This is reasonable in the light of the above discussion since $X = 0$ yields the greatest beat amplitude of the laser.

Comparison of our results for $N = 2$, $X = 0$ with the experimental (Lecompte et al., 1974) nonresonant values where $n = 11$ shows the experiment significantly below our result for the amplification factor. The experimental value (Lecompte et al., 1974) is about 18, ours for $X = 0$ is 171. This shows a nonzero value of X. In Table 2 we show the amplification factor for $F \sim (E^2)^{11}$. A value of X between 9 and 10 is indicated. This is a highly asymmetric two-mode laser. We should emphasize that mode locking (phase correlation between the two modes) cannot change the result, since for a two-mode laser the integral over the relative phase of the modes in Eq. (7) is equivalent to the time average so whether or not the modes are locked is irrelevant.

Table 2. Amplification Factor for $n = 11$, $N = 2$ as a Function of X

X	A
1	155.1
2	121.8
3	89.9
4	66.1
5	49.6
6	38.3
7	30.4
8	24.8
9	20.6
10	17.5

References

Agarwal, G. S. (1970). *Phys. Rev.*, **A1**, 1445.
Lecompte, C., Mainfray, G., Manus, C., and Sanchez, F. (1974). *Phys. Rev. Lett.*, **32**, 265.

44

Two-Photon Processes

R. Stephen Berry

The range of topics studied experimentally by two-photon processes, particularly through angular properties, extends from nuclear resonance through bound state spectroscopy to the photoelectric effect. We review the subject in a selective, topical way, to illustrate how a common theoretical frame underlies all these topics. Examples are taken from the study of angular correlations of gamma rays, of absorption spectroscopy and fluorescence with linear- and circular-polarized light, and of the angular distribution of photoelectrons.

1. Survey

Experimental studies of two-photon processes span a range from perturbed angular correlation studies in high-energy physics to two-photon spectroscopy of large molecules. Here, I shall try to give a sense of the range and present state of development of two-photon studies in atomic physics and in closely related studies in molecular physics by describing a few of these studies. I shall only be able to make passing reference to some of the others and, with apologies, pass over several of the experimental efforts in this field. I shall pay most attention to process in which *angular* information plays a key role. Even this relatively narrow part of the whole subject is already a lively field, with a considerable literature of its own.

Two-photon studies fall into overlapping categories, according to their objectives:

(i) Bound-state spectroscopy for the investigation of intrinsic atomic and molecular properties; two-photon spectroscopy, after its first demonstrations (Hughes and Grabner, 1950; Kaiser and Garrett, 1961; Abella,

---

R. Stephen Berry • Theoretical Chemistry Department, Oxford, England. Permanent address: Department of Chemistry, University of Chicago, Chicago, Illinois 60637, U.S.A.

1962; Yatsiv et al., 1965), has been used for high-resolution studies of atoms (Cagnac et al., 1973; Pritchard et al., 1974; Biraben et al., 1974; Levenson and Bloembergen, 1974; Bloembergen et al., 1974), and molecules (Dowley et al., 1967; Peticolas, 1967; Moore, 1971, R. Hochstrasser et al., 1973; 1974a,b; A. Hochstrasser, private communication), for the study of two-photon emission and excitation probabilities, (Richman and Chang, 1967; Lipeles et al., 1965; Prior, 1972; Kocher et al., 1972; Marrus and Schmieder, 1972; Bradley et al., 1972), and for a subject we shall discuss here, the identification of excited states from polarization studies (Hochstrasser et al., 1974a,b; A. Hochstrasser, private communication; Ovander, 1960; Inoue and Toyozawa, 1965; Chiu, 1968; Monson and McClain, 1970, 1972; McClain, 1971; Swofford and McClain, 1973; Harris and McClain, 1974; Drucker and McClain, 1974a,b,c).

(ii) Two-photon bound-state spectroscopy as a probe to study intermolecular processes (Freund et al., 1973; Gallagher et al., 1975).

(iii) Two-photon photodetachment and photoionization studies directed toward determining detachment and ionization thresholds and total two-photon transition probabilities of atoms (Hall et al., 1965; Hall, 1966; Rizzo and Klewe, 1966; Okuda et al., 1969; Kishi et al., 1970; Kishi and Okuda, 1971; Ambartzumian et al., 1971; Ambartzumian and Letokhov, 1972; Kogan et al., 1971; Fox et al., 1971), and molecules (Lineberger and Patterson, 1972; Collins et al., 1973).

(iv) Investigations of the relationship between photodetachment and photoionization probabilities and the polarization of the incident radiation (Fox et al., 1971; Lambropoulos, 1972; Mizuno, 1973).

(v) Studies directed toward the polarization properties of ionized electrons (Lambropoulos, 1973).

(vi) Studies of angular correlations of gamma rays to study molecular phenomena (Matthias et al., 1971; Shirley and Haas, 1972).

(vii) Measurements of angular distributions of electrons from two-photon photoionization (Edelstein et al., 1974).

Here, I shall review only a few topics from the list because of their appeal to my own tastes, and because of the relationship between these topics and the work of my friend and colleague, Ugo Fano. The general discussion of the theory is given by Melissa and Peter Lambropoulos (Chapter 42), and a discussion of electron polarization follows in the paper by H. D. Zeman.

2. *Spectroscopy: Symmetry Properties from Polarization Studies*

One of the most immediately attractive uses of two-photon methods has been in the study of bound states of atoms and molecules. The power of the method extends well beyond the obvious differences between selection rules for one- and two-photon electric-dipole absorption or emission processes, although this difference itself has proved useful for molecular spectroscopy. That is, it has been possible to observe and locate transitions to excited states that were too weak to be seen in one-photon spectroscopy because of the one-photon selection rule. Optical quenching of metastable (2s) hydrogen atoms is the simplest prototype of this use of the method. The two-photon mechanism simply provides an electromagnetic pathway between 1s and 2s levels. The method is applicable to H-like and He-like metastables (Lipeles et al., 1965; Prior, 1972; Kocher, 1972; Marrus and Schmieder, 1972).

The two-photon method becomes much richer when we admit the option of varying the polarization characteristics of the two-photons. Two-photon electronic spectroscopy of crystalline materials was analyzed first, primarily for molecular crystals and for linearly polarized light (Inoue and Toyozawa, 1965; Honig et al., 1967; Hernandez and Gold, 1966). The theoretical groundwork has been laid for studying two-photon spectra of atoms and molecules, even of low symmetry, in rigid media or in gases, with electric-dipole, electric-quadrupole and magnetic-dipole processes in any combination, with arbitrary propagation directions and with photons having arbitrary directions of linear polarization (Chiu, 1968). Most of the potential of this analysis is still unexploited, experimentally. However, the analysis I shall describe in some detail, because it has been used in the laboratory and because of its relevance to the polarization studies being described here, is that of McClain and his co-workers. This analysis deals with freely rotating molecules excited by two photons whose wavelength and polarization character could be varied independently (Manson and McClain, 1970, 1972; McClain, 1971).

The transition probability P_{of} for a two-photon electric-dipole process between bound states o and f is governed by a second-rank tensor $\mathbf{S}$. The components of $\mathbf{S}$ are induced by the polarization vectors of the two photons and the vectors of the two sets of transition dipoles that connect the initial state o with intermediate states i and the intermediate states i with the final state f, respectively. We use capital Roman letters to indicate (laboratory) coordinates for the photon polarization vectors—$(1, 0, 0)$, $(0, 1, 0)$, $2^{-1/2}(1, -i, 0)$, and $2^{-1/2}(1, i, 0)$ for x, y, right-circular and left-circular polarizations. We use lower-case Greek letters to indicate (molecule-frame) coordinates for the target atoms or molecules and $\mathbf{l}$ and $l_{\alpha\chi}$ to represent the coordinate transformation matrix from molecular to laboratory coordinates. Explicitly (Peticolas, 1967):

$$P_{of} = (2\pi)^2 \alpha^2 v_\lambda v_\mu g(v_\lambda + v_\mu) |\boldsymbol{\lambda} \cdot \mathbf{S}_{of} \cdot \boldsymbol{\mu}|^2$$
$$= K_{\lambda\mu} |\boldsymbol{\lambda} \cdot \mathbf{S}_{of} \cdot \boldsymbol{\mu}|^2 \tag{1}$$

where v_λ and v_μ are the photon frequencies and $g(v_\lambda + v_\mu)$ is a normalized line-shape function.

Monson and McClain (1970) expressed the components of the tensor $\mathbf{S}_{of}(\lambda, \mu)$ as the tensor products of three factors: the first is a second-rank tensor that is a function only of light polarization, and therefore of laboratory coordinates; the second is the (fourth-rank) tensor transforming bivectors in molecular coordinates to bivectors in laboratory coordinates, and the third is a second-rank tensor characteristic only of the atomic or molecular targets. Thus,

$$\mathbf{S}_{of}(\lambda, \mu) = \sum_{A,B} \sum_{\alpha,\beta} (\lambda_A \mu_B)(l_{A\alpha} l_{B\beta}) S_{\alpha\beta}^{of} \qquad (2)$$

where the transition-dipole term

$$\mathbf{S}_{\alpha\beta}^{of} = \sum_{\substack{\text{intermediate}\\\text{states } i}} \left[\frac{(o|r_\alpha|i)(i|r_\beta|f)}{E_i - E(\lambda)} + \frac{(o|r_\beta|i)(i|r_\alpha|f)}{E_i - E(\mu)} \right] \qquad (3)$$

The general expression for the predicted value of an observation characterized by the final state f and a detector described by a selection operator ε is, of course, expressed conveniently as the trace of the product of the density matrix ρ_f of the final state with the detector matrix ε (Fano, 1957):

$$P(f, \varepsilon) = \text{Tr}(\rho_f \varepsilon) \qquad (4)$$

We obtain ρ_f from ρ_0, the density matrix for the initial state, through $\mathbf{S}_{of}(\lambda, \mu)$:

$$\rho_f = \mathbf{S}_{of}^+ \rho_0 \mathbf{S}_{of} \qquad (5)$$

Monson and McClain calculate the probability that state f will be observed, so their detector operator ε is merely a projector onto the chosen final state. The initial distribution is taken as isotropic, with a simple rovibronic state (or, in solution, a single vibronic state), so that ρ_0 is a unit matrix with dimension equal to the degeneracy of the ground state. The crux of their treatment comes in their reduction of $\mathbf{S}_{of}(\lambda, \mu)$ for the case in which the targets are tumbling but undergo both steps of the two-photon absorption at the same instant. Here, they extended the averaging method of Pople and Buckingham (1955) to find the average over all orientations of the square modulus of $\mathbf{S}_{of}(\lambda, \mu)$:

$$\langle |\mathbf{S}_{of}(\lambda, \mu)| \rangle = \langle |\Sigma(\lambda_A \mu_B)(l_{A\alpha} l_{B\beta}) S_{\alpha\beta}^{of}|^2 \rangle$$
$$= \langle \Sigma(\lambda_A \mu_B \lambda_R^* \mu_S)(l_{A\alpha} l_{B\beta} l_{R\rho} l_{S\sigma})(S_{\alpha\beta}^{of} S_{\rho\sigma}^{+of}) \rangle \qquad (6)$$

The sum is taken over A, B, R, S, α, β, ρ, and σ, giving 3^8 terms in the sum. The averaging procedure supposes that the first and third factors, which do not depend on molecular orientation, can, in effect, be extracted and the average taken over the product of the four direction cosines to give

$$\langle |\mathbf{S}_{of}(\lambda,\mu)|^2 \rangle = \Sigma(\lambda_A \mu_B \lambda_R^* \mu_S^*)\langle l_{A\alpha} l_{B\beta} l_{R\rho} l_{S\sigma} \rangle (S_{\alpha\beta}^{of} S_{\rho\sigma}^{+of}) \tag{7}$$

The average represented by the second factor is

$$\langle l_\alpha l_\beta l_\rho l_\sigma \rangle = \tfrac{1}{30}[\delta_{AB}\delta_{RS}(4\delta_{\alpha\beta}\delta_{\rho\sigma} - \delta_{\alpha\rho}\delta_{\beta\sigma} - \delta_{\alpha\sigma}\delta_{\beta\rho})$$
$$+ \delta_{AR}\delta_{BS}(-\delta_{\alpha\beta}\delta_{\rho\sigma} + 4\delta_{\alpha\rho}\delta_{\beta\sigma} - \delta_{\alpha\sigma}\delta_{\beta\rho})$$
$$+ \delta_{AS}\delta_{BR}(-\delta_{\alpha\beta}\delta_{\rho\sigma} - \delta_{\alpha\rho}\sigma_{\beta\sigma} + 4\delta_{\alpha\sigma}\delta_{\beta\rho})] \tag{8}$$

Note the symmetry of the three terms. Any coordinate must appear 0, 2, or 4 times to give a nonvanishing contribution. The result is the expected simplification to a form in which $\langle |\mathbf{S}_{of}(\lambda,\mu)|^2 \rangle$ depends on three external (experimentally controllable) quantities F, G, and H, and three molecular parameters δ_F, δ_G, and δ_H. The two-photon absorption coefficient δ (cm^4 sec photon^{-1} molec^{-1}) analogous to the one-photon absorption cross section is

$$\langle \delta \rangle = K_{\lambda\mu}\langle |\mathbf{S}_{of}(\lambda,\mu)|^2 \rangle = F\delta_F + G\delta_G + H\delta_H \tag{9}$$

with

$$F = 4|\lambda \cdot \mu|^2 - 1 - |\lambda \cdot \mu^*|^2 \tag{10a}$$

$$G = -|\lambda \cdot \mu|^2 + 4 - |\lambda \cdot \mu^*|^2 \tag{10b}$$

$$H = -|\lambda \cdot \mu|^2 - 1 + 4|\lambda \cdot \mu^*|^2 \tag{10c}$$

and

$$\delta_F = \tfrac{1}{30}K_{\lambda\mu}\sum_{\alpha,\beta} S_{\alpha\alpha}S_{\beta\beta}^+ \tag{11a}$$

$$\delta_G = \tfrac{1}{30}K_{\lambda\mu}\sum_{\alpha,\beta} S_{\alpha\beta}S_{\alpha\beta}^+ \tag{11b}$$

$$\delta_H = \tfrac{1}{30}K_{\lambda\mu}\sum_{\alpha,\beta} S_{\alpha\beta}S_{\beta\alpha}^+ \tag{11c}$$

Alternatively,

$$\langle \delta \rangle = A + B_1|\lambda \cdot \mu^*|^2 + B_2|\lambda \cdot \mu|^2 \tag{12}$$

and

$$10\delta_F = A + B_1 + 3B_2 \tag{13a}$$

$$10\delta_G = 3A + B_1 + B_2 \tag{13b}$$

$$10\delta_H = A + 3B_1 + B_2 \tag{13c}$$

If one or both photons has a component of circular polarization, so that λ or μ is complex, then F, G, and H are independently variable and the *three* parameters δ_F, δ_G, and δ_H can be measured. If both photons are plane polarized, so that λ and μ are both real, then $F = H$, and only the parameters $(\delta_F + \delta_H)$ or $(B_1 + B_2)$ and δ_G can be determined from experiment. More specifically, if

$$|\lambda \cdot \mu|^2 = |\lambda \cdot \mu^*|^2 = \cos^2 \theta \tag{14}$$

then

$$\langle \delta \rangle = [4\delta_G - (\delta_F + \delta_H)] + [3(\delta_F + \delta_H) - 2\delta_G] \cos^2 \theta$$
$$= A + (B_1 + B_2) \cos^2 \theta \tag{15}$$

Monson and McClain point out that δ_F, proportional to $\sum_\alpha |S_{\alpha\alpha}|^2$, vanishes except when the first and second transitions have the same polarization. The parameter δ_G is proportional to the sum of the absolute squares of all nine transition moments $|S_{\alpha\beta}|^2$ and is therefore a measure of the total strength of the transition. The third parameter δ_H is similar in form to δ_G, except that the indices in one factor are transposed; δ_G and δ_H become equal when the first and second components of the transition $(o|r|i)$ and $(i|r|f)$ are parallel (for all intermediate states i) or when the frequencies v_λ and v_μ are equal.

McClain (1971) has worked out the general forms for the two-photon transition tensor $\mathbf{S}_{of}$ for all 32 point groups and for the three representations of lowest angular momentum of the groups $D_{\infty h}$ and $C_{\infty v}$ for the case in which the initial state o is totally symmetric—by far the most common case. From these patterns, he has shown how measurements giving δ_F, $\delta_G + \delta_H$, and $\delta_G - \delta_H$ provide the maximum possible information concerning the symmetry of final states following two-photon processes. (Previous authors cited by McClain had given related tables (Ovander, 1960; Inoue and Toyozawa, 1965), for oriented molecules only). The indicators δ_F, $\delta_G + \delta_H$, and $\delta_G - \delta_H$ are all either zero or positive; five combinations can arise. From these combinations alone, most, but not all, symmetry characters can be found for the first states. At worst, in the groups C_{2v}, D_2, and D_{2h} a totally symmetric state can be distinguished from a state having one of the other three symmetry characters, but the three cannot be distinguished among themselves.

In the laboratory, Monson and McClain (1972) carried out a complete determination of the plane defined by (12), in which $|\lambda \cdot \mu|^2$ and $|\lambda \cdot \mu^*|^2$ are the two independent variables. The apparatus (Figure 1) allowed the direct measurement of $\langle \delta \rangle$ in absorption with the polarizations of the two photons, from a ruby laser and a xenon flash lamp, independently variable from right circular through plane to left circular. The region of the plane of $|\lambda \cdot \mu|^2$, $|\lambda \cdot \mu^*|^2$ enclosed in the square defined by $0 \leq |\lambda \cdot \mu|^2, |\lambda \cdot \mu^*|^2 \leq 1$

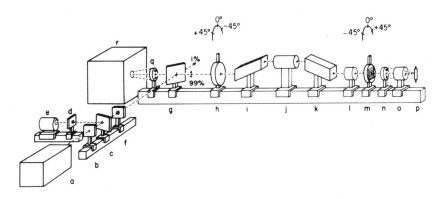

Figure 1. Apparatus used by Monson and McClain to carry out a polarization study of the two-photon absorption spectrum of 1-chloronaphthalene, and then other molecules (with permission of the authors and the *Journal of Chemical Physics*).

contains all experimental situations for simultaneous two-photon processes. The diagonal from $(0, 0)$ to $(1, 1)$ contains all cases in which one photon is linear-polarized, at $(0, 0)$. The two polarizations are orthogonal; at $(1, 1)$ both are linear and parallel. The corners $(0, 1)$ and $(1, 0)$ describe circular polarization of both photons with the same propagation and helicity and with the same sense and opposite helicity, respectively. The vertical axis $(1, |\lambda \cdot \mu^*|^2)$ describes the cases of two identical photons. This "experiment plane" is shown in Figure 2a.

Experimental results for the first system studied this way, the 1-chloronaphthalene molecule, are in Figures 2b,c. Monson and McClain measured the parameters at the four corners of the square of Figure 2a and found the cross-hatched planes as shown, with one ruby-laser photon and one photon of selected frequency from the xenon lamp. The lower figure corresponds to a transition to a totally symmetric upper state and is an allowed, expected process with $\delta_F > 0$, $\delta_G - \delta_H$ very small, and $\delta_G + \delta_H > 0$. The upper figure corresponds to a spectral region in which the polarization measurements show that $\delta_F > 0$, $\delta_G - \delta_H > 0$, and $\delta_G + \delta_H > 0$. Within the symmetry rules based on electronic excitation alone, this is forbidden for a two-photon transition in a molecule with the symmetry of naphthalene. The authors observe rapid variations in $\langle \delta \rangle$ with frequency of the incident light in this region, and infer that the two-photon process in the 32,000 cm^{-1} region is vibronically allowed and electronically forbidden. More recently, McClain and his co-workers have extended their polarization studies of two-photon absorption to diphenylbutadiene (Swofford and McClain, 1973), naphthalene (Drucker and McClain, 1947b), and biphenyl (Drucker

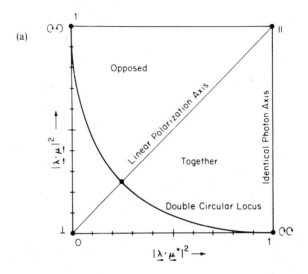

Figure 2. (a) The "experimental plane" defined by the axes $|\lambda \cdot \mu|^2$ and $|\lambda \cdot \mu^*|^2$; (b, c) perspective drawings showing best-fit planes of Eq. (12) for 1-chloronaphthalene at two different energies of the variable-energy photon (with permission of the authors and the *Journal of Chemical Physics*).

and McClain, 1974a), and also to include three-photon processes, including fluorescence. (See also Hochstrasser *et al.*, 1973, 1974a,b; Hochstrasser, private communication.) They have found it helpful to look not only at the δ parameters but also to plot the ratio $(3\delta_G - \delta_F)/(2\delta_G + \delta_F)$, the ratio of intensities of absorption of circular and linearly polarized light in a way precisely analogous to the ionization phenomena reported by Fox *et al.* (1971) and interpreted by P. Lambropoulos (1973) and Mizuno (1973).

In a recent extension of this line of investigation, Harris *et al.* (1974) have developed a phenomenology for treating polarization studies of two-photon processes when the target cannot be treated as small—specifically for the scattering of light from dilute polymer solutions, but the treatment is rather general. Here, by the way, the work is developed in the formalism of irreducible tensors, which makes the connection between the bound-

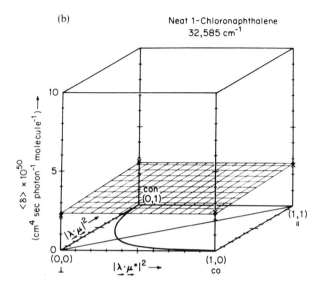

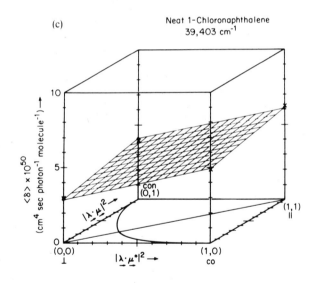

state and free-state work particularly clear. The treatment of scattering from polymers requires the explicit inclusion of cooperative scattering, and therefore makes it necessary to keep the exponential part of the matter–field Hamiltonian that was implicitly approximated as unity to obtain (3).

Let us put (6) into tensorial form. We have a tensor

$$\lambda_A \mu_B \lambda_R^* \mu_S^* = \zeta_{RABS} \tag{16}$$

describing only properties of the radiation field, the *polarization* tensor, and

$$S_{AB}^{of} S_{RS}^{+of} = \sum_{T_{ARBS}} l_{A\alpha} l_{B\beta} l_{R\rho} l_{S\sigma} S_{\alpha\beta}^{of} S_{\rho\sigma}^{+of} \tag{17}$$

the *molecular* tensor. Hence the product of interest is

$$\langle |\mathbf{S}_{of}(\lambda, \mu)|^2 \rangle = \Sigma\, \zeta_{RABS} T_{ARBS} \tag{18}$$

which takes a simple and useful form if ζ_{RABS} and T_{ARBS} are expressed in (spherical) irreducible tensors:

$$\langle |\mathbf{S}_{of}(\lambda, \mu)|^2 \rangle = \sum_{J=0}^{J} \sum_{m=-J}^{J} \sum_{g,h} (-1)^m \zeta_{-m}^J A_{gh}^J T_m^J(h) \tag{19}$$

The sum over J terminates with the maximum rank of the irreducible tensors in the series. A reducible fourth-rank tensor such as ζ_{RABS} or T_{ARBS}, the direct product of four vectors and therefore an 81-component tensor, contains no irreducible component of rank higher than 4. Hence the first sum runs from 0 through 4. The indices g and h refer to the different irreducible components of a given rank, if such exist, The factors A^J are 9-j symbols for the coupling of the four vectors labeled by the indices A, B, R, S. Harris and McClain couple the vectors of incoming polarizations λ_A and λ_R^* and of outgoing polarizations μ_B and μ_S^* to describe the polarization tensor in terms of the coupled vectors j_{AR}, j_{BS}; for the molecular tensor, they couple λ_R^* and μ_B, and λ_A and μ_S^* to define j_{BR}, j_{AS}. Then

$$\langle |\mathbf{S}_{of}(\lambda, \mu)|^2 \rangle = \sum_{J=0}^{4} \sum_{M=-J}^{J} \sum_{j_{AR}} \sum_{j_{BS}} \sum_{j_{AS}} \sum_{j_{BR}} (-1)^M \zeta_{-M}^J (j_{AR} j_{BS} | \lambda \mu)$$

$$\times \hat{j}_{AR} \hat{j}_{BS} \hat{j}_{AS} \hat{j}_{BR} \begin{Bmatrix} 1 & 1 & j_{AR} \\ 1 & 1 & j_{BS} \\ j_{AS} & j_{BR} & J \end{Bmatrix} T_M^J(j_{AS} j_{BR} | \mathbf{K}) \tag{20}$$

We let $\hat{j} = (2j+1)^{1/2}$. All that remains is to express the molecular tensor in the molecular coordinates, where we call it $\mathbf{l}_M^J$:

$$T_M^J(j_{AS} j_{BR} | \mathbf{H}) = \sum_{\eta=-J}^{J} D_{\eta M}^J(\omega) T_\eta^J(j_{AS} j_{BR} | \mathbf{K}) \tag{21}$$

The orientation averaging of (8) is equivalent to the orientation averaging of the rotation matrices $D_{\eta M}^J(\omega)$. (In case of polymers, or any other case in which the structure of the target is included, one more set of nontrivial factors enters, exponentials of the form exp $(i\mathbf{k} \cdot \mathbf{R})$, in which $\mathbf{k}$ refers to the light propagation and $\mathbf{R}$ is an intratarget vector. These factors must be included in the orientation averaging.)

3. Angular Correlations of Gamma Rays

The observation of angular correlations of gamma rays provides a useful tool for probing atomic and molecular properties, particularly when the decay is influenced by an external perturbation, such as the combined static and oscillatory fields of a magnetic resonance system. This topic, long established in nuclear physics, has been developed in the context of atomic and molecular properties (Matthias et al., 1971) in a way that exposes clearly its relationship to the bound-state work described in the previous section and the ionization processes we discuss in the next section. Shirley and Haas (1972) have reviewed the subject so recently that we need not go into details here. In essence, individual nuclei in an initially isotropic ensemble undergo the same 2γ cascade, and the two photons are received by separate but correlated detectors. Observation of one quantum defines its propagation axis along $\mathbf{k}_A$ and selects those nuclei for which the second photon, with momentum $\mathbf{k}_B$ will be measured. The angle of $\mathbf{k}_A$ with respect to $\mathbf{k}_B$ is called Θ. The photon angular distribution function $W(\Theta)$ is the desired quantity.

The natural method for developing $W(\Theta)$ is essentially a form of the preceding development for the case in which both photons are unpolarized (polarization is equal in both directions perpendicular to propagation, and uncorrelated), and the radiating nucleus has transition moments appropriate to a central-field system, but radiation is permitted to have multipolar forms higher than dipole. Apart from the addition of multipole—especially quadrupole—radiation, this much represents a simplification of (6) or (9), because the transition operator for molecules is, in general, a nontrivial tensor. The expressions for $W(\Theta)$ at the level of physical approximation of (6) or (9) is a simple sum of Legendre polynomials (Matthias et al., 1971; Shirley and Haas, 1972):

$$W(\Theta) = \sum_J C_J P_J(\cos \Theta) \qquad (22)$$

where the C_J's are essentially products of Clebsch–Gordan coefficients and J must be even.

The interesting enrichment of the theory comes when we introduce a time-dependent perturbation in the intermediate state. This may be an oscillatory process capable of assuming a frequency in resonance with the intermediate state of the emitter, or it may be a decay process. By writing the distribution function in the form of (4), with a common z axis for $\mathbf{k}_A$

and $\mathbf{k}_B$, and a *time-dependent* density matrix, one has, in general form and in irreducible tensors,

$$P(\Theta, t) = W(\mathbf{k}_A, \mathbf{k}_B; t) = \text{Tr}\,[\rho(\mathbf{k}_A; t)\varepsilon(\mathbf{k}_B)]$$

$$= \sum_{l,m} \rho_m^l(\mathbf{k}_A; t)\varepsilon_m^l(\mathbf{k}_B) \qquad (23)$$

The density matrix in the intermediate state $\rho(\mathbf{k}_A; t)$ evolves from $\rho(\mathbf{k}_A; 0)$ according to

$$\rho(\mathbf{k}_A; t) = \Lambda(t)\rho(\mathbf{k}_A; 0)\Lambda^+(t)$$

$$= \exp(-iH_{\text{pert}}t/\hbar)\rho(\mathbf{k}_A; 0)\exp(iH_{\text{pert}}t/\hbar) \qquad (24)$$

In the event that H_{pert} is a resonant oscillatory perturbation, e.g., field-split magnetic sublevels in the intermediate state, the resultant observation exhibits a regular oscillatory behavior in time if a timedependent observation is made while the resonant frequency is applied. When time-integrated measurements of $W(\Theta)$ are made, provided the decay lifetime of the intermediate state is comparable, the angular Larmor period (radians/second), the line shape—$W(\Theta)$ at a fixed Θ, as a function of the applied frequency in the vicinity of the resonant frequency—can show multiple peaks, analogous to the same phenomenon in optical pumping of atoms (Brossel and Bitter, 1952).

The case of a pure random decay is considered in the next section. Perturbed angular correlation has been used in this mode to measure collisional relaxation times for the decay of ^{111}Cd in dimethyl cadmium vapor in He, H_2, N_2, Ar, and Xe.

However, most of the experimental work concerning atoms and molecules based on nuclear two-photon processes has been done on solutions and solids.

4. Angular Distributions of Electrons from Two-Photon Ionization

The theory for this process for atomic absorbers, with no time dependence in the intermediate state and two identical photons, is essentially that of Zernik (1964) and Bebb (1966). It was reviewed and put in a general, orderly form by Peshkin (1971). A form for molecular absorbers, with unequal transition moments along different internal axes, was worked out by Tully *et al.* (1968). Among the more recent formulations of the relaxationless system, I might cite those of Arnous *et al.* (1973), the treatments of P. Lambropoulos (1974), a general resolvent development, and the papers already noted in the context of the differences between circular and linear-polarized light (P. Lambropoulos, 1973) and of production of polarized electrons (P. Lambropoulos, 1973). The angular distribution of two-photon photoelectrons was worked out by M. Lambropoulos and Berry (1973) in the case of atomic absorbers for transitions to a real intermediate level, which may or may not act as a real state, rather than a virtual state.

We developed explicit formulas for the case of two photons with the same polarization. This theory was intended specifically to include the effect of relaxation in the intermediate state. In other words, we began with (4) and (5), but, like (24), our formulation puts a time-dependent perturbation between the creation of the intermediate state and the transition to the final state. This is still an incomplete statement of the two-photon process because, unlike the approach of McClain and his co-workers, it does not make full use of the tensor character of the transition operator and the polarization operator. It explicitly omits reference to spin polarization. Moreover, the perturbation responsible for relaxation is assumed to be isotropic and incoherent. It does correspond to the experimental situation with which we shall conclude this discussion.

For atoms, the expression for the angular distribution of two-photon photoelectrons is a sum over initial states with angular moments a, intermediate states with angular momenta b and b', final states for all particles with angular momenta f and f' (because there are usually two open final channels), ion final states i, i' and angular momenta e, e' for the outgoing electron:

$$W(\omega) = \sum_{ff'} \sum_a \sum_{b,b'} \Big\{ (f|r|b)(b|r^\dagger|f')(b|r|a)(a|r^\dagger|b') \\ \times \sum_{e,e'} \sum_i \Xi(a, b, b', i, e, f, e', f') \Big\} \quad (25)$$

Like (9) and (18), Eq. (25) separates the "internal" dipole factors from the "external" polarization information. The factors Ξ contain the density matrices for the first and second photons and the initial state, the relaxation terms in the form of factors $(1 + \lambda_b \tau_b)^{-1}$, and the usual assortment of angular momentum factors to effect transformations from one set of axes to another. The presence of relaxation terms in the Ξ's makes the internal–external separation incomplete in (25). If the two photons have the same linear polarization, and only one intermediate and one final ion state occur, then the sums become rather simple and each Ξ has the form

$$\Xi = B_{00} + B_{02}(1 + \lambda\tau)^{-1} + [B_{20} + B_{22}(1 + \lambda\tau)^{-1}]P_2(\cos\theta) \\ + B_{42}(1 + \lambda\tau)^{-1}P_4(\cos\theta)$$

To extend this to the general case of molecular targets, one must replace the set of scalar matrix elements of r in (23) with a tensor, as in (18). The isotropic random relaxation process can be replaced with a more general intermediate-state evolution represented by an operator Λ as in (24). Note that for molecules the intermediate state may be represented as a stationary rotational state, as a rigidly oriented state (over which an ensemble average must be taken), or as a superposition, depending on the time scale of the excitation. To obtain maximum information from the angular measurements, one must also allow the polarizations to take on any combination of forms, as McClain has done. For arbitrary polarizations of the photons, the Ξ functions depend on $|\lambda \cdot \mu|^2$ and $|\lambda \cdot \mu^*|^2$. The "external" quantities governing the absorption are given by (9).

We have carried out two sets of experiments to measure the angular distribution of photoelectrons, thus far of atoms. The first set was done by Edelstein et al. (1974), with Ti atoms, for several reasons:

(1) foremost, Ti can be ionized via a resonant two-photon process with the N_2 laser (3371 Å);
(2) the intermediate level of interest fluoresces, so it can be monitored;
(3) the sample itself helps to keep the system clean.

There are also certain disadvantages—the high temperature required for

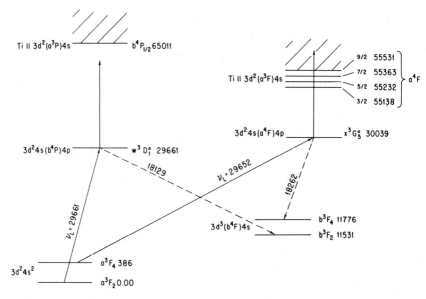

Figure 3. Part of the Grotrian diagram for the Ti atom, showing the laser-excited transition from a^3F_4 to x^3G_5 and ionization from this excited state. Spacing of the ion fine structure is exaggerated.

vaporization, and, far more important, the fact that the final ^4F term of the Ti$^+$ ion consists of four states ($J = \frac{3}{2}, \frac{5}{2}, \frac{7}{2}$, and $\frac{9}{2}$), giving rise to four groups of photoelectrons with kinetic energy separations too small to be resolved. Figure 3 shows the relevant part of the Grotrian diagram of Ti.

The apparatus consists of a nitrogen laser operating at 60 Hz, a polarizer, a half-wave plate rotating in synchrony with the laser to vary the direction of the axis of polarization, and a vacuum chamber containing an oven, two concentric equatorially grilled Cu spheres acting as Faraday cage and electron accelerating sphere, and a Bendix Channeltron electron multiplier detector. Signals were fed into a multichannel analyzer, also synchronized to advance with the laser pulse. Figure 4, is a schematic cutaway sketch of this apparatus. These measurements show that the photoelectron distribution is essentially isotropic for this system, and, incidentally, that the ionization cross section is of order 1×10^{-18} cm^{-2} for the $^3G_5$ intermediate state. Isotropy is perhaps less than thrilling when one looks for an effect whose interest lies in its anisotropy. However, our calculations show that, whatever the ratio of the amplitudes for outgoing s and d waves in this $4s \to 4p \to ks, kd$

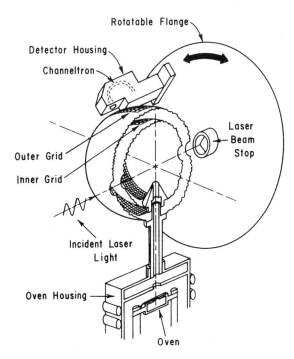

Figure 4. Schematic diagram of the apparatus used for studying two-photon ionization of Ti.

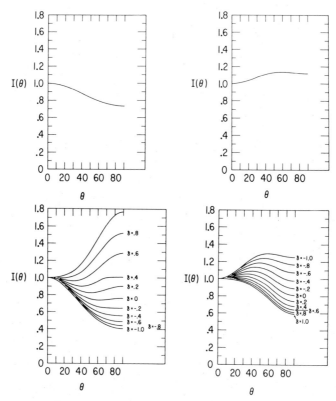

Figure 5. Calculated angular distributions for each of the final ion states $^2F_{3/2}$, $^2F_{5/2}$, $^2F_{7/2}$, and $^2F_{9/2}$. Only the d wave occurs for the first two, but both s and d waves can be associated with the states with $j = 7/2$ and $9/2$. The ratio of the amplitudes is denoted by δ and is considered an unknown parameter.

process, the sum of the intensities from all the final states of the system associated with the four final states of the Ti$^+$ ion gives an angular distribution that is nearly isotropic. The individual states are not this way, and if we could achieve electron energy resolution of order 100 cm^{-1}, we would expect to see angular distributions like those of Figure 5. The parameter δ in this figure and the one following is the ratio of transition amplitudes for outgoing s waves and outgoing d waves. We treated this as a variable parameter in the calculations. The summed contributions and the observations are shown in Figure 6.

We took some pains to validate the experiment. The rate of production of photoelectrons depends on the light intensities to the power 1.9 ± 0.2 when the light is attenuated by a factor of 50 below the normal operating

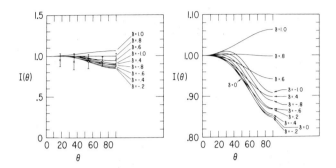

Figure 6. Summed calculated intensities from all the final levels for Ti$^+$ + e and barred circles showing the experimental results.

level; at the normal level, the exponent is close to unity because the first transition is nearly saturated (as it should be). No fluorescence is detected from any excited state of Ti except the one-quantum levels, particularly the $x^3G_5 \to b^3F_4$ line at 18,262 cm^{-1}. Note that we did use time gating of the detector to do a time-of-flight energy selection of the photoelectrons, choosing only those with kinetic energy of about 0.3 eV or more. This capability will be improved and used as the basis for photoelectron energy analysis.

The angular resolution of the apparatus was tested with a low-energy, narrow-angle electron gun, and found to be 7°, characteristic of our electron multiplier. Hence, we were assured that we could have detected anisotropy if it were present.

Our next step, still in progress, is a study of resonant two-photon ionization of Na. For this work, we need two photons of different energy, one to excite the Na($3p\,^2P_{1/2,3/2}$) term and another to ionize the atoms in these excited states. The apparatus to do this, constructed by John Duncanson, is shown schematically in Figure 7. The differences between this and the previous apparatus are the addition of a tunable dye laser to provide the NaD lines, and the fact that the polarization is varied for the N$_2$-laser light only.

The distribution of photoelectrons for ionization through the $^2P_{3/2}$ intermediate level with a fixed 45° angle between the detector and the dye-laser polarization axis, as a function of the direction of N$_2$-laser photon polarization axis, is shown in Figure 8. We are still carrying out both calculations and experiments on this system. Thus far, we are entirely reassured of our capability to detect anisotropy. We do know that the rate of production of photoelectrons depends on the first power of the intensity of both yellow and ultraviolet light over two orders of magnitude, so we are seeing a two-photon ionization. Whether we can obtain good enough resolution and

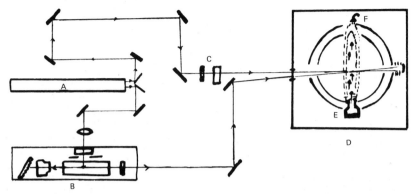

Figure 7. Schematic diagram of apparatus used to study angular distribution of two-photon photoelectrons from Na. (A) N_2 laser; (B) dye laser pumped by N_2 laser (with beam expander); (C) polarizer and rotator; (D) vacuum chamber; (E) oven and collimator; (F) channel multiplier (rotatable).

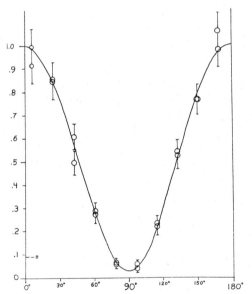

Figure 8. Observed angular distribution and best least-squares fit for two-photon ionization of Na through the $^2P_{3/2}$ level, with one resonant photon and one photon of $\lambda = 3371$ Å.

$$\tilde{W}(\theta) = a_0 + a_1 \cos^2(\theta_2) + a_2 \cos^4(\theta_2)$$

$$a_0 = 0.04 \pm 0.01$$

$$a_1 = 1.15 \pm 0.12$$

$$a_2 = -0.19 \pm 0.13$$

enough power to go a bit off resonance with our dye laser is still an open question. Naturally, we would like to be able to excite with circularly polarized light at the "Fano–Lambropoulos frequency," the frequency between the two components of the line-structure doublet at which two-photon ionization produces electrons with 100% spin polarization (P. Lambropoulos, 1973). Although one would want a direct measure of electron polarization to demonstrate this situation conclusively, measurement of the angular distribution of photoelectrons is tantamount to showing their degree of polarization when they leave their parent atoms.

Let me close by indicating the directions in which we see our future efforts in angular measurements of two-photon ionization. One can use this technique to probe the degree of homogeneous or inhomogeneous broadening of spectral lines in lasers, by exciting two closely spaced levels. This is a simple consequence of the ease with which this technique can distinguish whether interference terms between different intermediate states are influencing the angular distribution. Beyond this, we expect to use the method together with photoelectron energy analysis (not just selection) to investigate energy and angular momentum exchange between excited electrons and the ion cores to which they are bound, particularly in molecules. Lastly, we see this as a technique for probing those most difficult of molecular states, the dissociating excited states. With the time resolution of an N_2 laser, one can examine some predissociating states, but with the picosecond pulses of mode-locked lasers, one could look at ionization through directly dissociating states, at least when the leaving partners are both heavy and not terribly energetic.

Acknowledgments

The author would like to thank Drs. Robin Hochstrasser and W. Martin McClain for helpful comments and for providing unpublished material.

References

Abella, I. D. (1962). *Phys. Rev. Lett.*, **9**, 453.
Ambartzumian, R. V., and Letokhov, V. S. (1972). *Appl. Opt.*, **11**, 354.
Ambartzumian, R. V., Kalinin, V. P., and Letokhov, V. S. (1971). *Zh.E.T.P.Pis.Red.*, **13**, 305 (*JETP Lett.*, **13**, 217 (1971)).
Arnous, E., Klarsfeld, S., and Wane, S. (1973). *Phys. Rev.*, A7, 1559.
Bebb, H. B. (1966). *Phys. Rev.*, **149**, 25.
Bebb, H. B., and Gold, A. (1966). *Phys. Rev.*, **143**, 1.

Biraben, F., Cagnac, B., and Grynberg, G. (1974). *Phys. Rev. Lett.* **32**, 643.
Bloembergen, N., Levenson, M. D., and Salour, M. M. (1974). *Phys. Rev. Lett.*, **32**, 867.
Bradley, D. J., Hutchinson, M. H. R., and Koetser, H. (1972). *Proc. Roy. Soc. (Lond)*, **A329**, 105.
Brossel, J., and Bitter, F. (1952). *Phys. Rev.*, **86**, 308.
Cagnac, B., Grynberg, G., and Biraben, F. (1973). *J. Phys. (Paris)*, **34**, 56.
Chiu, Y-N., (1968). *J. Chem. Phys.*, **48**, 5702.
Collins, C. B., Johnson, B. W., Popescu, D., Musa, G., Pascoe, M. L., and Popescu, I. (1973). *Phys. Rev.*, **A8**, 2197.
Dowley, M. W., Eisenthal, K. B., and Peticolas, W. L. (1967). *J. Chem. Phys.*, **47**, 1609.
Drucker, R. P., and McClain, W. M. (1974a). *J. Chem. Phys.*, **61**, 2609.
Drucker, R. P., and McClain, W. M. (1974b). *Chem. Phys. Lett.*, **28**, 255.
Drucker, R. P., and McClain, W. M. (1974c). *J. Chem. Phys.*, **61**, 2616.
Edelstein, S., Lambropoulos, M. J., Duncanson, J., and Berry, R. S. (1974). *Phys. Rev.*, **A9**, 2459.
Fano, U. (1957). *Rev. Mod. Phys.*, **29**, 74.
Fano, U. (1969). *Phys. Rev.*, **178**, 131.
Fox, R. A., Kogan, R. M., and Robinson, E. J. (1971). *Phys. Rev. Lett.*, **26**, 1416.
Freund, S. M., Johns, J. W. C., McKellar, A. R. W., and Oka, T. (1973). *J. Chem. Phys.*, **59**, 3445.
Gallagher, T. F., Edelstein, S. A., and Hill, R. M. (1975). *Phys. Rev. Lett.*, **35**, 644.
Hall, J. L. (1966). IEEE, *J. Quant. Electron.*, QE-2, 361.
Hall, J. L., Robinson, E. J., and Branscomb, L. M. (1965). *Phys. Rev. Lett.* **14**, 1013.
Harris, R. A., McClain, W. M., and Sloane, C. F. (1974). *Molec. Phys.*, **28**, 381.
Hernandez, F. J., and Gold, A. (1966). *Phys. Rev.*, **156**, 26.
Hochstrasser, A. M. Private communication: a resolution of over 10^6 has been achieved in obtaining the two-photon absorption spectrum of NO.
Hochstrasser, R. M., Sung, H. N., and Wessel, J. E. (1973). *J. Chem. Phys.*, **58**, 4694.
Hochstrasser, R. M., Sung, H. N., and Wessel, J. E. (1974a). *Chem. Phys. Lett.*, **24**, 7.
Hochstrasser, R. N., Wessel, J. E., and Sung, H. N. (1974b). *J. Chem. Phys.*, **60**, 317.
Honig, B., Jortner, J., and Szöke, A. (1967). *J. Chem. Phys.*, **46**, 2714.
Hughes, V. W., and Grabner, L. (1950). *Phys. Rev.*, **79**, 314. 826.
Inoue, M., and Toyozawa, Y. (1965). *J. Phys. Soc. Japan*, **20**, 363.
Kaiser, W., and Garrett, C. G. B. (1961). *Phys. Rev. Lett.*, **7**, 229.
Kishi, K., and Okuda, T. (1971). *J. Phys. Soc. Japan*, **31**, 1289.
Kishi, K., Sawada, K., Okuda, T., and Matsuoka, Y. (1970). *J. Phys. Soc. Japan*, **29**, 1053.
Kocher, C. A., Clendinin, J. E., and Novick, R. (1972). *Phys. Rev. Lett.* **29**, 615.
Kogan, R. M., Fox, R. A., Burnham, G. T., and Robinson, E. J. (1971). *Bull. Am. Phys. Soc.*, **16**, 1416.
Lambropoulos, M., and Berry, R. S. (1973). *Phys. Rev.*, **A8**, 855.
Lambropoulos, P. (1972). *Phys. Rev. Lett.* **28**, 585.
Lambropoulos, P. (1973). *Phys. Rev. Lett.*, **30**, 413.
Lambropoulos, P. (1974). *Phys. Rev.*, **A9**, 1992.
Levenson, M. D., and Bloembergen, N. (1974). *Phys. Rev. Lett.* **32**, 645.
Lineberger, W. C., and Patterson, T. A. (1972). *Chem. Phys. Lett.* **13**, 40.
Lipeles, M., Novick, R., and Tolk, N. (1968). *Phys. Rev. Lett.*, **15**, 690.
Marrus, R., and Schmieder, R. W. (1972). *Phys. Rev.*, **A5**, 1160.
Matthias, E., Olsen, B., Shirley, D. A., and Templeton, J. E. (1971). *Phys. Rev.*, **A4**, 1626.
McClain, W. M. (1971). *J. Chem. Phys.*, **55**, 2789.
Mizuno, J. (1973). *J. Phys.. B.*, **6**, 314.

Monson, P. R., and McClain, W. M. (1970). *J. Chem. Phys.*, **53**, 29.
Monson, P. R., and McClain, W. M. (1972). *J. Chem. Phys.*, **56**, 4817.
Moore, C. B. (1971). *Ann. Rev. Phys. Chem.*, **22**, 387.
Okuda, T., Kishi, K., and Sawada, K. (1969). *App. Phys. Lett.*, **15**, 181.
Ovander, L. N. (1960). *Opt. Spectry.*, **9**, 302.
Peshkin, M. (1971). *Adv. Chem. Phys.*, **18**, 1.
Peticolas, W. (1967). *Ann. Rev. Phys. Chem.*, **18**, 233.
Pople, J., and Buckingham, A. D. (1955). *Proc. Roy. Soc. (Lond)*, **A68**, 905.
Prior, M. H. (1972). *Phys. Rev. Lett.*, **29**, 611.
Pritchard, D., Apt, J., and Ducas, T. W. (1974). *Phys. Rev. Lett.*, **32**, 641.
Richman, I., and Chang, N. C. (1967). *Appl. Phys. Lett.*, **10**, 218.
Rizzo, J. E., and Klewe, R. C. (1966). *Br. J. Appl. Phys.*, **17**, 1137.
Shirley, D. A., and Haas, H. (1972). *Ann. Rev. Phys. Chem.*, **23**, 385.
Swofford, R. L., and McClain, W. M. (1973). *J. Chem. Phys.*, **59**, 5740.
Tully, J. C., Berry, R. S., and Dalton, B. J. (1968). *Phys. Rev.*, **176**, 95.
Yatsiv, S., Wagner, W. G., Picus, G. S., and McClung, F. J. (1965). *Phys. Rev. Lett.*, **15**, 614.
Zernik, W. (1963). *Phys. Rev.*, **132**, 320.
Zernik, W. (1964). *Phys. Rev.*, **135**, A51.

45

Electron Spin Polarization from Multiple Photoionization Processes

Herbert D. Zeman

> The two-step photoionization of Cs vapor with a tunable blue dye laser is proposed as a source of spin-polarized electrons and compared with other pulsed sources of polarized electrons, especially the Fano effect. Its advantages over the Fano effect are described, as well as its own special problems. Measurements of the absolute photoionization cross sections from the excited 7P states of Cs are reported. At 4593 Å, the $7^2P_{1/2}$ state had $\sigma_{PI} = (6.2 \pm 0.5) \times 10^{-18}\,cm^2$. At 4555 Å, the $7^2P_{3/2}$ state had $\sigma_{PI} = (8.8 \pm 1.6) \times 10^{-18}\,cm^2$. Preliminary measurements made in collaboration with Koch and co-workers at the University of Bonn gave an electron spin polarization of 50%, a pulse intensity of 10^7 epp, and a pulse repetition rate of 2 pps, by using the $7^2P_{3/2}$ intermediate state of Cs (laser tuned to 4555 Å). Plans for further measurements are described.

For scattering experiments with spin-polarized electrons, it can be shown that the experimental uncertainty still remaining after a given measurement time is proportional to $1/P^2I$, where P is the polarization and I the current in the polarized beam (Zeman, 1971). Thus, for those experiments in which no limit is placed on the incoming polarized beam current, a low value of P can be compensated by a higher value of I. However, there are classes of scattering experiments (e.g., high-energy scattering from polarized targets) where I cannot be raised above a particular value without either destroying the target or overloading the detector system (Drachenfels et al., 1974). For these experiments, a source of polarized electrons with a large P (i.e., $P \geq 60\%$) is essential.

Herbert D. Zeman • Universität Münster, Münster, Germany. Present address: Stanford Research Institute, Menlo Park, California, U.S.A.

One of the most successful methods for producing large currents of highly polarized electrons, and the first method to be applied as a source of polarized electrons for a high-energy electron accelerator, has been the photoionization of polarized atomic beams (Hughes et al., 1972; Baum and Koch, 1969). For this experiment a beam of alkali atoms is polarized with a six-pole magnet and photoionized with unpolarized light. Fano (1969) was the first to realize that this method could be "reversed," at least as far as the experimental technique is concerned, and that under certain conditions an unpolarized alkali atomic beam photoionized with circularly polarized light would also yield a high degree of electron spin polarization. To polarize an atomic beam or a light beam, a high degree of collimation is needed, since the polarizer used in either case can accept only a limited solid angle. However, the six-pole magnet used to polarize an atomic beam has a considerably smaller acceptance than the polarizing prism or filter that can be used to polarize the light from a conventional lamp. Furthermore, if a laser is used, the light beam is already so well collimated that practically no intensity is lost in polarizing the light beam. Hence, the Fano effect has, at least in theory, great technical advantages over the photoionization of polarized atoms. These theoretical advantages have been borne out, at least in part, at the University of Bonn, where the first intense source of highly polarized electrons using the Fano effect has been developed (Drachenfels et al., 1974). They have already produced more electrons per pulse (3×10^9 epp) at higher polarization ($P = 90\%$) than has been achieved by the photoionization of polarized atoms to date (2×10^8 epp, $P = 78\%$, 10 pps) (Hughes et al., 1972), but up till now at a lower pulse repetition rate (1 pps).

Although the Fano effect seems better as a method for producing polarized electrons than the photoionization of polarized atoms, it has serious technical problems of its own. The highest polarization (100%) is achieved at a wavelength (2900 Å) where the photoionization cross section is very low (1×10^{-19} cm^2) and where no really intense sources of light are available. So far a frequency-doubled dye laser has been used (Drachenfels et al., 1974) which is costly and not very efficient.

It would be nice if one could perform the Fano effect with a visible-light laser at a wavelength where the photoionization cross section were, say, two orders of magnitude higher, and the electron polarization were still over 60%. This might, in fact, be achieved, but only in a two-step process.

In two-step photoionization, a tunable laser is tuned to resonance with an intermediate atomic state, and the laser line width and intensity are chosen so that just enough resonance light is present to produce excitation saturation, i.e., the population in the excited state is essentially equal to that in the ground state throughout the laser pulse. Atoms in the excited state are photoionized by light from the exciting laser itself, or from a second laser.

Electron Spin Polarization

Since the photoionization cross section from excited states is normally much larger than from the ground state, and since half the atoms are always in the excited state in this two-step photoionization process, a considerable improvement in efficiency should be achievable over the Fano effect. The electron spin polarization in two-step photoionization can also be as high as from the Fano effect.

This electron spin polarization in two-step photoionization results, as in the Fano effect, from spin–orbit coupling for the outermost electron in the atom. However, in the Fano effect, the spin–orbit coupling in the transition from the ground state to the continuum determines the photoelectron polarization, whereas, in the two-step process, the spin–orbit splitting of the discrete levels in the atom plays a more important role. This can be seen best with a few examples.

Figure 1 shows the angular-momentum levels which play a role in the two-photon photoionization of Cs vapor with a single blue laser. If we consider the case where laser light (4593 Å) resonant with the $7^2P_{1/2}$ state is used, the selection rules

$$\Delta j = \pm 1, 0$$
$$\Delta l = \pm 1$$

and

$$\Delta m_j = +1 \quad \text{(for right circularly polarized light)}$$

allow only a single path to be followed (arrow 1 in Figure 1), ending in a

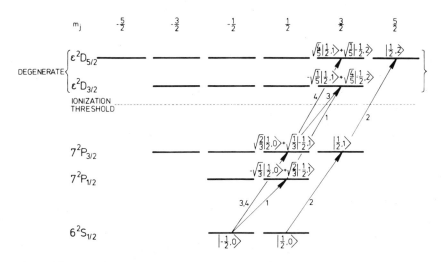

Figure 1. Resonant two-photon photoionization transitions in atomic Cs.

single angular momentum substate in the continuum. From the Clebsch–Gordan coefficients given in Figure 1 alone, one can calculate a spin polarization for this final state of -60%. One does not have to know anything about the transition probabilities between the various substates in the intermediate state and the continuum to get this polarization value. It comes directly from the angular momentum algebra. For the $7^2P_{3/2}$ intermediate state (i.e., laser light of 4555 Å), the various transition probabilities are important in calculating the polarization. If one assumes that they are all equal, a value of $+82\%$ results (Lambropoulos, 1973). It is interesting to note that if one uses the correct wavelength between 4555 Å and 4593 Å that 100% polarized photoelectrons will result (Lambropoulos, 1973). However, the yield would be so low that such an off-resonance two-photon photoionization would never be able to compete with the Fano effect as a source of polarized electrons.

Photoelectrons that are 100% polarized can be achieved in two other schemes, as shown in Figures 2 and 3. In the first, a three-step process, at least two lasers are needed, one right circularly polarized and one left circularly polarized. Na vapor is used and the $3^2P_{1/2}$ and $5^2S_{1/2}$ intermediate states are reached (Lambropoulos, 1974). In the second, a two-step process, an atom with three valence electrons, one of which has a $^2P_{1/2}$ ground state, is chosen (Farago and Walker, 1973). As a source of polarized electrons, these methods suffer either from the necessity of having a second tunable laser or from the difficulty of producing a dense beam of a high melting-point element. Hence, the rest of this paper will deal with two-photon photoionization of Cs with a single blue laser.

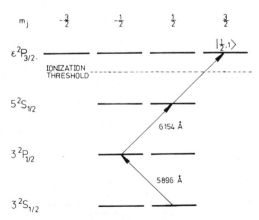

Figure 2. Resonant three-photon photoionization transitions in atomic Na.

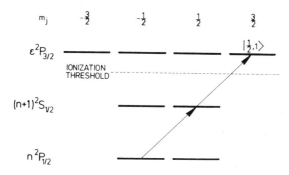

Figure 3. Resonant two-photon photoionization transitions in an atom with three valence electrons.

At Münster we have measured the absolute photoionization cross section from the $7^2P_{1/2}$ and the $7^2P_{3/2}$ excited states of cesium.* The experimental arrangement is shown in Figure 4.† Light from a flashtube-pumped coumarin-dye laser‡ is focused with a lens to a beam 1 mm in diameter between parallel flat electrodes 10 mm apart, between which a potential difference of up to 3500 V can be set. The electrons emitted from the Cs vapor between the electrodes during a laser pulse pass through a hole in one electrode and are separated from electrons from the electrodes themselves with an electron spectrometer. The photoelectrons, on leaving the spectrometer, are collected in a Faraday cup and integrated in a vibrating-reed electrometer. The laser pulse energy is measured with a calibrated thermopile. Since we did not know the vapor density of the Cs between the electrodes, we had to measure it indirectly by going to such high laser intensities (5 mJ) that essentially all of the atoms in the laser beam (90%) were photoionized in a single laser pulse. In this intensity region the number of photo electrons detected per pulse (I_E) is given by

$$I_E = fN_A[1 - \exp(-\sigma_{PI}I_L/2)] \quad (1)$$

where N_A is the number of atoms in the laser beam, I_L is the number of photons per unit area in the laser pulse, σ_{PI} is the photoionization cross section, and f is the fraction of the emitted photoelectrons that reach the collector. The factor of 2 comes from the fact that only half of the atoms are

---

* This measurement was performed by U. Heinzmann, D. Schinkowski, and H. Zeman and is described in D. Schinkowski, Staatsexamenarbeit, University of Münster, 1974.
† This apparatus was a modification of that described in U. Heinzmann, J. Kessler, and J. Lorenz, Z. Physik (1970), **240**, 42.
‡ This laser is from Carl Zeiss, Oberkochen, W. Germany. A smaller version of this laser is described in J. Kuhl, G. Marowski, P. Kunstmann, and W. Schmidt in Z. Naturforschung (1972), **27a**, 601.

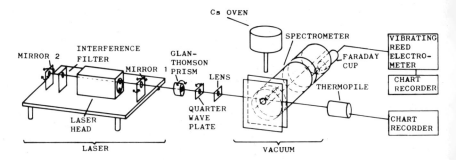

Figure 4. The experimental arrangement for the measurement of the absolute photoionization cross sections from the 7^2P excited states of Cs.

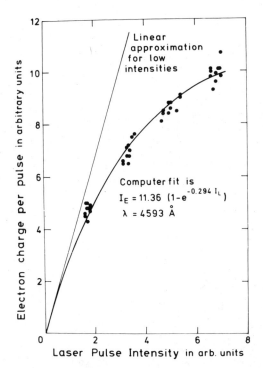

Figure 5. One set of measurements of photoelectron charge per pulse vs. laser pulse energy used to measure the absolute photoionization cross section from the $7^2P_{1/2}$ state. A computer fit to the data is shown.

in the excited state at any one time. This equation is valid only when the atoms in the Cs atomic beam do not move appreciably during the laser pulse. This criterion is fairly well satisfied for our laser pulse length of $\frac{1}{2}$ μsec.

Figure 5 shows one set of measurements I_E vs. I_L. To calculate σ_{PI}, a computer fitting program was used to fit Eq. (1) to the experimental data.* To get an absolute scale for I_L in photons per cm^2 per pulse, we had to measure exactly the laser beam cross-sectional area. This was done by cutting the laser beam off progressively with a knife edge, the position (x) of which was measured with a dial gauge, and measuring the laser beam energy after the knife edge with our thermopile. Assuming the laser beam had a circular cross section, we used a computer fitting program to fit the expected experimental curve,

$$I_L = \frac{A}{2}\left[x(B^2 - x^2)^{1/2} + B^2\left(\frac{\pi}{2} + \arcsin\frac{x}{B}\right)\right]$$

where A and B are the parameters to be adjusted, with our measured points (see Figure 6). The validity of this measurement procedure was checked by photographing a magnified image of the laser beam. The photographs verified that the beam had a circular cross section and gave the same value of cross-sectional area within experimental uncertainty. Also, horizontal and vertical scans with the knife yielded the same results for the beam diameter. The photoionization cross-section measurements yielded the following values: at 4593 Å the $7^2P_{1/2}$ state had $\sigma_{PI} = (6.2 \pm 0.5) \times 10^{-18}$ cm^2, and at 4555 Å the $7^2P_{3/2}$ state had $\sigma_{PI} = (8.8 \pm 1.6) \times 10^{-18}$ cm^2, where the experimental uncertainties given are purely statistical. The value at 4593 Å agrees with an extrapolation of photoionization cross sections calculated from the theoretical recombination cross sections of Norcross and Stone (1966) (see Figure 7).

When one considers that these photoionization cross sections are about 50 times larger than those from the ground state of cesium, and that at the present state of the art, an intense blue laser is easier and less expensive to build than a UV-producing laser of the same output, one would gather that this two-photon photoionization process would make a much more suitable source of polarized electrons than the Fano effect. However, this two-step process turns out to have some very special problems of its own.

The first problem plays a role when a two-photon process is performed with a single-light source. If one assumes that the laser line is infinitely narrow and tuned at the middle of a Doppler-broadened 7P absorption line of Cs, then the intensity needed to ensure that the number of Cs atoms

---

*The fitting programs were written by J. Brink-Spallink, Münster.

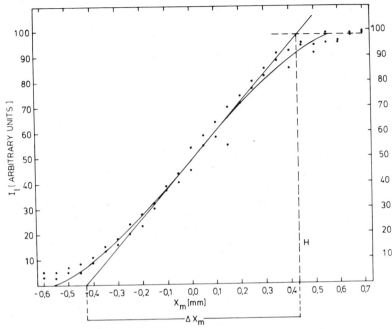

Figure 6. One set of measurements of the laser beam diameter using the movable knife technique. A computer fit to the data as well as a graphical fitting procedure is shown. To arrive at a value of B (the beam radius) graphically one needs the relationship $\Delta x_m = \pi B/2$. The line is a tangent to the curve at its point of greatest slope.

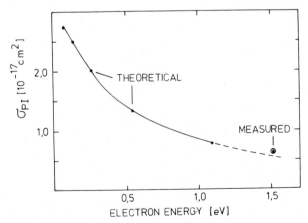

Figure 7. Our measured photoionization cross section for the $7^2P_{1/2}$ state of Cs at 4593 Å, compared with an extrapolation of the theoretical values of Norcross and Stone. The extrapolation is against photoelectron energy.

in the 7P state remains equal to the number of atoms in the ground state during the laser pulse (excitation saturation) is at least 10^5 times less than that needed to ensure that all of the atoms in the laser beam will be photoionized during a single laser pulse of $\frac{1}{2}$ μsec duration. Furthermore, a near total photoionization of the atoms in the laser beam is very desirable, since it would give the largest electron optical brightness in the output polarized electron beam obtainable from a given density of Cs vapor. However, with 10^5 times more light intensity than needed to produce excitation saturation, it would be impossible to achieve a polarized intermediate state with circularly polarized light. Even with our 99.9% right circularly polarized laser beam, 0.05% of the light is left circularly polarized, and that would be 50 times more than needed to saturate the wrong transition (i.e., $\Delta mj = -1$). Hence, one would really want a laser line 10^5 times broader than the Doppler width (5 mÅ) in order to compensate for this very large ratio in transition probabilities between the first and second step of the photoionization process. However, such a laser line (500 Å wide) would be 12 times broader than the separation (38 Å) between the two 7P levels, making it impossible to separate one from the other (NB, one gives -60% and the other $+82\%$ polarization).

The only way to overcome this difficulty with a single laser is to use a laser line narrow enough (ours has a 2.5 Å FWHM) to resolve the two 7P states and then to tune the laser slightly off resonance so that the Cs absorption line is excited only by the tail of the laser energy distribution. This method will only work if the laser has a smoothly varying spectral distribution and is very stable from pulse to pulse, because the tail of the spectral distribution tends to vary much more from pulse to pulse, or even during a single pulse, than the peak. When two separate lasers are used, the laser used for the excitation step can be given a very narrow line and tuned exactly on resonance, so that the entire laser intensity is within the Doppler width of the absorption line. Then one could reduce its intensity with neutral density filters to the required value. Hence, the above problem is easy to overcome with a two-laser system, as long as the light polarization is high enough.

The requirement for high light polarization is made more stringent when the laser pulse is of the same order as the lifetime of the intermediate state (as with our laser) or longer. In the case of the $7^2P_{1/2}$ intermediate state, only those ground state atoms in the $m_j = -\frac{1}{2}$ substate can be excited by right circularly polarized light (see Figure 1). However, excited atoms can spontaneously radiate into the $m_j = +\frac{1}{2}$ substate of the ground state, which will then increase in population. Hence, the atoms available to be excited by the small residual intensity of left circularly polarized light will actually increase during the laser pulse. This is the well-known process of optical pumping, which in this case is detrimental. The circular polarization of the resonant light from the laser can be increased, however, by passing it through a dense atomic beam of Cs. With the right Cs density, the unwanted left circularly polarized resonance light will be absorbed, while the right circularly polarized light can excite all the $m_j = -\frac{1}{2}$ atoms in the beam and still have enough intensity on leaving the beam to excite the atoms used for photoionization.

We have observed these effects at Münster by measuring the ratio of photoelectron current produced with linearly polarized light to that with circularly polarized light. We used a laser beam with 1 mm diameter and a 5 mJ output, so that 90% of the atoms excited by the laser beam were photoionized. If the light polarization had been 100% and the pulse length infinitely short, one would expect a ratio of 2, because all of the atoms in the ground state can be excited by linearly polarized light, and only those with $m_j = -\frac{1}{2}$ by right circularly polarized light. With a light polarization of 98% and a laser line width of about 0.1 Å we measured a ratio of 1, which clearly indicates that our light polarization was too low and that we had much too much resonant light. With a light polarization of 99.9% and a line width of 2.5 Å we measured a ratio of 1.5, a considerable improvement. By tuning 4 Å off resonance we measured a ratio of 2, which resulted from a cancellation of the effect due to too-low light polarization, which would make the ratio lower, and the effect due to optical pumping, which would make the ratio higher. By increasing the atomic beam density we then measured a ratio of 2.5, clearly demonstrating the optical pumping effect and the ability of a denser atomic beam to absorb falsely polarized light before the laser beam reaches the photoelectron extraction region.

If one does not need the highest electron optical brightness, one can reduce the need for very high light polarization when using a single laser to produce polarized electrons by two-photon photoionization, without reducing the total photoelectron current, by giving the laser beam a larger diameter. In that way one can reach that intensity where the right circularly polarized light will just be strong enough to saturate its transition and the left circularly polarized light will be too weak to saturate the opposite transition without tuning off resonance. We investigated this effect at the Fano-effect apparatus at the University of Bonn (Drachenfels et al., 1974). N. Ernst and H. Zeman from Münster collaborated with W. V. Drachenfels, U. T. Koch, R. D. Lepper, and T. M. Müller from the University of Bonn in these measurements. We used our blue laser from Münster with a beam diameter of 6 mm instead of 1 mm and with an intensity of 2 mJ instead of 5 mJ. By measuring the photoelectron current vs. the laser line central wavelength, we were able to see whether excitation saturation was occurring. Figure 8 shows the resulting "tuning curve" for three different laser

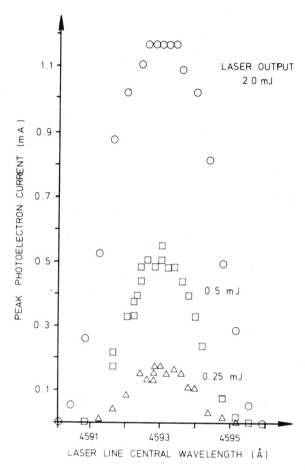

Figure 8. A measurement of the photoelectron charge per pulse vs. the laser line central wavelength for the two-photon photoionization of Cs vapor through the $7^2P_{1/2}$ intermediate state. Such a "tuning curve" was used to calibrate the wavelength scale of our laser.

intensities. Note that the saturation exhibition in the upper curve is absent from the two lower curves and that the heights of the two lower curves vary quadratically with laser energy. Note also the very high peak currents achieved there (1 mA = 3×10^8 epp). The width of the lower curve represents essentially the width of the laser line. We have measured a similar "tuning curve" with the new apparatus at Münster. It yielded a laser line width of 2.7 ± 0.1 Å. Currents of up to 3×10^7 epp have been achieved there so far.

During the collaborative experiment at Bonn, we discovered a serious difficulty with the two-step photoionization process. This difficulty is a result of one of the largest advantages of the two-photon process, namely, the very high photoionization cross section of the excited Cs atoms in comparison with ground-state atoms. A higher photoionization cross section usually means a higher electron impact ionization cross section. Hence, the gas of excited atoms tends to form an electron avalanche when the Cs pressure is too high. At the Bonn apparatus, where the Cs atomic beam pressure is about 10^{-5} torr, and where the electron beam is extracted and accelerated through the Cs vapor along the laser beam (see Figure 9), we had to limit the laser intensity and the Cs beam density so that not more than 10^7 photoelectrons were emitted per pulse in order to prevent the avalanche from totally overpowering the genuine photoelectrons. However, the avalanche was delayed with respect to the photoelectron pulse by several microseconds and it might be possible to eliminate it with a pulsed photoelectron extraction field. At the reduced current of 10^7 epp, we were able to

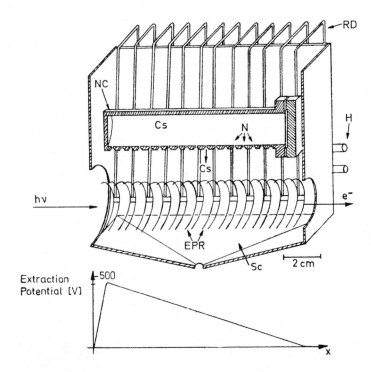

Figure 9. The electron extraction system and Cs oven for the Fano-effect apparatus at Bonn.

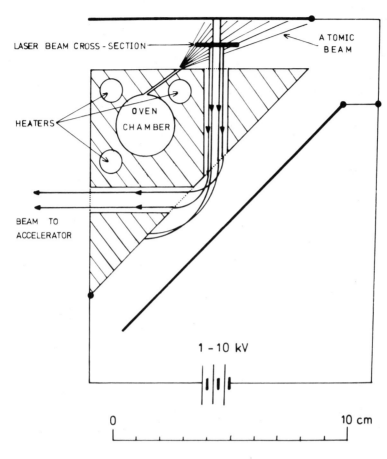

Figure 10. The new electron extraction system and Cs oven for the two-photon photoionization apparatus at Münster.

achieve 50% electron spin polarization through the $7^2P_{3/2}$ state, where 82% is expected. We hope that our transverse geometry (see Figure 10) at Münster will be less sensitive to avalanche, and that an exact comparison with theoretical predictions will be possible there.*

Acknowledgments

I would like to thank Professor Joachim Kessler for his support and encouragement during this work and I would also like to acknowledge the

---
*These measurements will be continued by Herbert Kaminski at Münster.

support of the National Science Foundation and the Alexander von Humboldt Foundation.

References

Baum, G., and Koch, U. (1969). *Nucl. Instr. and Meth.*, **71**, 189.
Drachenfels, W. von, Koch, U. T., Lepper, R. D., Müller, T. M., and Paul, W. (1974). *Z. Physik*, **269**, 387. Also, contribution to this volume.
Fano, U. (1969). *Phys. Rev.* **178**, 131.
Farago, P. S., and Walker, D. (1973). *J. Phys. B.*, **6**, L280.
Hughes, V. W., Long, R. L., Lubell, M. S., Posner, M., and Raith, W. (1972). *Phys. Rev.*, *A*, **5**, 195.
Lambropoulos, P. (1973). *Phys. Rev. Lett.*, **30**, 413.
Lambropoulos, P. (1974). *J. Phys. B.*, **7**, L33.
Müller, T. (1973). Diplomarbeit, University of Bonn.
Norcross, D. W., and Stone, P. M. (1966). *J. Quant. Spec. and Rad. Trans.*, **6**, 277.
Zeman, H. D. (1971). Doctoral thesis, Stanford University.

46

Total Cross Section for Elastic Scattering of Fast Charged Particles by a Neutral Atom

Michio Matsuzawa and Mitio Inokuti

> A compact formula valid asymptotically at high incident energies is derived. Our treatment expresses the total elastic-scattering cross section directly in terms of a few atomic ground-state expectation values, and therefore enables one to circumvent the standard two-step procedure (of first calculating the atomic form factor and then evaluating an integral over momentum transfer).

1. Introduction

Let us assume that an incident particle of charge z is structureless and is so fast that the Born approximation applies. Then the total cross section σ for elastic scattering by a neutral ground-state atom of atomic number Z is (p. 246 of Bethe and Jackiw, 1968)

$$\sigma = 8\pi z^2 k^{-2} \int_0^{2k} |Z - F(K)|^2 K^{-3} \, dK \tag{1}$$

where k is the incident momentum, $F(K)$ is the atomic form factor (as defined below) as a function of the magnitude K of momentum transfer $\mathbf{K}$, and all quantities are expressed in atomic units. (Specifically, the above σ is in units of the squared Bohr radius.)

Michio Matsuzawa and Mitio Inokuti • Argonne National Laboratory, Argonne, Illinois 60439, U.S.A.   M. Matsuzawa's present address is University of Electrocommunications, Chofu-shi, Tokyo, Japan.

The only atomic property that enters Eq. (1) is $F(K)$, which is defined as a ground-state expectation value:

$$F(K) = \left\langle \sum_{j=1}^{Z} \exp(i\mathbf{K} \cdot \mathbf{r}_j) \right\rangle \tag{2}$$

where $\mathbf{r}_j$ is the position of the jth atomic electron. Equivalently, we may write

$$F(K) = \int \exp(i\mathbf{K} \cdot \mathbf{r}) \rho(r) \, d\mathbf{r} \tag{3}$$

where

$$\rho(r) = \left\langle \sum_{j=1}^{Z} \delta(\mathbf{r}_j - \mathbf{r}) \right\rangle \tag{4}$$

is the one-electron density at position $\mathbf{r}$. Throughout the present paper we presume that the atoms are either spherical or randomly oriented, and imply that all expectation values are averages over magnetic sublevels. Thus, we may consider $\rho(r)$ as a real-valued function of the scalar variable r and $F(K)$ as a real-valued function of the scalar variable K.

An asymptotic expansion of σ for sufficiently large k is easily obtained by splitting the integral into two terms

$$\int_0^\infty - \int_{2k}^\infty$$

and then using some general properties of $F(K)$, as shown by Inokuti and McDowell (1974). The result is

$$\sigma = \pi z^2 k^{-2}(A + Bk^{-2} + Ck^{-4} + Dk^{-6} + \cdots) \tag{5}$$

where

$$A = 8 \int_0^\infty |Z - F(K)|^2 K^{-3} \, dK \tag{6}$$

$$B = -Z^2 \tag{7}$$

$$C = 0 \tag{8}$$

$$D = -\tfrac{1}{3}\pi Z \, d\rho(0)/dr \tag{9}$$

The coefficient A of the leading term is clearly the most important, and yet is the most complicated in its structure. The standard treatment by Inokuti and McDowell (1974), for example, takes two steps, i.e., the calculation of $F(K)$ from an atomic wave function and the evaluation of the integral of Eq. (6). We shall point out below a new expression that gives A in terms

Elastic Scattering of Fast Charged Particles   597

of a few ground-state expectation values *directly*, i.e., without use of $F(K)$ and its integral. In other words, our expression identifies the leading term of σ as a simple atomic property.

2. Theory

We substitute Eq. (3) into Eq. (6), recall that $F(K)$ is a function of the scalar variable K, and write

$$A = 8 \int_0^\infty dK\, K^{-3}(4\pi)^{-1} \int d\Omega \int d\mathbf{r}\, \rho(r)[\exp(i\mathbf{K}\cdot\mathbf{r}) - 1]$$

$$\times \int d\mathbf{r}'\, \rho(r')[\exp(-i\mathbf{K}\cdot\mathbf{r}') - 1] \tag{10}$$

where Ω represents the direction of the vector $\mathbf{K}$. Next we carry out the integration over Ω to obtain

$$A = 8 \int_0^\infty dK\, K^{-3} \int d\mathbf{r} \int d\mathbf{r}'\, \rho(r)\rho(r')$$

$$\times \left[\frac{\sin(KR)}{KR} - \frac{\sin(Kr)}{Kr} - \frac{\sin(Kr')}{Kr'} + 1 \right] \tag{11}$$

where $R = |\mathbf{r} - \mathbf{r}'|$.

Let us now consider the quantity

$$A(\varepsilon) = 8 \int_\varepsilon^\infty dK\, K^{-3} \int d\mathbf{r} \int d\mathbf{r}'\, \rho(r)\rho(r')$$

$$\times \left[\frac{\sin(KR)}{KR} - \frac{\sin(Kr)}{Kr} - \frac{\sin(Kr')}{Kr'} + 1 \right] \tag{12}$$

where $\varepsilon > 0$. Performing the integration over K first, we may write

$$A(\varepsilon) = 8 \int d\mathbf{r} \int d\mathbf{r}'\, \rho(r)\rho(r') J(\mathbf{r}, \mathbf{r}', \varepsilon) \tag{13}$$

where

$$J(\mathbf{r}, \mathbf{r}', \varepsilon) = \int_\varepsilon^\infty dK\, K^{-3} \left[\frac{\sin(KR)}{KR} - \frac{\sin(Kr)}{Kr} - \frac{\sin(Kr')}{Kr'} + 1 \right] \tag{14}$$

Through repeated partial integrations, we can show that

$$\int_\varepsilon^\infty dK\, K^{-3} \frac{\sin(Ks)}{Ks} = \frac{1}{2\varepsilon^2} + \left(-\tfrac{11}{36} + \tfrac{1}{6}\gamma + \tfrac{1}{6}\ln\varepsilon\right)s^2 + \tfrac{1}{6}s^2 \ln s + O(\varepsilon^2) \tag{15}$$

for small ε. Here we have used also the expansion of the cosine integral function

$$Ci(x) = -\int_x^\infty \frac{\cos t}{t} dt = \gamma + \ln x + \sum_{n=1}^\infty \frac{(-1)^n x^{2n}}{(2n)(2n)!}$$

and γ is the Euler constant (p. 348 of Magnus et al., 1966).

Substitution of formula (15) into Eq. (14) yields

$$J(\mathbf{r}, \mathbf{r}', \varepsilon) = (-\tfrac{11}{36} + \tfrac{1}{6}\gamma + \tfrac{1}{6}\ln\varepsilon)(R^2 - r^2 - r'^2)$$
$$+ \tfrac{1}{6}(R^2 \ln R - r^2 \ln r - r'^2 \ln r') + O(\varepsilon^2) \tag{16}$$

Because $\rho(r)$ is spherically symmetric by assumption, the term with

$$R^2 - r^2 - r'^2 = -2(\mathbf{r} \cdot \mathbf{r}')$$

gives a vanishing contribution to $A(\varepsilon)$ of Eq. (13). Then, after substitution of Eq. (16) into Eq. (13), we obtain our final result

$$A = \lim_{\varepsilon \to +0} A(\varepsilon)$$

$$= \tfrac{4}{3} \int d\mathbf{r} \int d\mathbf{r}' \, \rho(r)\rho(r')(R^2 \ln R - r^2 \ln r - r'^2 \ln r')$$

$$= \tfrac{4}{3} \int d\mathbf{r} \int d\mathbf{r}' \, \rho(r)\rho(r') R^2 \ln R - \tfrac{8}{3} Z \int d\mathbf{r} \, \rho(r) r^2 \ln r \tag{17}$$

Taking the limiting process after the integration over K is justified because the integral in Eq. (6) or (10) is completely well-defined with respect to small K. Indeed, power-series expansion of $\exp(i\mathbf{K} \cdot \mathbf{r})$ in Eq. (3) readily shows that $F(K) - Z = O(K^2)$ so that the integrand in Eq. (6) is analytic at $K = 0$. The introduction of nonzero ε was necessary merely for separate consideration of each term in the square brackets of Eq. (11).

The term with $R^2 \ln R$ in Eq. (17) is certainly a property of the atomic ground state. Nevertheless, it is bilinear in $\rho(r)$ and therefore is not strictly an expectation value in any real atomic state. However, we may consider that term as an expectation value in the following extended sense. Let $H(\mathbf{r}_1, \mathbf{r}_2, \ldots, \mathbf{r}_Z)$ be the real atomic Hamiltonian involving electron coordinates $\mathbf{r}_1, \mathbf{r}_2, \ldots, \mathbf{r}_Z$, and consider the direct sum

$$H(\mathbf{r}_1, \mathbf{r}_2, \ldots, \mathbf{r}_Z) \oplus H(\mathbf{r}'_1, \mathbf{r}'_2, \ldots, \mathbf{r}'_Z) = H_D.$$

Then the ground state of the "dual-atom" Hamiltonian H_D is the product

$$u_0(\mathbf{r}_1, \mathbf{r}_2, \ldots, \mathbf{r}_Z) u_0(\mathbf{r}'_1, \mathbf{r}'_2, \ldots, \mathbf{r}'_Z),$$

where u_0 is the ground state (presumed nondegenerate) of H. Then, we may regard the $R^2 \ln R$ term of Eq. (17) as an expectation value in that ground state of H_D.

The rigorous interpretation of the expectation value is only of conceptual importance. In practice, we can compute A by means of Eq. (17) as soon as we know $\rho(r)$ from an atomic wavefunction or otherwise. The $R^2 \ln R$ term can be reduced to a sum of one-electron integrals by use of an elementary expansion

$$\ln R = \ln r - \sum_{n=1}^{\infty} n^{-1} \cos(n\theta)(r'/r)^n \quad \text{(for } r' < r\text{)} \tag{18}$$

where $\cos \theta = (\mathbf{r} \cdot \mathbf{r}')/rr'$. Another method of treatment is to use the Hylleraas coordinates $r + r'$, $r - r'$, and R as integration variables. The merit of Eq. (17) lies in its precise precription that relates A to $\rho(r)$.

To verify the correctness of our theory, we have evaluated A for atomic hydrogen through Eq. (17). The result $A = 7/3$ agrees with the value obtained from Eq. (6) with the well-known form factor $F(K) = (1 + \frac{1}{4}K^2)^{-2}$.

3. Concluding Remarks

Although we have restricted our discussion to elastic scattering, we can easily generalize it to inelastic scattering. Actually, for inelastic scattering resulting in an *optically forbidden transition* of the atom, the asymptotic cross section is of the same analytic form as Eq. (5) (cf. Section 4.1 of Inokuti, 1971), and the coefficient of the leading term is given by an integral of a structure similar to Eq. (6). Therefore, it is possible to derive a formula analogous to Eq. (17); the only difference is that one should replace $\rho(r)$ with a transition density

$$\left\langle n \left| \sum_{j=1}^{Z} \delta(\mathbf{r}_j - \mathbf{r}) \right| 0 \right\rangle$$

where $|0\rangle$ and $|n\rangle$ are the initial state and the final state of the atom. For inelastic scattering resulting in an *optically allowed transition* of the atom, we need first to separate the dominant asymptotic term of the form $k^{-2} \ln k$ (whose coefficient is simply the dipole matrix element squared), and then to treat the second term of the form k^{-2}. The coefficient of that second term (i.e., $\ln c_n$ in the notation of Inokuti, 1971) then can be discussed in the same way as in Section 2. We shall give a full account of that procedure in a later publication.

Applications of Eq. (17) to some atoms such as He are in progress, and will be reported soon.

We should mention some published work closely related to ours. Indeed, various relations involving integrals of elastic- and inelastic-scattering form factors with respect to momentum transfer K have been pointed out (Silverman and Obata, 1963; Tavard and Roux, 1965; Tavard et al., 1965; Kohl and Bonham, 1967; Bonham, 1967; Sahni and Krieger, 1972; Lassettre and Dillon, 1973; O'Connell and Lightbody, Jr., 1975). All those reported relations pertain to integrals, including powers of K higher than those we have treated in the present paper, and therefore are mathematically much simpler.

Our final comment concerns a distinction between (a) all those relations involving integrals with respect to K and (b) standard sum rules involving a sum (and an integral) over excited states at fixed K, most notably the Bethe sum rule (Bethe and Jackiw, 1968; and Inokuti, 1971). The derivations of class (b) relations usually rely upon certain general properties of the Hamiltonian, such as its potential-energy part, dependent on r_j only, and its kinetic-energy part, proportional to the square of atomic-electron momenta. The relations of class (a) in contrast presume nothing specific about the atomic Hamiltonian, and therefore are much less restrictive. Indeed, they are merely mathematical identities that hold for any atomic model regardless of how much detailed dynamics it describes.

Acknowledgment

This work was performed under the auspices of the U.S. Energy Research and Development Administration.

References

Bethe, H. A., and Jackiw, R. W. (1968). *Intermediate Quantum Mechanics*, 2nd ed. New York: W. A. Benjamin.
Bonham, R. A. (1967). *J. Phys. Chem.*, **71**, 856.
Inokuti, M. (1971). *Rev. Mod. Phys.*, **43**, 297.
Inokuti, M., and McDowell, M. R. C. (1974). *J. Phys. B.*, **7**, 2382.
Kohl, D. A., and Bonham, R. A. (1967). *J. Chem. Phys.*, **47**, 1634.
Lassettre, E. N., and Dillon, M. A. (1972). *J. Chem. Phys.*, **59**, 4778.
Magnus, W., Oberhettinger, F., and Soni, R. P. (1966). *Formulas and Theorems for Special Functions of Mathematical Physics*. New York: Springer-Verlag.
O'Connell, J. S., and Lightbody, J. W. Jr. (1975). *Nucl. Phys.*, **A237**, 309.
Sahni, V., and Krieger, J. B. (1972). *Phys. Rev.*, **A6**, 928.
Silverman, J. H., and Obata, Y. (1963). *J. Chem. Phys.*, **38**, 1254.
Tavard, C., and Roux, M. (1965). *Compt. rend.*, **260**, 4460 and 4933.
Tavard, C., Rouault, M., and Roux, M. (1965). *J. Chim. Phys.*, **62**, 1410.

47
Recent Developments in Variational Principles and Variational Bounds: A Road Map

ROBERT BLAU, A. R. P. RAU, L. ROSENBERG, AND
LARRY SPRUCH

> Developments in the construction and use of general variational principles and bounds, particularly as applied to scattering lengths and matrix elements between bound states, are reviewed. Formal questions, such as the avoidance of near-singularities in the use of such variational principles, seem to be now in hand and the methods are ready for numerical application. Possibilities for extension to other scattering problems are indicated.

1. Introduction

Conference proceedings provide a welcome alternative to the possibilities provided in the usual journals, where motivation and physical insight must be sacrificed at the altars of brevity and the need for detailed proofs. Review articles should, properly, be exhaustive but are more often exhausting—to author and reader alike. This discussion, as an earlier one (Spruch, 1968), tries to provide a guide to what has recently been accomplished and what remains to be done (see also Spruch, 1973). The few proofs noted are just sketched, only the simplest cases are considered, and we limit ourselves largely to the work of our own group. Though the nature of the subject is not always conducive to such an approach, an effort is made to get to the heart of the matter, to strip the discussion of all but the basic physics; no effort could more properly be dedicated to Professor Fano.

ROBERT BLAU, L. ROSENBERG, and LARRY SPRUCH • Department of Physics, New York University, New York, New York 10003, U.S.A.   A. R. P. RAU • Department of Physics and Astronomy, Louisiana State University, Baton Rouge, Louisiana 70722, U.S.A.

At the time of the earlier discussion (Spruch, 1968), the variational principles (VPs) and variational bounds (VBs) which existed were very limited. VPs could be very successfully used for the evaluation of energies (diagonal matrix elements of the Hamiltonian H) and of the transition matrix elements (off-diagonal elements of H) describing scattering processes. VPs had not been very successful in determining bound-state properties of systems, represented by diagonal and off-diagonal matrix elements of the appropriate operator W. There were no VBs on any but an *extremely* restricted class of bound state matrix elements, and VBs on scattering problems were limited to targets with precisely known ground-state wave functions, and therefore to hydrogen-like targets. The situation has changed completely in the last year and a half.

2. The Construction of Variational Principles for Anything

Some papers have specified techniques for the construction of VPs, but for the most part the results were obscured by the use of perturbation-theoretic approaches, limited in generality, or remiss in not recognizing the need for a complete formal specification, including the phase, when necessary, of the conditions uniquely specifying the unknown functions that appear in the quantity of interest. Until recently, it was apparently not realized that for any well-defined problem in mathematical physics, a general procedure exists for constructing VPs for any quantity of interest without any restrictions on the equations involved. The equations need not be homogeneous or linear or self-adjoint; they may be differential or integral or integrodifferential; the system need not be invariant under rotation or space inversion or time reversal. The general procedure applies without regard to these aspects and is quite trivial. Suppose, for example, there is only one unknown function, ϕ, known to be real, which is specified by a set of conditions $B_i(\phi) = 0$, $i = 1$ to N; thus, we might have $B_1(\phi) \equiv (\phi, \phi) - 1$ as a normalization condition and $B_2(\phi) \equiv (H - E)\phi$ as together uniquely defining a bound-state wave function of energy eigenvalue E. If we are attempting to calculate a specified functional $F(\phi)$, which may, for instance, be a matrix element, $(\phi, W\phi)$, of some operator W, our VP is

$$F_v(\phi) = F(\phi_t) + \Sigma_i \mathscr{L}_{it}^{\dagger} B_i(\phi_t) \tag{1}$$

where ϕ_t is a trial function and the $\mathscr{L}_{it}$ are trial Lagrange multipliers. Depending on the nature of $F(\phi)$ and $B_i(\phi)$, the $\mathscr{L}_{it}$ may be numbers (λ), functions (L), operators etc., and, correspondingly, the dagger in Eq. (1) may stand for ordinary multiplication or a more complicated inner product. Equation

(1) has been written to satisfy the natural requirement that $F_v(\phi) \to F(\phi)$ when $\phi_t \to \phi$. Next, the $\mathscr{L}_i$ are determined by the requirement that $\delta^{(2)}F \equiv F - F_v$ vanish to first order in $\delta\phi \equiv \phi_t - \phi$, so that we do indeed have a VP. With some algebraic manipulations, which are normally straightforward and specific to the problem of interest, well-defined equations for $\mathscr{L}_i$ follow easily; the equations are necessarily linear by construction. Many examples are provided in Gerjuoy et al. (1972, 1973). The trial functions $\mathscr{L}_{it}$ should be chosen to be reasonably close to the $\mathscr{L}_i$. By construction, we then have

$$\delta^{(2)}F = O(\delta\phi^2) + \Sigma_i O(\delta\phi\delta\mathscr{L}_i) \qquad (2)$$

where $\delta\mathscr{L}_i \equiv \mathscr{L}_{it} - \mathscr{L}_i$. Such an incorporation of the equations of the system through Lagrange multipliers, which are defined so that $F_v(\phi)$ is a variational estimate, can be considered a "variational definition" of the functional $F(\phi)$ in the space of functions ϕ_t. In such a space, the expression in Eq. (1) coincides with $F(\phi)$ for $\phi_t = \phi$ and, for small departures around the exact ϕ, it differs from the exact value by second-order terms. This essentially pedestrian but general procedure has been shown (Gerjuoy et al., 1972, 1973) to lead to many familiar and new VPs.

3. Removal of the Near-Singularity

The formal construction of any VP is straightforward but the results are a bit of a fake, the burden of the problem having been shifted onto the $\mathscr{L}_i$. For the procedure to be significant practically, a method for getting reasonable $\mathscr{L}_{it}$ has to be provided. Of course, for any variational procedure, the question of the form of the trial functions chosen is important but, with something like the familiar Rayleigh–Ritz VB for the ground-state energy of a system, very broad considerations such as symmetry are normally sufficient when picking ϕ_t; indeed, one has to be clever to pick a ϕ_t which does not give good results! The problem for the general VPs of picking an $\mathscr{L}_{it}$ is, however, much more serious because of the very real dangers of the development of singularities in $\delta\mathscr{L}$ which make the VP worthless. This is because, typically, $\mathscr{L}$ obeys an inhomogeneous equation; for example, in looking for matrix elements involving the ground state ϕ_1, we have

$$(H - E_1)L = q(\phi_1) \qquad (3)$$

where $q(\phi_1)$ is a specified functional $[= -W\phi_1 + (\phi_1, W\phi_1)\phi_1$ when one seeks a VP for $(\phi_1, W\phi_1)]$ which is necessarily orthogonal to ϕ_1 for Eq. (3) to be meaningful. One *cannot* (Krieger and Sahni, 1972) get the corresponding L_t by solving the equation [with $E_{1t} \equiv (\phi_{1t}, H\phi_{1t})$]

$$(H - E_{1t})L_t = q(\phi_{1t}) \qquad (4)$$

Thus, since there is no homogeneous solution of $(H - E_{1t})$, a singularity develops in L_t in inverting this operator as $\phi_{1t} \to \phi_1$ with the attendant $E_{1t} \to E_1$; $\delta\mathcal{L}$ becomes *larger* as ϕ_{1t} is improved! When a homogeneous solution exists, this difficulty can be overcome by making L_t orthogonal to it (just as L can be chosen to be orthogonal to ϕ_1). Thus, the removal of the difficulty calls for the replacement of H in Eq. (4) by a modified Hamiltonian, $H_{\text{mod},t}$, which has ϕ_{1t} as its homogeneous solution. Until recently, the only prescriptions (Schwartz, 1959; Dalgarno and Stewart, 1960; Delves, 1967) for doing this were extremely restrictive, working only with nodeless ϕ_{1t} or with ϕ_{1t}, an unsymmetrized product of single particle orbitals. For this reason, no useful applications of VPs for matrix elements to any system other than He existed, save for one (Mueller et al., 1974) with a simple ϕ_{1t} to the VP for $\langle r^2 \rangle$ of heavier atoms; in fact, many formal developments in the construction of general VPs seem to have been aborted before reaching the stage of practical usefulness. The picture has changed, however, with the recent realization (Gerjuoy et al., 1974) that a general procedure of appropriately modifying the Hamiltonian existed in the literature (Rosenberg et al., 1960), albeit in a different context. In Eq. (4), $H - E_{1t}$ is replaced by

$$H_{\text{mod},t} - E_{1t}(1 - P_{1t}) \equiv H - \frac{HP_{1t}H}{E_{1t}} - E_{1t}(1 - P_{1t}) \qquad (5)$$

where $P_{1t} \equiv \phi_{1t}\phi_{1t}^{\dagger}$ projects out the trial ground state. (Note that matrix elements of $H_{\text{mod},t}$ involve only matrix elements of H and not of the more difficult H^2.) In going from the spectrum of H to that of $H_{\text{mod},t}$ the lowest eigenvalue, E_1, is shifted up to zero energy while the others are shifted up slightly by terms of second order. Therefore, if ϕ_{1t} is sufficiently good so that E_{1t} lies lower than an energy very close to the first excited state, E_2, of the system, the singularity difficulties are overcome. Furthermore, $H_{\text{mod},t} - E_{1t}(1 - P_{1t})$ can be shown to be a nonnegative operator so that an extremum principle can be formulated for obtaining L_t. Details are given in Gerjuoy et al. (1974), and numerical applications are now underway by Dr. Robin Shakeshaft. We note that, as one goes to heavier atoms, $E_2 - E_1$ becomes an increasingly smaller fraction of $E_1(\sim Z^{-7/3})$ and, therefore, ϕ_{1t} has to be fairly sophisticated for $E_{1t} \lesssim E_2$ to be satisfied.

4. Bounds and Variational Bounds

VPs provide stationary estimates that differ from the exact results only in second order, but they leave open the sign of the error. Bounds on properties which provide such information may be classified into the following scheme in order of increasing merit: (i) simple bounds, (ii) linear bounds, (iii) stationary bounds, also called VBs or extremum principles. Simple bounds do not involve trial functions for their evaluation but are expressions in terms of known properties of the system, such as sum rules and the position of low-lying energy levels. They are bounds, therefore, that are evaluated once and for all. Some simple bounds on atomic systems, such as those on the polarizability or $\langle r^2 \rangle$ for the ground state, have been known for a long time (Rebane and Braun, 1969) and, more recently, Blau et al. (1973a) developed a host of such bounds on $\langle W^2 \rangle$ in a fairly systematic way in terms of the spectrum of H and commutators of W with H, for W some power of the radial or momentum coordinate. These simple bounds may be expected to be roughly accurate since they incorporate some of the physics of the system under study. Even though rough, their merit lies in that they are often necessary inputs into the more sophisticated linear and stationary bounds.

Linear bounds on $\langle W \rangle$ are expressions involving trial functions ϕ_t which have errors of first order in $\delta\phi \equiv \phi_t - \phi$. Such bounds were first derived by Aranoff and Percus (1967); somewhat better results can be obtained (Weinhold, 1970; Blau et al., 1973a) by using the nonnegativity of the Gram determinant G, the determinant of the $n \times n$ matrix whose elements are the inner products of n functions. Let

$$\langle W \rangle \equiv (\phi, W\phi), \quad \langle W \rangle_t \equiv (\phi, W\phi_t) = (\phi_t, W\phi), \quad \langle W \rangle_{tt} \equiv (\phi_t, W\phi_t)$$

Choosing the three functions ϕ, ϕ_t, and $W\phi$, two of the six independent inner products are unity, a third is the overlap, $S \equiv (\phi, \phi_t)$, and we have

$$G(\phi, \phi_t, W\phi) \equiv \tilde{G}(\langle W^2 \rangle, S, \langle W \rangle, \langle W \rangle_t) \geq 0 \qquad (6)$$

Choosing ϕ, ϕ_t, and $W\phi_t$, we have on the other hand,

$$G(\phi, \phi_t, W\phi_t) \equiv \tilde{G}(\langle W^2 \rangle_{tt}, S, \langle W \rangle_t, \langle W \rangle_{tt}) \geq 0 \qquad (7)$$

These inequalities are quadratic functions of $\langle W \rangle$ and $\langle W \rangle_t$, respectively, and, therefore, yield *both* bounds on $\langle W \rangle$, provided we have bounds on S and on $\langle W^2 \rangle$. Bounds on S are well known. The need for bounds on $\langle W^2 \rangle$ (actually, only the upper bound) might appear to be a serious flaw in the procedure, but $\langle W^2 \rangle$ appears in Eq. (6), multiplied by a second-order quantity, $1 - S^2$. Thus, we need only use a simple bound on $\langle W^2 \rangle$ to get the more accurate linear bounds on $\langle W \rangle$.

To go very much beyond linear bounds, Blau et al. (1973b) emphasized the useful trick of not directly bounding the matrix element of interest but of first writing down the "variational identity" (Gerjuoy et al., 1972) $F(\phi) = F_v(\phi) + \mathscr{S}$ that corresponds to the VP in Eq. (1) with $\mathscr{S}$ of second order. The significance of the identity is that it specifically exhibits the second-order terms not contained in the VP. Since F_v is calculable, if we can bound $\mathscr{S}$ to some order of accuracy, more accurate bounds on the matrix element can be obtained. The second-order terms in $\mathscr{S}$ have the structure $(\delta\phi, \delta\phi)$ and $(\delta\phi, [H - E]\delta L)$. The former is just $2(1 - S)$; to bound the latter, one may use the Schwarz inequality or consider the Gram determinant of $\delta\phi$, $(H - E)\delta L$, and ϕ. Our first such attempt (Blau et al., 1973b) with first-order bounds on $\mathscr{S}$ led to quasi-stationary bounds (intermediate between linear and stationary bounds, with errors of $\delta\phi^{3/2}$) on matrix elements of arbitrary W. True stationary lower bounds were also obtained when W was positive-definite, in which case we could use the operator inequality,

$$W \geqslant \frac{W\phi_t \rangle \langle \phi_t W}{(\phi_t, W\phi_t)} \tag{8}$$

More recently, the approach was further refined (Blau et al., 1974) to give second-order bounds on $\mathscr{S}$ and, in turn, stationary upper and lower bounds for many operators of interest. Thus, for a wide variety of Ws, it is now possible to bracket the true value by upper and lower stationary bounds which can be progressively brought closer together by using progressively better ϕ_t.

5. Variational Bound on Scattering Length for Target Ground-State Wave Function Imprecisely Known

The above VBs on bound-state matrix elements can also be applied to zero-energy scattering (Blau et al., 1974), because VBs on the scattering length, A (assuming that the target ground-state wave function is given) are known (Spruch and Rosenberg, 1959). Assume, for example, that a positron is incident on an atom and that the e^+ + atom cannot form a composite bound state. (The necessary modification when they can is straightforward.) We then have the upper VB

$$A \leqslant A_t + (m/2\pi\hbar^2)(\Psi_t, [H - E_1]\Psi_t) \tag{9}$$

where Ψ_t, the trial zero-energy scattering function, satisfies the boundary condition

$$\Psi_t \sim \phi_1 \times (A_t - r)/r$$

r is the coordinate of the positron, and ϕ_1 and E_1 characterize the target ground state. Since ϕ_1 and E_1 are the only unknowns in the upper VB on A, the scattering problem has been reduced to the bound-state problem for which upper and lower VBs are available; only the upper VB is useful in the present context since, unfortunately, we do not have the equivalent of Eq. (9) in lower VB form.

The situation is only slightly more complicated for electron–atom scattering (Blau et al., 1975). With Ψ_t appropriately antisymmetrized, Eq. (9) remains valid. However, a new form of matrix element arises. For e^-–He scattering, for example, we have the "exchange" term:

$$J \equiv (\phi_1(1,2), U(1,2,3)\phi_1(2,3))$$

where ϕ_1 is the helium ground-state wave function and U is a known function or operator containing potential or kinetic energy operators and parts of the (known) trial function describing the e^-–He relative motion. J is perfectly well defined since $\phi_1(1,2)$ and $\phi_1(2,3)$ between them provide decay factors in all three coordinates, but for most Us of interest, one cannot immediately separate J into two factors, as one does in the use of the Schwarz inequality, each of which is square integrable in the space of all three coordinates. One can proceed as follows: As $r_2 \to \infty$, $\phi_1(2,3)$ decays exponentially. We can then rewrite J as

$$J = (\phi_1(1,2)F^{-1}(3)F(1), F(3)F^{-1}(1)U(1,2,3)\phi_1(2,3))$$

where $F^{-1}(3)$ decays rapidly enough to cause the first factor, involving $\phi_1(1,2)$, to be quadratically integrable, but $F(3)$ increases less rapidly than $\phi_1(2,3)$ decreases; a similar remark obtains for $F(1)$. At this stage, the preceding techniques can be used to obtain an upper VB on A.

We might remark at this point that the technique of accurately determining the scattering length, when used to get an equivalent determination of $A - A_B$, where A_B is the scattering length in the Born approximation, is useful in the analysis of higher-energy scattering as well, since the forward dispersion relation expresses $A - A_B$ as the integral of the total cross section over all momenta if no composite bound states exist. We write $\Psi_t = \Psi_B + \Theta$, where Ψ_B is the Born approximation wave function and Θ contains a scattering component. Since

$$A_B \equiv (m/2\pi\hbar^2)(\Psi_B, [H - E_1]\Psi_B)$$

a VB on $(A - A_B)$ follows from Eq. (9). When composite bound states exist, the dispersion relation involves pole terms whose residues must be bounded in order to preserve the bound on the cross-section integral. We have not succeeded in doing this yet, but we are hopeful.

The method leading to VBs on A can be extended to the positive energy single-channel scattering problem. However, the computation is considerably more difficult than for zero-energy scattering since exact numerical solutions of an equivalent one-body problem are required.

Acknowledgments

This work was supported at New York University by the U.S. Army Research Office, Durham, Grant No. DA-ARO-D-31-124-72-G92, by the Office of Naval Research, Grant No. N00014-67-A-0467-0007, and by the National Science Foundation.

References

Aranoff, S., and Percus, J. K. (1967). *Phys. Rev.*, **162**, 878.
Blau, R., Rau, A. R. P., and Spruch, L. (1973a). *Phys. Rev.*, **A8**, 119.
Blau, R., Rau, A. R. P., and Spruch, L. (1973b). *Phys. Rev.*, **A8**, 131.
Blau, R., Rosenberg, L., and Spruch, L. (1974). *Phys. Rev.*, **A10**, 2246.
Blau, R., Rosenberg, L., and Spruch, L. (1975). *Phys. Rev.*, **A11**, 200.
Dalgarno, A., and Stewart, A. L. (1960). *Proc. R. Soc. Lond.*, **A257**, 534.
Delves, L. M. (1967). *Proc. Phys. Soc. Lond.*, **92**, 55.
Gerjuoy, E., Rau, A. R. P., and Spruch, L. (1972). *J. Math. Phys.*, **13**, 1797.
Gerjuoy, E., Rau, A. R. P., and Spruch, L. (1973). *Phys. Rev.*, **A8**, 662.
Gerjuoy, E., Rau, A. R. P., Rosenberg, L., and Spruch, L. (1974). *Phys. Rev.*, **A9**, 108.
Krieger, J. B., and Sahni, V. (1972). *Phys. Rev.*, **A6**, 919 and 928.
Mueller, R., Rau, A. R. P., and Spruch, L. (1974). *Phys. Rev.* **A10**, 1511.
Rebane, T. K., and Braun, P. A. (1969). *Opt. Spectrosc.*, **27**, 486.
Rosenberg, L., Spruch, L., and O'Malley, T. F. (1960). *Phys. Rev.*, **118**, 184.
Schwartz, C. (1959). *Ann. Phys.*, **2**, 156 and 170.
Spruch, L. (1968). In *The Physics of Electronic and Atomic Collisions: Leningrad, July, 1967*. Ed. L. Branscomb, p. 89. Boulder: University of Colorado.
Spruch, L. (1973). *Fundamental Interactions in Physics*, New York: Plenum Press, p. 213.
Spruch, L., and Rosenberg, L. (1959). *Phys. Rev.*, **116**, 1034.
Weinhold, F. (1970). *Phys. Rev. Lett.*, **25**, 907.

48

Backscattering of Slow Electrons by Positive Ions

G. Drukarev

> The scattering cross section of slow electrons by positive ions at scattering angle 180° is considered as a function of energy. At this angle there will be no oscillations due to the factor $exp\,[i(z/k)\,\ln\sin(\theta/2)]$ in the Coulomb part of the amplitude. The shape of the cross section vs. energy curve depends on the non-Coulomb part of the interaction. The possibility of a deep minimum on this curve is indicated and the conditions of its existence are examined.

1. The subject of this paper is a general consideration of the energy dependence of the elastic scattering of an electron by a positive ion with charge Ze at 180°. The scattering amplitude can be expressed as a sum,

$$f = f_c + f' \tag{1}$$

where f_c is the Coulomb amplitude and f' the contribution of all other interactions. In the units $\hbar = m = e = 1$, f_c and f' have the form

$$f_c = \frac{Z}{2k^2\sin^2(\theta/2)}\exp 2i\left[\frac{Z}{k}\ln\sin\frac{\theta}{2} + \eta_0\right] \tag{2}$$

$$f' = \frac{1}{2ik}\sum_l (2l+1)\,e^{2i\eta_l}(e^{2i\delta_l} - 1)P_l(\cos\theta) \tag{3}$$

where

$$\eta_l = \arg\Gamma\left(l + 1 - i\frac{Z}{k}\right) \tag{4}$$

G. Drukarev • Leningrad State University, Leningrad, USSR.

is the Coulomb phase and δ_l is an additional phase shift due to other interactions in the presence of the Coulomb field.

Putting

$$f = \frac{1}{k} e^{2i\eta_0} F(k, \theta) \tag{5}$$

we have the following expression for the cross section:

$$\sigma = \left[\frac{Z}{2k^2 \sin^2 (\theta/k)} \cos \left(\frac{2Z}{k} \ln \sin \frac{\theta}{2} \right) + \frac{1}{k} \operatorname{Re} F \right]^2$$

$$+ \left[\frac{Z}{2k^2 \sin^2 (\theta/2)} \sin \left(\frac{2Z}{k} \ln \sin \frac{\theta}{2} \right) + \frac{1}{k} \operatorname{Im} F \right]^2 \tag{6}$$

From the term $(1/k) \ln \sin (\theta/2)$ there will be oscillations in the energy dependence at any given θ.

The only exception is $\theta = \pi$. Here all oscillations will vanish and the cross section will be

$$\sigma = \frac{Z^2}{4k^4} + \frac{Z}{k^3} \operatorname{Re} F(k, \pi) + \frac{1}{k} |F(K, \pi)|^2 \tag{7}$$

2. Let us consider $F(k, \pi)$ in detail. According to (3), (4), and (5), we have

$$F(k, \pi) = \sum_l (2l + 1) e^{i(\phi_l + \delta_l)} \sin \delta_l \tag{8}$$

Here

$$\phi_l = 2(\eta_l - \eta_0) + l\pi = 2 \sum_{m=1}^{l} \tan^{-1} \left(\frac{mk}{Z} \right) \tag{9}$$

Separating real and imaginary parts in (8) and denoting

$$\operatorname{Re} F(k, \pi) = R(k); \qquad \operatorname{Im} F(k, \pi) = I(k) \tag{10}$$

we have

$$R(k) = \sum_l (2l + 1) \cos (\phi_l + \delta_l) \sin \delta_l$$

$$I(k) = \sum_l (2l + 1) \sin (\phi_l + \delta_l) \sin \delta_l \tag{11}$$

The phase shifts δ_l at low energies behave like

$$\delta_l(k) = \delta_l(0) + \delta_l'(0)k^2 + \cdots \tag{12}$$

It follows from (9), (11), and (12) that at $k = 0$

$$R(0) = \tfrac{1}{2} \sum_l (2l + 1) \sin 2\delta_l(0)$$

$$R'(0) \equiv \left(\frac{dr}{dk}\right)_{k=0} = -\frac{1}{Z} \sum_l (2l + 1)(l + 1)l \sin^2 \delta_l(0) \qquad (13)$$

$$I(0) = \sum_l (2l + 1) \sin^2 \delta_l(0)$$

According to Seaton, at low energies $\delta_l(k)$ can be found by the extrapolation of the quantum defects $\mu_l(n)$ on the positive energies. In particular,

$$\delta_l(0) = \pi\mu_l(\infty) \qquad (14)$$

One should note that the addition of an integral multiple of π to the phase δ_l does not change F. Consequently, only the decimal part of μ_l matters. Usually the most important contribution at small k comes from the first three phases, $\delta_0, \delta_1, \delta_2$.

3. To get an approximate expression for the cross section at $k \ll 1$, we will replace $\mathrm{Re}\, F$ in (7) by $R(0) + KR'(0)$, and $|F|$ by $R^2(0) + I^2(0)$. Denoting

$$\begin{aligned} R(0) &= a \\ R^2(0) + I^2(0) + ZR'(0) &= b \end{aligned} \qquad (15)$$

we have

$$\sigma = \frac{Z^2}{4k^4} + \frac{Za}{k^3} + \frac{b}{k^2} \qquad (16)$$

In order to be able to neglect the term $\sim 1/k$, the value b should not be too small.

To prevent (16) from being negative one should have

$$b > a^2 \qquad (17)$$

We will assume that this is the case.

If $a < 0$, then σ can have a minimum at the point

$$k_0 = \frac{3}{4} \frac{|a|Z}{b} \left[1 - \frac{8b^{1/2}}{9a^2} \right] \qquad (18)$$

From this expression one can deduce that the minimum will exist under the condition

$$\frac{b}{a^2} < \frac{9}{8} \tag{19}$$

which does not contradict (17).

The value of σ at the point k_0 will be

$$\sigma_{min} = \frac{Z|a|}{2k_0^3} - \frac{Z^2}{4k^4}$$

If $b = a^2$, then $k_0 = Z/(2|a|)$ and $\sigma_{min} = 0$. Of course, the exact cross section will not reach zero because of the terms $\sim 1/k$ neglected in (16). Instead, there will be a deep minimum which is an analogue of the Ramsauer effect. At larger values of k, the neglected terms can become important. In this region one can expect smooth oscillations of cross section.

It gives me great pleasure to have this opportunity to express my great respect and admiration for Professor Ugo Fano.

49
Quantum Defect Theory of Excited $^1\Sigma_u^+$ Levels of H_2

O. ATABEK AND C. JUNGEN

> Vibration-electron coupling in Rydberg states of molecular hydrogen is treated by quantum defect methods. One single quantum defect function, $\mu_{p\sigma}(R)$, of simple algebraic form is used to fit the recently observed spectrum of $J = 0$ $np\sigma$ $^1\Sigma_u^+$ levels, extending from $n = 4$ up to the H_2^+ ($v = 1$) ionization threshold. The calculations reproduce with high accuracy all the strong perturbations taking place between Rydberg series of different vibrational quantum number, $0 \leqslant v \leqslant 5$. The resulting function $\mu_{p\sigma}(R)$ is directly related to the Born–Oppenheimer potential energy curve of the electronic states involved. The results are discussed with reference to the processes of vibrational preionization and vibrational predissociation.

A few years ago, Fano (1970) introduced a new concept into the theories of electron–atom collisions and multichannel interactions between Rydberg series (Seaton, 1966). He showed that a unified treatment of the full spectrum of electron energies, from the lowest Rydberg levels up into the continuous energy range, could be obtained by formulating the theory with simultaneous reference to two alternative extreme coupling cases. These are the two situations: A, strongly coupled motions of the excited electron and of the core particles, and B, completely independent motions of the excited electron and the core particles. The extreme situations A or B are characterized, respectively, by large or small differences $|\Delta\varepsilon|$ between electronic levels compared to the energy level structure in the residual core.

O. ATABEK • Laboratoire de Photophysique Moléculaire, Université Paris-Sud, 91405 Orsay, France.   C. JUNGEN • Centre de Mécanique Ondulatoire Appliquée, 23 rue de Maroc, 75019 Paris, and Laboratoire de Photophysique Moléculaire, Université Paris-Sud, 91405 Orsay, France.

This contribution deals with a molecular example, namely, with the interaction between vibrational and electronic motion in the $np\sigma$ Rydberg states of H_2. Fano's concepts are particularly useful in the theory of coupled nuclear and electronic motions in molecules. First of all, because of the close spacings of rotational and vibrational fine structure in molecular cores, levels belonging to situation A are actually observed in the spectrum, at low principal quantum numbers n^*, while a transition toward situation B occurs as n^* increases. Furthermore, in the case of vibration–electron interaction, there is even an overlap in the spectrum of the energy regions in which situations A and B hold approximately. This arises because there is an infinite number of vibrational channels: when the vibrational quantum number v assumes a high value, the vibrational energy $G(v)$ becomes comparable to the electron energy $|\varepsilon|$, while at the same time situation A still holds because of the small vibrational spacings $\Delta G(v)$.

In the special case of vibration–electron interaction, Fano's situation A is equivalent with the situation implied by the Born–Oppenheimer approximation. In the Born–Oppenheimer limit the electronic motion is fast compared to the vibrational motion of the nuclei. At the other extreme, the nuclei move much faster than the distant Rydberg electron in situation B where the Born-Oppenheimer approximation is entirely invalid. Quantum defect theory accounts for both extreme situations, including intermediate cases, by means of one single R-dependent quantum defect function $\mu_{p\lambda}(R)$ (R is the internuclear distance) (Herzberg and Jungen, 1972; Dill et al., 1973; Atabek et al., 1974). In the Born–Oppenheimer limit the Rydberg electron contributes to the electronic potential energy curve for vibrational motion, $U^{np\lambda}(R)$, according to

$$U^{np\lambda}(R) = U^+(R) - \frac{Ry}{[n - \mu_{p\lambda}(R)]^2} \quad (1)$$

where $U^+(R)$ is the potential energy curve for the H_2^+ core and Ry is the Rydberg constant. Conversely, in the limit opposite to the Born–Oppenheimer approximation a nondiagonal reaction matrix is obtained by averaging the R-dependent quantum defect over the vibrational motion of the core:

$$M_{vv'} = \int \chi_v^+(R) \tan[\pi\mu_{p\lambda}(R)] \chi_{v'}^+(R)\, dR \quad (2)$$

The functions $\chi_v^+(R)$ are the vibrational wave functions of the H_2^+ core.

Atabek et al. (1974) reported very recently the first application of quantum defect methods to vibration–electron interaction. They calculated the spectrum of $J = 1$ Rydberg levels of $^1\Pi_u^-$ symmetry in H_2 corresponding to excitation of a $np\pi$ electron, based on an R-dependent quantum defect

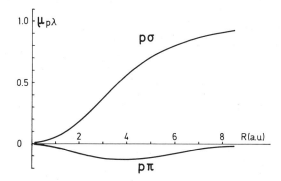

Figure 1. Quantum defect curves $\mu_{p\sigma}(R)$ and $\mu_{p\pi}(R)$. The curve $\mu_{p\sigma}(R)$ corresponds to Eq. (3) with the parameters $a = 2.062 \times 10^7$, $\alpha = -5.0$, $\beta = 7.863$, $\gamma = 2.089 \times 10^7$. The curve $\mu_{p\pi}(R)$ is derived from the *ab initio* potential energy curve for the $2p\pi$ C state of Kołos and Wolniewicz (1965).

$\mu_{p\pi}(R)$ which had been obtained from the accurate *ab initio* potential energy curve for the $2p\pi C$ state of Kołos and Wolniewicz (1965). This quantum defect curve $\mu_{p\pi}(R)$ is shown in Figure 1: the R dependence is small and the effects of vibration–electron interaction observed in the spectrum of $^1\Pi_u^-$ levels are accordingly limited (Herzberg and Jungen, 1972). In contrast, Herzberg and Jungen (1972) observed strong perturbations due to vibration–electron interaction in the $J = 0$ Rydberg levels of H_2 with $^1\Sigma_u^+$ symmetry corresponding to excitation of a $np\sigma$ electron, and in the following we shall interpret these perturbations on the basis of a quantum defect curve $\mu_{p\sigma}(R)$, which depends strongly on the internuclear distance R. Unlike $\mu_{p\pi}(R)$, the curve $\mu_{p\sigma}(R)$ cannot yet be derived from an *ab initio* potential energy curve,* but it has to be determined from the accurate spectroscopic data of Herzberg and Jungen (1972).

The qualitative behavior of the quantum defect $\mu_{p\sigma}(R)$ can be inferred from consideration of the dissociation behavior of the $np\sigma$ $^1\Sigma_u^+$ electronic states of H_2. A correlation diagram relating united-atom to separated-atom Rydberg orbitals has been given by Weizel (1931). It shows the $np\sigma$ electron to be "promoted," that is, having an n value higher by one unit in the united atom than in the separated atoms. This correlation characterizes an antibonding electron and implies a quantum defect curve $\mu_{p\sigma}(R)$ of the shape shown in Figure 1; the quantum defect is virtually zero in the united atom He, but it gradually increases as the nuclei move apart, until it approaches

---

*The $2p\sigma$ B state, for which an *ab initio* potential energy curve exists, is not a "pure Rydberg" state (see below).

unity at large internuclear distances. Strictly speaking, Weizel's correlation diagram holds only for the (separable) one-electron system of cylindrical symmetry. In H_2, with its second electron present in the core, interaction between the two electrons causes modifications in the dissociation behavior of $^1\Sigma_u^+$ states for low n (Mulliken, 1966), but these are of no importance in the present calculations, which exclude $n = 2$ and 3. For example, no molecular $^1\Sigma_u^+$ state correlates with two separated $H(1s)$ atoms. The $2p\sigma\ B^1\Sigma_u^+$ state, therefore, cannot behave according to the correlation diagram for Rydberg orbitals; this state loses its Rydberg character when R increases and becomes ionic.

For our purpose of fitting the experimental $J = 0$ levels using quantum defect theory, it seemed desirable to build the anticipated overall behavior of the quantum defect into the treatment from the beginning by expressing $\mu_{p\sigma}(R)$ in a suitable functional form. We have chosen the expression

$$\mu_{p\sigma}(R) = 1 - \frac{a}{(R - \alpha)^\beta + \gamma} \qquad (3)$$

which for $a \sim \gamma$ has the desired form and, while depending on only four parameters to be fitted, possesses a high degree of flexibility. The calculations have been performed in much the same way as described by Atabek et al. (1974). For a given set of parameters, a, α, β, γ reaction matrix elements were calculated according to Eq. (2) using the accurate ab initio potential energy curve for the core H_2^+. These elements were then used, together with the theoretical ionization potential of H_2 and the theoretical rotation–vibration energy levels of H_2^+, for the solution of the quantum defect eigenvalue problem (Seaton, 1966), which gave the calculated energy level positions directly. The best fit of the experimental data was obtained with the curve $\mu_{p\sigma}(R)$, shown in Figure 1 (the detailed results will be given in a later publication). A comparison of some of our results concerning the energy region between the $v = 0$ and $v = 1, N = 1$ ionization limits with the experimental data is made in the quantum defect plot shown in Figure 2. For each level the quantum defect $n - n^*$ (n^* is the effective principal quantum number) with respect to the $v = 1, N = 1$ limit has been plotted vs. the energy $T^+ - T$ measured from the same limit. Quantum defect theory predicts the theoretical quantum defects (circles) lying on a smooth curve representing $n - n^*$ as a function of $T^+ - T$. The calculated values agree with the observed ones (bars) in almost all cases, and they reproduce the numerous perturbations very well. The perturbations are so strong that nowhere can one see unperturbed behavior, that is, a constant (n-independent) quantum defect. The strongest perturbations, of the type $\Delta v = 1$, involve the members $n = 6$ and 7 of the series converging to the $v = 2$ ionization limit. Two weaker perturbations, of the type $\Delta v > 1$, involve the levels $5p\sigma, v = 3$ and $4p\sigma$,

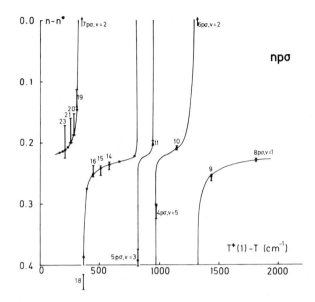

Figure 2. Quantum defect plot of the $np\sigma$, $v = 1$, $J = 0$ Rydberg series of H_2.

$v = 5$. While the $\Delta v = 1$ perturbations had been treated previously in the framework of situation B by Herzberg and Jungen (1972) (although less accurately than here), the $\Delta v > 1$ perturbations are for the first time accounted for in the present unified treatment. The $4p\sigma B''$ state is close to situation A (because of its low n value), and the gross positions of the vibrational levels are governed by their behavior according to this limiting situation. In this description, the vibrational frequency of the B'' state is about 130 cm^{-1} smaller than in H_2^+ owing to the antibonding effect exerted by the $np\sigma$ electron according to Eq. (1). From the point of view according to situation B, therefore, the level $4p\sigma$, $v = 5$ is shifted from its "unperturbed" position by several hundreds of cm^{-1}. Figure 2 shows that the quantum defect treatment accounts for both the gross position of $4p\sigma$, $v = 5$ and its interaction with the level $11p\sigma$, $v = 1$.

The quantum defect curve $\mu_{p\sigma}(R)$ extracted in this way from the spectroscopic data is related to the fixed-nuclei potential energy curves of the $np\sigma\ ^1\Sigma_u^+$ ($n \geq 4$) electronic states by Eq. (1), and it will serve for the comparison between experiment and *ab initio* theory once precise theoretical potential energy curves of these states become available. The present work has been undertaken with a much wider aim, however. We plan to treat by quantum defect methods vibrational preionization and predissociation, based on the quantum defect curve obtained from precise data of the discrete part of the spectrum.

The Rydberg levels represented in Figure 2 all lie in the $v = 0$ ionization continuum of H_2. In the absorption spectrum the transition to the level $5p\sigma$, $v = 3$ is by far the strongest of all transitions to $J = 0$ levels that appear between the $v = 0$ and $v = 1$ ionization limits (in the photoionization spectrum the same line is very weak). On the other hand, the Rydberg levels that interact most strongly with the $v = 0$ ionization continuum are those which, at least in a first approximation, belong to the $np\sigma$, $v = 1$ series. Consequently, it becomes clear that the photoionization process in H_2 cannot be understood simply in terms of the interaction of *one* discrete Rydberg series with the ionization continuum. Rather, it is characterized by the interplay of channel interactions among *several* discrete series, some of which contribute primarily oscillator strength, while others provide the coupling with the continuum. The interplay of interactions is further complicated by the fact that different channels belong to different coupling regimes, A and B. This complexity is borne out by the earlier model calculations of Berry and Nielsen, (1970), who did not use quantum defect theory. The present close fit of the experimental data by quantum defect theory demonstrates the capability of this method to account quantitatively for such complex behavior.

All the levels represented in Figure 2 are also subject to predissociation by the $2p\sigma B\ ^1\Sigma_u^+$ and $3p\sigma B'\ ^1\Sigma_u^+$ states. This interaction of the $np\sigma, v, J = 0$ levels with the vibrational continua of the two lowest members of the same Rydberg series is just another aspect of the vibration–electron interaction (Dill *et al.*, 1973). Striking experimental evidence supporting this point of view has been obtained very recently by Roncin *et al.* (1974). These authors observed the *emission* spectrum of H_2 under high resolution from levels near and above the ionization energy. Their emission spectrum has a much simpler structure than the corresponding portion of the absorption spectrum since it consists of $Q(J)$ lines (i.e., $\Delta J = 0$ transitions) only, while $R(J)$ and $P(J)$ lines (i.e., $\Delta J = +1$ and -1 transitions) are entirely absent. For these latter types, both preionization and predissociation have been established in the absorption spectrum by line broadening and asymmetric line shape. The symmetries of the levels involved in the transitions are such that only the $Q(J)$ lines correspond to pure excitation of a $np\pi$ electron, while the ($J = 0$) upper levels of the $P(1)$ lines have $np\sigma\ ^1\Sigma_u^+$ symmetry and those of the other P and R lines are mixtures of $np\sigma\ ^1\Sigma_u^+$ and $np\pi\ ^1\Pi_u^+$ due to rotation–electron interaction (l uncoupling). The striking point here is that there appears to be a close relationship between preionization and predissociation: in the $np\pi\ ^1\Pi_u^-$ upper levels of the $Q(J)$ lines, *both* processes must be very slow in order that fluorescence can occur. The quantum defect

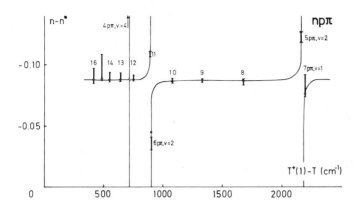

Figure 3. Quantum defect plot of the $np\pi^-$, $v = 1$, $J = 1$ Rydberg series of H_2.

theory of vibration–electron interaction predicts just this behavior since it relates preionization and predissociation of a $^1\Pi_u^-$ level to the *same* quantum defect curve, $\mu_{p\pi}(R)$. The contrasting behavior of the curves $\mu_{p\sigma}(R)$ and $\mu_{p\pi}(R)$ shown in Figure 1 has already been emphasized. The very different magnitude of the ensuing perturbations between discrete Rydberg series is illustrated by the comparison of Figure 2 with Figure 3, which has been taken from the paper of Atabek *et al.* (1974) and represents the $np\pi\ ^1\Pi_u^-$, $J = 1$ Rydberg series converging to the same limit as the $np\sigma\ ^1\Sigma_u^+$, $J = 0$ series shown in Figure 2. We conclude, finally, that it is the curve $\mu_{p\sigma}(R)$ whose R dependence governs the rates of vibrational preionization and vibrational predissociation in the levels of H_2 which are optically accessible from the ground electronic state.

References

Atabek, O., Dill, D., and Jungen, C. (1974). *Phys. Rev. Lett.*, **33**, 123.
Berry, R. S., and Nielsen, S. E. (1970). *Phys. Rev.*, **A1**, 395.
Dill, D., Chang, E. S., and Fano, U. (1973). In Abstracts of Papers, *Proceedings of the Eighth International Conference on the Physics of Electronic and Atomic Collisions, Belgrade*, p. 536. Eds. B. C. Dobič and M. V. Kurepa. Belgrade: Institute of Physics.
Fano, U. (1970). *Phys. Rev.*, **A2**, 353.
Herzberg, G., and Jungen, C. (1972). *J. Mol. Spectrosc.*, **41**, 425.
Kolos, W., and Wolniewicz, L. (1965). *J. Chem. Phys.*, **43**, 2429.
Mulliken, R. S. (1966). *J. Am. Chem. Soc.*, **88**, 1849.
Roncin, J. Y., Damany, H., and Jungen, C. (1974). In *Vacuum Ultraviolet Radiation Physics*. Eds. E. Koch, R. Haensel, and C. Kunz. Braunschweig: Vieweg.

Seaton, M. J. (1966). *Proc. Phys. Soc. Lond.*, **88**, 801.
Weizel, W. (1931). In *Handbuch der Experimentalphysik, Supplement*, Vol. 1, p. 42. Eds. W. Wien and F. Harms. Leipzig: Akademische Verlagsgesellschaft.

50
Auger Spectroscopy of Foil-Excited Beryllium Ions

R. BRUCH, G. PAUL, J. ANDRÄ, AND B. FRICKE

We have measured prompt and delayed emission spectra of electrons from foil-excited Be, Be^+, and Be^{2+} ions at 300 keV. On the basis of recently calculated eigenvalues we identified two lines in the prompt Be^+ spectrum as transitions from $2s^2 2p$ and $2s 2p^2$. The delayed Be spectrum indicates that transitions from highly excited quintet states occur. We propose radiationless deexcitation with one excited spectator electron not involved in the transition.

We have used the interaction of fast ionic beams (300 keV) with thin solid carbon targets (7 $\mu g/cm^2$) to populate highly excited ionic states in beryllium (Martinson, 1974), which subsequently decay either radiatively by electric dipole transitions or via autoionization to an adjacent electron continuum. The channels for autoionization of these states depend on the extent of configuration interaction with states with the same L, S, and J values and parity. Auger transitions allowing the variation of L, S, or J are only possible when the weaker spin–orbit (H_{so}), spin–other-orbit (H_{soo}), spin–spin (H_{ss}), and hyperfine coupling (H_{hf}) are involved (Das, 1973; Sevier, 1972).

One advantage of the beam–foil technique is the possibility for time-resolved observation of the ion beam, which permits one to separate the prompt electron emission ($\tau < 10^{-10}$ sec) from the delayed spectra ($\tau > 10^{-10}$ sec) characteristic of levels decaying by H_{so}, H_{soo}, H_{ss}, and H_{hf}. We have used a cylindrical electron spectrometer (Sellin, 1973) for accepting electrons emitted at $\theta = 42.3° \pm 1°$ ($2\Delta\theta = 2°$; beam diameter $\rho = 1.5$ mm) from

---

R. BRUCH, G. PAUL, AND J. ANDRÄ • Institut für Atom- und Festkörperphysik A, 1 Berlin 33, Boltzmannstrasse 20, Germany.   B. FRICKE • Gesamthochschule Kassel, 35 Kassel, Heinrich-Plett-Strasse 40, Germany.

a small beam volume on the downstream side of a thin carbon foil. This foil could be moved relative to the viewing region of the electron spectrometer to detect both the prompt and delayed spectra. The electron energies measured in the laboratory system are transformed to center-of-mass energies by standard calculations (Risley, 1974). The spread in beam energy and angular spread of the beam after foil excitation, together with the finite acceptance angle of the spectrometer, produces an effective energy resolution of 1.7% in the Be spectra, compared to the spectrometer resolution of 0.3% when applied to a stationary point source of electrons.

We obtained complex discrete electron spectra characteristic of multiply-excited levels in various charge states, superimposed on a continuum electron background produced by direct collision processes. In spite of this complexity, with the help of theoretical estimates of energy eigenvalues, we were able to interpret many of the observed structures (Bruch et al., 1975a,b). We note that a single level can often decay to several different final levels, giving rise to several spectral features of different intensities. On the basis of the electron configurations, we are able to compare the most probable decays with the observed lines. We have also measured decay curves for the spectrum near 102.4 eV and can resolve it into two decays, one of

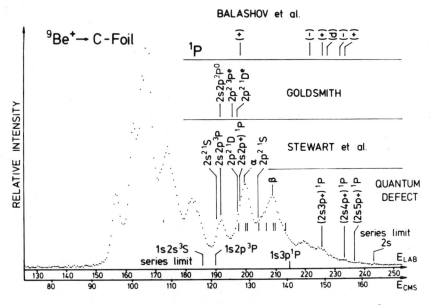

Figure 1. Electron spectrum of prompt autoionizing foil-excited Be, Be$^+$, and Be^{2+} states. (An absolute energy calibration ± 0.2 eV in the cms system was achieved using the delayed $1s2s2p(^4P_{5/2}) \rightarrow 1s^2\varepsilon f(^2F_{5/2})$ transition; $1s2s2p(^4P) = {}^4P^0(1)$; see Figure 2.)

3.1 ± 0.4 nsec lifetime and the other of 132 ± 50 nsec. The shorter decay is mainly due to the optical dipole transition $1s2p^2(^4P^e) \rightarrow 1s2s2p(^4P^o)$ measured by Hontzeas et al. (1973).

A prompt and a delayed spectrum is shown in Figures 1 and 2, respectively. The high-energy region of the prompt spectrum ($E_{cms} \geq 120$ eV) is attributed to Be II and Be III transitions. This part of the spectrum is compared with: (a) transitions $2lnl'$ ($n \geq 2) \rightarrow 1s\varepsilon l''$, calculated from theoretical two-electron autoionization energies (Balashov et al., 1970; Chan and Stewart, 1967; Perrott and Stewart, 1968); (b) two-electron energy values deduced from optical dipole transitions, as observed in the far-ultraviolet Be III spectrum, produced by a vacuum spark (Goldsmith, 1969); and (c) transition energies of the Rydberg series $2snp$ ($n \geq 3) \rightarrow 1s\varepsilon p$ converging to the series limit $2s\infty p \rightarrow 1s\varepsilon p$, calculated with zero quantum defect for $3p$, $4p$, and $5p$ electron orbitals.

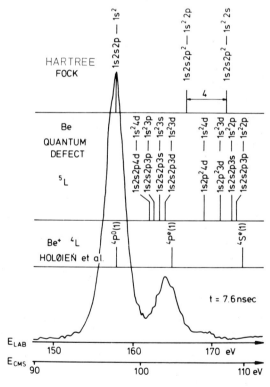

Figure 2. Delayed electron spectrum from foil-excited Be ions excited by impact on carbon foils.

For Be II, series limits of the $(1snln'l' \to 1s^2\varepsilon l'')$ transitions are given in Figure 1 at $E_{cms} = 118.6$ eV $[1s2s(^3S)]$, $E_{cms} = 121.9$ eV $[1s2p(^3P)]$, and $E_{cms} = 140.4$ eV $[1s3p(^1P)]$. In the energy region of interest, only the dominant lines of the $1snln'l' \to 1s^2\varepsilon l''$ series with n and $n' \geq 3$ should appear. No theoretical calculations are available yet for these series. However, in analogy to recent measurements of the $(3l3l')$ H$^-$ (Risley, 1974) and Li$^+$ (Bruch et al., 1975c) decays, one can assume that Be II transitions populating the $1s2s$ and $1s2p$ final ionic states should be much stronger than those decaying to the $1s^2$ ground state. Due to this branching only small contributions from $1snln'l'$ $(n, n' \geq 3) \to 1s^2\varepsilon l''$ transitions are expected in the energy range above $E_{cms} = 120$ eV. The prominent peaks α and β in the spectrum must therefore be assigned differently. They coincide with transition energies calculated recently (Lipsky and Ahmed, 1974) for the decay of $2s^22p(^2P^0)$ and $2s2p^2(^2S^e)$ states to $1s2s(^{1,3}S)$ and $1s2p(^{1,3}P)$ final ionic states, respectively. To our knowledge, no such promptly decaying three-electron autoionizing levels of positive ions have been observed previously (for the decay of negative ions see review article of Schulz, 1973). For the energy region $E_{cms} \leq 120$ eV, no complete comparison with theoretical calculations is yet possible. From first estimates, however, it is expected that the prominent features stem from three- and four-electron configurations (Bruch et al., 1975b).

Figure 2 shows part of the delayed Be spectrum 7.4 nsec after foil excitation and some theoretical energies estimated for highly excited three- and four-electron states. The Be$^+$ three-electron transitions among quartet states are assigned following the notion of Holøien and Geltman (1967). The $^4P^0(1)$, $^4P^e(1)$, and $^4S^e(1)$ states decay to the Be^{2+} state $1s^2$. We believe that the discrepancy between the calculated position for $1s2p^2(^4P^e) = {}^4P^e(1)$ ($E_{cms} = 103.1$ eV) and the center of the experimental peak ($E_{cms} = 102.4$ eV) is significant, in view of the reproducibility $\Delta E = \pm 0.2$ eV of the measurement. The theoretical estimate for the $^4P^e(1)$–$^4P^0(1)$ energy separation (Holøien and Geltman, 1967) should be good to at least the same accuracy, and in fact fits data for quartet levels in Be$^+$ (Hontzeas et al., 1973) within experimental uncertainty. Be is of particular interest here, since it is the lowest Z atom for which quintet states (^5L) occur. From the metastable spectrum it is likely that three electron quartet states alone are insufficient to explain the data. Accordingly, we propose metastable quintet states as explanation, which decay via a channel with the orbit of a spectator electron ($3s$, $3p$, $3d$, $4d$) remaining unchanged. This decay mechanism is illustrated in Figure 3 for a typical Be I four-electron configuration $1s2s2p3s$ decaying to four possible final ionic states $1s^22s(^2S)$, $1s^22p(^2P)$ (Junker and Bardsley, 1973), $1s2s^2(^2S)$, and $1s^23s(^2S)$. Since the $2s$–$2p$ correlation is supposedly stronger than the $2s$–$3s$ and $2p$–$3s$ one, the dominant decay channel leads

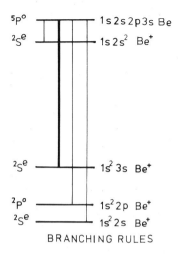

Figure 3. Decay scheme of the Be configuration (1s2s2p3s) with four open shells. (The scale is arbitrary; the Be$^+$ states shown are the final ionic states.)

probably to the $1s^2 3s(^2S)$ final state with the $3s$ spectator electron unchanged (Krause et al., 1971).

The transition energies have been estimated using the quantum defect method (Rudd and Macek, 1974), with $\delta = 0$ for $3d$ and $4d$ and $\delta \simeq 0.5$ for $3s$ and $3p$ orbitals (see Figure 2). Furthermore, Hartree–Fock calculations neglecting configuration interaction have been performed for the $1s2s2p^2 \rightarrow 1s^2 2p$ and $1s2s2p^2 \rightarrow 1s^2 2s$ transitions, where the $1s2s2p$ parent configuration has been used for energy calibration of the results.

All these delayed quintet transitions, in particular, the group between 100.5 and 103.0 eV, can very well contribute to the observed spectrum. However, more accurate theoretical values for the transition energies, transition probabilities, and branching ratios of such four-electron transitions are needed for a clear-cut interpretation of the spectrum. Together with more advanced experimental techniques, such as ion–electron coincidence measurements for projecting out a definite final charge state, accurate calculations could help to classify a lot of yet unidentified lines.

Acknowledgments·

We are indebted to Professor J. Macek for many stimulating discussions. The authors wish to thank M. Ahmed and L. Lipsky for communication of their unpublished data. This work was supported by the Sonderforschungsbereich 161 der Deutschen Forschungsgemeinschaft.

References

Balashov, V. V., Grishanova, S. J., Kruglowa, H. M., and Senashenko, X. (1970). *Opt. Spectr.*, **28**, 466.
Bruch, R., Paul, G., and Andrä, J. (1975a). *J. Phys. B.*, **8**, L253.
Bruch, R., Paul, G., Andrä, J., and Fricke, B. (1975b). *Phys. Lett.*, **53A**, 293.
Bruch, R., Paul, G., Andrä, J., and Lipsky, L. (1975c). *Phys. Rev.*, **A12**, 1808.
Chan, Y. M. C., and Stewart, A. L. (1967). *Proc. Phys. Soc.*, **90**, 619.
Das, T. P. (1973). *Relativistic Quantum Mechanics of Electrons*, 153 and 190.
Goldsmith, S. (1969). *J. Phys. B.*, **2**, 1075.
Holøien, E., and Geltman, S. (1967). *Phys. Rev.*, **153**, 81.
Hontzeas, S., Martinson, I., Enman, P., and Buchta, R. (1973). *Nucl. Instr. Meth.*, **110**, 51.
Junker, B. R., and Bardsley, J. N. (1973). *Phys. Rev.*, **A8**, 1345.
Krause, M. O., Carlson, T. A., and Moddeman, W. E. (1971). *J. Phys. (Paris) Colloq.*, **C.4**, 139.
Lipsky, L., and Ahmed, M. (1974). Private communication.
Martinson, I. (1974). *Physica Scripta*, **9**, 281.
Perrott, R. H., and Stewart, A. L. (1968a). *J. Phys. B.*, Ser. 2, **1**, 381.
Perrott, R. H., and Stewart, A. L. (1968b), *J. Phys. B.*, Ser. 2, **1**, 1226.
Risley, J. S., Edwards, A. K., and Geballe, R. (1974). *Phys. Rev.*, **A9**, 115.
Rudd, M. E., and Macek, J. H. (1974). *Case Studies in Atomic Physics*, **III**, 49.
Schulz, G. J. (1973). *Rev. Mod. Phys.*, **45**, 403.
Sellin, I. A. (1973). *Nucl. Instr. Meth.*, **110**, 477, and other references quoted there.
Sevier, K. D. (1972). *Low Energy Electron Spectrometry*, 145.

51

The Perturbed Series, $2p^5(^2P_{3/2,\,1/2})ns$, of Ions along the Ne I Isoelectronic Sequence

K. T. Lu and M. W. D. Mansfield

Eigenchannel quantum defect theory describes perturbed Rydberg series of an atom or ion in terms of a set of parameters. We calculate these parameters by means of a Hartree–Fock program developed by Froese-Fischer for ions. The perturbed spectra of ions along an isoelectronic sequence can be studied by analyzing these parameters as functions of effective nuclear charge z.

An initial attempt is being made here to study the Ne-like ion spectra isoelectronically by the eigenchannel quantum defect theory (Fano, 1970; Lu, 1971).

Photoabsorption of a neutral Ne atom from its ground state leads to $2p^5(^2P)ns$ and $2p^5(^2P)nd$, $J = 1$, five series in the final states. The configuration interaction between ns series and nd series is very weak and can be neglected entirely (Starace, 1973). The same situation is expected for Ne-like ions.

We thus analyze only the two-channel series, $2p^5(^2P_{3/2})ns$ and $2p^5(^2P_{1/2})ns'$, $J = 1$. For each ionized atom with effective nuclear charge z, $z = Z - (N - 1)$, we assign (Lu and Fano, 1970) two alternative effective quantum numbers v_1 and v_2 to each energy level E_n of these two series by the equations

$$E_n = I_1 - \frac{z^2}{2v_1^2} = I_2 - \frac{z^2}{2v_2^2} \quad \text{(in a.u.)} \tag{1a}$$

K. T. Lu • Physics Department, Imperial College, London, SW7, England. M. W. D. Mansfield • Physikalisches Institut der Universität Bonn, Germany.

or

$$\Delta I = I_2 - I_1 = \frac{z^2}{2}\left(\frac{1}{v_2^2} - \frac{1}{v_1^2}\right) \tag{1b}$$

where I_1 and I_2 are the ionization potentials corresponding to the $2p^5(^2P_{3/2,1/2})$ levels of the ion. We also assign to each level E_n a single quantum defect $\mu = n - v_1$, where n is an integer. Theory (Fano, 1970; Lu, 1971) requires all the levels to lie on a smooth curve of the form

$$F(\mu, v_2) = \sin^2\theta \sin\Pi(\mu - \mu_1)\sin\Pi(v_2 + \mu_2)$$
$$+ \cos^2\theta \sin\Pi(\mu - \mu_2)\sin\Pi(v_2 + \mu_1) = 0 \tag{2}$$

Here μ_1 and μ_2 are two eigenvalues belonging to the eigenstate α of a reaction matrix (Seaton, 1966), and θ is the angle of the orthogonal transformation matrix $U_{i\alpha}$ which diagonalizes this reaction matrix. $U_{i\alpha}$ transforms from eigenstate α of the close-coupled electron–ion compound to eigenstate i of the electron–ion scattering system. These three parameters, μ_α, $\alpha = 1, 2$, and θ, together with the doublet P core splitting ΔI (Edlén, 1964), completely determine the character of the two mutually perturbing Rydberg series of a given ion.

To analyze this two-channel problem along the NeI isoelectronic sequence, we study the z dependence of the parameters, θ, and the difference of the eigenvalues $\delta_\alpha = \mu_1 - \mu_2$. The scattering eigenstate i is defined to be jj coupled. The eigenstates α are determined by interaction in the small region of r close to the nucleus, in which electrostatic interactions are stronger than spin–orbit interactions. This is true for the lower ionized atoms of the sequence. We expect the eigenstates α to be LS coupled. The angle defining LS-jj orthogonal transformation is $\theta_{LS} = \cos^{-1}\sqrt{\frac{2}{3}} = 35.263°$. However, the situation changes as the spin–orbit interaction of the $2p^5$ core increases its strength relative to the electrostatic interaction between the optical electron and the core for the more highly ionized atoms of the sequence. Thus the character of the eigenstates α are intermediate and eventually tend toward being jj coupled. By allowing a spin–orbit interaction in the secular equations, we obtain δ_α and θ as

$$\delta_\alpha = \frac{2v_1^3}{z^2}(1 - \tfrac{1}{2}x + \tfrac{9}{16}x^2)^{1/2}G_1/3 \tag{3}$$

and

$$\theta = \theta_{LS} + \tfrac{1}{2}\tan^{-1}\left(\frac{x/\sqrt{2}}{1 - x/4}\right) \tag{4}$$

where x is the ratio of the spin–orbit energy ζ_p to the Slater exchange integral $G_1/3$. A computer program of Froese-Fischer (1969) was used which calculates

Hartree–Fock wave functions for an isoelectronic sequence and uses these to form Slater integrals G_1, and the spin–orbit parameter ζ_p and therefore the parameters δ_α and θ can be calculated.

In order to study the interactions between these two-channel series, we rewrite Eq. (2), following Fano (1970), as

$$\tan \Pi(\mu - \bar{\mu}) - \tan \Pi(\Delta_p - \bar{\mu}) = -\frac{\Gamma}{\tan \Pi(v_2 + \bar{\mu}) - \tan \Pi(\Delta_p - \bar{\mu})} \quad (5)$$

which gives μ explicitly as a function of v_2. In (5) we have introduced the notations

$$\bar{\mu} = \tfrac{1}{2}(\mu_1 + \mu_2) \quad (6a)$$

$$\tan \Pi(\Delta_p - \bar{\mu}) = \tan \tfrac{1}{2}\Pi\delta \cos 2\theta \quad (6b)$$

$$\Gamma = \tan^2 \tfrac{1}{2}\Pi\delta \sin^2 2\theta \quad (6c)$$

where Δ_p indicates a "plateau value" of μ, Γ indicates an "interaction strength," as discussed below, and $\bar{\mu}$ is the average eigenvalue which can be calculated by the Hartree–Fock program. Since each term in (5) involves only trigonometric functions, both μ and v_2 are periodic and have period 1. The curve represented by (5) can be plotted on a unit square, i.e., μ versus v_2 (modulus 1), as shown in Figure 1 for selected ions Ne I, Al IV, Cl VIII, and

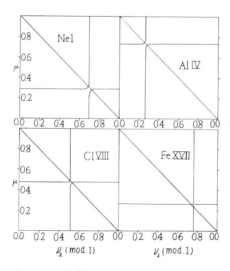

Figure 1. Solid curve is the μ versus v_2 (mod. 1) plot of Eq. (5), and the diagonal line is $\mu + v_2 = 1$. Examples are shown for Ne I, Al IV, Cl VIII, and Fe XVII.

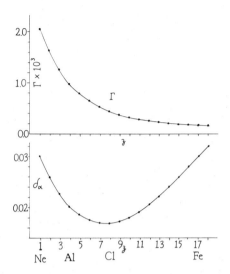

Figure 2. Calculated values of Γ and δ_α as a function of z.

Fe XVII, with the parameters calculated from Eqs. (3), (6a), and (6c). Figure 2 shows the calculated values of δ_α and Γ from equations (3) and (6c), respectively.

Equation (5) shows that μ would vary by a sharp step in the limit $\Gamma \equiv \tan^2 \frac{1}{2}\Pi\delta \sin^2 2\theta \to 0$, which corresponds to "vanishing interaction" between the two channels. For weak interaction, μ would remain approximately equal to its plateau value Δ_p for all values of $v_2 \sim \Delta_p - 2\bar{\mu}$. At this special value of v_2, μ would rise stepwise by 1. This vanishing interaction limit could occur for either of two reasons: (a) $\delta_\alpha \sim 0$, implying that the electrostatic interaction is zero between the p^5 ion core and the optical s electron, or (b) $\theta \sim 0$, implying that the eigenchannels of α and i coincide. Reason (b) applies to the more highly ionized atoms in the sequence, as seen from the values of Γ in Figure 2. An example is given for Fe XVII in Figure 1. The two channels, $2p^5(^2P_{3/2})ns$ and $2p^5(^2P_{1/2})ns'$, are almost decoupled. Under this circumstance configuration interaction can be neglected. Thus single configuration treatment with intermediate coupling gives oscillator strength of Fe XVII ion resonance lines that agree well with observed intensity ratio in the corona (Kastener et al., 1967). On the other hand, configuration interaction must be taken into account in the calculation for lower ionized atoms in the sequence.

Note that μ_1 and μ_2 coincide (modulo 1) with the values of μ at the interactions of the curve represented by Eq. (5) with the plot's diagonal

$\mu + \nu_2 = 1$ in Figure 1. The values of μ_1 and μ_2 are thus determined conveniently from an experimental plot of μ against ν_2, as suggested elsewhere (Lu and Fano, 1970).

In conclusion, the eigenchannel quantum defect theory describes the perturbed Rydberg series of an atom or ion in terms of a set of parameters. The study of these parameters as functions of effective nuclear charge z makes it possible to analyze the spectra of the ion isoelectronically.

References

Edlén, B. (1964). In *Encyclopedia of Physics*, **27**, Ed. S. Flugge. Berlin: Springer-Verlag.
Fano, U. (1970). *Phys. Rev.*, A**2**, 353.
Froese-Fischer, C. (1969). *Comp. Phys. Commun.*, **1**, 151.
Kastener, S. O., Omidvar, K. and Underwood, J. H. (1967). *Astrophys. J.*, **148**.
Lu, K. T. (1971). *Phys. Rev.*, A**4**, 579.
Lu, K. T. and Fano, U. (1970). *Phys. Rev.*, A**2**, 81.
Seaton, M. J. (1966). *Proc. Phys. Soc. Lond.*, **88**, 815.
Starace, A. F. (1973). *J. Phys. B.*, **6**, 76.

52
Effect of LS Term Dependence on Some Rare Gas Transition Probabilities

P. L. ALTICK AND JACK R. WOODYARD, SR.

> *A comparison is made between the standard semiempirical method of treating rare gas spectra and an ab initio frozen-core Hartree–Fock calculation which uses an LS coupled basis set. For the 5p-4s J = 1 to J = 1 transitions in argon some large discrepancies in the transition probabilities are shown to exist between the two methods due to the fact that in the semiempirical approach, only one radial dipole matrix element is used, while in the other calculation there are different radial dipole matrix elements for each LS term.*

The visible and near-visible spectra of the rare gases have been studied since the early days of spectroscopy, both experimentally and theoretically. More recently these spectra have played important roles in plasma diagnostics. Yet even today the overall availability of reliable data on transition probabilities for these spectra is poor. Part of the difficulty on the theoretical side is the proper characterization of the levels. It was realized very early that neither pure LS coupling nor pure jj coupling were appropriate. For example, the two $3p^54s$ levels in Ar are separated by 0.0076 a.u. The splitting of the $3p^5$ core may be taken as measure of the spin–orbit interaction and is 0.0042 a.u. The difference in Hartree–Fock eigenvalues for $3p^54s\,^1P$ and ^3P may be used as a measure of the Coulomb interaction and is 0.0058 a.u., so, in this instance, the two interactions have comparable

---

P. L. ALTICK and JACK R. WOODYARD, SR. • Physics Department, University of Nevada, Reno, Nevada 89507, U.S.A.

magnitudes. For higher members of the Rydberg series, the jK coupling, first suggested by Racah (1942), becomes quite good, but for lower levels, the coupling is intermediate and must be found by a diagonalization of the energy matrix.

The purpose of this paper is to point out that, in a theoretical calculation, even obtaining the proper coupling does not, by itself, ensure accurate transition probabilities. This will be illustrated by a comparison of the standard semiempirical approach found, for example, in Condon and Shortley (1970) with some recent *ab initio* calculations (Woodyard and Altick, 1974).

The theoretical foundation for the semiempirical approach is the central field approximation, i.e., all the orbitals are to be computed in one suitable central potential, and then the residual Coulomb interaction and the spin–orbit interaction are treated by perturbation theory. The first-order formulas for energy levels contain a number of parameters, Slater integrals, and spin–orbit integrals, which are adjusted to give the best overall fit to the experimental spectrum. This procedure has been in use for about 40 years (Shortley, 1935) and, until about a year ago, was the only theoretical method applied to the rare gas spectra.

Not only energy levels, but other atomic parameters also can be generated semiempirically. The Landé g factors are essentially the expectation values of the z component of total spin and follow immediately from the zero-order wave functions in intermediate coupling. It should be especially noted that the Landé factors follow just from the angular momentum coupling; there are no radial integrals involved. The only property of the radial functions which enters is their orthonormality. To compute transition probabilities, however, a radial dipole matrix element arises which must be evaluated in some way or another. The usual procedure is to find this element by using the method of Bates and Damgaard (1949). Typically, one (Murphy, 1968) or two (Johnston, 1967) different values of dipole matrix elements are used to normalize the transition probabilities between two different configurations.

The *ab initio* calculations, however, are performed by a frozen-core Hartree–Fock method (FCHF). This can best be explained by looking at a specific case so the method will be described for argon. The orbitals $1s$, $2s$, $2p$, $3s$, $3p$ are computed once in the configuration $1s^2 2s^2 2p^6 3s^2 3p^5$, and then held fixed. Valence orbitals ns, np, nd are then computed in LS coupling using the fixed core. The result is that there are no matrix elements of the Coulomb Hamiltonian between configurations involving different members of the same Rydberg series, i.e.,

$$\langle 3p^5 nl^{[S]}L|H|3p^5 n'l^{[S]}L\rangle = 0 \tag{1}$$

where $[S] = 2S + 1$. This basis is thus partially diagonalized and there are a number of radial wave functions for each configuration, unlike the semiempirical method. For example, in the $3p^5np$ configuration there are different radial functions for np^3S, np^3P, np^1P, np^3D. These various LS states are used to form an energy matrix which includes the spin–orbit interaction, and the eigenvalues of the matrix yield the intermediate coupling levels. Landé g factors follow as in the semiempirical case. To compute transition probabilities, however, there are several different dipole matrix elements associated with the different FCHF LS terms, and their differences can have important effects, as will be shown below.

As an example, the $5p$–$4s$, $J = 1$ to $J = 1$ transitions in argon will be taken. The energy discrepancies with experiment in the semiempirical method are extremely small, while they average about 10% in the *ab initio* calculation. Both approaches yield good values for the Landé g factors, as shown in Table 1, and this is interpreted to mean that both describe the intermediate coupling correctly. The transition probabilities are shown in Table 2, and one notes large discrepancies in the transitions involving the $3p_{10}$ level and less severe ones involving other levels. Investigating the matter, it is found that the radial dipole matrix elements among the various LS terms are quite different. They are displayed in Table 3, and it can be seen that the element involving $5p^3S$ even has a different sign than the others! The last four columns in Table 2 list numbers proportional to the line strengths, i.e., the transition energy has been factored out. The $3p_4 \to 1s_2$ transition is arbitrarily given the value 1, as only the ratios are important. The column labelled "AF" results from a computation using the FCHF eigenvectors, but only angular factors for the dipole matrix elements, i.e.,

Table 1. Landé g Factors for the $3p^55p$ and $3p^54s$ Levels in Ar

The column labeled HF gives the *ab initio* results. The column labeled SE gives the semiempirical results of Johnston (1967). The observed data (OBS) are also taken from Johnston.

Level (Paschen)	HF	SE	OBS
$1s_4$	1.390	1.404	1.399
$1s_2$	1.110	1.101	1.101
$3p_{10}$	1.895	1.90	1.892
$3p_7$	1.009	1.01	1.007
$3p_4$	0.647	0.61	0.645
$3p_2$	1.450	1.45	1.456

Table 2. Transition Probabilities and Line Strengths for the 5p–4s $J = 1$ to $J = 1$ Transitions in Argon

The labels are the same as in Table 1 except for AF, which signifies a calculation using only angular factors for the dipole matrix elements.

Transition	Aij (10^7 sec^{-1})			Line strength (arb. units)			
	HF	SE	OBS*	HF	SE	AF	OBS*
$3p_{10} \to 1s_4$	0.00001	0.039	0.0015	0.0002	0.47	0.52	0.048
$\to 1s_2$	0.00723	0.00028	0.011	0.215	0.004	0.0022	0.44
$3p_7 \to 1s_4$	0.111	0.17	0.084	2.41	1.9	2.03	2.5
$\to 1s_2$	0.0141	0.027	0.0102	0.386	0.39	0.40	0.39
$3p_4 \to 1s_4$	0.00390	0.015	0.00271	0.0713	0.15	0.11	0.070
$\to 1s_2$	0.0438	0.083	0.0312	1.0	1.0	1.0	1.0
$3p_2 \to 1s_4$	0.00579	0.024	0.00438	0.105	0.24	0.269	0.113
$\to 1s_2$	0.0620	0.10	0.0387	1.40	1.2	1.56	1.23

*Wiese et al., 1969.

the differences in radial dipole elements, are ignored. It is seen that the semi-empirical calculation agrees nicely with AF, as it should if the coupling is the same in both cases, but both differ significantly from the FCHF calculation and observation for the $2p_{10}$ transitions, and the FCHF calculation is much better at reproducing the trend of the experimental results.

The 4p–3s transition in Ne was also analyzed in this way, and similar discrepancies between using the full dipole elements and just the angular factors were found. Unfortunately, there is neither experimental nor semi-empirical data to compare with.

It seems reasonable to conclude that standard semiempirical calculations of transition probabilities should be treated with caution, especially

Table 3. Radial Dipole Matrix Elements (in a.u.) among Various LS Terms in Argon

The configurations are $3p^55p$ and $3p^54s$.

	$3p^54s$	
$3p^55p$	$3p$	$1p$
^3S	0.0492	—
^3P	-0.575	—
^1P	—	-0.403
^3D	-0.346	—

when the radial dipole element is small. The above example shows that no single value of the radial dipole element can reproduce the experimental results for the transition probabilities and that additional flexibility, such as is given by LS-term dependence, is needed.

The effects of LS-term dependence on the calculation of theoretical quantities has also been pointed out by Dill et al. (1973), who were studying angular distribution asymmetry parameters, and by Hansen (1973), who was concerned with energy levels in atoms and ions.

During the last few years, Fano and co-workers have developed a more general semiempirical method based on quantum defect theory (e.g., Lu and Fano, 1970) and have now developed it to the point where completely *ab initio* calculations are possible (Fano and Lee, 1973; Lee, 1974). This formulation allows for different radial dipole matrix elements for different channels and is thus not subject to the problems of the older method.

References

Bates, D. R., and Damgaard, A. (1949). *Phil. Trans. Roy. Soc.*, A**242**, 101.
Condon, E. U., and Shortley, G. H. (1970). *The Theory of Atomic Spectra*. London: Cambridge Univ. Press.
Dill, D., Manson, S. T., and Starace, A. F. (1973). *Bull. Am. Phys. Soc.*, **18**, 1514.
Fano, U., and Lee, C. M. (1973). *Phys. Rev. Lett.*, **31**, 1573.
Hansen, J. E. (1973). *J. Phys. B.*, **6**, 1751.
Johnston, P. D. (1967). *Proc. Phys. Soc. Lond.*, **92**, 896.
Lee, C. M. (1974). *Phys. Rev.*, **A10**, 584.
Lu, K. T., and Fano, U. (1970). *Phys. Rev.*, A**2**, 81.
Murphy, P. W. (1968). *J. Opt. Soc. Am.*, **58**, 1200.
Racah, G. (1942). *Phys. Rev.*, **61**, 537.
Shortley, G. H. (1935). *Phys. Rev.*, **47**, 295.
Wiese, W. L., Smith, M. W., and Miles, B. M. (1969). *Atomic Transition Probabilities*, NSRDS-NBS 22 V II. Washington, D.C.: U.S. Govt Printing Office.
Woodyard, J. R. Sr., and Altick, P. L. (1974). *J. Phys. B.*, **7**, 2298.

53
Resonances and Cusps in Electron Impact on Atoms

D. E. GOLDEN

If a resonance exists below an excitation threshold, the analytic formulation of the resonance shape can be used to obtain the shape of the cusp at the excitation threshold by extrapolation. The detection of cusp effects allows the accurate calibration of electron energy scales and thus the accurate determination of resonance energies, and it also gives information about threshold laws for excitation. Measurements of cusps and resonances can be used to help place absolute scales on scattering cross section measurements without the absolute measurement of pressure or beam intensity. Furthermore, while it is well known that spin-polarization effects are large when the scattering amplitude is rapidly varying, it has been explicitly shown that such will be the case in the vicinity of threshold cusps in electron scattering.

A long time ago Wigner (1948) showed that all scattering cross sections exhibit cusp behavior at thresholds of new scattering channels, and then he went on to add a footnote to the effect that probably these effects were not important.

It is true that cusp effects observed thus far are for the most part small. It is only because of the advances in the technology of high-energy resolution electron scattering measurements that these effects can be observed.

The detection of cusp effects allows the accurate calibration of electron energy scales and gives information about threshold laws for excitation. Measurements of cusps and resonances can be used to help place absolute scales on scattering cross-section measurements without the absolute measurement of pressure. The absolute pressure measurement is the most difficult measurement necessary to obtain absolute cross sections for

---

D. E. GOLDEN • Department of Physics and Astronomy, University of Oklahoma, Norman, Oklahoma 73069, U.S.A.

experiments which detect the scattered electrons. Furthermore, while it is well known that spin-polarization effects are large when the scattering amplitude is rapidly varying (Franzen and Gupta, 1965), Burke and Mitchell (1974) have recently shown explicitly that such will be the case in the vicinity of threshold cusps in electron scattering from one-electron atoms.

The elastic cross section for electrons on cesium as calculated by Burke and Mitchell (1974), with and without spin-orbit coupling, is shown in Figure 1, and the resulting spin polarization in Figure 2. The results indicate that the spin polarization rises to a maximum between the thresholds and dies away on either side.

There is some inconclusive evidence for the existence of cusp behavior in the electron transmission experiments in helium of Kuyatt *et al.* (1965) and Golden and Zecca (1970). However, it is only recently that Sanche and Schulz (1972) have classified the structure at the 2^3S threshold in their transmission experiment in helium as being due to the combination of a

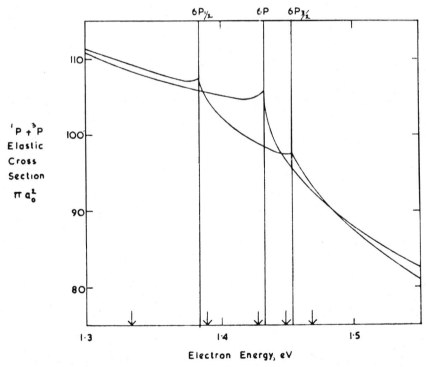

Figure 1. The ^1P and ^3P contribution to the elastic electron–cesium atom cross section. The energies where spin-polarization calculations have been carried out are indicated by arrows. (After Burke and Mitchell, 1974.)

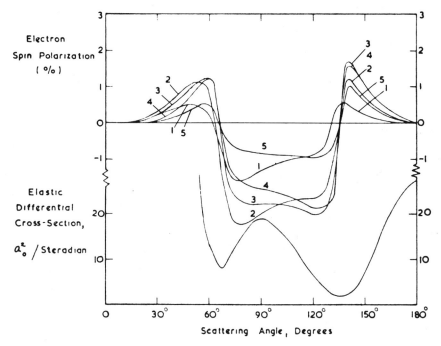

Figure 2. The electron spin polarization for unpolarized electrons scattered from unpolarized cesium atoms near the $6P_{1/2}$ and $6P_{3/2}$ thresholds. Curve 1, 1.332 eV, curve 2, 1.389 eV, curve 3, 1.427 eV, curve 4, 1.448 eV, curve 5, 1.468 eV. The elastic differential cross section at 1.427 eV is also shown.

"Wigner cusp" in the elastic cross section at the threshold and the rise of the inelastic cross section above threshold. At about the same time, Andrick et al. (1972) succeeded in making the first unambiguous measurement of a cusp in differential electron–atom scattering. Figure 3 shows the differential elastic electron–sodium scattering cross section at various angles in the vicinity of the 3^2P threshold obtained by Andrick et al. (1972). The shape and magnitude of the structure are in general agreement with the independent close-coupling calculation of Moores and Norcross (1972), who find a 20% effect in the total elastic cross section. Figure 4 shows the total elastic cross section given by Moores and Norcross (1972) compared with the recent absolute measurements of Rubin (1972) and the older results of Perel et al. (1962) and Brode (1929), normalized to the calculation at 5 eV. As can be readily seen from the figure, the error bars on the measurements are too large to show the cusp in the total cross section given by the theory.

In general, the differential elastic cross section may either increase or decrease both above and below the opening of a new scattering channel.

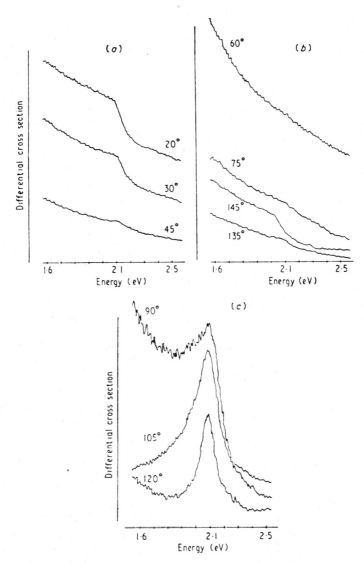

Figure 3. Differential cross section of elastic e^-–Na scattering versus energy at various scattering angles. The scales of (b) and (c) are enlarged against the scale of (a) by a factor 3.7 and 25 respectively. The curves are calibrated against each other at 1.6 eV using theoretical data of Moores and Norcross (1971).

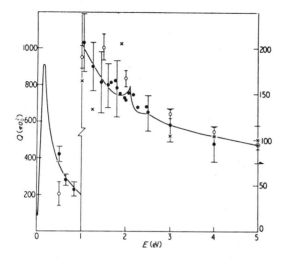

Figure 4. Total cross section for scattering of electrons by sodium. —— present results; ×, experiment of Brode (1929), divided by two; ○, experiment of Perel *et al.* (1962), normalized to theory at 5 eV; ● experiment of Rubin (1972).

See, for example, Baz *et al.* (1962). Thus, the joining of these possibilities at threshold leads to four possible shapes, which we refer to as positive or negative cusps or positive or negative steps. The total elastic cross section many only decrease above the opening of a new scattering channel, although it may either increase or decrease below the threshold. Therefore, only a positive cusp or a negative step is possible in this case. One many think of the decrease in the total scattering cross section above threshold as due to the increase in the inelastic cross section. Thus the form of the drop in the elastic cross section is directly related to the form of the rise in the inelastic cross section, although the two effects are not necessarily equal.

It has recently been shown by Cvejanovič *et al.* (1974) that near an inelastic threshold of energy E_{th}, the differential scattering cross section is given by

$$\frac{d\sigma(\theta)}{d\Omega} = \frac{d\sigma_{\text{th}}(\theta)}{d\Omega} - bk^{-1}|k_2|\,\text{Re}\,(h) \qquad E < E_{\text{th}}$$
$$\frac{d\sigma(\theta)}{d\Omega} = \frac{d\sigma_{\text{th}}(\theta)}{d\Omega} + bk^{-1}k_2\,\text{Im}\,(h) \qquad E > E_{\text{th}} \tag{1}$$

where
$$h = f_{th}(\theta)\exp(-2i\delta_{0_{th}})$$
$$\hbar k = (2mE)^{1/2}$$
$$\hbar k = [2m(E_{th} - E)]^{1/2} \quad (2)$$
$$\hbar k_2 = [2m(E - E_{th})]^{1/2}$$

and the scattering amplitude is given by

$$f(\theta) = f_{th}(\theta) + \tfrac{1}{2}bk^{-1}k_2 \exp(2i\delta_{0_{th}}) \quad (3)$$

Cvejanovič et al. (1974) have used the above equations to fit their differential elastic electron–helium scattering cross sections for 90° scattering near the 2^3S threshold. At 90° the P-wave phase shift is zero, the D-wave phase shift was taken to be 0.05 (see Knowles and McDowell, 1973), and all higher phase shifts were neglected. Thus b, δ_0, and the position of the cusp were used as parameters in the fitting procedure. The value of b leads immediately to the threshold dependence of the excitation cross section for the 2^3S level from

$$\sigma(1^1S \to 2^3S) = 2\pi bk^{-2}k_2 \quad (4)$$

while δ_0 gives the S-wave contribution to the elastic cross section. Figure 5 shows the differential elastic scattering cross section for helium at 90° for

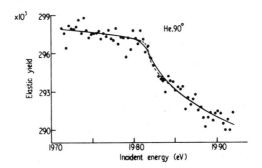

Figure 5. Differential cross section for elastic electron–helium scattering at 90°, in the neighborhood of the $2(^3S)$ threshold at 19.818 eV. The full curve is the best fit obtained with Eqs. (10) and (11). The broken curve is this fit before convolution with the apparatus function (which has a FWHM equal to 18 meV).

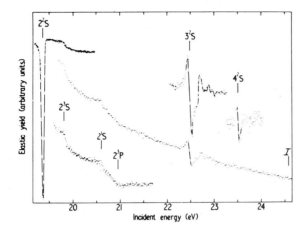

Figure 6. Differential cross section for elastic electron–helium scattering at 90°, from 19.1 to 24.7 eV. The different spectra shown in the figure have the same energy scale but different yield scales and may have different sloping backgrounds.

an energy resolution of 18 meV. The points are the experimental values measured by Cvejanovič et al. (1974), while the line is the best fit to the data using the three-parameter theory discussed above. Once this fit has been made, one knows where 19.818 eV is on the raw energy scale, and the energy scale is calibrated. Figure 6 shows the elastic differential spectrum of helium run at an energy resolution of 38 meV in order to obtain better statistics. The energy scale has been placed by the 2^3S cusp calibration and yields 19.37 eV for the position of the lowest ^2S resonance.

In a recent electron transmission experiment in helium, Golden et al. (1974) have obtained the analytic structure of the scattering matrix in the vicinity of the 2^3S threshold from the analytic structure of the ^2S resonance below the 2^3S threshold using effective range theory as used, for example, by Burke (1965). The matrix element corresponding to elastic scattering near a resonance is taken as

$$S = e^{i\delta_0}\frac{(E - E_r - i\Gamma/2)}{(E - E_r + i\Gamma/2)}e^{i\delta_0} \qquad (5)$$

where δ_0 is the nonresonant phase shift and Γ is the resonance width with Wigner form $\Gamma = k\gamma^2\kappa$. The scattering matrix is continued across the threshold according to the prescription $i\kappa \to k_2$. Since $\gamma^2 k$ is effectively constant in the vicinity of the threshold, we may take $\Gamma = \bar{\Gamma}\kappa$, where $\bar{\Gamma}$ is a constant

Then the elastic scattering cross section is given by

$$\sigma \propto \left[\sin^2 \delta_0 + \frac{\bar{\Gamma}^2 \kappa^2/4}{(E - E_r)^2 + \bar{\Gamma}^2 \kappa^2/4} \right.$$

$$\left. - \frac{\bar{\Gamma}\kappa}{2} \left(\frac{(E - E_r) \sin 2\delta_0 + \bar{\Gamma}\kappa \sin^2 \delta_0}{(E - E_r)^2 + \bar{\Gamma}^2 \kappa^2/4} \right) \right] \quad E < E_{\text{th}} \quad (6)$$

$$\sigma \propto \left[\sin^2 \delta_0 + \frac{\bar{\Gamma}^2 k_2^2/4}{(E - E_r + \bar{\Gamma}k_2/2)^2} - \frac{\bar{\Gamma}^2 k_2^2 \sin^2 \delta_0}{(E - E_r + \bar{\Gamma}k_2/2)^2} \right] \quad E_{\text{th}} < E$$

The cusp arises due to the third term in Eq. (6), which predicts a decrease in the elastic scattering cross section above threshold and an increase or decrease below threshold depending upon the sign of $\sin 2\delta_0$. The sine of $2\delta_0$ is negative (see Ehrhardt et al., 1968), so that the Wigner cusp here corresponds to a rapid drop in the elastic scattering cross section at the 2^3S threshold. The derivative of the transmitted current then shows a peak centered approximately at the 2^3S threshold. Golden et al. (1974) have used Eq. (6) to fit experimentally differentiated electron transmission data between 19.2 and 20 eV. In this fitting procedure the parameters E_r, δ_0, and Γ were varied in order to obtain agreement between calculated and measured differentiated transmitted currents. Thus the energy scale of the experiment

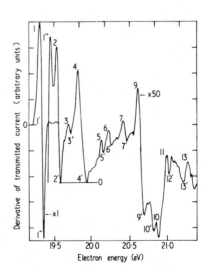

Figure 7. Measured energy differentiated output signals from 19.2 to 20.2 eV.

Table 1. Positions and Relative Peak Heights for the 2^2S State of He$^-$ and the Threshold of 2^3S State of He

	P_0(eV)	P_1(eV)	P_2(eV)	$\dfrac{\Delta I(P_1)}{\Delta I(P_0)}$	$\dfrac{\Delta I(P_2)}{\Delta I(P_0)}$	Γ(eV)	δ_0(deg)	E_r(eV)
Expt.	19.34	19.43	19.825	0.014	0.011			
Calc.	19.34	19.42	19.825	0.015	0.012	0.013	102	19.35

was established. Figure 7 shows the measured differentiated transmitted electron spectrum of He near the vicinity of the $n = 2$ threshold. The feature 1–1'–1''–1''' is due to the lowest ^2S resonance, now placed at 19.35 eV, and the cusp at the 2^3S threshold is labeled 4. The peak labeled 1''' is due to the asymmetry in the 2^2S resonance. It is this resonance which gives rise to the cusp at the 2^3S threshold. Table 1 shows a comparison between the experimentally measured positions and peak height ratios of features 1 and 4 compared with calculations of these features using Eq. (6). As one can see, the agreement between the measured and calculated positions and peak height ratios is excellent. The values of Γ, δ_0, and E_r determined from this fitting procedure are listed in the last three columns of Table 1. It should be pointed out again that the S-wave contribution to the elastic scattering cross section is determined solely from the shape of the resonance and cusp, independently of any cross-section measurement.

The separation between 1''' and 2 is only 80 meV, and an energy resolution of the order of 25 meV was required in order to resolve these two features. The position of 2–2' is in excellent agreement with the previous results of Kuyatt et al. (1965), Gibson and Dolder (1969), and Golden and Zecca (1970), but in disagreement with the more recent result of Sanche and Schulz (1972). By studying the results of Sanche and Schulz (1972), we have obtained an estimate of 63 meV for their instrumental energy resolution, which we would argue is insufficient to separate the 1''' peak from 2. In addition, the experiments of Sanche and Schulz (1972) have been done in an axial magnetic field, which has been shown by Golden (1973) to decrease the sensitivity for the detection of elastic scattering. There is a similar cusp at the 2^1S threshold (9). There are additional structures between He$^-$ 2^2S and He 2^1S, in agreement with the previous results of Golden and Zecca (1970), which must still be classified as UFOs (unidentified fysical objects) because no calculation has yet accounted for them. However, I would point out that these UFOs are extremely narrow features (less than 100 μeV*), and similar features have been observed in electron excitation of some of the more highly excited

Table 2. He $^-2^2S$

Names	Year	Position (eV)	Width (meV)	δ_0	δ_1	δ_2
Schulz	1963	19.3	—	—	—	—
Simpson and Fano	1963	—	∼10	—	—	—
Golden and Bandel	1965	19.29	—	—	—	—
Kuyatt, Simpson, and Mielczarek	1965	19.31	—	—	—	—
Andrick and Ehrhardt	1966	—	15–20	100	25	4
Ehrhardt, Langhans, and Linder	1968	—	12	109	20	3
Gibson and Dolder	1969	—	8	111	17	3
Golden and Zecca	1971	19.30	8	—	—	—
Sanche and Schulz	1972	19.35	—	—	—	—
Cvejanovic, Comer, and Read	1974	19.37	9	106	—	—
Golden, Schowengerdt, and Macek	1974	19.35	13	102	—	—
Burke, Cooper, and Ormonde	1966	19.33	39	—	—	—
Young	1968	19.67	—	—	—	—
Perkins	1969	19.69	—	—	—	—
Weiss and Krauss	1970	19.37	—	—	—	—
Temkin, Bhatia, and Bardsley	1972	19.36	14	—	—	—
Sinfailam and Nesbet	1972	19.4	15	129	9	1
Ormonde and Golden	1973	19.38	11.5	103	—	—

states of He (see Heddle, 1974). Table 2 shows a summary of the results regarding the quantitive information concerning the 2^2S He$^-$ state in historical fashion. The latest calibrations of the position of this feature using the 2^3S cusp by Sanche and Schulz (1972), Cvejanovič et al. (1974), and Golden et al. (1974) move the position of this feature up by about 50 meV so that it is now in excellent agreement with the most recent independent calculations by Temkin et al. (1972), Sinfailam and Nesbet (1972), and Ormonde and Golden (1973). The width measurements agree very well with the calculations. Of course, Professor Fano's rough estimate of this width 11 years ago (Simpson and Fano, 1963) is still holding up. The other information obtained from the measurements is the S-wave phase shift at the resonance energy, which is also in good agreement with the independent measurements of this quantity by Andrick and Ehrhardt (1966), Ehrhardt et al. (1968), and Gibson and Dolder (1969).

*The width estimates given by Golden et al. (1974) are too large. Revised width estimates for these UFOs show that they are sufficiently narrow so that one should consider spin–orbit effects leading to the formation and decay of quartet negative ion states in subsequent resonance calculations.

References

Andrick, D., and Ehrhardt, H. (1966). *Z. Physik*, **192**, 99.
Andrick, D., Eyb, M., and Hofmann, M. (1972). *J. Phys. B.*, **5L**, 15.
Baz, A. I., Puzikov, L. D., and Smordinskii, Ya. A. (1962). *Sov. Phys. JETP*, **15**, 865.
Brode, R. B. (1929). *Phys. Rev.*, **34**, 673.
Burke, P. G. (1965). *Adv. Phys.*, **14**, 521.
Burke, P. G., and Mitchell, J. F. B. (1974). *J. Phys. B.*, **7**, 214.
Burke, P. G., Cooper, J. W., and Ormonde, S. (1966). *Phys. Rev. Lett.*, **17**, 345.
Cvejanovic, S., Comer, J., and Reed, F. H. (1974). *J. Phys. B.*, **7**, 468.
Ehrhardt, H., Langhans, L., and Linder, F. (1968). *Z. Physik*, **214**, 179.
Franzen, W., and Gupta, R. (1965). *Phys. Rev. Lett.*, **15**, 819.
Gibson, J. R., and Dolder, K. T. (1969). *J. Phys. B.*, **2**, 741.
Golden, D. E., and Bandel, H. W. (1965). *Phys. Rev.*, **138**, A14.
Golden, D. E. (1973). *Rev. Sci. Instrum.*, **44**, 1339.
Golden, D. E., and Zecca, A. (1970) *Phys. Rev.*, **A1**, 241.
Golden, D. E., Schowengerdt, F. D., and Macek, J. (1974). *J. Phys. B.*, **7**, 478.
Heddle, D. W. O. (1974). In *International Symposium on Electron and Photon Interactions with Atoms*.
Knowles, M., and McDowell, M. R. C. (1973). *J. Phys. B.*, **6**, 300.
Kuyatt, C. E., Simpson, J. A., and Mielczarek, S. R. (1965). *Phys. Rev.*, **138**, A385.
Moores, D. L., and Norcross, D. W. (1972). *J. Phys. B.*, **5**, 1482.
Ormonde, S., and Golden, D. E. (1973). *Phys. Rev. Lett.*, **31**, 1161.
Perkins, J. F. (1969). *Phys. Rev.*, **178**, 89.
Perel, J., Englander, P., and Bederson, B. (1962). *Phys. Rev.*, **128**, 1148.
Rubin, K. (1972). Unpublished.
Sanche, L., and Schulz, G. J. (1972). *Phys. Rev.*, **A5**, 1672.
Schulz, G. J. (1963). *Phys. Rev. Lett.*, **10**, 104.
Simpson, J. A., and Fano, U. (1963). *Phys. Rev. Lett.*, **11**, 158.
Sinfailam, A. L., and Nesbet, R. K. (1972). *Phys. Rev.*, **A6**, 2118.
Temkin, A., Bhatia, A. K., and Bardsley, J. N. (1972). *Phys. Rev.*, **A5**, 1663.
Weiss, A., and Krauss, M. (1970). *J. Chem. Phys.*, **52**, 4263.
Wigner, E. P. (1948). *Phys. Rev.*, **73**, 1002.
Young, A. D. (1968). *J. Phys. B.*, **1**, 1073.

54

High-Resolution Studies of Electron–Atom Collisions

F. H. READ

The various factors which limit the attainable energy resolution in low-energy electron–atom scattering experiments are discussed, and examples of recent high-resolution studies are described.

1. Introduction

Until recently the smallest widths of peaks in published spectra of low-energy electron–atom scattering experiments have been about 40 meV (FWHM), but in the last two or three years technical improvements have led to better resolutions, and it has now become possible to resolve new features and to see previously known features with greater clarity.

The observed widths of peaks are limited by the energy resolutions of the energy selection and analysis systems, as well as by the effects of Doppler broadening and the broadening caused by the presence of stray electrostatic fields in the target region (which we shall refer to as "field broadening"), and as the selector and analyzer resolutions have been improved, so have the effects of Doppler and field broadening become more and more evident. It is the purpose of the present paper to discuss these various limitations and the ways in which they combine to give the widths of observed spectral peaks, and also to describe recent progress in minimizing them. The discussion will be limited to those experiments in which the incident particles are

---

F. H. READ • Laboratoire de Physique et Optique Corpusculaires, Université de Paris VI, France. Permanent address: Physics Department, Schuster Laboratory, University of Manchester, England.

electrons having energies in the range from about one to a few hundred electron volts. Some of the results are well known, but it will be necessary to summarize them before presenting the new results and ideas.

2. The Widths of Observed Spectral Lines

Suppose that the probability distribution of the energies E_i of the electrons in the incident beam is $f_S(E_i - E_S)$, where the function f_S and the mean energy E_S are determined by the energy selector S and the beam transport system between the selector and target and, similarly, that the probability of detecting a scattered electron of energy E_f is $f_A(E_f - E_A)$, where f_A and E_A are determined by the analyzer A and by the beam transport systems before and after this analyzer. Suppose also that the effective energy changes E_T, which are caused by the thermal motion of the target atoms, have a probability distribution $f_T(E_T)$, and that the energies E_V caused by field broadening have a probability distribution $f_V(E_V)$.

These four distributions f_S, f_A, f_T, and f_V combine in different ways for different types of scattering experiments. A detailed discussion of this, and also further information about Doppler broadening, has been given by Read (1975). Only the essential features will be summarized here, and in order to do this we shall make the approximation that all four distributions are nearly Gaussian in shape, with full widths at half maximum given by $\Delta_S, \Delta_A, \Delta_T$, and Δ_V, respectively, and that they are independent and uncorrelated. Three types of spectra will be considered, namely, energy-loss spectra, resonance-scattering spectra, and threshold-excitation spectra.

In the measurement of energy-loss spectra (in which E_S is kept constant while E_A is varied), the widths $\Delta_{EL}(\text{FWHM})$ are given, with the above assumptions, by

$$\Delta_{EL}^2 \approx \Delta_S^2 + \Delta_A^2 + \Delta_T'(\theta)^2 \tag{1}$$

When the target velocities have Maxwellian distribution in the reaction plane, $\Delta_T'(\theta)$ is given by

$$\Delta_T'(\theta) = \left(\frac{E}{E_i}\right)^{1/2} \Delta_T \tag{2}$$

(see, for example, Chantry, 1971), where

$$\Delta_T = 4(\ln 2)^{1/2} \left(\frac{kTmE_i}{M}\right)^{1/2} \tag{3}$$

and

$$E = \tfrac{1}{2} m \Delta v^2 = \tfrac{1}{2} m(v_i^2 + v_f^2 - 2v_i v_f \cos\theta) \tag{4}$$

and where m and M are the electron and target mass, T is the target temperature, v_i and v_f are the incident and final electron velocities, θ is the scattering angle, and therefore Δv is the change in electron velocity. For a Maxwellian distribution at 290°K, Δ_T has the value $(E_i/A)^{1/2} \times 12$ meV, where E_i is measured in eV and A is the atomic mass number of the target atom. When the target atoms are collimated to form a beam the range of velocities in the reaction plane is reduced, and the value of Δ_T is therefore also reduced (see below).

When the atomic state being excited has a natural width Γ which is not negligible, then this will also contribute to the observed Δ_{EL}, although natural line shapes have Lorentzian or Fano–Beutler profiles in general, and since the widths of such profiles do not add in quadrature with the widths of Gaussian profiles, other methods of combining the widths must be used (Comer and Read, 1973).

The width Δ_{EL} is not affected by field broadening, provided that the inelastic scattering cross section is not quickly varying, since any potential energy gained by an incident electron on approaching the point of scattering is lost again on leaving the target region after the scattering event. The Doppler width $\Delta'_T(\theta)$ is smallest at 0°, and at this angle it decreases as the incident energy is increased at high incident energies, as pointed out by Chantry (1971). The dependence on θ has been noted by Hall et al. (1973).

A second common type of scattering experiment is that in which resonances are studied by varying the selector and analyzer energies together. The combination of the distributions f_S, f_A, f_T, and f_V is more complicated for this type of experiment, but with the assumptions stated above one finds that for resonance peaks having a narrow natural width, the observed width Δ_{RES} is given by

$$\Delta_{RES}^2 \approx \frac{\Delta_S^2 \Delta_A^2}{\Delta_S^2 + \Delta_A^2} + \Delta_V^2 + g(\theta)\Delta_T^2 \tag{5}$$

where

$$g(\theta) = \frac{(\alpha\Delta_S^2 + \Delta_A^2)^2 + (\beta\Delta_S^2)^2 + \beta^2(\Delta_S^2 + \Delta_A^2)\Delta_T^2}{(\Delta_S^2 + \Delta_A^2)[(\Delta_S^2 + \Delta_A^2) + \{(1-\alpha)^2 + \beta^2\}\Delta_T^2]} \tag{6}$$

and where αv_i and βv_i are the components of v_f parallel and perpendicular to the direction of v_i. That is, $\tan\theta = \beta/\alpha$ and $v_f^2 = v_i^2(\alpha^2 + \beta^2)$. As in the case of energy-loss spectra, Δ_T is reduced when an atomic beam is used. When the natural width of the resonance is comparable with Δ_{RES}, the two shapes must be convoluted together of course, although, as in the case of energy-loss spectra, the widths cannot in general be added in quadrature.

The factor $g(\theta)$ is always less than or equal to unity, and it effectively attenuates the width Δ_T. This attenuation increases as θ increases, and $g(\theta)$

is smallest for backward scattering. For example, when $\Delta_S = \Delta_A$, and when the scattering is elastic, $g(\theta)$ reduces to $\cos^2(\theta/2)$, which means that g is unity at 0°, decreases to $\frac{1}{2}$ at 90°, and actually becomes zero at 180°. It appears that this effect has not yet been observed or exploited.

The essential reason for the attenuation of Δ_T can be understood as follows. The electrons have a high probability of passing the combined energy windows of the selector and analyzer if they suffer small thermal energy changes in the collision. In fact, the magnitudes of the energy changes increase with the scattering angle and with the component of the atomic velocity v_T in the direction of the velocity change Δv (Chantry, 1971; Hall et al., 1973). Therefore, when they are scattered through large angles the electrons are collected with high probability only if they collide with target atoms having a small velocity component in the direction Δv, and this consequently provides a selection mechanism for the target atoms and results in an effective decrease in the Doppler width. It also results, incidentally, in a reduction in the measured yield. In the case of transmission experiments in which the scattered electrons are not observed, there is no reduction in the Doppler width, and one has

$$(\Delta_{RES}^{trans})^2 = \Delta_S^2 + \Delta_V^2 + \Delta_T^2 \tag{7}$$

The third type of experiment to be considered is that in which threshold excitation spectra are obtained by keeping the analyzer energy E_A constant, usually at a very small value, while the selector energy E_S is varied. If it can be assumed that the excitation cross section has a peak at threshold with a Gaussian shape and a width Δ_σ, then one has

$$\Delta_{TH}^2 \approx \Delta_S^2 + \Delta_T^2 + \frac{\Delta_A^2(\Delta_V^2 + \Delta_\sigma^2)}{\Delta_A^2 + \Delta_V^2 + \Delta_\sigma^2} \tag{8}$$

Although the assumed shape of the excitation cross section is not very realistic, this expression nevertheless has some usefulness when considering the widths of threshold excitation peaks.

3. Comparison and Minimization of Peak Widths

In the absence of thermal broadening Δ_T and field broadening Δ_V the narrowest peaks should be those observed in resonance spectra, since for these the selector and analyzer functions are multiplied together rather than convoluted with each other. For example, the ratio of Δ_{RES} to Δ_{EL} is given by

$$\frac{\Delta_{RES}}{\Delta_{EL}} \approx \frac{\Delta_S \Delta_A}{\Delta_S^2 + \Delta_A^2} \leq \frac{1}{2} \tag{9}$$

the equality sign holding only when $\Delta_S = \Delta_A$. The fact that ratios as small as this are not obtained in practice is an obvious indication of the existence of other broadening effects.

Thermal broadening can be reduced by using an atomic beam rather than a gas cell, and clearly it is important to do this in high-resolution work with light atoms or molecules. The reduction in $\Delta'_T(\theta)$ and Δ_T depends on the type of gas beam, but if the beam has an intensity distribution with a full angular width at half-maximum equal to Δ_B radians, the reduction factor is about $\frac{1}{2}\Delta_B$.

Even when the thermal width Δ_T has been reduced by using atomic beams or by studying heavy atoms, the ratio of Δ_{RES} to Δ_{EL} given by Eq. (9) does not seem to have been obtained, which is no doubt due to the existence of the field broadening Δ_V. In practice it is difficult to control or reduce stray potential variations, which are often caused by nonuniform surface potentials of the target chamber or target shield, but clearly it is most important to do so. The presence of field broadening is probably the major contribution to the observed widths of resonant features observed in the highest-resolution experiments (see below).

The selector and analyzer resolutions also limit the widths of observed peaks of course, and although it is possible in principle to design selection systems having very small values of Δ_S, this can only be done at the expense of greatly reducing the incident electron current. It has been shown (Read *et al.*, 1974) by extensive computer calculations in which account has been taken of the effects of space charge spreading, Boersch energy spreading, lens and selector aberrations, and gun and cathode limitations, that the maximum current I is related to the energy spread Δ_S by

$$I \approx c\Delta_S^{2.5} \tag{10}$$

where c depends on the type and size of selector (for example, c is 0.027 nA · meV$^{-2.5}$ for a hemispherical selector of mean diameter 10 cm). In practice most laboratories have obtained up to about 5% of this theoretical maximum, and it may therefore be possible in the future to obtain currents up to 20 times larger (at the same Δ_S), or, alternatively, to obtain values of Δ_S up to about three times smaller (for the same I), although this would need careful design, especially with regard to lens limitations.

4. Examples of High-Resolution Spectra

The narrowest peaks in published energy-loss spectra for low-energy (~ 1 to 100 eV) electron scattering appear to be those of Joyez *et al.* (1973a), who obtained a width of 17 meV (FWHM) in the rotational excitation of H_2 at incident energies in the range from 10 to 12 eV. This width would imply from Eq. (1) that either or both Δ_S and Δ_A are less than about 12 meV. In the absence of thermal and field broadening this would imply that the widths of resonance peaks should be less than about 8 meV, apart from the natural peak widths, whereas the observed widths of the resonances in H_2 were all greater than 19 meV. Since there is reason to believe that some of these resonances have natural widths which are much less than 19 meV (and also the helium resonance at 19.4 eV which was observed at the same time was found to have a width of 19 meV, which is greater than its known natural width of 9 meV), and since the thermal widths for these energy-loss and resonance-scattering spectra are probably comparable, one can only conclude that the field-broadening width Δ_V is greater than about 17 meV, and that it is in fact the major limitation to the resolution in these resonance spectra.

These resolutions were obtained (i) by using an atomic beam to reduce Δ_T, (ii) by carefully shielding and baking the target region to reduce Δ_V, and (iii) by using hemispherical energy selectors and analyzers of the general design of Simpson (1964) to give low values of Δ_S and Δ_A. These methods have of course been used, in whole or part, by many other experimenters. It is possible, although certainly not proven, that field broadening is the major limitation in the resolution obtainable in resonance experiments when atomic beams are used, and that the differences between observed resolutions are essentially caused by differences in the field broadening Δ_V. Experiments in which gas cells are used are somewhat different in that the Doppler width Δ_T is usually the most important limitation when light gases are being studied. In general, there seems to be a choice between gas cell experiments in which Δ_V can be made small but Δ_T is large, and gas beam experiments in which Δ_T is reduced but Δ_V is increased, but perhaps the "backscattering" method of Burrow and Sanche (1972) will offer a means of combining the best features and of exploiting the reduced value of the constant $g(\theta)$ at backward angles.

The resolution obtained by Joyez et al. (1973a) enabled individual rotational transitions $J_i \rightarrow N_r \rightarrow J_f$ to be studied, where J_i, N_r, and J_f are the rotational quantum numbers of the target, resonant, and final molecule, respectively. By these means it was possible to establish that the symmetry classifications of series a and c of H_2^- are $^2\Sigma_g^+$ and $^2\Pi_u$, respectively. More recently Wong and Schulz (1974) have obtained very similar resolutions in energy-loss spectra for the rotational excitation of H_2 which have enabled them to study the dependence on the rotational quantum numbers of the differential cross sections. Resonance spectra using this same apparatus have not been published.

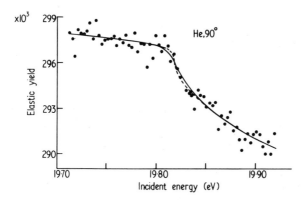

Figure 1. Differential cross section for elastic electron–helium scattering at 90°, in the neighborhood of the 2^3S threshold (Cvejanović et al., 1973).

The apparatus used by Joyez et al. (1973a) has also been used (Cvejanović et al., 1974) to study the 2^2S resonance and the 2^3S cusp in electron–helium scattering. Figure 1 shows that at the energy of the 2^3S threshold the elastic scattering cross section has a discontinuity in the form of a rounded step. The resolution in these experiments was 18 meV (FWHM), and, once again, it is clear from Eq. (5) (which applies to cusps as well as to resonances) that the field broadening width Δ_V is probably the major limitation to the observed resolution. The analysis of these experimental results has shown (i) that the central energy of the 2^2S resonance is 19.367 ± 0.009 eV, (ii) that this resonance has a natural width 9 ± 1 meV, and (iii) that the S-wave phase shift at the resonance energy is $106 \pm 3°$. Another recent high-resolution study of this resonance and cusp is that of Golden et al. (1974), in which the major limitation to the resolution seems to be the Doppler width Δ_T, since a gas cell was used.*

Progress of a different nature has recently been made in studies of threshold excitation spectra, and it has become possible (see, for example, Joyez et al., 1973b) to obtain peak widths as low as 35 meV. This has been achieved (i) by using an atomic beam to decrease Δ_T, and (ii) by using the achromaticity of a conventional lens and analyzer system to make Δ_A small for low-energy electrons. It can be seen from Eq. (8) that if Δ_A is not small, the threshold peak widths are limited by the field broadening Δ_V and by the natural width of the excitation peak, if such a peak exists.

---

*Two very recent studies, using a time-of-flight technique [J. E. Land and W. Raith, *Phys. Rev.*, **A9**, 1592–1602 (1974)] and a photoionization source [A. C. Gallagher and G. York, *Rev. Sci. Inst.*, **45**, 662–668 (1974)] respectively, have resulted in energy resolutions of better than 10 meV for resonance structures in transmission experiments.

In a similar technique (Cvejanović and Read, 1974a,b), an atomic beam is also used, and a weak but nonuniform potential distribution is deliberately introduced into the target region to give an enhanced detection probability for the lowest-energy electrons. This potential distribution has a saddle point at the scattering center, and therefore the electrostatic field is zero at this point and Δ_V is not increased appreciably, but the fields surrounding this region focus low-energy electrons towards the analyzers and detectors. This results in a value of Δ_A which is estimated to be about 15 or 20 meV. Figure 2 shows an example of a threshold excitation spectrum for helium obtained by this method (Cvejanović and Read, 1974b).

It can be seen from Figure 2 that a cusp exists at the ionization energy. This cusp appears to be a manifestation of the angular and spatial correlations which exist between the two outer electrons when they both have very low energies (Fano, 1974). Similar cusps have been observed in Ar and N_2 (Cvejanović and Read, 1974a).

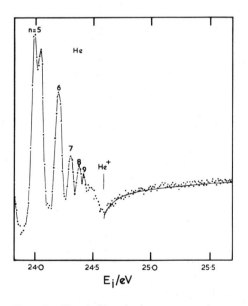

Figure 2. Threshold excitation spectrum showing the yield of very-low-energy electrons ($\lesssim 50$ meV) resulting from electron–helium scattering in the neighborhood of the ionization threshold (Cvejanović and Read, 1974b).

5. Conclusions

Although very much has already been learned from low-energy electron impact spectroscopy, it is clear that some aspects of this spectroscopy will become truly fruitful only when higher-energy resolutions are available. This is particularly true in cases of very narrow resonances and cusps, and also in the case of rotational and vibrational excitations of molecules. As higher resolutions become available it becomes more and more important to understand, and, if possible, to exploit, the ways in which various factors contribute to observed resolutions, and it has been the purpose of this report to discuss this and to give some examples of the progress which has been made.

References

Burrow, P. D., and Sanche, L. (1972). *Phys. Rev. Lett.*, **28**, 333.
Chantry, P. J. (1971). *J. Chem. Phys.*, **55**, 2746.
Comer, J., and Read, F. H. (1973). *J. Elec. Spectroscop.*, **2**, 87.
Cvejanović, S., and Read, F. H. (1974a). *J. Phys. B.*, **7**, 1180.
Cvejanović, S., and Read, F. H. (1974b). *J. Phys. B.*, **7**, 1841.
Cvejanović, S., Comer, J., and Read, F. H. (1974). *J. Phys. B.*, **7**, 468.
Fano, U. (1974). *J. Phys. B.*, **7**, L401.
Golden, D. E., Schowengerdt, F. D., and Macek, J. (1974). *J. Phys. B.*, **7**, 478.
Hall, R. I., Joyez, G., Mazeau, J., Reinhardt, J., and Schermann, C. (1973). *J. de Physique*, **34**, 827.
Joyez, G., Comer, J., and Read, F. H. (1973a). *J. Phys. B.*, **6**, 2427.
Joyez, G., Hall, R. I., Reinhardt, J., and Mazeau, J. (1973b). *J. Elec. Spectroscop.*, **2**, 183.
Read, F. H., Comer, J., Imhof, R. E., Brunt, J. N. H., and Harting, E. (1974). *J. Elec. Spectroscop.*, **4**, 293.
Read, F. H. (1975). *J. Phys. B.*, **8**, 1034.
Simpson, J. A. (1964). *Rev. Sci. Instr.*, **35**, 1698.
Wong, S. F., and Schulz, G. J. (1974). *Phys. Rev. Lett.*, **32**, 1089.

55
Electron Excitation of Xenon near Threshold

N. Swanson, R. J. Celotta, and C. E. Kuyatt

> An electron energy-loss spectrum and excitation functions for the four lowest excited states in Xe have been measured at a scattering angle of 45° in the near-threshold region (8–14 eV). Four peaks in the energy-loss spectrum above the $^2P_{3/2}$ ionization limit can be fitted to a $5p^5np'$ Rydberg series. The excitation functions for the two lowest states at 8.32 and 8.44 eV show large peaks due to decay of the $5p^56s^2(^2P_{1/2})$ resonance, as well as other resonance peaks which correlate well with previous transmission measurements.

1. Introduction

Discovery of lasing action in dimers of Kr and Xe has stimulated interest in the excitation of these gases. A previous study of Kr (Swanson et al., 1973) showed that the near-threshold excitation of the lower-lying states is dominated by resonant effects.

We have measured an energy-loss spectrum for Xe which shows both optically allowed and optically forbidden states and includes peaks in the autoionizing region between the $^2P_{3/2}$ and $^2P_{1/2}$ ionization limits. Several of these peaks can be fitted to a $5p^5np'$ Rydberg series ($n = 7, 8, 9, 10$). We have also measured excitation functions for the four lowest excited states belonging to the $5p^56s$ configuration. These excitation functions show marked structure near threshold, including both broad and narrow peaks which can be identified with $5p^56s^2$ and $5p^56s5d$ negative ion resonances.

N. Swanson, R. J. Celotta, and C. E. Kuyatt • National Bureau of Standards, Washington, D.C. 20234, U.S.A.

2. Apparatus

The measurements were made on an electron monochromator–analyzer combination described in detail in Swanson et al. (1973). It consists of copper hemispherical energy selectors for the monochromator and analyzer and a cylindrical molybdenum target gas cell. The incident electron beam is monochromatized, adjusted in energy, and focused on the gas cell. The scattered electrons are energy analyzed and individually counted by an electron multiplier followed by a fast amplifier, discriminator, and scaler. The output pulse count for each channel is accumulated in an on-line minicomputer.

3. Energy Loss Spectrum

Figure 1 shows the energy loss spectrum obtained for Xe at 14 eV incident energy and 45° scattering angle with 45 meV resolution. The first three excited states at 8.32, 8.44, and 9.45 eV are clearly resolved. These states have configurations $5p_{3/2}^5 6s\,(J = 2, 1)$ and $5p_{1/2}^5 6s\,(J = 0)$, respectively. The fourth peak consists of the overlap of the $5p_{1/2}^5 6s\,(J = 1)$ and $5p_{3/2}^5 6p$ $(J = 1)$ states at 9.57 and 9.58 eV, respectively. All the peaks up to the $^2P_{3/2}$

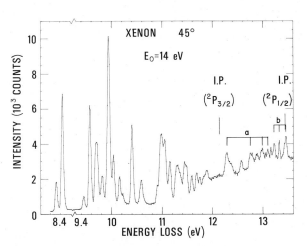

Figure 1. Energy loss spectrum for Xe at 14 eV incident electron energy and 45°, measured with 45 meV resolution. The data are an accumulation of 600 sweeps in 10 meV steps with 17 msec/point/sweep. The $^2P_{3/2}$ and $^2P_{1/2}$ ionization potentials are indicated.

ionization potential at 12.13 eV can be identified with single or overlapped spectroscopically known states, both electric dipole allowed and forbidden.

The region between the $^2P_{3/2}$ and $^2P_{1/2}$ ionization potentials contains discrete states with a $5p^5_{1/2}$ core which can autoionize to the $5p^5_{3/2}$ continuum. The s' and d' states were first observed by Beutler (1935) in optical absorption. The d' states are lifetime broadened, while the s' states are sharply peaked. Photoabsorption cross sections for these states have recently been measured by Carter and Hudson (1973), among others.

Peaks due to $8s'$ and $9s'$ excitation can be seen in Figure 1 at 12.58 and 12.89 eV, respectively. However, there are no observable peaks at energies of the d' levels. Instead a series of strong peaks with high-energy shoulders can be seen at 12.28, 12.75, 12.99, and 13.10 eV (Series a) together with three relatively intense peaks at 13.23, 13.33, and 13.46 eV (Series b) due to electrons ejected in the decay of autoionizing states.

In an ionization current measurement in this region, Marmet et al. (1972) were able to observe several peaks due to the md' transitions. They also observed three other peaks at 12.30, 12.67, and 12.92 eV, whose energies did not correspond to a Rydberg series and which they attributed to negative ion resonances.

Grasso (1969) measured the yield of ion current vs. electron energy in this region, and by taking derivatives of his data brought out structure corresponding to s' and d' states, as well as structure he assigned to p' states.

The reason peaks corresponding to s' transitions and not d' transitions can be seen in Figure 1 must result from the higher energy above threshold (~ 1.5 eV) and possibly the choice of scattering angle compared to the measurements of Marmet et al. (1972) and Grasso (1969). The similar shapes and regular spacings of the four peaks at 12.28, 12.75, 12.99, and 13.10 eV led us to calculate quantum defects assuming that the peaks were due to p' transitions, with results that can be seen in Table 1.

The peak positions fit a term series quite well. [The quantum defects for the $6p'$ ($J = 1$) levels at 10.96 and 11.07 eV are 3.66 and 3.60, respectively,

Table 1. Calculated Quantum Defects for the Tabulated Peaks in Xenon

Peak energy (eV)	Assumed configuration	Quantum defect
12.28	$7p'$	3.57
12.75	$8p'$	3.55
12.99	$9p'$	3.48
13.10	$10p'$	3.64

reasonably close to the tabulated values.] The variations in quantum defects in the table correspond to variations in peak energies of about 0.01 eV. The energies in Table 1 agree with those measured by Stebbings *et al.* (1974) for the $mp'[\frac{1}{2}]_1$ levels, where $m = 7, 8$.

Whether the peaks seen by Marmet *et al.* (1972) are really members of this series with a larger energy uncertainty, or are in fact resonances,

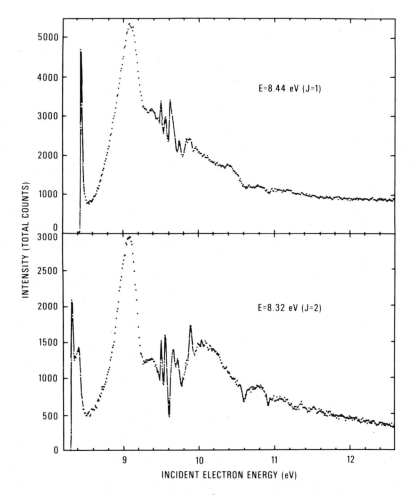

Figure 2. Excitation functions of the $5p^5_{3/2}6s$ states of Xe at 45°, measured with 30 meV resolution. The data in Figures 2 and 3 were accumulated in 400 computer-controlled sweeps with 10 meV/step and 17 msec/point/sweep. The sharp peaks at threshold in the two figures is at least partly an instrumental effect.

is not clear. Their energy scale is uncertain by at least one channel unit, or 0.024 eV. Sanche and Schulz (1972), in a transmission measurement on Xe, saw a weak resonance doublet at about 12.12 and 12.23 eV, and Kisker (1972) observed a resonance doublet at 12.13 and 12.27 eV in the optical excitation function for the $6p'$–$6s$ transition. In Figure 2 there may be a weak minimum at 12.33 eV in the 8.32 eV data, and in Figure 3 there may be maxima at 12.10 eV and 12.27 eV in the 9.57 eV data.

4. Excitation Functions

Resonances in the scattering of electrons by Xe were first seen by Kuyatt et al. (1965) in a transmission measurement, and later in more detail by Sanche and Schulz (1972) using a derivative technique. Measurements of elastic and inelastic scattering in Kr (Swanson et al. 1973) showed that resonances both dominated the near-threshold excitation of the lower lying states and produced marked differences in the excitation functions for states only a few tenths of an eV apart.

Figure 2 shows the electron excitation functions at 45° for the $5p_{3/2}^5 6s$ ($J = 2, 1$) states of Xe at 8.32 and 8.44 eV, respectively, measured with 30 meV resolution. Both excitation spectra possess a sharp peak at threshold, and a broad, very strong peak at 9.09 eV, produced by the decay of the $5p^5 6s^2(^2P_{1/2})$ resonance. The decay peak appears with comparable intensity in both channels, unlike the case in Kr, where decay to the $J = 1$ channel is more highly favored (Swanson et al., 1973).

Sanche and Schulz (1972) observed resonances due to the $5p^5 6s^2$ $^2P_{3/2}$ and $^2P_{1/2}$ negative ion states, as well as a series of sharp resonances between 9.5 and 10 eV and an isolated resonance at 8.48 eV. This resonant structure also appears in the two curves of Figure 2 as a succession of six peaks between 9.3 and 10 eV, four of which occur at identical energies in the two spectra, at 9.38, 9.50, 9.56, and 9.88 eV. Two minima at about 10.60 and 10.91 eV appear more strongly in the 8.32-eV data, which also include a peak at 8.42 eV and a broad maximum at 10.1 eV.

Figure 3 shows the electron excitation functions at 45° and 30 meV resolution for the $5p_{1/2}^5 6s(J = 0)$ state at 9.45 eV, and the unresolved $5p_{1/2}^5 6s(J = 1)$ and $5p_{3/2}^5 6p(J = 1)$ states at 9.57 and 9.58 eV. There is little similarity between the two traces other than the sharp threshold peak. The 9.45 eV data show sharp peaks as in Figure 2, at 9.52, 9.58, 9.73, and 9.80 eV, and a sharp step at 11.00 eV. The $J = 1$ data show smaller peaks at 9.75, 9.87, 10.50, and 11.07 eV, a broad maximum centered at about 10.3 eV, and minima at 10.60 and 10.88 eV.

In Kr, sharp structure above the $^2P_{1/2}$ resonance was interpreted as due to $5s4d$ resonances, based in part on a comparison of the scaled resonance

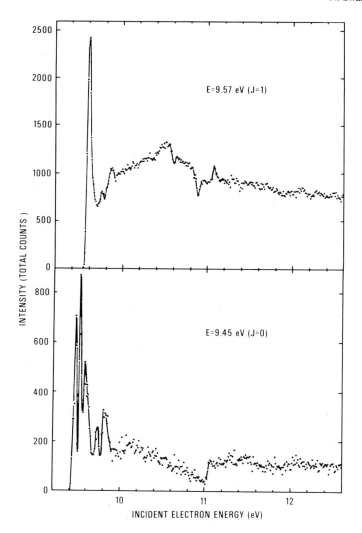

Figure 3. Excitation functions of the $5p^5_{1/2}6s$ states of Xe at 45°, measured with 30 meV resolution. The higher-energy state overlaps the $6p$ ($J = 1$) state at 9.58 eV.

energies with the corresponding Rb I spectrum (Swanson et al., 1973). A similar analogy can be made for the sharp Xe resonances using the Cs I spectrum as a guide. Cooper (1973) has pointed out that, based on a comparison of core splittings for K, Rb, and Cs with Ar, Kr, and Xe, Beutler (1934) has interchanged the lowest $5p^5_{3/2}6s5d$ and the $5p^5_{1/2}6s^2$ levels in Cs, and that the 6s5d level is between the two $6s^2$ levels. This identification implies that the resonance at 8.42 eV in Figure 2 has a 6s5d configuration.

Acknowledgment

We would like to thank John W. Cooper for many useful and stimulating discussions on the interpretation of resonance structure.

References

Beutler, H. (1934). *Z. Phys.*, **88**, 25.
Beutler, H. (1935). *Z. Phys.*, **93**, 177.
Carter, V. L., and Hudson, R. D. (1973). *J. Opt. Soc. Am.*, **63**, 733.
Cooper, J. W. (1973). Private communication.
Grasso, F. (1969). *J. Mass. Spect. Ion. Phys.*, **2**, 357.
Kisker, E. (1972). *Z. Phys.*, **257**, 51.
Kuyatt, C. E., Simpson, J. A., and Mielczarek, S. R. (1965). *Phys. Rev.*, **138**, A385.
Marmet, P., Bolduc, E., and Quemener, J. J. (1972). *J. Chem. Phys.*, **56**, 3463.
Sanche, L., and Schulz, G. J. (1972). *Phys. Rev.*, **A5**, 1672.
Stebbings, R. F., Dunning, F. B., and Rundel, R. D. (1974). *4th Int. Conf. Atomic Physics, Heidelberg.* Invited paper.
Swanson, N., Cooper, J. W., and Kuyatt, C. E. (1973). *Phys. Rev.*, **A8**, 1825.

56
Resonances in the Excitation of Ne by Electrons at Energies between 40 and 50 eV

H. G. M. HEIDEMAN AND T. VAN ITTERSUM

> A study has been made of the excitation of neon levels by electron impact in the energy range from 40 to 50 eV. Two weak resonances near 43 and 47 eV were observed. The conclusion seems justified that most of the structures, observed by Grissom et al. (1960) in a trapped electron experiment and attributed by them to Ne^- resonances, are in fact due to autoionizing states whose thresholds have shifted due to the displaced threshold phenomenon, recently reported by Hicks et al. (1973).

A study has been made of the excitation of neon levels by electron impact in the energy range from 40 to 50 eV. Earlier experiments by Grissom et al. (1969) indicated that in this energy region the inelastic scattering is largely affected by many resonance structures. Most of the structures were attributed to triply excited Ne^- states.

To investigate whether and to what extent these resonances also affect the excitation of neon states, and thereby to definitely distinguish between autoionizing and Ne^- states, is the main purpose of this work. Our method (Heideman, 1971) is based on the observation of the light emission following the electron impact excitation of the levels concerned.

Our measured excitation curve of the $3p'(\frac{1}{2})_0$ state (Paschen notation) exhibits much less structure between 40 and 50 eV than the trapped-electron spectrum of Grissom et al. (1969). Only two weak resonances near 43 and

---

H. G. M. HEIDEMAN and T. VAN ITTERSUM • Fysisch Laboratorium der Rijksuniversiteit, Utrecht, Holland.

47 eV, respectively, could be detected. A possible explanation of this apparent discrepancy is the following.

In the experiment of Grissom et al. (1969), both autoionizing (Ne**) and negative ion states (Ne$^-$) can be observed. The observed autoionizing states are excited just above their thresholds (this is inherent to the trapped electron method), and consequently the phenomenon of displaced thresholds, recently reported by Hicks et al. (1973), is involved. This phenomenon is due to a post collision interaction between the (slow) inelastically scattered and the ejected electron and causes an apparent shift in the threshold energy of an autoionizing state. It is therefore quite possible that most of the structures observed by Grissom et al. (1969), and which they ascribe to Ne$^-$ resonances, are in fact due to autoionizing states whose positions have shifted due to the aforementioned phenomenon.

References

Grissom, J. T., Garrett, W. R., and Compton, R. N. (1969). *Phys. Rev. Lett.*, **23**, 1011.
Heideman, H. G. M., van Dalfsen, W., and Smit, C. (1971). *Physica*, **51**, 215.
Hicks, P. J., Comer, J., Cvejanovič, S., and Read, F. H. (1973). Abstracts of papers, *VIII ICPEAC, Belgrade, Yugoslavia*, pp. 513.

57
Resonance Series in Helium

D. W. O. HEDDLE

> We discuss the classification of resonances observed in the optical excitation functions of helium and show that there are series of terms $1sns^2(^2S)$ and $1snsnp(^2P)$. Two other resonances are identified as $1s3p^2(^2D)$ and possibly $1s3p^2(^2S)$; these may also be members of series. Strong features in the 3^3D and 4^3D excitation functions are identified as shape resonances in the n^3X terms of helium, where X signifies the term having the maximum possible value of angular momentum. Some sharp and weak features are noted in both the 3S and 3D excitation functions at energies where Feshbach resonances cannot occur.

The occurrence of resonances in helium is well established and they have been recorded in experiments which observe elastic or inelastic scattering of electrons or the production of metastable or radiating states. They have been reviewed by Schulz (1973). From our observations of electron excitation to radiating states, and from published data at lower electron energies, we can divide the resonances into three groups. One group consists of series of Feshbach resonances, a second a series of near-threshold shape resonances, while a third group comprises a number of features at energies where one might expect shape resonances, but which are quite sharp and weak.

In order to examine the types of series we might expect, it is helpful to to consider the physical phenomea which enable us to detect a resonance. A necessary requirement is that the resonance should cause a change in the scattering cross section from some smoothly varying pattern. This requires that the lifetime of the resonance be not too short, which in turn requires that the motion of the two excited electrons be quite strongly correlated, particularly in respect to their radial motion. If the radial motion were

---

D. W. O. HEDDLE • Department of Physics, Royal Holloway College, University of London, Egham Hill, Egham, Surrey, England.

completely correlated we could express the potential (in a hydrogenic approximation) in terms of a single coordinate and find the energy eigenvalues for the system very easily. The potential is given by

$$V = \frac{-1}{4\pi\varepsilon_0}\left(\frac{2Ze^2}{r} - \frac{e^2}{2r}\right)$$

and the energies by

$$T = \frac{-R^*}{n^2}$$

where

$$R^* = R\frac{(4Z-1)^2}{8}$$

This expression ignores all magnetic interactions, but it is not at all bad in describing the H^- and He^{+-} terms.

In a real case there are magnetic interactions and also, in general, interactions with core electrons, but we would expect terms described with reasonable accuracy by

$$T = \frac{-RZ_{\text{eff}}^2}{(n-\delta)^2}$$

to occur and for Z_{eff} and δ to be adequately constant along a series. We can make a graphical examination of this assumption by plotting values of $(-R/T)^{1/2}$, where T is determined experimentally against a running integer n, which we identify as the principal quantum number of the two excited electrons. Resonances belonging to different series will lie along different lines. We shall use this representation in Figure 3.

The exact energy of a resonance is not always easy to determine, particularly if a second resonance is close by. Figure 1 shows the excitation function of the 4^1S state of helium (Heddle et al., 1973). There are plainly two major resonances, but their energies are not immediately clear. An extrapolation from the high-energy end is indicated in this figure, which suggests that the low-energy resonance may be nearly symmetric and have an energy of 23.83 eV, while the higher-energy resonance is a peak at 23.94 eV. We have suggested earlier (Heddle et al., 1974a) that these resonances could be described by configurations $1s5s^2$ and $1s5p^2$ because they appear equally clearly in the 4^1S and 4^3S excitation functions, though we noted that configurations such as $1s5s(^1S)\,5p$ and $1s5s(^3S)5p$ might be so close in energy that their decay into 4^1S and 4^3S states could appear to have a common origin.

Resonance Series in Helium

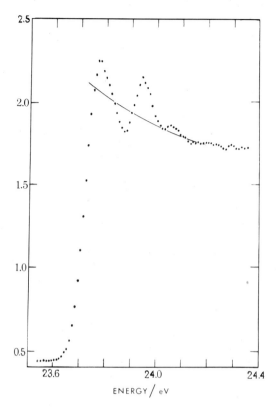

Figure 1. The excitation function of the 4^1S state, showing an extrapolation which suggests that the first maximum and minimum may be the extrema of an almost symmetric resonance.

Figure 2 shows the excitation functions of the 3^3S, 3^1S, 4^3S, 4^1S, 5^3S, 5^1S, and 6^1S states on a common energy scale, but with arbitrary ordinates. At the top of the figure we show four series which appear to fit the major features quite well. The labeling of the series depends in part on other measurements, notably the identification of the resonances at 19.37 eV and 22.47 eV as ^2S and that at 22.65 eV as ^2P. The terms have been arranged in series from study of Figure 3. The ^2P series and the lower-energy ^2S series are quite clear in this figure. There is no real indication of energy differences between the ^2P terms, which result from the three configurations $1snsnp$ having different spin orientations.

The very large peak at 22.94 eV in the 3^3S excitation function does not fit into either series. We would expect the next most loosely bound configuration to be $1snp^2$, which has terms ^2D and ^2S, of which ^2D would be

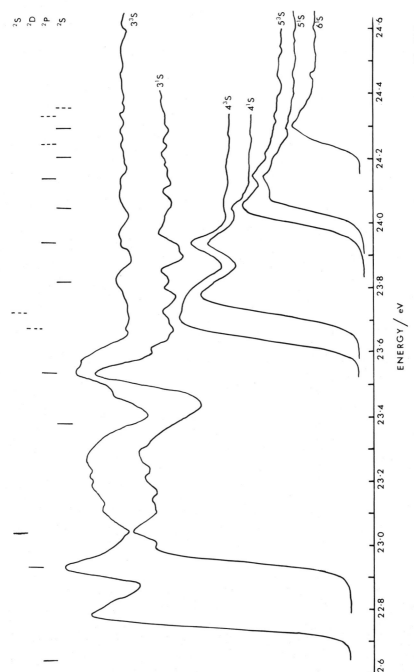

Figure 2. The excitation functions of seven S states. The vertical scales are arbitrary. Series of resonances are indicated at the top. The higher members, indicated by broken lines, have been obtained using Figure 3.

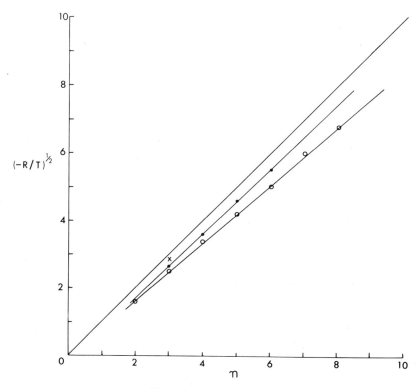

Figure 3. Values of $(-R/T)^{1/2}$, where T is the binding energy of the resonance state, for many of the resonances shown in Figure 2. The first member of the ^2P series and the first two members of the ^2S series lie below the 3^3S threshold and have been taken from published data. ○—$1sns^2(^2S)$ series; ●—$1snsnp(^2P)$ series; ×—$1s3p^2(^2D)$. The diagonal line represents the limit of bound states.

more tightly bound. Preliminary observations of electron scattering with excitation of the 3^3S state (J. M. Foster, private communication) indicate that the 22.94 eV feature, while prominent in scattering at 90°, vanishes in observations made at 55°. This leads us to suggest that this peak is due to the $1s3p^2(^2D)$ term. It is shown in Figure 3 by ×. The energy of the next member of the series may be estimated by drawing a line of suitable slope through this point and it is shown on Figure 2 by a broken line. There is no clear indication of a resonance in the $3S$ excitation functions at this energy, but the sharp rise at the onset of the 4^3S excitation function may be due to this. The initial rise of the 3^1S excitation function appears to be due to the $1s3p^2(^2D)$ resonance. The peak in this function at 23.04 eV is not a threshold effect, but may be the $1s3p^2(^2S)$ resonance. We have not marked this on Figure 3: it would lie between the ^2D point and the diagonal line drawn across this figure. The next member of this series, estimated as before, is shown on Figure 2 by a broken line, and there does seem to be a corresponding feature in the 3^1S excitation function. There is evidence from the near-threshold polarization of light from the helium D states (Heddle *et al.*, 1974b; Heddle, Keesing, and Parkin, unpublished) that there is a ^2S resonance at or just below these thresholds. This is in accord with the present identification.

The topmost diagonal line in Figure 3 corresponds to the n^1P terms of neutral helium which are the most lightly bound. Any Feshbach resonance must lie below this line. We would not expect any resonance to be more tightly bound than $1sns^2(^2S)$, and so the lowest line in Figure 3 represents a lower bound. There are clearly regions of energy which lie above the upper bound for a given value of n and below the lower bound for $(n + 1)$. For example, energies which correspond to

$$3.003 < (-R/T)^{1/2} < 3.37$$

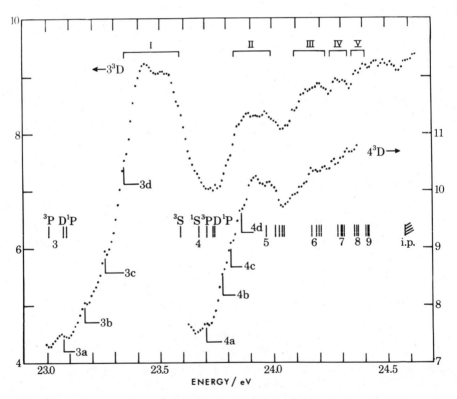

Figure 4. The excitation functions of the 3^3D and 4^3D states. The energies of features I to V should be compared with those of the neutral helium states.

i.e., 23.08 < E < 23.39 eV are not available to Feshbach resonances. Reference to Figure 2 shows that there are at least three weak and fairly sharp features in this energy range. We have no better description of these features as yet. They require further study by electron scattering techniques.

The excitation functions of the 3^3D and 4^3D states are shown in Figure 4. The small, barely resolved features 3b, 3c, and 3d are in this same "non-Feshbach" region, but the major features, labeled I–V, could well contain Feshbach resonances. Their form is much more complex than anything in the S-state excitation functions: each feature appears to contain two major resonances and some smaller ones. Features III, IV, and V extend in energy above the helium states with $n = 6, 7$, and 8, respectively, so they cannot be pure Feshbach resonances. We cannot, of course, rule out some Feshbach component, but we can describe these features in a consistent fashion which lends some support to a suggestion by Fano (1974).

We suggest that I is a shape resonance in the 3^3D state, II is a shape resonance in the 4^3D and 4^3F states which appears in the 3^3D excitation function as a result of cascade, and III is a shape resonance in the 5^3F and 5^3G states and is also seen as a result of cascade. Figure 5 shows the relevant

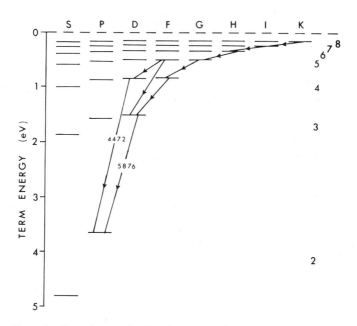

Figure 5. Term diagram for the triplet states of helium. Excitation of the states of maximum angular momenta leads to population of the ^3D state with $n = 3$ only. Excitation of states having less angular momentum can also populate ^3D states with $n > 3$.

part of the helium term diagram. Of the excitation to the 5^3F state, approximately one-third cascades to 4^3D and two-thirds to 3^3D. All the excitation to the 5^3G state cascades (in two steps) to the 3^3D state. Feature III is much more prominent in the 3^3D excitation function than in that of the 4^3D state, which suggests that the contribution from 5^3G is much greater than that from 5^3F. In a similar fashion, the appearance of feature IV suggests that it is predominantly the result of excitation to the 6^3H state, which can populate only 3^3D by cascade.

Our overall interpretation of these features is that each is the result of excitation to a state of principal quantum number greater by 2 than the number assigned to the feature and that the excitation is predominantly to the state having the largest angular momentum possible.

Acknowledgments

The optical excitation data discussed in this paper were obtained at the University of York, and much of it was taken by Dr. R. G. W. Keesing, Dr. Jelena M. Kurepa, and Dr. A. Parkin.

References

Fano, U. (1974). *J. Phys. B.*, **7**, L401.
Heddle, D. W. O., Keesing, R. G. W., and Kurepa, Jelena. M. (1973). *Proc. Roy. Soc. (Lond.)*, **A334**, 135.
Heddle, D. W. O., Keesing, R. G. W., and Kurepa, Jelena M. (1974a). *Proc. Roy. Soc. (Lond.)*, **A337**, 435.
Heddle, D. W. O., Keesing, R. G. W., and Watkins, R. D. (1974b). *Proc. Roy. Soc. (Lond.)*, **A337**, 443.
Schulz, G. J. (1973). *Rev. Mod. Phys.*, **45**, 378.

Index

A

Absolute cross sections, photoionization, 52–56
—— differential, for e–H, 333
Alignment of atoms, 465, 489ff, 517, 521
Analysis of rare gas spectra, 633–637
Angular correlation of γ rays, 569–570
Angular distribution of electrons
—— from electron impact ionization, 414
—— scattered by atoms, 309ff
Angular distribution of photoelectrons, 11, 61–65
—— asymmetry parameter β, 11, 61–63, 83–88, 92
—— for open-shell atoms, 83–88
—— in photodetachment, 119
—— in two-photon processes, 560, 570–577
Anisotropic interactions, 6
Anisotropic states of atoms, 485, 515
Anisotropy of quenching radiation, 339–346
Asymmetric surface interaction, 523
Asymmetry
—— in atomic collisions, 515–523
—— in electron scattering, 235–239
—— in emission of H_x, 365–373
—— in photoabsorption, see Asymmetric surface interaction
—— of radiation from metastable H, 339–345
Atomic form factor, 595
Auger spectroscopy, 621–626
—— transitions, 138
Autoionization, 35–38, 50–54, 75
—— and resonances, 36, 75
—— Fano profiles, 7–8, 56–57
—— levels in He, 7
—— postcollision interaction, 37

B

Backscattering of electrons, 609
Beam–foil excitation, 495, 516–523, 621
—— of Be and its ions, 621–626
—— tilted beam–foil geometry, 490, 513, 520–523
Boersch energy spreading, 655
Born approximation, 187, 245ff, 384, 397, 469
Branching ratios, in photodetachment and photoionization, 141–148

C

Classical models, 185–189
Coherence effects in beam–foil excited radiation, 520
Coherent impact excitation, 365, 458ff, 507, 520
Coincidence techniques
—— (e, 2e) coincidences, 149, 160, 394, 402ff
—— (e, e − $\hbar$) coincidences, 376, 455–483, 498
Collision-induced fluorescence, 488–490
Configuration interaction
—— close-coupling, 4, 416
—— effects on electron and photon interactions with atoms, 1–26
—— e–H scattering, 15
—— from (e, 2e) impulsive reactions, 149
—— in electron collisions, 29, 55ff
—— in resonance width calculations, 9
Correlation effects in electron–atom scattering, 419–420
—— in atomic structure, 18, 151
—— of loosely bound electrons, 125

Coulomb–Born approximation, 256–257, 387–395
—— for excitation of H, 390, 393
—— for excitation of He, 391–392
—— for ionization, 395
Cusps, in electron–atom scattering, 639, 657; see also Resonances

D

Density matrix, 486–488
—— and coincidence measurements, 502–504
—— and Fano parameters, 566–569
Dispersion relations, and phase shift analysis, 161–179
—— for e–He, 169–171
—— for e–Ne, 171–174
—— for positron, He, 167, 183
—— sum rule, 166–167
Doppler broadening in high-resolution studies of electron–atom scattering, 651ff

E

Eigenfunction expansions, 4, 248–249
Eikonal approximation, 245ff, 251–254
—— eikonal–Born series, 285–297
—— many-channel theory, 275–283
Elastic scattering of electrons, 185–189
—— and optical potential, 299–308
—— by H, 288, 309–338
—— by He, 187, 290, 304–305, 436–439, 644–649, 657–658
—— by Ne and Ar, 188, 306–307
—— by positive ions, 609–612
Electron–alkali scattering, 109–123
Electron impact ionization, 393–395, 397ff, 411ff
Electron–photon correlation experiments, 269–270, 437–438, 455–483
—— and optical polarization, 464–465
—— for He, 470–479
—— for Ly α, 261–262, 499
—— general theory, 501–514
Energy level analysis, 51
Energy loss spectra, 661–662
Excitation by electron impact, of light atoms, 245–273
—— of C, 27–34
—— of H, 15, 257–259, 280, 291–296, 325–334, 365–373

—— of He, 263–270, 281–282, 349–363, 671–678
—— of He^+, 17
—— of N, 21, 27–34
—— of N^+, 18
—— of Ne, 669–673
—— of O, 19–20, 27–34
—— of Xe, near threshold, 661–667

F

Fano effect, in photoionization, 241–243, 581–594; see also Asymmetry
Fano–Lambropoulos frequency, 577
Fano–Macek parameters for orientation and alignment, 465–446, 489, 504–506
Fano parameters of autoionization resonances, 7–8
Feshbach resonances, 6, 671
Field broadening in high-resolution studies of electron–atom collisions, 651
Forbidden transitions, 599
Forward scattering of electrons, 181–184, 451

G

Glauber approximation, 287, 384, 498
Green's function, in many-body theory, 255–256

H

Hanle effect, 366, 493
High-resolution studies of electron–atom collision processes, 317, 358ff, 639, 651ff, 661, 669, 671

I

Intense electromagnetic fields, 525, 533, 551
Ionization, triple differential cross section, 394–395, 397–410, 411–414

L

Lamb shift, 345–346
Laser excitation, of Na, 375–386
Lasers, as a source of intense electromagnetic fields, 525–551
LS-term dependence of transition probabilities, 633

Index

M

Many-body theory, 255–256, 435–444
—— and e–$\hbar$ coincidence measurements, 481
Matrix variational method, 5, 20
—— for electrons on C, N, O, 27–34
Metastable atoms, 339–347
Monochromators for synchrotron radiation, 48–50
Mott matrix element, 151
Mott scattering, 216, 241
Multiphoton processes, 133, 525–552, 553–558; *see also* Two-photon processes
Multipole radiation, 569
Multipole states of collisionally excited atoms, 376, 383

N

Nonstatistical branching ratios in atomic processes, 141ff

O

Optical potential, 249–250
—— for elastic scattering, 299–308
Optical pumping, 377ff, 491
Orientation of atoms, 455, 466, 477, 485ff, 521, 523

P

Phase shift analysis (and dispersion relations), 161–179
—— and absolute cross sections, 314–315
—— for e–Hg, 174–177
—— uniqueness, 163–166
Photodetachment
—— branching ratios, 141–148
—— K^-, 115–118
—— Li^-, 114
—— Na^-, 115–116
—— of alkali negative ions, 109–123
—— Rb^-, Cs^-, 120–122, 130
—— Se^-, 127
—— threshold behavior, 110, 125–132
—— two-electron, 131–132
Photoelectron spectroscopy, 57–60
Photoionization, 39–67; *see also* Angular distribution
—— and QDM, 99–107
—— branching ratios, 141–148
—— multiple, of rare gases, 133–139
—— multiphoton, 527–537
—— of Ar, 65, 91
—— of Ba, 69–81
—— of Be, 104–106
—— of H^-, 13
—— of He, 13, 96
—— of Ne, 92–93, 627–630
—— of rare gases, 13, 64–65
—— of S, 87
—— of Xe, 12, 64, 69–81
Photon sources, and synchrotron radiation, 42–48
Polarization, optical
—— and amplitude measurements, 464
—— of decay photons, 2
—— of impact radiation, 485, 676
—— of Ly α, 219–260
Polarized electrons, sources, 241–243; *see also* Spin polarization
—— in scattering by Hg, 173, 445–453
Polarized frozen core and target, 415–432
—— for e–N scattering, 429–430
Polarized orbital methods, 251
Positron–atom scattering, 181–184

Q

Quantum defect method (QDM), 99–107
—— applied to Xe, 663
—— for excited $^1\Sigma_u^+$ levels of H_2, 613–620
—— for the Ne I isoelectronic sequence, 627–631
Quenching radiation, anisotropy, 339–347

R

Ramsauer effect, 612
Random phase approximation (RPA), 5, 138
—— for Ar 3s, 14
—— for He, 352
Resonances
—— and cusps, in electron impact, 639–649
—— in elastic scattering by H, 318–319
—— in He, 675–682
—— in inelastic scattering, 328–330
—— in Ne, 229–233, 669–670
—— in Xe, 661–667
R-matrix, 4
Rydberg states and series, 18, 75, 613–620, 627

S

Spin–orbit coupling, 203, 235
Spin polarization
—— and conservation of total spin, 203–213
—— and multiphoton processes, 542–546, 581–594
—— and resonance scattering, 215–228
—— for Cs, 209–211, 238, 581–594, 640–641
—— for H, 209
—— in Ar, 217
—— in electron–atom scattering, 191–202, 235–239
—— in Hg, 220, 226–227, 445–454
—— in Kr, 218
—— in Li, 197
—— in Na, 194, 642–643
—— in Ne, 217, 222–225, 229–233
—— in Xe, 219
Stoke's parameters, 511–513, 521
Synchrotron radiation (and photoionization), 38–67

T

Tilted beam-foil geometry, 515, 521
Theory of the measurement
—— of electron–photon coincidence experiments, 458ff, 501ff
—— of impact radiation of the atoms, 485–500
—— of laser-excited atoms, 375ff
Threshold behavior
—— of excitation of He by electrons, 349–364
—— of multiple photoionization of rare gases, 134
—— of photodetachment, 110–113, 125–132
Time-of-flight technique for electron–atom collision processes, 657
Total elastic electron–atom cross section, 595–560
—— for Li, 196–197
—— for Na, 195–196
Transmission experiments for electron–atom collision processes, 657
Triple differential cross section of electron impact ionization, 395–396, 397
Two-photon processes, 559–579
—— and symmetry properties, 560–566
—— of Cs, as a polarized electron source, 581–594

V

Variational principles and bounds, 601–608

W

"Weak forces" in collisional excitations of atoms, 492